Stochastic Differential Equations: Theory and Applications

INTERDISCIPLINARY MATHEMATICAL SCIENCES

Published

Vol. 1: Global Attractors of Nonautonomous Dissipative Dynamical Systems
David N. Cheban

Vol. 2: Stochastic Differential Equations: Theory and Applications
A Volume in Honor of Professor Boris L. Rozovskii
eds. Peter H. Baxendale & Sergey V. Lototsky

Vol. 3: Amplitude Equations for Stochastic Partial Differential Equations
Dirk Blömker

Vol. 4: Mathematical Theory of Adaptive Control
Vladimir G. Sragovich

Vol. 5: The Hilbert–Huang Transform and Its Applications
Norden E. Huang & Samuel S. P. Shen

Vol. 6: Meshfree Approximation Methods with MATLAB
Gregory E. Fasshauer

Interdisciplinary Mathematical Sciences – Vol. 2

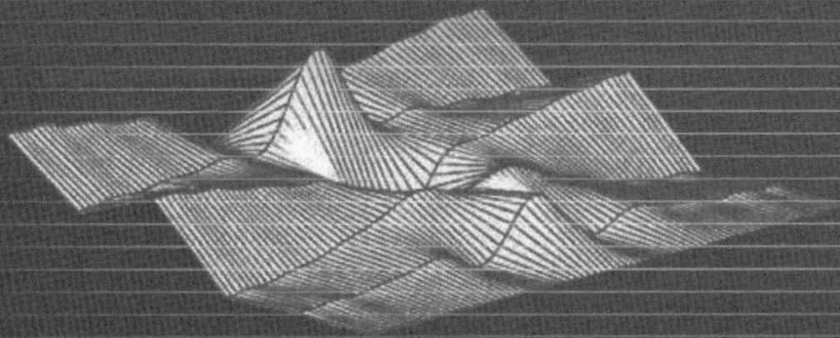

A Volume in Honor of Professor Boris L. Rozovskii

Stochastic Differential Equations: Theory and Applications

Editors

Peter H. Baxendale
Sergey V. Lototsky

University of Southern California, USA

World Scientific

NEW JERSEY • LONDON • SINGAPORE • BEIJING • SHANGHAI • HONG KONG • TAIPEI • CHENNAI

Published by

World Scientific Publishing Co. Pte. Ltd.
5 Toh Tuck Link, Singapore 596224
USA office: 27 Warren Street, Suite 401-402, Hackensack, NJ 07601
UK office: 57 Shelton Street, Covent Garden, London WC2H 9HE

British Library Cataloguing-in-Publication Data
A catalogue record for this book is available from the British Library.

Interdisciplinary Mathematical Sciences — Vol. 2
STOCHASTIC DIFFERENTIAL EQUATIONS: THEORY AND APPLICATIONS
A Volume in Honor of Professor Boris L. Rozovskii

ISBN-13 978-981-270-662-1
ISBN-10 981-270-662-3

Printed in Singapore by World Scientific Printers (S) Pte Ltd

Professor Boris L. Rozovskii

Preface

This festschrift volume is dedicated to Boris Rozovskii on his 60th birthday.

The first paper in the volume, *Stochastic Evolution Equations* by N. V. Krylov and B. L. Rozovskii, was originally published in Russian in 1979 (*Itogi Nauki i Tekhniki, Seriya Sovremennye Problemy Matematiki*, Vol, 14, pp. 71–146). The English translation was first published in the Journal of Soviet Mathematics, Vol. 14, pp. 1233–1277, 1981, ©Plenum Publishing Co. We are very grateful to the current copyright holder, Springer, for the permission to include the paper in the volume. After more than a quarter-century, this paper remains a standard reference in the field of stochastic partial differential equations (SPDEs) and continues to attract attention of mathematicians of all generations, because, together with a short but thorough introduction to SPDEs, it presents a number of optimal and essentially non-improvable results about solvability for a large class of both linear and non-linear equations.

The other papers in this volume were specially written for the occasion. The 14 contributions deal with a wide range of topics in the theory and applications of stochastic differential equations, both ordinary and with partial derivatives.

Eight of the contributions are related to stochastic partial differential equations. *D. Blömker and J. Duan* investigate behavior of the mean energy and correlation function for the Burgers equation with various types of random perturbation. *L. Borcea, G. Papanicolaou, and C. Tsogka* study asymptotics of the space-time Wigner transform for the stochastic Schrödinger equation and apply the results to broadband array imaging in random media. *A. de Bouard and A. Debussche* establish existence and uniqueness of a global square-integrable solution of the stochastic Korteweg-de Vries equation with multiplicative noise. *Z. Brzeźniak and L. Debbi* establish global existence and uniqueness of a mild solution for a fractional Burgers equation with multiplicative space-time white noise. *F. Flandoli and M. Romito* study regularity properties of a transition semigroup associated with the Navier-Stokes equation driven by a non-degenerate additive space-time noise. *I. Gyöngy and A. Millet* establish the rate of convergence of the implicit Euler scheme for a class of nonlinear SPDEs. *N.V. Krylov* proves the maximum principle for a large class of linear stochastic parabolic equations and uses the result to study spatial regularity of the solution for equations on the half-line. *G. Da Prato* investigates the Kolmogorov equation associated with a class of non-linear stochastic parabolic equations.

Six of the contributions are related to stochastic ordinary differential equations. *A. Cadenillas, J. Cvitanic, and F. Zapatero* study a stochastic control problem that models the use of stock options for executive compensation. *P. Chigansky and R. Liptser* establish the large deviations principle for a class of stochastic equations with rapidly growing coefficients and a possibly degenerate diffusion. *D. Crisan and S. Ghazali* study numerical approximation of the probability distribution of the solution at a fixed time for a class of stochastic equations in the Stratonovich form and apply the results to the nonlinear filtering problem. *L. Decreusefond and D. Nualart* show that the solutions of stochastic equation driven by a fractional Brownian motion with Hurst parameter bigger than 1/2 generate a flow of homeomorphisms. *Yu. A. Kutoyants* reviews recent results on statistical inference for a stochastic equation with delay when the delay parameter is unknown. *R. Mikulevicius and H. Pragarauskas* investigate a linear integro-differential equation connected with jump-diffusion processes.

Preparation of this volume was a joint effort by a number of people. We are very grateful to J. Duan, who initiated the project; to Rok Ting Tan and Yubing Zhai, who managed the project at *World Scientific*; to Inge Weijman and Berendina van Straalen, who processed our copyright clearance at *Springer*; to Abhinav Guru, who helped to prepare the TEX file of the paper by Krylov and Rozovskii. Our special gratitude goes to all the contributors to this volume and all the referees who carefully reviewed the submitted papers and assured the highest quality of all contributions.

P. Baxendale
S. Lototsky

Boris Rozovskii

In the life of every mathematician there are usually several key events that facilitate a successful career. For Boris Rozovkii, one such event happened in 1962, when, shortly after entering the Odessa State University, he got a special invitation to transfer to the Faculty of Mechanics and Mathematics (the famous MechMat) at the Moscow State University, and immediately found himself at the center of the Soviet and, to some extent, the world mathematics. The undergraduate program of study at MechMat leads to an analog of Master of Science degree and requires the students to engage in serious research. Boris started to work with I. Girsanov, of the Girsanov Theorem fame. After the tragic death of I. Girsanov in an avalanche, Boris became a student of A. N. Shiryev, and thus an "academic grandson" of A. N. Kolmogorov. While Kolmogorov did not participate directly in the upbringing of his "grandson," his influence through lectures and seminars was strong and beneficial.

From the start, Boris worked at the junction of stochastic analysis and statistics, and with considerable success. His undergraduate work on change detection in a Poisson process was praised by A. N. Kolmogorov, which is probably the highest reward a young mathematician could have received at the time. His Ph.D. dissertation, *On stochastic equations arising in filtering of Markov processes*, published in 1972, became one of the three works that formed the foundation of the modern theory of stochastic partial differential equations (SPDEs); the other two, by E. Pardoux and M. Viot, appeared in 1975 and were also Ph.D. dissertations.

In 1972 Boris became a faculty member at the Moscow Institute of Advanced Studies for Engineers and Managers in Chemical Industry (MIASCME). This institution had a strong statistical group, which Boris joined. Working with chemical engineers was a valuable experience, as it taught him how to effectively communicate mathematics to non-mathematicians and provided ample opportunity to learn about various applications of mathematics. This experience came in very handy later, when Boris became the director of the Center for Applied Mathematical Sciences at the University of Southern California (USC).

Meanwhile, Boris continued to work, both independently and with N. V. Krylov, on nonlinear filtering of diffusion processes and the general theory of SPDEs. The result was a complete L_2 theory for linear SPDEs, as well as several key results for nonlinear equations. The linear theory was summarized in the book *Stochastic Evolution Systems* (1983 in Russian, 1990 in English). While the Russian edition looked more like a brochure than a book, the English translation, with extra material on Malliavin Calculus, was in hard cover and had a much more respectable appearance. The inside of the English translation was a different story; because of a technical glitch, the text contained an enormous number of typos, some of them rather serious. In the end, though, this proved to be a blessing in disguise, as reading the book and correcting the typos became a rite of passage for every student interested in SPDEs; correcting all those typos makes the reading much more challenging and rewarding.

In 1984 Boris was granted the second doctoral degree, Doctor of Science in Physics and Mathematics, by the Vilnius State University. This degree was an important benchmark for a scholar in the former Soviet Union.

In 1988 Boris emigrated to the USA. After a short tenure at the University of North Caroline, Charlotte, he moved to Los Angeles, CA, to become director of the Center for Applied Mathematical Sciences (CAMS) at the University of Southern California. Nonlinear filtering and SPDEs are connected with a large number of mathematical problems, both applied and theoretical, and, while at USC, Boris explored these connections to the fullest. As a Director of CAMS, Boris was very successful in attracting extensive external research funds and became involved in numerous projects in computational fluid dynamics, finance, network security, physical oceanography, target recognition and tracking, and many other areas.

Applied problem not only help to write a successful grant proposal, but can also lead to new and unexpected theoretical developments. For example, physical oceanography, the subject of the first big grant from the Office of Naval Research Boris secured as the director of CAMS, lead to a new estimation theory for SPDEs, developed in his joint papers with M. Huebner, R. Khasminskii, S. Lototsky, and L. Piterbarg. Similarly, numerical computation of the optimal nonlinear filter, the key component of many successful proposals, lead to a new way of analyzing linear and nonlinear SPDEs using Wiener chaos decomposition. Yet another direction of his research at USC was a comprehensive theory of absolute continuity and singularity of measures generated by solutions of nonlinear SPDEs, developed jointly with R. Mikulevicius.

B. L. Rozovskii has published in over 25 different journals with over 40 collaborators. His Erdős number is 4, via several connections, such as A. N. Shiryaev→L. A. Shepp→P. Frankl; R. Z. Khasminskii→O. Zeitouni→P. Diaconis; P. Baxendale→D. Stroock→P. Diaconis.

Not long ago Boris became a grandfather (his granddaughter Mia was born in 2005), and shortly after that, an "academic grandfather." Still, he shows no signs of slowing down, publishing more papers in the past year than in the previous five. In fact, his recent move to the East Coast suggests that he is at the peak of his mathematical career.

Main landmarks of B. L. Rozovskii's scientific career

Boris L'vovich Rozovskii was born on June 8, 1945, in Odessa, Ukraine. Currently, he is Professor at the Division of Applied Mathematics, Brown University.

1968 Graduated from the Department of Mechanics and Mathematics, Moscow State University, with Master of Science degree in Probability and Statistics.
1972 Ph.D. in Physical and Mathematical Sciences, Moscow State University.
1984 Doctor of Science in Physics and Mathematics.

1985 Professor and Head of the Informatics Laboratory, Moscow Institute of Advanced Studies for Engineers and Managers in Chemical Industry.
1991 Professor, Department of Mathematics, USC.
1992 Director, Center for Applied Mathematical Sciences, University of Southern California.
2006 Professor, Division of Applied Mathematics, Brown University.

Graduate students of B. L. Rozovskii

M. Huebner (USC, 1993)
K. Owens (USC, 1994)
A. Fung (USC, 1995)
S. Lototsky (USC, 1996)
C. Rao (USC, 1998)
S. Kligys (USC, 1998)
A. Petrov (USC, 2000)
G. Yaralov (USC, 2000)

B. L. Rozovskii: Honors and Awards

Fellow of the Institute of Mathematical Statistics (1997)
Peter-the-Great Medal (International Academy of Natural and Social Sciences, 1997)
Kolmogorov Medal (Kolmogorov Centennial Conference, 2003)

Publications of B. L. Rozovskii

Books

1. *Stochastic evolution systems. Linear theory and applications to the statistics of random processes* (in Russian). Moscow: *Nauka*, 1983.
2. *Data analysis in chemical research. Statistical foundations* (in Russian). Moscow: *Khimija*, 1984.
3. *Stochastic evolution systems. Linear theory with applications to non-linear filtering.* Mathematics and its Applications (Soviet Series), vol. 35. Dordrecht: *Kluwer Academic Publishers*, 1990.

Papers

1. Wiener chaos solutions of linear stochastic evolution equations (with S. Lototsky). *Ann. Probab.*, 34 (2006), no. 2, 638–662.
2. Wiener chaos expansions and numerical solutions of randomly forced equations of fluid mechanics (with T. How et al.), *J. Comput. Phys.* 216 (2006), no. 2, 687–706.
3. Stochastic differential equations: A Wiener chaos approach (with S. Lototsky). In *From Stochastic Calculus to Mathematical Finance*, ed. Yu. Kabanov et al.,

pp. 433–506. Berlin: *Springer*, 2006.

4. Strong Solutions of Stochastic Generalized Porous Media Equations: Existence, Uniqueness and Ergodicity. (with G. Da Prato et al.) *Comm. Partial Dif. Eq.*, 31 (2006), no. 1-3, 277–291.
5. A novel approach to detection of intrusions in computer networks via adaptive sequential and batch-sequential change-point detection methods (with R. Blazek et al.), *IEEE Transactions on Signal Processing*, to appear.
6. Detection of intrusions in information systems by sequential change-point methods (with A. Tartakovsky et al.). *Stat. Methodol.*, 3 (2006), no. 3, 252–293.
7. Detection of intrusions in information systems by sequential change-point methods. Authors' response (with A. Tartakovsky et al.) *Stat. Methodol.* 3 (2006), no. 3, 329–340.
8. A filtering approach to tracking volatility from prices observed at random times (with J. Cvitanic et al). *Ann. Appl. Probab.*, 16 (2006), no. 3, 1633–1652.
9. Numerical estimation of volatility values from discretely observed diffusion data (with J. Cvitanic et al), *Journal of Computational Finance* (to appear).
10. A novel approach to detection of intrusions in computer networks via adaptive sequential and batch-sequential change-point detection methods (with R. Blazek et al.) *International Journal of Computing and Information Sciences.*
11. Global L_2-solutions of stochastic Navier-Stokes equations (with R. Mikulevicius). *Ann. Probab.*, 33 (2005), No. 1, 137–176.
12. Passive Scalar Equation in a Turbulent Incompressible Gaussian Velocity Field (with S. Lototsky), *Russian. Math. Surveys*, 59 (2004), no.2, 297–312.
13. Stochastic Navier-Stokes equations for turbulent flows (with R. Mikulevicius). *SIAM J. Math. Anal.* 35 (2004), no. 5, 1250–1310.
14. A diffusion model of roundtrip time (with S. Bohacek). Computational Statistics and Data Analysis, *Comput. Statist. Data Anal.*, 45 (2004) no. 1, 25–50.
15. On martingale problem solutions for stochastic Navier-Stokes equations (with R. Mikulevicius). In *Stochastic partial differential equations and applications*, ed. G. Da Prato and L. Tubaro. Lecture Notes in Pure and Applied Mathematics Series 227. New York: Marcel Dekker, 2002.
16. A note on Krylov's L_p-theory for systems of SPDEs (with R. Mikulevicius). *Electron. J. Probab.* 6 (2001), no. 12, 1–35.
17. On equations of stochastic fluid mechanics (with R. Mikulevicius). In *Stochastics in finite and infinite dimensions: in honor of Gopinath Kallianpur*, ed. T. Hida et al., 285–302. Trends Math. Boston: Birkhauser, 2001.
18. Stochastic Navier-Stokes equations: propagation of chaos and statistical moments (with R. Mikulevicius). *In Optimal control and partial differential equations: in honor of Professor Alain Bensoussan*, ed. J. L. Menaldi et al., 258–267. Amsterdam: IOS Press, 2001.
19. Approximation of the Kushner equation of nonlinear filtering (with K. Ito). *SIAM J. Control Optim.* 38 (2000), no. 3, 893–915.

20. Parameter estimation for stochastic evolution equations with non-commuting operators (with S. Lototsky). In *Skorokhod's ideas in probability theory*, ed. V. Korolyuk et al., pp. 271–280. Kiev: *Institute of Mathematics of the National Academy of Sciences of Ukraine*, 2000.
21. Fourier-Hermite expansions for nonlinear filtering (with R. Mikulevicius). *Teor. Veroyatnost. i Primenen.* 44 (1999), no. 3, 675–680. English translation *Theory Probab. Appl.* 44 (2000), no. 3, 606–612.
22. Spectral asymptotics of some functionals arising in statistical inference for SPDE's (with S. Lototsky). *Stochastic Process. Appl.* 79 (1999), no. 1, 69–94.
23. Recursive nonlinear filter for a continuous-discrete time model (with S. Lototsky). *IEEE Trans. Automatic Cont.* 48 (1998), no. 8. 1154–1158.
24. Martingale problems for stochastic PDE's (with R. Mikulevicius). In *Stochastic partial differential equations: six perspectives*, ed. R. Carmona and B. L. Rozovskii, pp. 243–325. Math. Surveys Monogr., vol. 64. Providence, RI: *American Mathematical Society*, 1998.
25. Normalized stochastic integrals in topological vector spaces (with R. Mikulevicius). In *Séminaire de Probabilités XXXII*, pp. 137–165. Lecture Notes in Math, vol. 1686. Berlin: *Springer*, 1998.
26. Linear parabolic stochastic PDE's and Wiener chaos (with R. Mikulevicius). *SIAM J. Math. Anal.* 29 (1998), no. 2, 452–480.
27. Weighted stochastic Sobolev spaces and bilinear SPDE's driven by space-time white noise (with D. Nualart). *J. Funct. Anal.* 149 (1997), no. 1, 200–225.
28. On asymptotic problems of parameter estimation in stochastic PDE's: discrete time sampling (with L. Piterbarg). *Math. Methods Statist.* 6 (1997), no. 2, 200–223.
29. Nonlinear filtering revisited: a spectral approach (with S. Lototsky and R. Mikulevicius). *SIAM J. Control Optim.* 35 (1997), no. 2, 435–461.
30. On asymptotic properties of an approximate maximum likelihood estimator for stochastic PDEs (with M. Huebner and S. Lototsky). In *Statistics and control of stochastic processes*, ed. Yu. M. Kabanov et al. pp. 139–155. River Edge, NJ: *World Scientific*, 1997.
31. Recursive multiple Wiener integral expansion for nonlinear filtering of diffusion processes (with S. Lototsky). In *Stochastic processes and functional analysis*, ed. J. Goldstein et al., pp. 199–208. Lecture Notes in Pure and App. Math., vol. 186. New York: Marcel Dekker, 1997.
32. Maximum likelihood estimators in the equations of physical oceanography (with L. Piterbarg). In *Stochastic modelling in oceanography*, ed. R. Adler et al., pp. 397–421. Progr. Probab., vol. 39. Boston: *Birkhauser*, 1996.
33. On asymptotic properties of maximum likelihood estimators for parabolic stochastic PDE's (with M. Huebner). *Probab. Theory Related Fields* 103 (1995), no. 2, 143–163.
34. On stochastic integrals in topological vector spaces (with R. Mikulevicius).

Stochastic analysis (Ithaca, NY, 1993), pp. 593–602. Proc. Sympos. Pure Math., vol. 57. Providence, RI: *American Mathematics Society*, 1995.

35. Estimates of turbulent parameters from Lagrangian data using a stochastic particle model (with A. Griffa et al.). Journal of Mar. Res. 53 (1995), no. 3, 371–401.
36. Statistics and physical oceanography (with D. B. Chelton et al.). Stat. Sci. 9 (1994), no. 2, 167–201.
37. Uniqueness and absolute continuity of weak solutions for parabolic SPDE's (with R. Mikulevicius). *Acta Appl. Math.* 35 (1994), no. 1–2, 179–192.
38. Soft solutions of linear parabolic SPDE's and the Wiener chaos expansion (with R. Mikulevicius). In *Stochastic analysis on infinite-dimensional spaces*, ed. H. Kunita and H.-H. Kuo, pp. 211–220. Pitman Res. Notes Math. Ser., vol. 310. Baton Rouge, LA: *Longman Sci. Tech, Harlow*, 1994.
39. Kinematic dynamo and intermittence in a turbulent flow. (with P. Baxendale). *Geophys. Astrophys. Fluid Dynam.* 73 (1993), no. 1-4, 33–60.
40. Two examples of parameter estimation for stochastic partial differential equations (with M. Huebner and R. Khasminskii). In *Stochastic processes. A festschrift in honor of Gopinath Kallianpur*, pp. 149–160. New York: *Springer*, 1993.
41. Some results on a diffusion approximation to the induction equation. In *Stochastic partial differential equations and applications (Trento, 1990)*, ed. G. Da Prato and L. Tubaro, pp. 268–281. Pitman Res. Notes in Math. Ser., vol. 268. Baton Rouge, LA: *Longman Sci. Tech, Harlow*, 1992.
42. A simple proof of uniqueness for Kushner and Zakai equations. In *Stochastic analysis*, ed. E. Mayer-Wolf, pp. 449–58. Boston: *Academic Press*, 1991.
43. Measure-valued solutions of second-order stochastic parabolic equations (with O.G. Purtukhiya, in Russian). In *Statistics and control of random processes*, ed. A. N. Shiryaev, pp. 177–79. Moscow: *Nauka*, 1989.
44. On the mathematical theory of a hydromagnetic dynamo in a random flow (in Russian). *Dokl. Akad. Nauk SSSR* 293 (1987), no. 6, 1311–1314.
45. On the statistic estimation of reliability of determining aqueous solution pH by acid-base indicator paper (with V.M. Ostrovskaja et al., in Russian). *J. Analit. Chem. USSR Acad. Sci.* 42 (1987), no. 9, Part 2, 1369–1371.
46. Nonnegative L_1-solutions of second order stochastic parabolic equations with random coefficients. In *Statistics and control of stochastic processes (Moscow, 1984)*, ed. N. V. Krylov et al. pp. 410–427. Transl. Ser. Math. Engrg., New York: *Optimization Software*, 1985.
47. Filtering interpolation and extrapolation of degenerate diffusion processes. Backward equations (in Russian). *Teor. Veroyatnost. i Primenen.* 28 (1983), no. 4, 725–737.
48. Stochastic partial differential equations and diffusion processes (with N. V. Krylov, in Russian). *Uspekhi Mat. Nauk* 37 (1982), no. 6, 75–95.

49. Characteristics of second-order degenerate parabolic Itô equations (with N. V. Krylov, in Russian). *Trudy Sem. Petrovsk.* 8 (1982), 153–168.
50. Smoothness of solutions of stochastic evolution equations and the existence of a filtering transition density (with A. Shimizu). *Nagoya Math. J.* 84 (1981), 195–208.
51. On the first integrals and Liouville equations for diffusion processes (with N. V. Krylov). In *Stochastic differential systems (Visegrad, 1980)*, pp. 117–125. Lecture Notes in Control and Information Sci., vol. 36. New York: *Springer*, 1981.
52. On complete integrals of Itô equations (with N.V. Krylov, in Russian). *Uspekhi Mat. Nauk*, 35 (1980), no. 4, 147.
53. A note on the strong solutions of stochastic differential equations with random coefficients. In *Stochastic differential systems. (Proc. IFIP-WG 7/1 Working Conference, Vilnius, 1978)*, pp. 287–296. Lecture Notes in Control and Information Sci., vol. 25. New York: *Springer*, 1980.
54. Conditional distributions of degenerate diffusion processes (in Russian). *Teor. Veroyatnost. i Primenen.* 25 (1980), no. 1, 149–154.
55. Itô equations in Banach spaces and strongly parabolic stochastic partial differential equations (with N. V. Krylov, in Russian). *Dokl. Akad. Nauk SSSR* 249 (1979), no. 2, 285–289.
56. Stochastic evolution equations (with N. V. Krylov, in Russian), *Current Problems in Mathematics*, vol. 14, pp. 71–147. Moscow: *Akad. Nauk SSSR, Vsesoyuz. Inst. Nauchn. i Tekhn. Informatsii*, 1979. English translation *J. Soviet Math.* 16 (1981), no 4, 1233–1277.
57. Fundamental solutions of stochastic partial differential equations and the filtering of diffusion processes (with L. G. Margulis, in Russian). Uspekhi Mat. Nauk 33, no. 2 (1978), 197.
58. Conditional distributions of diffusion processes (with N. V. Krylov, in Russian). *Izv. Akad. Nauk SSR Ser. Mat.* 42 (1978), no. 2, 356–378.
59. The Cauchy problem for linear stochastic partial differential equations (with N. V. Krylov, in Russian). *Izv. Akad. Nauk SSR Ser. Mat.* 41 (1977), no. 6, 1329–1347.
60. Stochastic partial differential equations (in Russian). *Mat. Sb. (N.S.)* 96 (1975), no. 138, 314–341.
61. Stochastic differential equations in infinite-dimensional spaces and filtering problems (in Russian). In *Proceedings of the School and Seminar on the Theory of Random Processes (Druskininkai, 1974), Part II*, pp. 147–194. Vilnius: *Inst. Fiz. i Mat. Akad. Nauk Litovsk. SSR*, 1975.
62. The Itô-Wentzell formula (in Russian). *Vestnik Moskov. Univ. Ser. I Mat. Meh.* 28 (1973), no. 1, 26–32.
63. On infinite systems of stochastic differential equations that arise in the theory of optimal nonlinear filtering (with A. N. Shiryaev, in Russian). *Teor. Verojatnost.*

i Primenen. 17 (1972), 228–237.

64. Stochastic partial differential equations that arise in nonlinear filtering problems (in Russian). *Uspekhi Mat. Nauk* 27 (1972), no. 3, 213–214.
65. The problem of "disorder" for a Poisson process (with L. I. Galtchuk, in Russian). *Teor. Verojatnost. i Primenen.* 16 (1971), 729–734.

Edited Volumes

1. *Applied Mathematics & Optimization.* Special issue on Approximation in Stochastic Partial Differential Equations, Guest editor B. Rozovskii, *Springer*, 2006.
2. *Stochastic partial differential equations: six perspectives.* Ed. R. Carmona and B. L. Rozovskii. Mathematical Surveys and Monographs, vol. 64. Providence, RI: *American Mathematical Society*, 1998.
3. *Statistics and control of stochastic processes.* The Liptser festschrift: papers from the Steklov Seminar (Moscow, 1995/1996). Ed. Yu. M. Kabanov, B. L. Rozovskii, and A. N. Shiryaev. River Edge, NJ: *World Scientific*, 1997.
4. *Stochastic modelling in oceanography.* Ed. R. Adler, P. Muller, and B. L. Rozovskii. Progress in Probability 39. Boston: *Birkhauser*, 1996.
5. *Stochastic partial differential equations and their applications.* Proceedings of the IFIP WG 7/1 International Conference (Charlotte, NC, 1991). Ed. B. L. Rozovskii and R. B. Sowers. Lecture Notes in Control and Information Sci., vol. 176. Berlin: *Springer*, 1992.

Selected Conference Proceedings

1. Chaos expansions and numerical solutions of randomly forced equations of fluid dynamics (with T. How et al.), Proceedings of the Sixth Helenic-European Conference on Computer Mathematics and its Applications, HERCMA 2003, Vol. 1, ed. E. A. Lipitakis, pp. 12–22.
2. Novel Approach to Detection of "Denial-of-Service" Attacks via Adaptive Sequential and Batch-Sequential Change-Point Detection Methods (with R. Blazek et al.). In *Proceedings of the 2nd Annual IEEE Systems, Man, and Cybernetics Information Assurance Workshop (West Point, NY, 2004).* New York: *Institute of Electrical and Electronics Engineers*, 2004.
3. A New Adaptive Batch and Sequential Methods for Rapid Detection of Network Traffic Changes with Emphasis on Detection of "Denial-of-Service" Attacks, (with R. Blazek and H. Kim). In *Proceedings of the 53rd Session of the International Statistical Institute (Seoul, 2001).* New York: *Physica-Verlag*, 2001.
4. Tracking Volatility (with J. Cvitanic and R. Liptser). In *Proceedings of the 39th IEEE Conference on Decision and Control, IEEE Control Systems Society (Sydney, 2000).* New York: *Institute of Electrical and Electronics Engineers*,

2000.

5. Interactive Banks of Bayesian Matched Filters (with R. Blazek and A. Petrov). In *SPIE Proceedings: Signal and Data Processing of Small Targets (Orlando, FL, 2000)*, vol. 4048, ed. O. E. Drummond. Bellingham, WA: *SPIE (The International Society for Optical Engineering)*, 2000.
6. Optimal nonlinear filtering for track-before-detect in IR image sequences (with A. Petrov). In SPIE Proceedings: Signal and Data Processing of Small Targets (Denver, CO, 1999), vol. 3809, ed. O. E. Drummond. Bellingham, WA: *SPIE (The International Society of Optical Engineering)*, 1999.
7. An adaptive Bayesian approach to fusion of imaging and kinematic data (with A.Tartakovsky and G.Yaralov). In *Proceedings of the 2nd International Conference on Information Fusion (Fusion '99, Sunnyvale, CA, 1999).* Madison, WI: *Omnipress*, 1999.
8. Matched filters and hidden Markov models with distributed observation (with S. Kligys). In *Proceedings of the Fourth Annual U.S. Army Conference on Applied Statistics (Aberdeen Proving Ground, MD, 1998)*, ed. Barry A. Bodt. ARL-SR-84. Aberdeen, MD: *Army Research Laboratory*, 1999.
9. State estimation in hidden Markov models with distributed observation (with S. Kligys). In *Theory and Practice of Control Systems: Proceedings of the 6th IEEE Mediterranean Conference (Alghero, Sardinia, 1998)*, ed. A. Tornambe et al. River Edge, NJ: *World Scientific*, 1998.
10. Splitting-up discretization for Kushner's equation of nonlinear filtering (with K. Ito). In *Proceedings of the 36th IEEE Conference on Decision and Control, IEEE Control Systems Society (San Diego, CA, 1997).* New York: *Institute of Electrical and Electronics Engineers*, 1998.
11. Solving hidden Markov problems by spectral approach (with C.P. Fung). In *Proceedings of the 3rd IEEE Mediterranean Symposium, (PLACE, 1995)*, vol. II. New York: *Institute of Electrical and Electronics Engineers*, 1995.
12. Separation of observations and parameters in nonlinear filtering (with R. Mikulevicius). In *Proceedings of the 32nd IEEE Conference on Decision and Control, (San Antonio, TX, 1993)*, vol.2. New York: *Institute of Electrical and Electronics Engineers*, 1993.
13. Statistics and physical oceanography (with A. Griffa et al.). *Report of the National Research Council.* Washington, D.C.: *National Academy Press*, 1993.
14. Nonlinear filtering revisited: A spectral approach II (with S. Lototsky and R. Mikulevicius). In *Proceedings of the IEEE & SIAM CDC 35th Conference on Decision and Control, (Kobe, Japan, 1996)*, vol. 4. Madison, WI: *Omnipress*, 1997.
15. On the kinematic dynamo problem in a random flow. In *Probability Theory and Mathematical Statistics: Proceedings of 5th Vilnius Conference on Probabilty Theory and Mathematical Statistics, (Vilnius, 1985)*, vol. II. Utrecht: *VNU Science Press*, 1987.

16. On the kinematic dynamo problem in a random flow. In *Probability Theory and Mathematical Statistics: Proceedings of the 4th Vilnius International Conference on Probability Theory and Mathematical Statistics (Vilnius, 1985)*. Vilnius: *Akad. Nauk Litovsk. SSR, Inst. Mat. i Kibernet*, 1985.
17. Filtering of degenerate diffusion type processes. Backward equations. In *Stochastic Optimization International Conf. Abstracts, (Kiev,1984)*, Part II.
18. Backward equations of conditional and unconditional diffusion. In *Proceedings of the 4th Soviet-Japan Symp. on Probab. Theor. and Math. Stat. Abstracts (Tbilisi, 1982)*.
19. Backward filtering equations. In *15th All-Union School-Colloquium on Probab. Theor. and Math. Stat. Abstracts (Bakuriani, 1981)*. Tbilisi: *Metsniereba*, 1981.
20. Liouville equations for a diffusion Markov process (in Russian). In *14th All Union School on Probab. Theor. and Math. Stat. Abstracts (Bakuriani, 1980)*. Tbilisi: *Metsniereba*, 1980.
21. On the first integral and Liouville equations. In Abstracts of 3rd Working Conference on Stochastic Differential Equations (Visegrad, 1980), Budapest: *SZAMKI*, 1980.
22. On the extrapolation of a signal with a martingale type noise (in Russian). In *5th International Symposium on Inform. Theory. Abstracts (Tbilisi, 1979)*.
23. Non-linear filtering of diffusion processes: an analytical approach. In *International Symposium on Stochastic Differential Equations. Abstracts of Communications (Vilnius, 1978)*. Vilnius: *Inst. Math. and Cybernet. Acad. Sci. Lithuanian SSR*, 1978.
24. On Itô equations in Hilbert spaces. In *2nd Vilnius Conference on Probability and Mathematical Statistics. Abstracts of Communications (Vilnius, 1977)*. Vilnius: *Inst. Mat. i Kibernet. Akad. Nauk Litovsk. SSR*, 1977.
25. On Cauchy problem for superparabolic stochastic differential equations (with N.V. Krylov). In *Proceedings of the Third Japan-USSR Symposium on Probability Theory (Tashkent, 1975)*. Lecture Notes in Mathematics, vol. 550. New York: *Springer*, 1976.
26. On stochastic differential equations in partial derivatives. In *International Conference on Probability Theory and Mathematical Statistics. Abstracts of Communications (Vilnius, 1973)*. Vilnius: *Akad. Nauk Litovsk. SSR*, 1973.
27. Reduced form of non-linear filtering equations (with A. N. Shiryaev). In *IFAC Symposium on Stochastic Control. Supplement of abstracts (Budapest, 1974)*.

Selected Technical Reports and Teaching Aids

1. Stochastic Navier-Stokes equations for turbulent flows (with R. Mikulevicius). Warwick Preprint: 21/ 2001.
2. Detection algorithms and track-before-detect architecture based on nonlinear filtering for infrared search and track systems (with S. Kligys and A. Tar-

takovsky). Technical Report CAMS-98.9.1, Center for Applied Mathematical Sciences, University of Southern California, 1998.

3. Domain pursuit method for tracking ballistic targets (with R. Rao and A.Tartakovsky). Technical Report CAMS-98.9.2, Center for Applied Mathematical Sciences, University of Southern California, 1998.
4. Parameter estimation for stochastic evolution equations with non-commuting operators (with S. V. Lototsky). IMA Preprint Series #1501, University of Minnesota, 1997.
5. Statistics and physical oceanography (with A. Griffa et al.). Report of the National Research Council, *National Academy Press*, 1993.
6. Lecture notes on stochastic partial differential equations. University of North Carolina, Charlotte, 1990.
7. Real time statistical quality control (with Yu. P. Adler) (in Russian). Moscow: *Znanije Publishing House*, 1984.
8. Studies in the theory of stochastic partial differential equations (in Russian). Doctor of Sci. Dissertation (Vilnius, Lithuania, 1984).
9. Statistical methods in chemical industry. Methods and instructions (with B. I. Pashko et al., in Russian). Center for Scientific Organization of Labor in Chemistry, Moscow 1983.
10. Mathematical design of experiments in textile industry (with Yu. P. Adler et al., in Russian). MIASCME, 1984.
11. Mathematics design of experiments. Methods and instructions (with Yu. P. Adler et al., in Russian). MIASCME, 1983.
12. Statistical methods in analytical chemistry. Instructions (with Yu. P. Alder et al., in Russian). MIASCME, 1981.
13. Statistical quality control. Methods and instructions (with Yu. P. Alder et al., in Russian). MIASCME, 1978.
14. Optimal design of experiments. Methods (with Yu. P. Alder et al., in Russian). MIASCME, 1978.
15. Lectures in probability theory (in Russian). MIASCME, 1974.
16. On stochastic equations arising in filtering of Markov processes (in Russian). Ph. D. Dissertation, Moscow State (Lomonosov) University, 1972.

Contents

Stochastic Differential Equations: Theory and Applications

Chapter 1

Stochastic Evolution Equations

Nicolai V. Krylov and Boris L. Rozovskii*

The theory of strong solutions of Itô equations in Banach spaces is expounded. The results of this theory are applied to the investigation of strongly parabolic Itô partial differential equations.

Contents

*This paper was originally published in Russian in 1979 (*Itogi Naukt i Tekhniki, Seriya Sovremennye Problemy Matematiki*, Vol, 14, pp. 71–146) The English translation was first published in the Journal of Soviet Mathematics, Vol. 14, pp. 1233–1277, 1981, © Plenum Publishing Co. The permission of the current copyright holder, Springer, to include the paper in this volume is gratefully acknowledged.

1. Introduction

1.1. *Itô Equations in Banach Spaces*

The theory of Itô stochastic differential equations is one of the most beautiful and most useful areas of the theory of stochastic processes. However, until recently the range of investigations in this theory had been, in our view, unjustifiably restricted: only equations were studied which could, in analogy with the deterministic case, be called ordinary stochastic equations. The situation has began to change in the last 10–15 years. The necessity of considering equations combining the features of partial differential equations and Itô equations has appeared both in the theory of stochastic processes and in related areas.

Such equations have appeared in the statistics of stochastic processes (filtering of diffusion process), statistical hydromechanics, population genetics, Euclidean field theory, classical statistical field theory, and other areas. Concrete examples of equations of this type are presented below. These equations describe the evolution in time of processes with values in function spaces or in other words, random fields in which one coordinate — the "time" — is distinguished.

The objective of the present work is to show how to create a unified theory which include both ordinary Itô equations and a rather broad class of stochastic partial differential equations.

We realize our program by considering equations of the Itô type in Banach spaces. More precisely we consider the equation

$$du\,(t,\omega) = A\,(u\,(t,\omega)\,,t,\omega)\,dt + B\,(u\,(t,\omega))\,dw\,(t)\,, \tag{1.1}$$

where $A(\cdot,t,\omega)$ and $B(\cdot,t,\omega)$ are families of unbound operators in Banach spaces, depending on the elementary outcome ω in a non-anticipating fashion, and $w(t)$ is a process with values in some Hilbert space and with independent (in time) increments. Such equations are usually called stochastic evaluation equations.

1.2. *Examples of Stochastic Evolution Equations*

1.2.1. *Linear Equation of Filtring of Diffusion Processes*

Filtering is one of the most important problems in statics of random processes. In essence, it consists of the following [1]. Consider a two-component process $Z = (x,y)$, e.g., the $(n+m)$-dimensional diffusion process

$$\begin{gathered} dx(t) = a(x(t),y(t),t)dt + b(x(t),y(t),t)dw(t), \\ dy(t) = g(x(t),y(t),t)dt + \sigma(y(t),t)dw(t), \\ x\,(0) = x_0, y(0) = y_0, \end{gathered}$$

where $\dim x = n$ and $w(t)$ is a standard $m+n$-dimensional Wiener process. It is assumed that the component x of the process z is nonobservable. Given a function

$f = f(x)$, it is required to find the best mean-square estimate of $f(x(t))$ on the basis of observations of the trajectory of the observable component y up to time t. In other words, this estimate is to be sought as a functional of the component y. It is well known that such an estimate is the conditional expectation of $f(x(t))$ given the σ-algebra $\mathcal{F}_t^y$ generated by the values of y up to time t, i.e., $\mathbb{E}[f(x(t))|\mathcal{F}_t^y]$. The filtering problem consists in computing this conditional expectation.

In Ref. [2] we showed that under broad assumptions

$$\mathbb{E}[f(x(t))|\mathcal{F}_t^y] = \int_{\mathbb{R}^n} f(x)\varphi_t(x)dx \left(\int_{\mathbb{R}^n} \varphi_t(x)dx\right)^{-1}, \tag{1.2}$$

where $\varphi_t(x)$ is the solution of the Cauchy problem

$$\begin{aligned} d\varphi_t(x) &= \{\frac{1}{2}t_r D_{xx}(bb^*\varphi_t(x)) - D_x(a\varphi_t(x))\}dt + [(\sigma\sigma^*)^{-1/2}g\varphi_t(x) \\ &\quad + D_x((\sigma\sigma^*)^{-1/2}\sigma b^*\varphi_t(x))](\sigma\sigma^*)^{-1/2}dy(t),^{\ddagger} \\ \varphi_0(x) &= P(x_0 \in dx)/dx, \end{aligned}$$

D_{xx} is the matrix of the second derivatives, and D_x is the vector of the first derivatives. This is a linear stochastic differential equation with unbounded operators of "drift" and "diffusion".

1.2.2. *Equations of Population Genetics*

One of the most important types of models in population genetics are the models with geographic structure. These are models in which the structure of population change not only in time but also in space (geographically). Various probabilistic models of this sort have been proposed by Bailey [3], Crow and Kinmura [4], Malecot [5], and others. All these models are discrete. Dawson [6] and Fleming [7] proposed continuous (in time and space) models which are limits of the corresponding discrete models. These works of Dawson and Fleming continue conceptually the well-known work of Feller [8].

The equation proposed by Dawson for the mass distribution $p(t,x)$ of the population has the form

$$dp(t,x) = a\Delta p(t,x)dt + c\sqrt{p(t,x)}\,dw(t,x), \tag{1.3}$$

while the equations of Fleming has the form

$$dp(t,x) = \big(\Delta p(t,x) + ap(t,x) - \beta\big)dt + \sqrt{\frac{p(t,x)(1-p(t,x))_+}{2}}\,dw(t,x) \tag{1.4}$$

In both cases Δ is the Laplace operator, α, β, c are constants, $(\alpha)_+ = \max(a,0)$, and $w(t,x)$ is a Wiener process with values in $L_2(\mathbb{R}^d)$, where d is the dimension of

‡ Here $*$ is the symbol for the conjugate; the arguments x, t, and $y(t)$ of the coefficients have been dropped.

x, and with a nuclear covariance operator. This means that $w(t,x)$ is a stochastic process with values in $L_2(\mathbb{R}^d)$, such that, for every function $e \in L_2(R^d)$,

$$w_e(t) = \int_{\mathbb{R}^d} w(t,x)e(x)dx$$

is a one dimensional Wiener process and

$$\mathbb{E}(w_{e_1}(t) - w_{e_1}(s))(w_{e_2}(t) - W_{e_2}(s)) = (t-s)(e_1, Qe_2)_{L_2(\mathbb{R}^d)},$$

where Q is a nuclear operator on $L_2(\mathbb{R}^d)$ and $(\cdot,\cdot)_{L_2(\mathbb{R}^d)}$ is the inner product in $L_2(\mathbb{R}^d)$; for more details, see Ref. [9]).

Wiener processes with values in Hilbert spaces are discussed in more detail in Section 2.2 (see, in particular, Definition 2.15).

1.2.3. *System of Navier-Stokes Equations with Random External Forces*

In the physics literature on the theory of turbulence (see, e.g. Novikov [10], Monin and Yagiom [11], Klyatskin [12] and the literature cited there) a model of the motion of an incompressible fluid is considered under the action of random external forces; the model is described by the following system of Navier-Stokes equations:

$$\begin{aligned} du_i(t,x) &= \left(\nu\Delta u_i(t,x) - \sum_{k=1}^{3} u_k \frac{\partial u_i(t,x)}{\partial x_k} - \frac{\partial p}{\partial x_i}\right) dt + dw_i(t,x), \\ \sum_{k=1}^{3} \frac{\partial u_k}{\partial x_k} &= 0. \end{aligned} \tag{1.5}$$

Here $u = (u_1, u_2, u_3)$ is the velocity vector, p is the pressure, ν, the viscosity, and $w_i(t,x)$ are independent Wiener processes with values in a function space. For equation (1.5) in a cylinder $(0,T) \times G$, where G is a domain in $\mathbb{R}^3$ with boundary Γ, the first boundary value problem has been considered:

$$u(t,x)\,|_{[0,T]\times\Gamma} = 0, \ u(0,x) = u_0(x).$$

1.2.4. *Equation of the Free Field*

Let $\mathfrak{L} = \mathfrak{L}(\mathbb{R}^{d+1})$ be the space of rapidly decreasing function on $\mathbb{R}^{d+1}$, and let $\mathfrak{L}'$ be the dual space of $\mathfrak{L}$, that is, the Schwartz space of slowly increasing generalized functions. We denote by $\mathfrak{G}$ the σ-algebra in $\mathfrak{L}'$, generated by the cylinder sets. On the measurable space $(\mathfrak{L}', \mathfrak{G})$, it is possible to construct a probability measure ν with the characteristic functional

$$C_\nu(\eta) = \int_{\mathfrak{L}'} e^{\sqrt{-1}\,\eta\omega}\nu(d\omega) = \exp\left\{-\left(\eta, \frac{1}{2}\left(-\Delta_{t,x} + m^2\right)^{-1}\eta\right)_{L_2(\mathbb{R}^{d+1})}\right\},$$

where $\eta \in \mathfrak{L}$, $\Delta_{t,x} = \sum_{i=1}^d \frac{\partial^2}{\partial x_i^2} + \frac{\partial^2}{\partial t^2}$, m is a number, and $\eta\omega$ is the value of the functional $\omega \in \mathfrak{L}'$ on $\eta \in \mathfrak{L}$.

It is known (see, e.g., the monograph of Simon [13]) that the free field ξ is one of the simplest objects of relativistic quantum mechanics; in the Euclidean model it can be interpreted as a canonical generalized random field on the probability space $(\mathfrak{L}', \mathfrak{G}, \nu)$: $\xi(\omega, t, x) = \omega(t, x)$ for each $\omega \in \mathfrak{L}'$.

Further, let $\dot{w}$ be a generalized white noise, i.e., the canonical generalized random field on the probability space $(\mathfrak{L}', \mathfrak{G}, \mu)$, where μ is the Gaussian measure with the characteristic operator

$$C_\mu(\eta) = \exp\left(-\frac{1}{2}\|\eta\|^2_{L_2(\mathbb{R}^{d+1})}\right);$$

see e.g. Ref. [9].

Hida and Strett [14] showed that the Euclidean free field $\xi(t,x)$ is a stationary in t solution of the equation

$$\frac{\partial \xi(t,x)}{\partial t} = -\sqrt{-\Delta_x + m^2}\,\xi(t,x) + \dot{w}(t,x), \tag{1.6}$$

where

$$\Delta_x = \sum_{i=1}^{d} \frac{\partial^2}{\partial x_i^2}$$

and the operator $\sqrt{-\Delta_x + m^2}$ is understood in the sense of the theory of generalized functions. Regarding this equation, see also the survey of Dawson [15]. The work of Albeverio and Hoegh-Krohn [16] is a good example of stochastic evolution equations in Euclidean field theory.

The above examples present only a small fraction of stochastic evolution equations considered in recent years. We have selected these examples because they have been studied in detail at a mathematical level of rigor.

Modern physics journals are an inexhaustible source of stochastic evolution equations of the most varied type which are studied only at a physical level of rigor.

1.3. *Stochastic Evolution Equations with Bounded Coefficients and Linear Stochastic Evolution Equations*

The impetus for the first mathematical investigations in the area of stochastic evolution equations were not, however, the demands of physics or biology but rather the inner requirements of mathematics, viz., of the theory of differential equations with variational derivatives. In the mid-sixties, Daletskii and Baklan [17–19] studied stochastic evolution equations with the goal of constructing a solution of the Cauchy problem for the Komogorov equation in variational derivatives

$$-\frac{\partial F(x,t)}{\partial t} = \frac{1}{2}tr[B^*(x,t)F''(x,t)B(x,t)] + A(x,t)F'(x,t),\ t \le T,\ F(x,t) = \phi(x).$$

In these works, a probabilistic method of constructing solutions of the Kolmogorov equation, which is well known in the finite-dimensional case, was used; the method consists in writing the solution in the form $F(x,t) = \mathbb{E}\,[\Phi(u(T))|u(t) = x]$, where $u(t)$ is a solution of the stochastic evolution equation (1.1). To carry out this program, it was necessary to study stochastic equations, and Refs. [17–19] were among the first to address this issue. In these works, B is assumed to be bounded (as an operator) and to satisfy a Lipschitz condition, while A is either the infinitesimal generator of some contractive semigroup or is also bounded and Lipschitz. This direction was further developed by the authors themselves and their students in Refs. [20–24].

Moreover, a number of works [25–32] appeared in which linear stochastic evolution equations were studied for other applications, such as control, filtering, and extrapolation of linear stochastic Itô partial differential equations. From the point of view of proving the existence of a solution, these works are very similar to Ref. [18]: the operator $A(s)$ is always assumed to generate a semigroup $T_{s,t}$, and the operator B is bounded and satisfies a Lipschitz condition (we remark that these conditions are often not satisfied for the filtering equations). In the above works, the following equations was considered:

$$u\,(t) = T_{0,t}u_0 + \int_0^t T_{s,t}B(u(s))dw(s). \tag{1.7}$$

The proof of existence and uniqueness of solution of this equation is accomplished simply by the method of contraction mappings, since $T_{s,t}$ is a bounded operator. It is then proved, or simply taken for granted, that, under additional conditions the solution of equation (1.7) belongs to the domain of the operator A, and equation (1.1) is equivalent to (1.7). It should be noted that the conditions for the equivalence of equations (1.1) and (1.7) obtained in these works are rather burdensome to be useful in application to stochastic Itô partial differential equations, and the reason is similar to the deterministic case. Indeed, the theorems on the solvability of inhomogeneous parabolic equations obtained e.g., by methods of potential theory (see Chapter 1 in Ref. [33]) are much stronger than their counterparts obtained by the theory of inhomogeneous semigroups for linear operator equations [34].

Methods of potential theory in application to linear equations of type (1.1) were used by Rozovskii and Margulis [35, 36]. In these works it is assumed that the "diffusion" operator has order zero, while the "principal part" of the "drift" operator does not depend on the elementary outcome. In the first work, the unique solvability of the Cauchy problem is proved, while in the second a fundamental solution of the equation is constructed, and precise estimates of it in the Hölder classes are obtained. By modifying the method of Ref. [35], Shimizu [37] proved the unique solvability of the second boundary-value problem for the same equation as in Ref. [35].

We mention several works of qualitative character. Markus [38] investigated stationary solutions of linear equations, Shimizu [39] proved some comparison theorems, and Mahno [40] studied the averaging principle for equations of type (1.7).

Bensoussan [41] used a completely different idea to construct solutions of a linear

equations of type (1.1) for $B \equiv 1$. In contrast to all the works cited above, he did not assumed that A generates a semigroup; this assumption was replaced by a coercivity condition. To construct the solution, Bensoussan utilized the method of time discretization. The coercivity condition ensures that the corresponding discrete equation is easily solvable and the solution satisfies the necessary estimates; a weak limiting procedure is then carried out. In the deterministic case, this method was applied earlier by Lions [42].

1.4. *Nonlinear Stochastic Evolution Equations*

The important program started by Bensoussan in Ref. [41] was continued in his joint paper with Temam [43], where the method of discretization was applied to an equation with a nonlinear "drift" operator. It was assumed that $B = 1$, while the operator A satisfies a monotonicity condition.

Theory of monotone operators is one of the most beautiful areas of modern nonlinear analysis. The foundations of this theory were laid in the works of Vainberg, Kachurovskii, Minty, and Browder [44–47]. Further development is reflected in the monographs [42, 48–50]. Application of this theory to nonlinear elliptic and parabolic equations are described in Ref. [51]. The work of Bensoussan and Temam [43] generalizes one of the methods from the monograph of Lions [42].

The method of monotonicity in application to stochastic evolution equations was further developed in the works of Pardoux [52, 53]. In these works a general stochastic evolution equation is considered with unbounded nonlinear operators of "drift" and "diffusion." As a special case, the result of Pardoux contains the results of [41, 43, 54], as well as the deterministic situation described by Lions [42, Section 2.1]. The results of [25–27, 29–32] are also essentially covered, although here a region where they are formally distinct can be indicated: the situation is analogous to the difference between the theory of differential equations in divergence and non-divergence forms.

The solution obtained in Ref. [53] belongs to the domains of the operators A and B and is measurable with respect to the σ-algebra generated by the Wiener process and the initial condition. The solution is constructed on a prescribed probability space; i.e., in correspondence with the terminology of stochastic ordinary differential equations, it is a strong solution.

It should be noted that equations (1.3), (1.4), (1.5), and (1.6) do not satisfy the assumptions of Paradox; in fact, solutions of the equations (1.3), (1.4), (1.5) are constructed in the sense of (probabilistic) distributions. The works [55–61] are devoted to an investigation of the Navier-Stokes equation (1.5), while Eqs. (1.3) and (1.4) are investigated in Refs. [15, 61]. In these works, a measure is sought which is supported on the trajectories of solutions and is a solution of either the Kolmogorov or the Hopf equation (formally) associated with the equation in question.

1.5. *Content and Organization of the Work*

In the present work, we generalize the results of Pardoux. In considering the same situation as in Ref. [53], we admit dependence of the coefficients on the elementary outcome in a nonanticipating way. We show that certain conditions of Pardoux are superfluous; in particular, this is the case of the local Lipschitz condition on the "diffusion coefficient." The Markov property of the solution (in t) is proved for equations with deterministic coefficients. On the basis of general results regarding the solvability of stochastic evolution equations, we investigate quasilinear stochastic Itô partial differential equations (of any order) which satisfy the so-called condition of strong probability. In the deterministic case ($B \equiv 0$) this condition coincides with the well-known condition of strong parabolicity of Vishik [62]. In the linear case, we introduced the condition of strong parabolicity [63] and systematically investigated it [64]. An analogous condition in the linear situation was introduced also by Pardoux [53].

We point out that in the finite-dimensional case our results generalize somewhat Itô's classical theorem on the strong solvability of stochastic equations with random coefficient satisfying Lipschitz conditions [34, 65] (see Example 3.15).

The paper is organized as follows. Aside from the introduction, it contains three big sections, each with its own introduction outlining the content and organization. Section 2 is devoted to the theory of stochastic integration in Hilbert spaces. In Section 2.2, the concept of a martingale and of a Wiener process in a Hilbert space is introduced, and stochastic integrals over these processes are described. In Sections 2.3 and 2.4, Itô's formula for the square of the norm of a semimartingale in a rigged Hilbert space is proved. This result plays an extremely important role in the entire theory.

Section 3 is devoted to the proof of the main theorems about solvability and regularity of solutions of stochastic evolution equations.

In Section 4, the results of Section 3 are applied to the Itô stochastic partial differential equations.

Sections 2.3 and 2.4, where Itô's formula for the square of the norm of a semimartinagle in a Hilbert space is establishes, and Section 3.4, where the finite-dimensional case is considered, may be of independent interest for some readers. The exposition and notations in these sections are independent of the remainder of the work. On the other hand, we go much further here than is required by the rest of the text, so that a reader interested in a rapid acquaintance with the ideas of Itô's stochastic partial differential equations should concentrate on the statements of the main results in Sections 2.3, 2.4, and 3.4, while skipping most of the proofs.

2. Stochastic Integration in Hilbert Spaces

2.1. *Introduction*

The theory of stochastic integration in infinite-dimensional spaces is a broard and rapidly developing area of the theory of stochastic processes. A complete survey of this area is beyond the purpose of the present work, and we will discuss only "selected" questions of the theory of stochastic integration in Hilbert spaces which we use directly in the sequel.

There are two groups of these "selected" questions:

a) construction of a stochastic integral over a square-integrable martingale with values in a Hilbert space;
b) derivation of Itô's formula for the square of the norm of a semimartinagale is a rigged Hilbert space.

In Section 2.2 we consider questions of group a) in a very simple situation — the integration with respect to a continuous martingale. This section is entirely of survey character and there are practically no proofs given. It is based on Refs. [32, 66, 67].

The problem formulated in part b) occupies a central spot and is solved in Sections 2.3 and 2.4. It is assumed that a Hilbert space H is rigged by a pair of Banach spaces V and V', i.e. there are the following dense embeddings: $V \subset H \subset V'$), and we consider a "semimartingale" of the form

$$v(t) = \int_0^t v'(s)\, ds + m(t),$$

where $v \in V$, $v' \in V'$, and $m(t)$ is a martingale in H. Itô's formula is derived for $\|v(t)\|_H^2$.

The derivation of Itô's formula for the square of the norm of a semimartingale with all components take value in a single Hilbert space differs little from the finite-dimensional case. However, in passing from this situation to the one described above, the jump in complexity is comparable to the jump in complexity in passing from bounded to unbounded operators. This comparison is more apt than is apparent at a first glance. We will see below that evolution equations with unbounded operators cannot be considered in a single space — it is necessary to separate the domain of the operators (the space V) from the range of the operators (the space V' or H). This situation was first considered by Pardoux [53]. Using entirely different methods, we generalize his result to non-reflexive and non-separable spaces and provide a much sorter proof.

Itô's formula for the square of the norm is essentially used in Section 3 to derive a priory estimates and to prove uniqueness and continuity of the solution.

To conclude the section we give a brief and, of course, incomplete survey of existing results related to stochastic integrals in Hilbert spaces.

The first important work in this direction was apparently the work of Daletskii [17], where he constructed a Wiener process with an identity covariance operator in a Hilbert space, or, more precisely, in a certain nuclear extension of this space, and defined a stochastic integral. Later, on the basis of the work of Gross [68, 69], Kuo [70] studied a stochastic integral with respect to an abstract Wiener process in a Banach space. The results of Daletskii were also extended in Refs. [71–73], and others.

Kunita [66] initiated the study of the problem of integrability with respect to a square-integrable martingale in a Hilbert space. Metivier and his student subsequently achieved considerable progress in this direction [31, 32, 40, 73–75]. The important work of Meyer [67] should also be mentioned.

The interested reader can systematically study the subject using Refs. [40, 67, 70, 74, 76].

2.2. *Stochastic Integrals in Hilbert Spaces*

We begin this section by recalling and discussing some very basic concepts and results about mappings of measurable spaces into Banach spaces.

Let (S, Σ, μ) be a complete measure space, and let $(X, \mathcal{X})$ be a Banach space with the σ -algebra $\mathcal{X}$ of Borel sets relative to the strong topology. We denote by X^* the Banach space dual to X and by xx^* the value of a functional $x^* \in X^*$ on $x \in X$.

A mapping (function) $x : S \to X$ is called measurable if, for each $\Gamma \in \mathcal{X}$, $\{s : x(s) \in \Gamma\} \in \Sigma$.

A mapping $x : S \to X$ is called weakly measurable if for each $x* \in X^*$, $x(s)x^*$ is a measurable mapping of S into $\mathbb{R}$.

A mapping $x : S \to X$ is called strongly measurable if there exists a sequence of measurable simple functions converging to $x(s)$, μ-almost surely.

It is clear that the concept of strong measurability coincides with the concept of measurability if X is separable.

Theorem 2.1 (Pettis). *A mapping $x : S \to X$ is strongly measurable if and only if it is weakly measurable and there exists a set $B \subset \Sigma$ such that $\mu(B) = 0$, while the set of values of $x(s)$ on $S \setminus B$ is separable. In particular, if X is separable, then the concept of weak and strong measurability coincide.*

Proof. See Yosida [51]. □

Let $(\Omega, \mathcal{F}, P)$ be a probability space with an expanding system of σ-algebras $\mathcal{F}_t \subset \mathcal{F}$, $t \in \mathbb{R}_+ = [0, \infty)$. We assume that $\mathcal{F}_0$ is complete with respect to the measure P.

Definition 2.2. A random variable in X is a measurable mapping of (Ω, F, P) into X.

Definition 2.3. We say that $x = x(t, \omega)$ is an $\mathcal{F}_t$-consistent stochastic process in X if, for every $t \geq 0$, $x(t, \cdot)$ is a measurable mapping of $(\Omega, \mathcal{F}_t, P)$ into X.

Definition 2.4. Completely measurable sets are the subsets of $[0, \infty) \times \Omega$ that are elements of the smallest σ-algebra relative to which all real $\mathcal{F}_t$-consistent right continuous process without discontinuities of the second kind are measurable in (t, ω). A mapping (process) $x : [0, \infty) \times \Omega \to X$ is called completely measurable if for any Borel set $\Gamma \subset X$ the set $\{(t, \omega) : \ x(t, \omega) \in \Gamma\}$ is completely measurable.

In this paper we must sometimes consider random variables with values in embedded Banach spaces. Let V and H be two separable Banach spaces, where V is a subset of H and the embedding operator assigning to an element $v \in V$ the corresponding element $v \in H$ is a continuous operator from V to H.

Lemma 2.5. *a) If x is a random variable with values in V (relative to the Borel σ-algebra in V), then x is a random variable with values in H.*

b) If x is a random variable with values in H, then $\{\omega : x(\omega) \in V\} \in \mathcal{F}$.

Proof. Assertion a) of the lemma is a consequence of the measurability of a superposition of measurable mappings. Assertion b) follows from the completeness of $\mathcal{F}$ and the fact that the continuous image of a Borel (in V) set of V is analytic and is hence universally measurable in H; cf. Ref. [77]. □

We now define a martingale with values in a real Hilbert space H and the stochastic integral over such a martingale. We will consider only martingales $m = m(t)$ that are strongly continuous in t and have a separable range. We temporarily denote by H_1 the closed linear hull of the range of $m(t, \omega)$, $t \geq 0$, $\omega \in \Omega$.

If $h(t, \omega) \perp H_1$, then, for all (t, ω), it is natural to set $\int_0^t h\,(s)\,dm\,(s) = 0$. Therefore, in view of the orthogonal decomposition of H into H_1 and H_1^L, it suffices to study integration of functions with values in H_1. These arguments make the following assumption natural, and we adopt it to the end of the section: H is a separable Hilbert space, identified with its dual in the natural way.

For $h_1, h_2 \in H$ we denote by $h_1 h_2$ the scalar product of h_1, h_2; $h_1^2 = h_1 h_1, |h_1| = (h_1^2)^{1/2}$.

For random variables in H with finite expectation of the norm, it is possible to define the conditional expectation in complete analogy to the finite-dimensional case. Namely, let $\mathcal{G}$ be a sub-σ-algebra of $\mathcal{F}$ and let x be a random variable in H with $\mathbb{E}|x| < \infty$.

Definition 2.6. The conditional expectation of x given $\mathcal{G}$ is the random variable $\mathbb{E}(x|\mathcal{G})$ in H such that, for every $y \in H$,

$$y\mathbb{E}(x|\mathcal{G}) = \mathbb{E}(yx|\mathcal{G}) \quad \text{(a.s.)}.$$

It is clear that the random variable so defined in unique (a.s.).

Definition 2.7. A stochastic process x in H is a martingale relative to the family $\{\mathcal{F}_t\}$ if

a) x is $\mathcal{F}_t$-consistent,

b) $\mathbb{E}|x(t)| < \infty$ for all $t \geq 0$,

c) $\mathbb{E}[x(t)|\mathcal{F}_s] = x(s)$ $(a.s.)$ for all $0 \leq s \leq t$.

The next theorem follows immediately from Definitions 2.6 and 2.7, and the equivalence of the three concepts of measurability in separable spaces.

Theorem 2.8. *A stochastic process $x(t)$ in H with $\mathbb{E}|x(t)| < \infty$ for all $t \geq 0$ is a martingale relative to the family $\mathcal{F}_t$ if and only if, for every $y \in H$, the stochastic process $yx(t)$ is a one-dimensional martingale relative to $\mathcal{F}_t$.*

Definition 2.9. As in the finite-dimensional case, we say that a stochastic process $x(t)$ in H is a local martingale[a] and write $x \in \mathfrak{M}_{loc}(\mathbb{R}_+, H)$ if there exists a sequence of stopping times $\tau_n \uparrow \infty\,(a.s.)$ such that the process $x(t \wedge \tau_n)$ is a martingale for each n. We call the sequence $\{\tau_n\}$ a localizing sequence.

For simplicity we henceforth consider only (strongly) continuous martingales and local martingales.

We denote by $\mathfrak{M}^c_{loc}(\mathbb{R}_+, H)$ the collection of all continuous local martingales in H issuing from zero. It is easy to see that the following results holds.

Theorem 2.10. *If $x \in \mathfrak{M}^c_{loc}(\mathbb{R}_+, H)$, then there exists a sequence $\{\tau'_n\}$ of stopping times localizing x for which $\mathbb{E}\sup_{t\geq 0}|x(t \wedge \tau'_n)|^2 < \infty$. If $\mathbb{E}|x(t)|^2 < \infty$ for some $t \geq 0$, then $\mathbb{E}\sup_{s<t}|x_s|^2 \leq 4\mathbb{E}|x_t|^2$.*

We assume from now on that, for every local martingale, the localizing sequence $\{\tau_n\}$ from Definition 2.9 coincides with the sequence $\{\tau'_n\}$ from Theorem 2.10.

The next theorem plays an important role in the construction of a stochastic integral with respect to $x \in \mathfrak{M}^c_{loc}(\mathbb{R}_+, H)$.

Theorem 2.11. *If $x \in \mathfrak{M}^c_{loc}(\mathbb{R}_+, H)$, then $|x(t)|^2$ is a local submartingale.*

Proof. This follows from the equality

$$\mathbb{E}\big(|x(t \wedge \tau_n) - x(s \wedge \tau_n)|^2|\mathcal{F}_s\big) = \mathbb{E}\big[(|x(t \wedge \tau_n)|^2|\mathfrak{F}_s\big) - |x(s \wedge \tau_n)|^2,$$

which, in turn, follows immediately from the martingale property of the Fourier coefficient of x in a basis in H. □

Definition 2.12. An increasing process $\langle x\rangle_t$ for $x \in \mathfrak{M}^c_{loc}(\mathbb{R}_+, H)$ is the increasing process in the Doob-Meyer decomposition of $|x(t)|^2$.

From the Doob-Meyer theorem it follows that $\langle x\rangle_t$ is a.s. uniquely defined and continuous in t.

[a]Here and below we consider martingales and local martingales only relative to the family $\{\mathcal{F}_t\}$.

As in the finite-dimensional case, if $x, y \in \mathfrak{M}^c_{loc}(\mathbb{R}_+, H)$, then we set

$$\langle x, y\rangle_t = \frac{1}{4}\Big(\langle x+y\rangle_t - \langle x-y\rangle_t\Big).$$

It is easy to verify that if the first assertion of Theorem 2.10 holds with the same sequence τ_n for both $x(t)$ and $y(t)$, then, for each $t, s \geq 0$, $t \geq s$, and each n, we have

$$\begin{aligned}\mathbb{E}\big((x(t\wedge\tau_n) - x(s\wedge\tau_n))(y(t\wedge\tau_n) - y(s\wedge\tau_n))|\mathcal{F}_s\big)\\ = \mathbb{E}\big(\langle x, y\rangle_{t\wedge\tau_n} - \langle x, y\rangle_{s\wedge\tau_n}|\mathcal{F}_s\big) \quad (a.s.)\end{aligned}$$

Theorem 2.13 (Burkholder's Inequality). *If $x \in \mathfrak{M}^c_{loc}(\mathbb{R}_+, H)$ and τ is an a.s. finite stopping time, then*

$$\mathbb{E}\sup_{t\leq\tau}|x(t)| \leq 3\mathbb{E}\langle x\rangle_\tau^{1/2}.$$

A proof of this theorem can be found, e.g., in Ref. [77].

In H we fix an orthogonal basis $\{h_i,\ i \geq 1\}$ and set $x^i(t) = h_i x(t)$. It is known that, almost surely for each i, j, the measure on $[0, \infty)$ generated by the process $\langle x^i, x^j\rangle_t$ is absolutely continuous with respect to the measure generated by $\langle x\rangle_t$.

For further exposition of the theory of martingales, we need the following construction. Let E be a separable Hilbert space which is naturally identified with its dual, let $\{e_i\, i \geq 1\}$ be an orthonormal basis in E, let $\mathcal{L}(H, E)$ be the space of continuous linear operators from H to E, and let $\mathcal{L}_2(H, E)$ be subspace of $\mathcal{L}(H, E)$ consisting of all Hilbert-Schmidt operators. It is known that $\mathcal{L}_2(H, E)$ is a separable Hilbert space with the norm

$$\|B\| = \left(\sum_i |Bh_i|^2\right)^{1/2} = \left(\sum_{i,j}(e_j, Bh_i)^2\right)^{1/2},$$

and $\|B\|$ does not depend on the choice of bases in H and E.

Given a symmetric nonnegative nuclear operator Q in $\mathcal{L}(H, H)$, we denote by $\mathcal{L}_Q(H, E)$ the set of all linear (bounded or unbounded) operators B defined on $Q^{1/2}H$, taking $Q^{1/2}H$ into E and having the property $BQ^{1/2} \in \mathcal{L}_2(H, E)$.

For $B \in \mathcal{L}_Q(H, E)$ we define $|B|_Q = \|BQ^{1/2}\|$. It is known that if $B \in \mathcal{L}_2(H, E)$, then $|B| \leq \|B\|$,

$$B \in \mathcal{L}_Q(H, E), \text{ and } |B|_Q \leq |B|(trQ)^{1/2}.$$

We return to $x \in \mathfrak{M}^c_{loc}(\mathbb{R}_+, H)$. It can be shown that there exists a completely measurable process $Q_x = Q_x(t, \omega)$ with values in $\mathcal{L}_2(H, H)$ such that, for all (t, ω), the operator $Q_x(t, \omega)$ is symmetric, nonnegative, and nuclear, $trQ_x(t, \omega) = 1$, and

$$h_i Q_x(t, \omega) h_j = \frac{d\langle x^i, x^j\rangle_t}{d\langle x\rangle_t} \quad (dP \times d\langle x\rangle_t \text{ a.s.})$$

for all i, j and any basis $\{h_i\}$, where $dP \times d\langle x \rangle_t$ is the differential of the measure

$$\mu(A) = \mathbb{E} \int_0^\infty \chi_A(t, \omega) d\langle x \rangle_t,$$

defined on the product of $\mathcal{F}$ and the Borel σ-algebra on $[0, \infty)$. We call the process Q_x the correlation operator of x.

If $B(t)$ is a completely measurable process in $\mathcal{L}_2(H, E)$ and

$$\mathbb{E} \int_0^t \|B(s)\|^2 d\langle x \rangle_s < \infty \quad \text{(a.s.)}$$

for any $t \geq 0$, then there exists a square-integrable martingale $y(t)$ in E which is strongly continuous in t and such that, for every orthonormal basis $\{h_i\}$ and every $v \in E$, $T \geq 0$,

$$\lim_{n \to \infty} \mathbb{E} \sup_{t \leq T} \left| v\, y(t) - \sum_{i=1}^n \int_0^t vB(s) h_i d(h_i x(s)) \right|^2 = 0.$$

Two processes possessing this property obviously coincide a.s. for all t. It is therefore correct to write

$$y(t) = \int_0^t B(s) dx(s). \tag{2.1}$$

To compute $y(t)$ we fix bases in H, E and set

$$y(t) = \sum_{i=1}^\infty e_i y^i(t), \qquad y^i(t) = \sum_{j=1}^\infty \int_0^t e_i B(s) h_j d(h_j x(s)),$$

where the series converges in $L_2(\Omega, C([0,T], E))$, i.e. in the mean-square, uniformly on each finite time interval. By direct computation,

$$\langle y \rangle_t = \int_0^t |B(s)|^2_{Q_s(s)} d\langle x \rangle_s, \tag{2.2}$$

and this together with the inequality $|B|_{Q_x} \leq \|B\|$ leads to an extension of the stochastic integrals (2.1) to completely measurable functions $B(s)$ such that, for all $t \geq 0$,

$$\int_0^t \|B(s)\|^2 d\langle x \rangle_s < \infty.$$

This extension is done in the usual way and preserves the property (2.2). The resulting stochastic integral is continuous in t and is a local martingale. The following theorem shows that it is possible to extend the concept of a stochastic integral to an even larger class of processes $B(s)$.

Theorem 2.14. *Let $B = B(s, \omega) \in \mathcal{L}_{Q_x(s,\omega)}(H, E)$ be such that $BQ_x^{1/2}$ is a completely measurable process in $\mathcal{L}_2(H, E)$, and for each t the right-hand side of (2.2)* *is a.s. finite. Then the sequence*

$$y_n(t) = \int_0^t B(s)Q_x^{1/2}(s)\left(\frac{1}{n} + Q_x^{1/2}(s)\right)^{-1} dx(s)$$

converges in probability, uniformly with respect to t, to a limit denoted by $y(t)$. Moreover, $y(t) \in \mathfrak{M}^c_{loc}(\mathbb{R}_+, E)$ and equality (2.2) holds.

Proof. Note that $B_n(s) = B(s)Q_x^{1/2}(s)(1/n + Q_x^{1/2}(s))^{-1}$ is a completely measurable process with values in $\mathcal{L}_2(H, E)$ and

$$\|B_n(s)\| \le n|B(s)|_{Q_x(s)},$$

$$|B_n(s) - B_m(s)|_Q^2 = \sum_{i\ge1} |B(s)Q_x^{1/2}(s)e_i(s)|^2 \left(\frac{a_i(s)}{\frac{1}{n} + a_i(s)} - \frac{a_{i(s)}}{\frac{1}{m} + a_i(s)}\right)^2,$$

where $e_i(s)$ are the eigenvectors of $Q_x(s)$ and $a_i^2(s)$ are the corresponding eigenvalues. It is evident from this that $|B_n(s)-B_m(s)|_Q \le |B(s)|_Q$ and $|B_n(s)-B_m(s)|_Q \to 0$ as $n, m \to \infty$. Hence,

$$\langle y_n - y_m \rangle_t = \int_0^t |B_n(s) - B_m(s)|^2_{Q_x(s)} d\langle x \rangle_s \to 0,$$

and the assertions of the theorem are deduced in the well-known way. □

Stochastic integrals for functions in $\mathcal{L}_{Q_x}(H, E)$ have previously been defined in Ref. [32], but the construction presented here is somewhat different.

By definition, the process $y(t)$ in Theorem 2.14 is taken equal to the right-hand side of (2.1). If X is a separable Hilbert space and $A \in \mathcal{L}(E, X)$, then for all t (a.s.)

$$Ay(t) = \int_0^t AB(s)dx(s). \tag{2.3}$$

We choose an element $e \in E$ and by means of it define an operator $\widehat{e} \in \mathcal{L}(E, \mathbb{R})$ by the formula $\widehat{e}y = ey$, where ey is the scalar product in E. From (2.3) we then have

$$ey(t) = \int_0^t \widehat{e}B(s)dx(s).$$

We observe that the operator $\widehat{e}B(s)$ acts on $Q_x^{1/2}H$ by the formula $h \to eB(s)h$, while the latter is equal to (B^*e, h) if $B \in \mathcal{L}(H, E)$. Finally, if $h(s) \in H$ is completely measurable, and for all $t \ge 0$,

$$\int_0^t |\widehat{h}(s)|^2_{Q_x(s)} d\langle x \rangle_s = \int_0^t |Q_x^{1/2}(s)h(s)|^2 d\langle x \rangle_s < \infty \quad (a.s.),$$

then we write

$$\int_0^t h(s)dx(s) = \int_0^t \widehat{h}(s)dx(s).$$

We now introduce the concept of a Wiener process in H.

Definition 2.15. Let Q be a nuclear symmetric nonnegative operator on H with $trQ < \infty$. A Wiener process relative to $\{\mathcal{F}_t\}$ in H with covariance operator Q is a continuous martingale $w(t)$ with values in H, correlation operator $(trQ)^{-1}Q$, and such that $w(0) = 0$, $\langle w \rangle_t = trQ \cdot t$.

It is known that, for every suitable probability space and every nuclear symmetric nonnegative operator Q on H with $trQ > 0$, there exists a Wiener process with covariance operator Q. Clearly, $\mathbb{E}w^2(t) = trQ \cdot t$. Stochastic integrals with respect to a Wiener process possess especially good properties. For example, they are defined not only for completely measurable $B(s)$ but also for operators measurable in (s, ω) that are $\mathcal{F}_s$-consistent and satisfy $\int_0^t |B(s)|_Q^2 ds < \infty$ (a.s.) for all $t \geq 0$.

We conclude this section with a remark that, in place of an infinite time interval above, we could consider a segment of the form $[0, T]$. To have the possibility of doing this, it suffices to extend the processes in question to $t \geq T$ by setting them equal to the value they assume at $t = T$.

2.3. *Itô's Formula for the Square of the Norm*

Let V be a Banach space, let V^* be the dual space of V, and let H be a Hilbert space (we assume that all three are real spaces). If $v \in V$ ($h \in H$, $v^* \in V^*$), then $|v|$ (resp. $|h|, |v^*|$) denotes the norm of v (h, v^*) in V (H, V^*); if $h_1, h_2 \in H$, then $h_1 h_2$ denotes the scalar product of h_1, h_2; the result of the action of a functional $v^* \in V^*$ on an element $v \in V$ is written as either vv^* or v^*v. Let Λ be a bounded, linear operator acting from V to H such that ΛV is dense in H.

We consider three processes $v(t, \omega) \in V$, $h(t, \omega) \in H$, $v^*(t, \omega) \in V^*$ defined for $t \geq 0$ on some complete probability space $(\Omega, \mathcal{F}, P)$ and connected with some expanding family of complete σ-algebras $\mathcal{F}_t \subset \mathcal{F}$, $t \geq 0$, in the following way:

(1) $v(t, \omega)$ is strongly measurable in (t, ω) and is weakly $\mathcal{F}_t$-measurable in ω for almost all t;
(2) for every $g \in V$ the quantity $gv^*(t, \omega)$ is $\mathcal{F}_t$- measurable in ω for almost every t and is measurable in (t, ω);
(3) $h(t, \omega)$ is strongly continuous in t, is strongly measurable in ω relative to $\mathcal{F}_t$ for each t, and is a local semimartingale: there exist strongly $\mathcal{F}_t$-measurable continuous processes $A(t), m(t)$ in H such that $m(t)$ is a local martingale, the trajectories $A(t, \omega)$ for each ω have a finite variation on bounded time intervals, and $h(t) = A(t) + m(t)$.

We fix $p \in (1, \infty)$ and set $q = p/(p-1)$. We assume that $|v(t)| \in L_p([0, T])$ $(a.s.)$ for all $T \geq 0$ and there exists a function $f(t, \omega)$ measurable in (t, ω) such that $f(t) \in L_q([0, T])$ $(a.s.)$ for all $T \geq 0$ and $|v^*(t)| \leq f(t)$ for all (t, ω). Regarding the last condition, it is useful to note that $|v^*(t)|$ is, generally speaking, not measurable. This norm is measurable, e.g., if V is separable, and in this case $|v^*(t)| \in L_p([0, T])(a.s.)$ for all $T \geq 0$.

We formulate the main result regarding Itô's formula.

Theorem 2.16. *Let τ be a stopping time and suppose that, for every $g \in V$ almost everywhere on the set $\{(t,\omega) : t < \tau(\omega)\}$,*

$$(\Lambda g)(\Lambda v(t)) = \int_0^t g v^*(s) ds + (\Lambda g) h(t). \tag{2.4}$$

Then there exist a set $\Omega' \subset \Omega$ and a function $\tilde{h}(t)$ with values in H such that

a) $P(\Omega' = 1)$, $\tilde{h}(t)$ is strongly $\mathcal{F}_t$- measurable on the set $\{\omega : t < \tau(\omega)\}$ for any t, $\tilde{h}(t)$ is continuous in t on $(0, \tau(\omega))$ for every ω, and $\Lambda v(t) = \tilde{h}(t)$ (a.s. on $\{(\tau,\omega) : t < \tau(\omega)\}$;

b) for $\omega \in \Omega'$ and $t < \tau(\omega)$

$$\tilde{h}^2(t) = h^2(0) + 2\int_0^t v(s)v^*(s)ds + 2\int_0^t \tilde{h}(s)dh(s) + \langle m\rangle_t; \tag{2.5}$$

c) if V is separable, then, for $\omega \in \Omega'$, $t < \tau(\omega)$, $g \in V$

$$(\Lambda g)\tilde{h}(t) = \int_0^t g v^*(s) ds + (\Lambda g) h(t); \tag{2.6}$$

d) if V is separable and (2.4) is satisfied for some $t \geq 0$ and each $g \in V$(a.s.) on $\{\omega : t < \tau(\omega)\}$, then $\Lambda v(t) = \tilde{h}(t)$ (a.s.) on $\{\omega : t < \tau(\omega)\}$.

We take up the proof of this theorem after discussing its hypotheses and assertions.

The stochastic integral in (2.5) exists if $\tilde{h}(s)$ is completely measurable and, for $t < \tau(\omega)$,

$$\int_0^t |\tilde{h}(s)| d\|A\|_s + \int_0^t |\tilde{h}(s)|^2 d\langle m\rangle_s < \infty,$$

where

$$\|A\|_s = \lim_{n\to\infty} \sum_{k=0}^{\infty} \left| A\left(s \wedge \frac{k+1}{2^n}\right) - A\left(s \wedge \frac{k}{2^n}\right)\right|.$$

Both these conditions are satisfied because $\tilde{h}(s)$ is continuous in s and is $\mathcal{F}_s$-consistent, while $\langle m\rangle_t + \|A\|_t < \infty$. We point out that by a stochastic integral we always understand a continuous (for all ω) process.

Further, since $v(s)$ is strongly measurable, while $v^*(s)$ is weakly measurable, $v(s)v^*(s)$ is measurable in (s,ω), and by our assumptions it is locally integrable in s (a.s.) All expressions in (2.5) are therefore meaningful.

Assertion d) is a simple corollary of the preceding assertions. Indeed, from (2.4) and (2.6) we have $(\Lambda g)\tilde{h}(t) = (\Lambda g)\Lambda v(t)$ (a.s.) on $\{\omega : t < \tau(\omega)\}$ for every $g \in V$. But V is separable, and therefore $(\Lambda g)\tilde{h}(t) = (\Lambda g)\Lambda v(t)$ for all $vg \in V$, a.s. on $\{\omega : \tau < \tau(\omega)\}$. Since ΛV is dense in H, assertion d) follows.

We now show that it is enough to prove the theorem for separable H and V Indeed, since process $v(t)$ is strongly measurable in (t, ω), there exists a process $v'(t)$ which coincides with $v(t)$ for almost all (t, ω) and has its range in some separable subspace $V' \subset V$. It is shown similarly that on a set of full probability Ω'' all the values of $h(t, \omega)$, $t \geq 0$, $\omega \in \Omega''$ lie in some separable subspace $H' \subset H$. We may assume with no loss of generality that $\Omega'' = \Omega$. By hypothesis, ΛV is dense in H. Hence, there exists a separable subspace $V'' \subset V$ such that $H' \subset \overline{\Lambda V''}$. Suppose now that V_1 is the closed space spanned by $V' \cup V''$ and H_1 is the closed space spanned by ΛV_1. The spaces V_1, H_1 are separable, and ΛV_1 is dense in H_1. Further, $v'(t) \in V_1$, $h(t) \in H_1$, while the functionals $v^*(t)$ on V are also functionals on V_1. It may therefore be assumed that $v^*(t) \in V_1^*$ and the norm of $v^*(t)$ in V_1^* is no larger than $|v^*(t)|_{V^*}$. Relation (2.4) is preserved for every $g \in V_1$ (even for $g \in V$), because $v(t) = v'(t)$ a.s. in (t, ω). Hence, if Theorem 2.16 is true for separable V and H, then we obtain the fist assertion in the general case by applying it to V_1, H_1.

We may thus assume with no loss of generality until the end of the section that V and H are separable.

We will now explain why (2.5) is called Itô's formula for the square of the norm. For this purpose we place all processes $v(t)$, $h(t)$, $v^*(t)$ in a single space.

In those cases where the same vector belongs to various spaces we equip its norm with the symbol of the space in which it is considered.

Suppose that the space V is a (possibly, non-closed) subspace of H, is dense in H in the norm of H, and $|\varphi|_H \leq N|\varphi|_V$ for all $\varphi \in V$, where N does not depend on v. Suppose that H is, in turn, a subspace of some Banach space V' and that H is dense in V'. Then

$$V \subset H \subset V'. \tag{2.7}$$

We assume that the scalar product in H possesses the following property: if $\varphi \in V$, $\psi \in H$, then $|\varphi\psi| \leq |\varphi|_V\,|\psi|_{V'}$. Since the embeddings in (2.7) are dense, it is possible to uniquely define $\varphi\psi$ for $\varphi \in V$, $\psi \in V'$ as $\lim_{n\to\infty} \varphi\psi_n$, where $\psi_n \in H$ and $|\psi - \psi_n|_{V'} \to 0$. Obviously, for a fixed $\psi \in V'$, the mapping $\varphi \mapsto \varphi\psi$ is a bounded linear functional on V. We suppose that for $\psi \in V'$ the equality $\varphi\psi = 0$ for all $\varphi \in V$ implies that $\psi = 0$. Then the mapping which, to every $\psi \in V'$ assigns the corresponding functional $\varphi \mapsto \varphi\psi$ on V, is a one-to-one mapping of V' into some subset of the dual space V^* of V. We now make an additional assumption that every bounded linear functional on V has the form $\varphi \mapsto \varphi\psi$ for some $\psi \in V'$, and then the mapping $V' \to V^*$ mentioned above becomes both one-to-one and onto so that V' can be identified with V^* if desired. We note that under this identification the norms of ψ as an element of V' and as an element of V^* are, in general, different, and in this section we will not identify V' with V^*[b]. Finally, we suppose that that there is a function $v'(t, \omega)$ with values in V' such that, for every

[b]On the contrary, in the following two sections we will make this identification.

$g \in V$, $gv'(t)$ is measurable in (t, ω) and, for every $t \geq 0$, is $\mathcal{F}_t$-measurable in ω, and also $|v'(t)|_{V'} \leq f(t)$ for all (t, ω), where the function f is the same as on page 16. Then, for every $g \in V$, $t \geq 0$, we have

$$\left| \int_0^t gv'(s)ds \right| \leq |g|_V \int_0^t f(s)ds.$$

Therefore, from the properties of $f(t)$, it follows that there exists a set $\Omega' \subset \Omega$ such that $P(\Omega') = 1$ and for $\omega \in \Omega'$, $t \geq 0$, the mapping $g \mapsto \int_0^t gv'(s)ds$ is a bounded linear functional on V. Under our assumptions, this functional can be written in the form $g \mapsto g\psi(t)$, where $\psi(t) \in V'$, and, for $\omega \in \Omega'$ we have

$$\psi(t) = \int_0^t v'(s)ds.$$

Theorem 2.16 then takes the following form, where, for simplicity of the formulation, we take $\tau = \infty$; generalization to the case of arbitrary τ is obvious.

Theorem 2.17. *For $\omega \in \Omega'$ and $t \geq 0$ we define*

$$\tilde{h}(t) = \int_0^t v'(s)ds + h(t), \tag{2.8}$$

and suppose that $\tilde{h}(t) = v(t)$ for almost all (t, ω). Then there exists a set $\Omega'' \subset \Omega'$ such that $P(\Omega'') = 1$ and, for $\omega \in \Omega''$, the function $\tilde{h}(t)$ takes values in H, is continuous in H with respect to t, is strongly $\mathcal{F}_t$-measurable with respect to ω for each t (as a function with values in H), and

$$\tilde{h}^2(t) = h^2(0) + 2\int_0^t v(s)v'(s)ds + 2\int_0^t \tilde{h}(s)dh(s) + \langle m \rangle_t. \tag{2.9}$$

We note that relation (2.9) is obtained if the rules for computing the stochastic differential $d\tilde{h}^2(t)$ for the process $\tilde{h}(t)$ defined in (2.8) are applied. One of the difficulties in justifying (2.9) is that it is generally not clear why $\tilde{h}(t) \in H$ (in other words, why $\tilde{h}^2(t)$ exists), because equation (2.8) only defines $\tilde{h}(t)$ as a process with values in V', and it is generally not true that $\tilde{h}(t) \in V$ for all (t, ω).

Proof. [of Theorem 2.17 using Theorem 2.16] We take for Λ the identify operator and use the fact that the formulation of Theorem 2.17 does not contain the process $v^*(t)$. We construct a process $v^*(t)$ on the basis of $v'(t)$ as the process corresponding to $v'(t)$ under the mapping $V' \to V^*$. Then $v^*(t)$ satisfies the assumptions preceding Theorem 2.16, and (2.4) follows from (2.8). On the basis of Theorem 2.16, we construct a set $\Omega'' \subset \Omega'$ and a process $\tilde{h}^1 = \tilde{h}^1(t)$ such that $P(\Omega'')$ =1, $\tilde{h}^1$ is continuous in H with respect to t, is an $\mathcal{F}_t$-consistent process, and $\tilde{h}^1(t) = v(t)$ for almost all (t, ω); for $\omega \in \Omega''$ formula (2.9) holds if $\tilde{h}$ is replaced by $\tilde{h}^1$, and for

$\omega \in \Omega''$, $g \in V$, $t \geq 0$,

$$g\tilde{h}^1(t) = \int_0^t gv'(s)ds + gh(t) = g\tilde{h}(t).$$

From the equality of the extreme terms, it follows that $g(\tilde{h}^1(t) - \tilde{h}(t)) = 0$ for all $g \in V$, and hence $\tilde{h}^1 = \tilde{h}$ for $t \geq 0$, $\omega \in \Omega''$; from the properties of $\tilde{h}^1$ we now obtain all the required properties of $\tilde{h}$. The proof of the theorem is complete. □

We proceed to prepare for the proof of Theorem 2.16. We have already agreed to consider sparable V and H. Further, if in place of τ in Theorem 2.16 we take $\tau \wedge n$, prove the theorem for $\tau \wedge n$ and then let $n \to \infty$, then we obtain the proof of the theorem for τ. It may therefore be assumed that τ is a bounded stopping time, and then a non-random change of time reduces everything to the case $\tau \leq 1$.

We further note that both $|v(s)|$ and $|v^*(s)|$ are $\mathcal{F}_s$-measurable for almost all s and are measurable in (s, ω). Hence, the process

$$r(t) = h^2(0) + \|A\|_t + \langle m \rangle_t + \int_0^t |v(s)|^p \, ds + \int_0^t |v^*(s)|^q \, ds$$

is $\mathcal{F}_t$-measurable for each t and is continuous in t (a.s). This implies that for every $n \geq 0$ $\tau_n = \inf\{t \geq 0 : r(t) \geq n\} \wedge \tau$ is a stopping time. Since $\tau_n \uparrow \tau$, it suffices to prove Theorem 2.16 with τ in its formulation replaced by τ_n. Moreover, process $r(t)$ is bounded in (t, ω) on $\{(t, \omega) : t \leq \tau_n(\omega)\}$, and it may therefore be assumed in the proof of Theorem 2.16 that the process $r(t)$ is bounded on $\{(t, \omega) : t \leq \tau(\omega)\}$. For $t > \tau$, we set $v(t) = 0$, $v^*(t) = 0$, $h(t) = h(\tau)$, thus ensuring that the process $r(t)$ is bounded on $[0, \infty) \times \Omega$. After this, multiplying (2.4) by a suitable constant ε and replacing v, v^*, h by εv, εv^*, εh, we reduce the matter to the case where $r(t) \leq 1$.

These arguments show that it suffices to prove Theorem 2.16 in the special case considered below.

2.4. *Proof of Theorem 2.16*

In this section we prove Theorem 2.16 under the additional assumptions that V and H are separable, $\tau \leq 1$, and, for all ω,

$$\int_0^1 |v(s)|^p \, ds + \int_0^1 |v^*(s)|^q \, ds + \|A\|_1 + \langle m \rangle_1 + h^2(0) \leq 1 \tag{2.10}$$

It was shown above that these additional assumptions do not lead to any loss of generality.

Lemma 2.18. *There exist (a) a sequence of nested partitions*

$$0 = t_0^n < t_1^n < \ldots < t_{k(n)+1}^n = 1$$

of the interval $[0,1]$ *with diameter tending to zero, and (b) a set* $\Omega' \subset \Omega$ *with the following properties:*

1) $P(\Omega') = 1$ *and, if* $\omega \in \Omega'$ *and* $t \in I = \{t_i^n;\ i = 1, \dots, k(n),\ n \geq 1\}$ *are such that* $t < \tau(\omega)$, *then equality* (2.4) *is satisfied for all* $g \in V$ *and, for every* $s \in I$, *the quantity* $v(s)$ *is* $\mathcal{F}_s$*-measurable;*
2) *Define the processes* $v_n^1(t)$ *and* $v_n^2(t)$ *by* $v_n^{(1)}(t) = v(t_i^n)$ *for* $t \in [t_i^n, t_{i+1}^n)$, $i = 1, \dots, k(n)$, $v_n^{(1)}(t) = 0$ *for* $t \in [0, t_1^n)$; $v_n^{(2)}(t) = v(t_{i+1}^n)$ *for* $t \in [t_i^n, t_{i+1}^n)$, $i = 1, \dots, k(n) - 1$, $v_n^{(2)}(t) = 0$ *for* $t \in \left[t_{k(n)}^n, 1\right)$. *Then, for* $j = 1, 2$

$$\mathbb{E} \sup_{t\in[0,1]} \left|v_n^{(j)}(t)\right|^p < \infty, \qquad \lim_{n\to\infty} \mathbb{E} \int_0^1 \left|v(t) - v_n^{(j)}(t)\right|^p dt = 0. \tag{2.11}$$

Proof. Let $[a]$ be the integer part of the number a, (for $a > 0$, $[a]$ is the largest integer smaller than a) and let $\varkappa^1(n,t) = 2^{-n}[2^n t]$, $\varkappa^2(n,t) = 2^{-n}[2^n t] + 2^{-n}$, $v(t) = 0$ for $t \in [0,1]$. Standard arguments of Doob show that there exists a sequence of integers $r_n \to \infty$ such that, for $j = 1, 2$ and almost all $s \in [0,1]$,

$$\lim_{r_n\to\infty} \mathbb{E} \int_0^1 |v(t) - v(\varkappa^j(r_n, t+s) - s)|^p dt = 0 \tag{2.12}$$

Further, it follows from Fubini's theorem and separability of V that there exists a set $T \subset [0,1]$ of unit Lebesgue measure such that, for all $t \in T$ and all $g \in V$, equality (2.4) is satisfied (a.s) on $\{\omega : t < \tau(\omega)\}$, and the quantity $v(t)$ is $\mathcal{F}_t$-measurable. It is clear that, for every $s \in [0,1]$, all values of the functions $\varkappa^j(r_n, t+s) - s$ for $t \in [0,1]$, $j = 1, 2$, $n \geq 1$, lying in $[0,1]$ also belong to T. We fix a suitable s so that (2.12) is also satisfied; we define $\{t_i^n\}$ as the set of values of $\varkappa^1(r_n, t+s) - s$ for $t \in [0,1]$ which lie in $[0,1]$, to which we add the points 0 and 1, and we denote by Ω' the set of ω for which equation (2.4) is satisfied for all $g \in V$, $t = t_i^n < \tau(\omega)$, $i = 1, \dots, k(n)$, $n \geq 1$. All assertions of the lemma are then valid except possibly for the first inequality in (2.11). We note, however, that by virtue of the second inequality in (2.11), for sufficiently large n,

$$\mathbb{E} \int_0^1 \left|v_n^{(j)}(t)\right|^p dt < \infty$$

This inequality is equivalent to the first inequality in (2.11), which is thus valid for large n. For small n it is clearly valid since our partitions are nested. The proof of the lemma is complete. □

Lemma 2.19. *For* $\omega \in \Omega'$, $t, s \in I$, $s \leq t \leq \tau(\omega)$

$$|\Lambda v(t)|^2 - |\Lambda v(s)|^2 = 2 \int_s^t v(t) v^*(u) du + 2\Lambda v(s)(h(t) - h(s)) + |(h(t) - h(s)|^2 \\ - |\Lambda(v(t) - v(s)) - (h(t) - h(s))|^2, \tag{2.13}$$

$$|\Lambda v(t)|^2 = 2\int_s^t v(t)v^*(u)du + h^2(t) - |\Lambda(v(t) - (h(t)|^2 . \tag{2.14}$$

The proof of this lemma is based on using (2.4) with $g = v(t)$ or $g = v(s)$ and simple algebraic transformations, which we leave to the reader while suggesting that (2.14) be derived first and (2.13) then proved by subtracting the appropriate equalities (2.14).

Lemma 2.20.

$$\mathbb{E} \sup_{t\in I,\, t<\tau} |\Lambda v(t)|^2 < \infty.$$

Proof. From (2.14) and (2.11), with $t \in I$,

$$\mathbb{E}\chi_{_{t<\tau}} |\Lambda v(t)|^2 \le \mathbb{E}h^2(t) + \frac{2}{p} t\,\mathbb{E}\,|v(t)|^p + \frac{2}{q}\mathbb{E}\int_0^t |v^*(s)|^q\, ds < \infty. \tag{2.15}$$

Further, by (2.13) and (2.14), with $t = t_i^n < \tau(\omega)$, $\omega \in \Omega'$

$$\begin{aligned}\left|\Lambda v_n^1(t)\right|^2 &= h^2(0) + 2\int_0^t v_n^2(s)v^*(s)ds + 2\int_0^t \Lambda v_n^1(s)dh(s) + 2h(0)(h(t_i^n) - h(0))\\ &+ \sum_{j=1}^{i-1} \left|h(t_{j+1}^n) - h(t_j^n)\right|^2 - \sum_{j=0}^{i-1} \left|\Lambda\big(v(t_{j+1}^n) - v(t_j^n)\big) - (h(t_{j+1}^n) - h(t_j^n))\right|^2,\end{aligned} \tag{2.16}$$

where in the last sum the term corresponding to $j = 0$ is taken equal to $|\Lambda(v(t_1^n) - h(t_1^n)|^2$. To estimate the second therm on the right side of (2.16), we apply (2.10) and the Burkholder inequality. We then obtain

$$\begin{aligned}\mathbb{E}\sup_{i\ge 1} \chi_{_{t_1^n<\tau}} \left|\int_0^{t_1^n} \Lambda v_n^1(u)dh(u)\right| &\le \mathbb{E}\sup_{t\le 1} \left|\int_0^t \chi_{_{u<\tau}} \Lambda v_n^1(u)dh(u)\right|\\ &\le \mathbb{E}\sup_{i\ge 1} |\Lambda v(t_i^n|\, \chi_{_{t_i^n<\tau}} \|A\|_1 + 3\mathbb{E}\left(\int_0^1 \chi_{_{u<\tau}} \left|\Lambda v_n^1(u)\right|^2 d\langle m\rangle_u\right)^{1/2}\\ &\le 4\mathbb{E}\sup_{i\ge 1} |\Lambda v(t_i^n|\, \chi_{_{t_i^n<\tau}} \le \frac{1}{4}\mathbb{E}\sup_{i\ge 1} |\Lambda v(t_i^n|^2\, \chi_{_{t_i^n<\tau}} + 16.\end{aligned}$$

We note that it follows from (2.15) that the last expression is finite. Moreover, from (2.10)

$$\mathbb{E}\sum_{j=0}^{k(n)} \left|h(t_{j+1}^n) - h(t_j^n)\right|^2 \le 4\mathbb{E}\,\|A\|_1^2 + 2\mathbb{E}m^2(1) \le 6.$$

From this and from (2.16)

$$\mathbb{E}\sup_{i\ge 1} |\Lambda(v(t_i^n)|\, \chi_{_{t_i^n<\tau}} \le \frac{4}{p}\mathbb{E}\int_0^1 \left|v_n^{(2)}(t)\right|^p dt + \frac{4}{q} + 100.$$

Letting here $n \to \infty$ and using the fact that by (2.11) the right side is bounded in n, we obtain the assertion of the lemma. The proof of the lemma is complete. □

Lemma 2.21. *Let*

$$\Omega'' = \Omega' \cap \left\{ \omega : \sup_{t \in I,\ t < \tau} |\Lambda v(t)| < \infty \right\}.$$

Then there exists a function $\tilde{h} = \tilde{h}(t)$ with values in H, which, for $\omega \in \Omega''$ and $t \in I$, $t < \tau(\omega)$, is weakly continuous in H with respect to t and satisfies (2.6) for all $g \in V$ Moreover, $\tilde{h}(t)$ is $\mathcal{F}_t$-measurable on $\{\omega : t < \tau(\omega)\}$, $\tilde{h}(t) = \Lambda v(t)$ (a.s.) on $\{(t, \omega) : t < \tau(\omega)\}$, $\tilde{h}(t) = \Lambda v(t)$ for $\omega \in \Omega''$, $t \in I$, $t < \tau(\omega)$.

Proof. For $\omega \in \Omega''$, $t \in I$, $t < \tau$, $g \in V$ the function $(\Lambda g)\Lambda v(t)$ coincides with the right side of (2.6), which is continuous in t. Hence, for $s < \tau$ there exists $\lim_{\substack{I \ni t \to s \\ t < \tau}} (\Lambda g)\Lambda v(t) < \tau$.

Since $|\Lambda v(t)|$ is bounded on $I \cap \{t < \tau\}$, this means that for $s < \tau$, $I \ni t \to s$ there exists the weak limit of $\Lambda v(t)$, which we denote by $\tilde{h}(s)$. It is clear that $\tilde{h}(t)$ satisfies (2.6) for all $\omega \in \Omega''$, $t < \tau(\omega)$ $g \in V$. The assertions of the lemma follow from this in an almost obvious way. The proof of the lemma is complete. □

We set $\tilde{h}(t) = \tilde{h}(\tau)$ for $t \geq \tau$, $\tilde{h}(t) = 0$ for $\omega \notin \Omega''$, and we note that for $\omega \in \Omega''$

$$\sup_{t \leq 1} \left| \tilde{h}(t) \right| = \sup_{t \in I,\ t < \tau} |\Lambda v(t)| < \infty. \tag{2.17}$$

Moreover, for $t \leq 1$, the integral $\int_0^t \left| \tilde{h}(s) \right| ds$ is well defined, because $\tilde{h}(s)$ is completely measurable due to separability of H and continuity of $\pi \tilde{h}(s)$ for $\omega \in \Omega''$, where π is the operator of projecting onto some finite-dimensional subspace of H.

Lemma 2.22. *Define $\tilde{h}_n(t) = \tilde{h}(t_i^n)$ for $t \in [t_i^n, t_{i+1}^n)$, $i = 1, \dots, k(n)$. Then*

$$\lim_{n \to \infty} \sup_{t \leq 1} \left| \int_0^t \tilde{h}_n(s) dh(s) - \int_0^t \tilde{h}(s) dh(s) \right| = 0 \tag{2.18}$$

in probability.

Proof. Let $h_1, \dots, h_r, \dots$ be an orthonormal basis in H, and let π_r be the operator projecting H onto the space spanned by $h_1, \dots, h_r$. Since $\pi_r \tilde{h}(s)$ is continuous on Ω'', it follows that (a.s.)

$$\lim_{n \to \infty} \left\{ \int_0^t \left| \pi_r \tilde{h}_n(s) - \pi_r \tilde{h}(s) \right| d \langle m \rangle_s + \int_0^t \left| \pi_r \tilde{h}_n(s) - \pi_r \tilde{h}(s) \right| d \|A\|_s \right\} = 0.$$

It therefore suffices to prove that, for every $\varepsilon > 0$,

$$\lim_{r \to \infty} \sup_n P \left\{ \sup_{t \leq 1} \left| \int_0^t (1 - \pi_r) \tilde{h}_n(s) dh(s) \right| > 2\varepsilon \right\} = 0 \tag{2.19}$$

and

$$\lim_{r\to\infty} P\left\{\sup_{t\le 1}\left|\int_0^t (1-\pi_r)\tilde{h}(s)dh(s)\right| > 2\varepsilon\right\} = 0.$$

We shall prove only the first equality. The second is proved similarly. Noting that π_r is a self-adjoint operator, for every N, $\delta > 0$ we estimate the probability in (2.19) as follows:

$$\begin{aligned} &P\left\{\sup_{t\le 1}\left|\int_0^t \tilde{h}_n(s)d((1-\pi_r)h(s))\right| > 2\varepsilon\right\} \le \frac{\delta}{\varepsilon^2} + P\left\{\int_0^t \left|\tilde{h}_n(s)\right| d\left\|(1-\pi_r)A\right\|_s > \varepsilon\right\} \\ &+ P\left\{\int_0^t \left|\tilde{h}_n(s)\right|^2 d\left\langle(1-\pi_r)m\right\rangle_s > \delta\right\} \le \frac{\delta}{\varepsilon^2} + 2P\left\{\sup_{s\le 1}\left|\tilde{h}(s)\right| > N\right\} \\ &+ \frac{N}{\varepsilon}\mathbb{E}\|(1-\pi_r)A\|_1 + \frac{N}{\delta}\mathbb{E}\left|(1-\pi_r)m(1)\right|^2. \end{aligned} \tag{2.20}$$

We denote by h^i the i-th coordinate of $h \in H$ in the basis $\{h_i\}$, and we set $a^i(t) = dA^i(t)/d\|A\|_t$. Then as $r \to \infty$

$$\|(1-\pi_r)A\|_1 = \int_0^1 \left(\sum_{i>r}\left(a^i(t)\right)^2\right)^{1/2} d\|A\|_t \to 0,$$

$$|(1-\pi_r)m(1)|^2 = \sum_{i>r}\left(m^i(1)\right)^2 \to 0.$$

From this and from (2.20) and (2.10) we see that the left side of (2.19) does not exceed

$$\frac{\delta}{\varepsilon^2} + 2P\left\{\sup_{s\le 1}\left|\tilde{h}(s)\right| > N\right\}.$$

Since δ, N are arbitrary and (2.17) is satisfied, the last expression can be made arbitrarily small. The proof of the lemma is complete. □

We now define the set Ω''' which will play the role of Ω' in Theorem (2.16). It is possible to find a sequence $\{n'\}$ along which the left side of (2.17) tends to zero (a.s.). It may be assumed with no loss of generality that the original sequence has this property. Moreover, we set $\pi_0 = 0$; then, as is well know, in probability for $r > 0$, $t \in [0,1]$,

$$\lim_{n\to\infty}\sum_{t^n_{j+1}\le t}\left|(1-\pi_r)(h(t^n_{j+1}) - h(t^n_j)\right|^2 = \langle(1-\pi_r)m\rangle_t. \tag{2.21}$$

Therefore, there exists a subsequence along which the last equality, understood in the sense of pointwise convergence, is true for all $r > 0$, $t \in I$ almost surely. To simplify the notation, we assume that this subsequence also coincides with the original sequence. We set

$$Q_1 = \left\{\omega : \lim_{n\to\infty}\sup_{t\le 1}\left|\int_0^t \tilde{h}_n(s) - \tilde{h}(s))dh(s)\right| = 0\right\},$$

$$Q_2 = \bigcap_{r=0}^{\infty} \bigcap_{t \in I} \left\{ \omega : \lim_{n \to \infty} \sum_{t_{j+1}^n \le t} \left| (1-\pi_r)(h(t_{j+1}^n) - h(t_j^n) \right|^2 = \langle (1-\pi_r) m \rangle_t \right\},$$

$$Q_3 = \left\{ \omega : \lim_{n \to \infty} \int_0^1 \left| v_n^{(i)}(s) - v(s) \right|^p ds = 0, \ i = 1, 2 \right\},$$

$$Q_4 = \left\{ \omega : \lim_{r \to \infty} \langle (1-\pi_r) m \rangle_i = 0 \right\},$$

$$Q''' = Q'' \bigcap \bigcap_{i=1}^{4} Q_i,$$

and set $\tilde{h}(t,\omega) = 0$ on the complement of Q'''.

From what has been said above, $P(Q_i) = 1$, $i = 1, 2$. From Lemma 2.18 it follows that it may be assumed that $P(Q_3) = 1$. Since the sequence $\langle (1-\pi_r)m \rangle_1$, $r \ge 0$, is decreasing (a.s.) and $\mathbb{E}\langle (1-\pi_r)m \rangle_1 = \mathbb{E}\left|(1-\pi_r)m(1)\right|^2 \to 0$, it follows that $P(Q_4) = 1$. Thus, $P(Q''') = 1$.

Lemma 2.23. *For $\omega \in Q'''$, $t, s \in I$, $s < t < \tau(\omega)$*

$$\begin{aligned} \left| \tilde{h}(t) - \tilde{h}(s) \right|^2 &= 2 \int_s^t (v(u) - v(s))\, v^*(u) du \\ &\quad + 2 \int_s^t \left(\tilde{h}(u) - \tilde{h}(s) \right) dh(u) + \langle m \rangle_t - \langle m \rangle_s, \end{aligned} \tag{2.22}$$

$$\left| \tilde{h}(t) \right|^2 = h^2(0) + 2 \int_0^t v(u) v^*(u) du + 2 \int_0^t \tilde{h}(u) dh(u) + \langle m \rangle_t. \tag{2.23}$$

Proof. We first prove (2.23). We fix $\omega \in Q'''$, $t \in I$, $t < \tau(\omega)$. From (2.16) and Lemma 2.21 for n such that t is a point of the partition $t_0^n < t_1^n < \ldots t_{k(n)+1}^n$, we have

$$\begin{aligned} \tilde{h}^2(t) &= h^2(0) + 2 \int_0^t v_n^{(2)}(u) v^*(u) du + 2 \int_0^t \tilde{h}_n(u) dh(u) \\ &\quad + \sum_{t_{j+1}^n \le t} \left| h(t_{j+1}^n) - h(t_j^n) \right|^2 - \sum_{t_{j+1}^n \le t} \left| \left(\tilde{h}(t_{j+1}^n) - \tilde{h}(t_j^n) \right) - \left(h(t_{j+1}^n) - h(t_j^n) \right) \right|^2. \end{aligned}$$

Letting $n \to \infty$, we find

$$\tilde{h}^2(t) = h^2(0) + 2 \int_0^t v(u) v^*(u) du + 2 \int_0^t \tilde{h}(u) dh(u) + \langle m \rangle_t - J,$$

where

$$J = \lim_{n \to \infty} \sum_{t_{j+1}^n \le t} \left| \left(\tilde{h}(t_{j+1}^n) - \tilde{h}(t_j^n) \right) - \left(h(t_{j+1}^n) - h(t_j^n) \right) \right|^2,$$

and the last limit exists and is finite. For the proof of (2.23) it suffices to show that $J = 0$. It is convenient to assume that the basis $\{h_i\}$ is formed from elements of the set ΛV. Since the latter is dense in H, this last assumption involves no loss of generality. Moreover, it is obvious that for every $r \geq 1$ there exists a function $\tilde{v}_r(t)$ continuous in the norm of V such that $\Lambda\tilde{v}_r(t) = \pi_r h(t)$ for all (t,ω). On the basis of the function $\tilde{v}_r$ we construct functions $\tilde{v}_{r,n}^{(i)}$, $i = 1,2$ just as in Lemma 2.18 the functions $\tilde{v}_r^{(i)}$ are constructed on the basis of v. Noting that by (2.4) and Lemma 2.21 for $t_{j+1}^n \leq t$ and every $\varphi \in V$ there is the equality

$$\left((\tilde{h}(t_{j+1}^n) - \tilde{h}(t_j^n)\right) - \left(h(t_{j+1}^n) - h(t_j^n))\right) \Lambda\varphi = \int\limits_{t_j^n}^{t_{j+1}^n} \varphi v^*(u)du,$$

we find easily for any $r \geq 0$

$$J = \lim_{n\to\infty} \int_0^t (v_n^{(2)}(u) - (v_n^{(1)}(u))v^*(u)du - \lim_{n\to\infty} \int_0^t (\tilde{v}_{r,n}^{(2)}(u) - (\tilde{v}_{r,n}^{(1)}(u))v^*(u)du$$
$$- \lim_{n\to\infty} \sum_{t_{j+1}^n \leq t} \left((\tilde{h}(t_{j+1}^n) - \tilde{h}(t_j^n)) - (h(t_{j+1}^n) - h(t_j^n))\right)(1-\pi_r)(h(t_{j+1}^n) - h(t_j^n)).$$

Here the first limit is equal to zero; the second is equal to zero, because $\omega \in Q_3$ and the continuity of $\tilde{v}_r(u)$ implies $\lim_{n\to\infty} \left|\tilde{v}_{r,n}^{(2)}(u) - \tilde{v}_{r,n}^{(1)}(u)\right| = 0$ uniformly in u. Hence,

$$J \leq \lim_{n\to\infty} \left(\sum_{t_{j+1}^n \leq t} \left| \left((\tilde{h}(t_{j+1}^n) - \tilde{h}(t_j^n)\right) - (h(t_{j+1}^n) - h(t_j^n))\right|^2 \right)$$
$$\times \left(\sum_{t_{j+1}^n \leq t} \left|(1-\pi_r)\left((h(t_{j+1}^n) - h(t_j^n)\right)\right|^2 \right)^{1/2} = J^{1/2}\langle(1-\pi_r)m\rangle_t^{1/2}.$$

As $r \to \infty$ this implies that $J = 0$. Equality (2.23) is proved. Equality (2.22) is deduced from (2.23) by means of the relations $(a-b)^2 = a^2 - b^2 - 2b(a-b)$ and

$$-2\tilde{h}(s)(\tilde{h}(t) - \tilde{h}(s)) = -2\int_s^t v(s)v^*(u)du - 2\int_s^t \tilde{h}(s)dh(u).$$

The proof of the lemma is complete. □

We now finish the proof of Theorem 2.16. Because of (2.21) and (2.23) it remains for us to prove the strong continuity of $\tilde{h}(t,\omega)$ in t for $t < \tau(\omega)$, $\omega \in \Omega'''$. Since a weakly continuous function with a continuous norm is strongly continuous, it suffices to prove (2.23) for $t < \tau(\omega)$, $\omega \in \Omega'''$. For $t = 0$ (2.23) is obvious. We fix $t > 0$, $t < \tau(\omega)$, $\omega \in \Omega'''$.

For all sufficiently large n it is possible to define $j = j(n)$ such that $0 < t_j^n \leq t < t_{j+1}^n$. We set $t(n) = t_{j(n)}^n$ and note that $t(n) \uparrow t$,

$$\lim_{n\to\infty} \int_{t(n)}^t |v(u) - v(t(n))| \cdot |v^*(u)|\, du \leq \lim_{n\to\infty} \int_0^1 \left|v(u) - v_n^{(1)}(u)\right| \cdot |v^*(u)|\, du = 0,$$

$$\lim_{n\to\infty} \sup_{s\leq t} \left| \int_{t(n)}^s \tilde{h}(u) - \tilde{h}(t(n)) dh(u) \right| \leq 2 \lim_{n\to\infty} \sup_{s\leq 1} \left| \int_0^s \tilde{h}(u) - \tilde{h}_n(u)) dh(u) \right| = 0,$$

$$\lim_{n\to\infty} \left(\langle m \rangle_t - \langle m \rangle_{t(n)} \right) = 0.$$

Therefore, there exists a subsequence $n(k)$ such that for $s(k) = t(n(k))$ we have

$$\sum_{k=1}^{\infty} \left\{ \left(\int\limits_{s(k)}^{s(k+1)} |v(u) - v(s(k))| \cdot |v^*(u)|\, du \right)^{1/2} + \left(\langle m \rangle_{s(k+1)} - \langle m \rangle_{s(k)} \right)^{1/2} \right.$$
$$\left. + \left(\int\limits_{s(k)}^{s(k+1)} \left(\tilde{h}(u) - \tilde{h}(s(k)) dh(u) \right)^{1/2} \right) \right\} < \infty.$$

From (2.22) we then find that

$$\sum_{k=1}^{\infty} \left| \tilde{h}(s(k+1)) - \tilde{h}(s(k)) \right| < \infty.$$

Therefore, $\tilde{h}(s(k))$ for $k \to \infty$ has a strong limit. Since $s(k) \to t$, it follows that $\tilde{h}(s(k))$ converges weakly to $\tilde{h}(t)$. Thus, $\tilde{h}(s(k)) \to \tilde{h}(t)$ strongly in H, and, substituting the numbers $s(k)$ in place of t in (2.23), and passing to the limit $k \to \infty$, we obtain (2.23) for the t chosen. This complete the proof of Theorem 2.16.

3. Itô Stochastic Equations in Banach Spaces and the Method of Monotonicity

3.1. *Introduction*

In this chapter we consider the Itô equations

$$v(t) = u_0 + \int_0^t A(v(s), s) ds + \int_0^t B(v(s), s) dw(s) \tag{3.1}$$

in Banach spaces. The coefficients $A(v, s)$, $B(v, s)$ of "drift" and "diffusion" are generally assumed to be unbounded non-linear operators. They may depend on the elementary outcome in a nonanticipatory fashion. By w we understand a Wiener process with values in some Hilbert space.

An existence and uniqueness theorem will be proved for the solution of an equation slightly more general than (3.1) and certain qualitative results on the solution will be obtained.

A solution is understood to be a trajectory with values in the domain of the operators $A(\cdot, t)$, $B(\cdot, t)$ (the domain of the operators A and B is assumed to be independent of t) that satisfies (3.1) and is consistent with the same system of σ-algebras as $w(t)$, $A(\cdot, t)$, and $B(\cdot, t)$. This system is assumed to be given together with the original probability space and the Wiener process. A solution is thus understood in the "strong" sense.

The main conditions on A an B are the conditions of monotonicity and coerciveness; see (A_2), (A_3) in Section 3.2. The following equations satisfy these assumptions in spaces of Sobolev type:

$$\begin{aligned} du(t,x) &= a(t,\omega)(-1)^{m+1}\frac{\partial^m}{\partial x^m}\left(\left|\frac{\partial^m}{\partial x^m}u(t,x)\right|^{p-2}\frac{\partial^m}{\partial x^m}u(t,x)\right)dt \\ &\quad + b(t,\omega)\left|\frac{\partial^m}{\partial x^m}u(t,x)\right|^{p/2}dw(t), \end{aligned} \tag{3.2}$$

$p > 1$, $x \in G \subset \mathbb{R}$, $-2(p-1)a + 4^{-1}p^2b^2 \leq -\varepsilon$, $\varepsilon > 0$, and $w(t)$ is a Wiener process with values in $\mathbb{R}$;

$$\begin{aligned} du(t,x) &= \sum_{i,j=1}^{n}\frac{\partial}{\partial x^i}\left(a_{ij}(t,x,\omega)\frac{\partial}{\partial x^j}u(t,x)\right)dt \\ &\quad + \sum_{j=1}^{m}\sum_{i=1}^{n}b_{ij}(t,x,\omega)\frac{\partial}{\partial x^j}u(t,x)dw^j(t), \end{aligned} \tag{3.3}$$

$x \in \mathbb{R}^n, a_{ij}, b_{ij}$ are bounded measurable functions such that, for some $\lambda > 0$,

$$\sum_{i,j=1}^{n} a_{ij}\xi^i\xi^j - \sum_{i,j=1}^{m}\sum_{\ell=1}^{n}\frac{1}{2}b_{i\ell}b_{j\ell}\xi^i\xi^j \geq \lambda\sum_{i=1}^{n}(\xi^i)^2$$

for all t, x, ω and every vector $\xi \in \mathbb{R}^n$; $w^i(t)$ are independent Wiener process with values in $\mathbb{R}$, and ω is an elementary outcome.

These and other stochastic partial differential equations are considered in detail in Section 4.

The results of the present section are a refinement of the results of Pardoux [52, 53] which, in turn, generalize the results of Bensoussan and Temam [43].

As already mentioned above, we have succeeded in showing that certain conditions of Pardoux are superfluous, in particular, the local Lipschitz condition for the operator B.

The method of proving the existence theorem (the most difficult and important part of this section) has been borrowed from Pardoux and corresponds to a Galerkin scheme: a finite-dimensional analog of equation 3.1 is considered (Section 3.3), estimates of the solution independent of the dimension are obtained (Section

3.4), and then (by the method of monotonicity) a passage to the limit is realized (Section 3.5). The basic improvements which make it possible in the final analysis to generalize the results of Pardoux are made at the first step in Section 3.3, where a theorem is obtained generalizing a well-known theorem of Itô on the existence and uniqueness of strong solutions of a stochastic equation with random coefficients satisfying Lipschitz conditions.

3.2. *Assumptions and the Main Results*

Let $(\Omega, \mathcal{F}, P)$ be a complete probability space with an expanding system of σ-algebras $\{\mathcal{F}_t\}$ $(t \in [0,T], T < \infty)$ imbedded in $\mathcal{F}$. We assume that the family $\{\mathcal{F}_t\}$ is complete with respect to the measure P.

Further, let H and E be real separable Hilbert spaces, naturally identified with their duals H^* and E^*; let $w(t)$ be a Wiener process in E with nuclear covariance operator Q (see Section 2.2), and let $z(t)$ be a square-integrable martingale in H. We also consider a real, separable, reflexive Banach space V and its dual space V^*.

As in Section 2, if v is an element of V and v^* is an element of V^*, then vv^* denotes the value of v^* on v; $|\cdot|_X$ and $(\cdot,\cdot)_X$ denote, respectively, the norm in the space X and the scalar product in the space X if X is a Hilbert space. In Section 3.3, where finite-dimensional spaces are considered, this notation is simplified; special mention is made of this.

As before, $\mathcal{L}_Q(E,H)$ is the space of all linear operators Φ defined on $Q^{1/2}E$ and taking $Q^{1/2}E$ into H such that $\Phi Q^{1/2} \in \mathcal{L}_2(E,H)$ (the space of Hilbert-Schmidt operators from E to H). Recall that $\mathcal{L}_Q(E,H)$ is a separable Hilbert space relative to the scalar product $(\Phi,\Psi)_Q = tr\Phi Q^{1/2}\left(\Psi Q^{1/2}\right)^*$; we write $|\cdot|_Q$ to denote the norm in this space.

The following assumptions are henceforth used:

a) $V \subset H \equiv H^* \subset V^*$;
b) V is dense in H (in the norm of H);
c) there exists a constant c such that, for all $v \in V$, $|v|_H \le c|v|_V$;
d) $vv^* = (v,v^*)_H$ if $v^* \in H$.

An important example of spaces possessing properties a)– d) is the Sobolev space $W_p^m(G)(= V)$ and $L_2(G)(= H)$ where G is a bounded domain in $\mathbb{R}^d$, and $dp \ge 2(d - mp)$. These spaces are discussed in more detail in Section 4. Some other triples of spaces possessing properties a)–d) are also presented there. We recall also that in Section 2 triples of spaces V, H, V' connected by less rigid assumptions have already been considered, and the "implication" of assumptions a) – d) was discussed in some detail; in particular, the possibility of identifying V' with V^* by means of $(\cdot,\cdot)_H$ was discussed.

We fix numbers p and q, $p \in (1,\infty)$, $q = \dfrac{p}{p-1}$.

Suppose that, for each $(v, t, \omega) \in V \times [0, T] \times \Omega$,

$$A(v, t, \omega) \in V^*, \quad B(v, t, \omega) \in \mathcal{L}_Q(E, H).$$

We assume that, for each $v \in V$, the functions $A(v, t, \omega)$, $B(v, t, \omega)$ are measurable in (t, ω) relative to the measures $dt \times dP$ and are $\mathcal{F}_t$-consistent, i.e., for each $v \in V$, $t \in [0, T]$ they are $\mathcal{F}_t$-measurable in ω.

We recall that, since V^* and $\mathcal{L}_Q(E, H)$ are separable, the concepts of strong and weak measurability coincide, and we will speak simply of measurability. Suppose further that on Ω there is given an $\mathcal{F}_0$-*measurable* function u_0 with values in H, while on $[0, T] \times \Omega$ there is given a nonnegative function $f(t, \omega)$ measurable in (t, ω) and $\mathcal{F}_t$-consistent.

We assume that, for some constants K, $\alpha > 0$ and for all $v, v_1, v_2 \in V$, $(t, \omega) \in [0, T] \times \Omega$, the following conditions are satisfied.

(A_1) Semicontinuity of A: the function $vA(v_1 + \lambda v_2)$ is continuous in λ on $\mathbb{R}$.

(A_2) Monotonicity of (A, B):

$$2(v_1 - v_2)(A(v_1) - A(v_2)) + |B(v_1) - B(v_2)|_Q^2 \leq K\,|v_1 - v_2|_H^2 .$$

(A_3) Coercivity of (A, B):

$$2vA(v) + |B(v)|_Q^2 + \alpha\,|v|^p \leq f + K\,|v|_H^2 .$$

(A_4) Boundedness of the growth of A:

$$|A(v)|_{V^*} \leq f^{1/q} + K\,|v|_V^{p-1} .$$

(A_5)

$$\mathbb{E}\,|u_0|_H^2 < \infty, \quad \mathbb{E}\int_0^T f(t)dt < \infty.$$

Under these assumptions we consider on $[0, T] \times \Omega$ the stochastic evolution equation

$$v(t, \omega) = u_0(\omega) + \int_0^t A(v(s, \omega), s, \omega)ds + \int_0^t B(v(s, \omega), s, \omega)dw(s, \omega) + z(t, \omega). \quad (3.4)$$

Definition 3.1. A solution (or a V-solution) of equation (3.4) is a function $v(t, \omega)$ with values in V defined on $[0, T] \times \Omega$, measurable in (t, ω), $\mathcal{F}_t$-consistent, satisfying the inequality

$$\mathbb{E}\int_0^T \left(|v(t)|_V^p + |v(t)|_H^2\right) dt < \infty, \quad (3.5)$$

and satisfying equation (3.4) in the sense of equality of elements of V^* for almost all $(t, \omega) \in [0, T] \times \Omega$.

In this definition it is implicitly assumed that integrals in (3.4) and (3.5) are meaningful. This assumption is, however, superfluous in view of what follows.

By Lemma 2.5 the function v is measurable not only as a function with values in V but also as a function with values in H. Therefore, $|v(t)|_V$, $|v(t)|_H$ are measurable with respect to (t, ω) and condition (3.5) is meaningful.

We show further that the functions $A(v(t), t)$, $B(v(t), t)$ are measurable in (t, ω) and $\mathcal{F}_t$-measurable in ω. We first consider $A(v(t), t)$ and use the same sort of lemma as in the theory of monotone operators (see e.g. Ref. [42]).

Lemma 3.2. *If a sequence v_n converges strongly in V to v, then the sequence $A(v_n)$ converges weakly to $A(v)$ in V^*.*

Proof. Because of assumption (A_4), for every subsequence $\{\mu\}$ of natural numbers, the sequence $A(v_\mu)$ is bounded in V^*, and therefore there exists a subsequence $\{\eta\}$ of the sequence $\{\mu\}$ along which $A(v_\eta)$ converges weakly to some $A_\infty \in V^*$. We will now show that $A_\infty = A(v)$.

Let u be an arbitrary element of V. By assumption (A_2)

$$(u - v_\eta)(A(u) - A(v_\eta)) - K\,|u - v_\eta|_H^2 \leq 0.$$

Passing to the limit $\eta \to \infty$ in this equality, we obtain

$$(u - v)(A(u) - A_\infty) - K\,|u - v|_H^2 \leq 0. \tag{3.6}$$

We now suppose that $u = v + \lambda y$ where $\lambda \in \mathbb{R}_+$ and y is some element of V; from (3.6) it then follows that

$$y(A(v + \lambda y) - A_\infty) - K\lambda\,|y|_H^2 \leq 0.$$

Passing to the limit $\lambda \downarrow 0$ in this equality and using the semicontinuity of A (assumption (A_1),) we obtain

$$y(A(v) - A_\infty) \leq 0.$$

Since y is arbitrary, it follows that $A_\infty = A(v)$ and, since the subsequence $\{\mu\}$ was arbitrary, the proof of the lemma is complete. □

It follows from the lemma that the function $v \to uA(v, t, \omega)$ is continuous for each $u \in V$, $(t, \omega) \in [0, T] \times \Omega$. As a result, if $v = v(t, \omega)$ is measurable in (t, ω) and $\mathcal{F}_t$-consistent, then so is $A(v(t, \omega), t, \omega)$. Moreover, in view of (A_4), (A_5), $A(v(t, \omega), t, \omega)$ is a summable function of t for almost all ω if (3.5) is satisfied.

The problem of the measurability of $B(v(t, \omega), t, \omega)$ is easier, since it is obvious that by assumptions (A_2) and (A_4), $B(\cdot, t, \omega)$ for all t, ω is a strongly continuous function from V to $\mathcal{L}_Q(E, H)$.

From assumptions (A_3) and (A_4) it follows that, for all u, t, ω,

$$|B(u, t, \omega)|_Q^2 \leq c\big(f(t, \omega) + |u|_H^2 + |u|_V^p\big). \tag{3.7}$$

Therefore,

$$\mathbb{E}\int_0^T |B(v(t),t)|_Q^2\,dt < \infty,$$

and the stochastic integral in (3.4) is defined and is a square-integrable martingale in H.

Remark 3.3. It is obvious (see the clarifications of Theorem 2.14) that v is a solution of (3.4) if and only if, for some set $Y \subset V$ dense in V (in the norm of V) and for every $u \in Y$, the equality

$$(u, v(t))_H = (u, u_0)_H + uA(v(s), s)ds + \int_0^t \tilde{u}B(v(s), s)dw(s) + (u, z(t))_H$$

holds a.e. (t, ω).

Definition 3.4. An H-solution of (3.4) is a solution $u(t,\omega)$ with values in H defined on $[0,T] \times \Omega$, strongly continuous in H with respect to t, $\mathcal{F}_t$-consistent, and such that

(a) $u \in V$ a.e. in (t,ω) and

$$\mathbb{E}\int_0^t \left(|u(t)|_V^p + |u(t)|_H^2\right) dt < \infty.$$

(b) there exists a set $\Omega' \subset \Omega$ of probability one on which, for all $t \in [0,T]$,

$$u(t) = u_0 + \int_0^t A(u(s), s)ds + \int_0^t B(u(s), s)dw(s) + z(t), \tag{3.8}$$

where the equality is understood as an equality of elements of V^*.

In clarification of (3.8) we note that, by Lemma 2.5, the function $\chi_V(u(t))$ is measurable in (t,ω), $\mathcal{F}_t$-measurable in ω, equal to one [a.e. in (t,ω)] by condition (a), and the integrals in (3.8) are therefore understood as

$$\int_0^t \chi_V(u(s))A(u(s), s)ds, \quad \int_0^t \chi_V(u(s))B(u(s), s)dw(s).$$

Definition 3.5. An H-solution u is called a continuous modification of a solution v in H if $u(t,\omega) = v(t,\omega)$ a.e. in (t,ω).

We now formulate the main results of this Section. All the assumptions above are assumed to be satisfied.

Theorem 3.6. *A solution v of equation* (3.4) *exists.*

From this theorem and Theorem 2.17 we immediately obtain the following result.

Corollary 3.7. *There exists a continuous modification of the solution v of equation* (3.4) *in H.*

Theorem 3.4 is proved in Section 3.5. The next theorem contains the uniqueness assertion for the solution of equation (3.4).

Theorem 3.8. *Let $v^n(t)$, $n = 0, 1, 2, \ldots$, be solutions of equation* (3.4) *with initial data $u_0 = u_0^n$, where $\mathbb{E}\,|u_0^n|_H^2 < \infty$ and $\mathbb{E}\,\left|u_0^n - u_0^0\right|_H^2 \to 0$ as $n \to \infty$. Suppose that $u^n(t)$ are continuous modifications of $v^n(t)$ in H. Then, for every $\varepsilon > 0$,*

$$\lim_{n\to\infty} \left\{ \sup_{t\le T} \mathbb{E} \left|u^n(t) - u^0(t)\right|_H^2 \right\} + P\left\{ \sup_{t\le T} \left|u^n(t) - u^0(t)\right|_H \ge \varepsilon \right\} = 0.$$

Remark 3.9. We will see from the proof of the theorem that the result is valid if, in the definition of the solution, condition (3.5) is dropped and only condition (A_2) is required of the coefficients of (3.4).

Theorem 3.10. *If v is a solution of* (3.4) *and u is its continuous modification in H, then*

$$\mathbb{E} \sup_{t\le T} |u(t)|_H^2 + \mathbb{E}\int_0^T |v(t)|_V^p \, dt \le c \left(\mathbb{E}\,|u_0|_H^2 + \mathbb{E}\int_0^T f(t)dt + \mathbb{E}\,|z(t)|_H^2 \right),$$

where c depends only on K, p, T, and α.

Theorems 3.8 and 3.10 are proved in Section 3.4. The following theorem on the Markov property of solutions of (3.4) also belongs to the basic results of this section. This theorem is proved at the end of Section 3.5.

Theorem 3.11. *Suppose that A and B do not depend on ω, $z(t) \equiv 0$, $v = v(t)$ is a solution of equation* (3.4), *and $u = u(t)$ is its continuous modification in H. Then $u(t)$ is a Markov random variable.*

Remark 3.12. With no loss of generality, we can take $K = 0$ in conditions (A_2) and (A_3). Indeed, if v is a solution of (3.4), then $v(t)e^{-Kt}$ is a solution of an equation of the type (3.4) with A replaced by $e^{-Kt}(A - KI)$, where I is the identity operator, and with B replaced by $e^{-Kt}B$, and these new A and B satisfy conditions (A_2) and (A_3) with $K = 0$.

Remark 3.13. Our assumption that the spaces in question are real is not essential and may be relaxed if, in conditions (A_2)–(A_3), in place of

$$vA(v_1 + \lambda v_2), \ (v_1 - v_2)(A(v_1) - A(v_2)), \ vA(v)$$

we write

$$\operatorname{Re}\big(vA(v_1 + \lambda v_2)\big), \ \operatorname{Re}\big((v_1 - v_2)(A(v_1) - A(v_2))\big), \ \operatorname{Re}\big(vA(v)\big).$$

3.3. *Itô Equations in* $\mathbb{R}^d$

Let $\mathbb{R}^d$ be Euclidean space of dimension d with a fixed orthonormal basis, let x^i be the $i-th$ coordinate of a point $x \in \mathbb{R}^d$, let $(\Omega, \mathcal{F}, P)$ be a complete probability space, and let $\{\mathcal{F}_t\}$ be an expanding family of complete σ-algebras $\mathcal{F}_t \subset \mathcal{F}$. Let $m(t)$ be a d_1-dimensional continuous local martingale relative to $\{\mathcal{F}_t\}$, with $m(0) = 0$, and let $A(t)$ be a continuous, real, nondecreasing $\mathcal{F}_t$-consistent process with $A(0) = 0$. Suppose that, for $t \geq 0$, $x \in \mathbb{R}^d$, $\omega \in \Omega$, a $d \times d_1$ matrix $b(t, x)$ and a d-dimensional vector $a(t, x)$ are defined. We assume that, for each $x \in \mathbb{R}^d$, $a(t, x)$ and $b(t, x)$ are completely measurable relative to $\{\mathcal{F}_t\}$ and are continuous in x for each (t, ω). Let x_0 be a d-dimensional $\mathcal{F}_0$-measurable random vector.

We consider the following equation:

$$x(t) = x_0 + \int_0^t a(s, x(s))dA(s) + \int_0^t b(s, x(s))dm(s). \tag{3.9}$$

Equation (3.9) will be considered under certain additional conditions on a, b, A, and m whose formulation requires the following notation. By the Doob-Meyer theorem there exists a continuous increasing process $\langle m \rangle_t$ for which $(m^2(t)) - \langle m \rangle_t)$ is a local martingale relative to $\{\mathcal{F}_t\}$ and $\langle m \rangle_0 = 0$. For $i, j = 1, \ldots, d_1$ we further define by means of the Doob-Meyer theorem continuous processes $\langle m^i, m^j \rangle_t$ having locally bounded variation in t for which every process $m^i(t)m^j(t) - \langle m^i, m^j \rangle_t$ is a local martingale and $\langle m^i, m^j \rangle_0 = 0$. We recall that the matrix $(\langle m^i, m^j \rangle_t)$ is nonnegative definite and

$$\langle m \rangle_t = \sum_{i=1}^{d_1} \langle m^i, m^i \rangle_t$$

for all t (a.s.). We fix a continuous real nondecreasing $\mathcal{F}_t$-consistent process V_t such that, for each ω, the measures on the t axis generated by the functions $A(t)$, $\langle m \rangle_t$ are absolutely continuous relative to the measure corresponding to V_t (for example, we can take $V_t = A(t) + \langle m \rangle_t$.)

We define

$$c^{ij}(t) = \frac{d\langle m^i, m^j \rangle_t}{dV_t}, \qquad C(t) = \left(c^{ij}(t)\right),$$

$$\alpha(t, x) = a(t, x)\frac{dA(t)}{dV_t}, \ \beta(t, x) = b(t, x)C_t^{1/2}.$$

We assume that the following conditions are satisfied, in addition to those enumerated above: for each $x \in \mathbb{R}^d$, $T > 0$

$$\int_0^T |\alpha(t, x)| \, dV_t = \int_0^T |a(t, x)| \, dA(t) < \infty \ (a.s.); \tag{3.10}$$

for every $R > 0$ there exists a nonnegative completely measurable process $K_t(R)$ such that,for all $T > 0$,

$$\int_0^T K(R)dV_t < \infty \text{ (a.s.)},$$

and, for each z, x, $y \in \mathbb{R}^d$ such that $|x|$, $|y| \le R$,

$$2(x-y)(\alpha(t,x) - \alpha(t,y)) + \|\beta(t,x) - \beta(t,y)\|^2 \le K_t(R)(x-y)^2, \tag{3.11}$$

$$2z\alpha(t,z) + \|\beta(t,z)\|^2 \le K_t(1+z^2) \tag{3.12}$$

for almost all t relative to the measure dV_t, where $K_t = K_t(1)$, $\|\gamma\|$ for a matrix γ means $(tr\,(\gamma\gamma^*))^{1/2}$, and, for ε, $\delta \in \mathbb{R}^d$, $\delta\varepsilon$ denotes the scalar product of δ, ε.

Theorem 3.14. *There exists a continuous, $\mathcal{F}_t$-consistent process $x(t)$ for which (3.9) is satisfied for all t with probability one. If $x(t)$, $y(t)$ are two continuous $\mathcal{F}_t$-consistent processes satisfying (3.9) for each t (a.s.), then*

$$P\left\{\sup_{t\ge 0} |x(t) - y(t)| > 0\right\} = 0.$$

The present theorem is a generalization of Itô's classical theorem on the existence of a strong solution of a stochastic equation of type (3.9) with random coefficients. In the present theorem we have avoided the Lipschitz condition and replaced it by condition (3.11) which we shall call the monotonicity condition.

Example 3.15. Let $w(t)$ be a one-dimensional Wiener process, let $p \in (1,2)$, and let $a(t,\omega)$ and $b(t,\omega)$ be completely measurable functions such that

$$\int_0^T a(t,\omega)dt < \infty \text{ (a.s.)}$$

and

$$-2(p-1)a(t,\omega) + \frac{p^2}{4}b^2(t,\omega) \le 0 \quad \text{(a.s. in } (t,\omega)).$$

We consider the equation

$$dx(t) = -|x(t)|^{p-1} \operatorname{sgn} x(t)a(t)dt + |x(t)|^{p/2} b(t)dw(t).$$

It is clear that for $p < 2$ the coefficients of this equation do not satisfy Lipschitz conditions, but satisfy the monotonicity condition. Indeed, by the formula of Hadamard,

$$\begin{aligned}
-2a\cdot(|x|^{p-1}\operatorname{sgn} x - |y|^{p-1}\operatorname{sgn} y)(x-y) &+ b^2\cdot(|x|^{p/2} - |y|^{p/2})^2 \\
&= -2a\cdot(x-y)^2\int_0^1 (p-1)\,|x+\tau(y-x)|^{p-2}\,d\tau \\
+b^2\cdot(x-y)^2 &\left(\int_0^1 \frac{p}{2}\,|x+\tau(y-x)|^{\frac{p}{2}-1}\operatorname{sgn}(x+\tau(y-x))d\tau\right)^2 \\
&\le (x-y)^2\left[\frac{p^2}{4}b^2 - 2(p-1)a\right]\int_0^1 |x+\tau(y-x)|^{p-2}\,d\tau \le 0.
\end{aligned}$$

If we take $V_t = t$ then from this inequality we obtain the monotonicity condition with $K_t = 0$ and also condition (3.12), since in the present case $\alpha(t,0) = \beta(t,0) = 0$.

Thus, condition (3.11) is weaker than the Lipschitz condition. However, extremely "bad" functions may not satisfy (3.11). We remark without proof that (3.11) implies differentiability of $\alpha(t,x)$ with respect to x for almost all x and that the first generalized derivatives of $\beta(t,x)$ with respect to x are locally square-summable.

We note that in contrast to Itô's work we consider the equation over a semimartingale. Generalizations of Itô's results to stochastic equations over a semimartingale have been obtained by Kazamaki [78], Doléans-Dade [79], Protter [80], Gal'chuk [81], Lebedev [54] and others. The results of these works are a special case of our assertion as far as continuous semimartingales are concerned. It seems to us that the extension of Theorem 3.14 to the case of discontinuous semimartingales is an accessible problem, but is beyond our present interest.

As always in similar situations, the uniqueness assertion in Theorem 3.14 is very easy to prove. Indeed, let $x(t)$, $y(t)$ be two solutions of (3.9). By the continuity with respect to t of $x(t)$, $y(t)$, and the integrals in (3.9), the processes $x(t)$, $y(t)$ satisfy (3.9) for all t on the same set of probability one. Next, define

$$\Psi_t(R) = \exp\left(\int_0^t K_s(R)dV_s\right),$$

apply Itô's formula to $|x(t)-y(t)|^2\,\Psi_t(R)$ and use (3.11). We then find that, (a.s.) for all t,

$$\begin{aligned}&|x(t\wedge\tau(R)) - y(t\wedge\tau(R))|^2\,\Psi_{t\wedge\tau}(R)\\&\le 2\int_0^{t\wedge\tau(R)} \Psi_s(R)\,(x(s)-y(s))\,(b(s,x(s)-b(s,y(s)))dm(s) \equiv m_t'(R),\end{aligned}$$

where $\tau(R)$ is the first exit time of $\max\big(|x(t)|,|y(t)|\big)$ from $[0,R)$. We see that the local martingale $m_t'(R)$ is nonnegative, and hence $m_t'(R)$ is a supermartingale. Since $m_0'(R) = 0$, it follows that $m_t'(R) = 0$ (a.s.), so that $|x(t\wedge\tau(R)) - y(t\wedge\tau(R))| = 0$ (a.s.) for all t, and, since $|x(t)-y(t)|$ is continuous in t and R is arbitrary, it follows that $\sup_t |x(t)-y(t)| = 0$ (a.s.) as required.

The existence assertion in Theorem 3.14 will be proved after a number of auxiliary propositions. In Theorem 3.14 it is asserted, in particular, that the right side of (3.9) exists for some process $x(t)$. This fact for any $\mathcal{F}_t$-consistent continuous process follows immediately from the following lemma.

Lemma 3.16. *For every $T, R > 0$ (a.s.)*

$$\int_0^T \sup_{|x|\le R} |\alpha(t,x)|dV(t) < \infty, \quad \int_0^T \sup_{x\le R} \|\beta(t,x)\|^2\,dV_t < \infty.$$

Proof. Let $\{y_i\}$ be a countable dense subset of $\{x : |x| \leq R + 2\}$. We define $\varphi_n(x, z) = \max((x - y_i)z, i = \{1, \dots, n\}, x, z \in \mathbb{R}^d$. It is clear that each $\varphi_n(x, z)$ is a continuous function and $\varphi_n(x, z) \uparrow xz + |z|(R+2)$ as $n \to \infty$. The last function is continuous and greater than two if $|x| \leq R$, $|z| = 1$. By Dini's theorem there exists an n_0 such that $\varphi_{n_0}(x, z) \geq 1$ for $|x| \leq R$, $|z| = 1$. This implies that $\varphi_{n_0}(x, z) \geq |z|$ for $|x| \leq R$ and all $z \in \mathbb{R}^d$. Substituting the points y_i into (3.11) in place of y and computing the upper bounds over i for $|x| \leq R$, we find (with $\alpha(t, x)$ in the place of z) that

$$2|\alpha(t, x)| \leq 4(R + 2)^2 K_t(R + 2) + 2(R + 2) \sum_{i=1}^{n_0} \alpha(t, y_i).$$

Because of (3.10) this proves the first assertion of the lemma. The second assertion follows in an obvious way from the first and from (3.12). The proof of the lemma is complete. □

Lemma 3.17. *Let $f(x)$ be a real locally bounded function on $\mathbb{R}^d$, let $n > 0$, and let $N = \sup\{|f(x)| : |x| \leq n\}$. Then on $\mathbb{R}^d$ there exists a real function $g(x)$ such that $g(x) = f(x)$ for $|x| \leq n$ and $g(x) = 0$ for $|x| \geq n + 1$, $|g(x)| \leq |f(x)|$, and*

$$|g(x) - g(y)|^2 \leq |f(x) - f(y)|^2 + N^2 (x - y)^2 \tag{3.13}$$

for all $x, y \in \mathbb{R}^d$. Moreover, if $f(x)$ is continuous in x and depends in a measurable way on several parameters, then $g(x)$ is continuous in x and is measurable with respect to these parameters.

Proof. Define[c] $h(x) = (n + 1)N - N|x|$ and

$$\begin{aligned} g_{(+)}(x) &= \max(\min(h(x), f_+(x)), 0), \\ g_{(-)}(x) &= \max(\min(h(x), f_-(x)), 0), \\ g(x) &= g_{(+)}(x) + g_{(-)}(x). \end{aligned}$$

We will prove that this function g satisfies all the requirements. The last assertion of the lemma is obvious, since N is a measurable function of those parameters on which f depends, while the continuity of g follows from the continuity of f and (3.13).

Further, $g_{(+)} \leq f_+$, $g_{(-)} \leq f_-$. Therefore, $g_\pm = g_{(\pm)}$, $|g| \leq |f|$, and $g(x)$ has the same sign as $f(x)$. Moreover, it is obvious that $h(x) \leq 0$ for $|x| \geq n + 1$. Therefore, $g(x) = 0$ for $|x| \geq n + 1$. Since $h(x) \geq N \geq |f(x)|$ for $|x| \leq n$, it follows that $g(x) = f(x)$ for $|x| \leq n$.

It remains to prove (3.13). We fix points x, y. If $f(x)$ and $f(y)$ have different signs, then the same is true of $g(x)$ and $g(y)$, and

$$|g(x) - g(y)| = \big| |g(x)| + |g(y)| \big| \leq \big| |f(x)| + |f(y)| \big| = |f(x) - f(y)|.$$

[c]This method of constructing g was suggested to us by A. D. Wentzell.

We now consider the case where $f(x)$ and $f(y)$ have the same sign. Suppose, to be specific, that $f(x) \geq 0$, $g(x) \geq 0$; then $g(x) = g_{(+)}(x)$, $g(y) = g_{(+)}(y)$. Since the absolute value of the difference of the upper (lower) bounds does not exceed the upper (lower) bound of the absolute value of the difference, it follows that

$$\begin{aligned}|g(x) - g(y)| &\leq |\min(h(x), f(x)) - \min(h(y), f(y))| \\ &\leq \max\left(|h(x) - h(y)|, |f(x) - f(y)|\right),\end{aligned}$$

which implies (3.13). The proof of the lemma is complete. □

Lemma 3.18. *For every $n > 0$, there exist processes $\tilde{a}(t,x)$, $\tilde{b}(t,x)$, N_t such that $\tilde{a}(t,x) \in \mathbb{R}^d$, $\tilde{b}(t,x)$ is a $d \times d_1$ matrix, N_t is a real process, $\tilde{a}$, $\tilde{b}$, N are defined for all $x \in \mathbb{R}^d$, $t \geq 0$, $\omega \in \Omega$, are continuous in x and completely measurable, $a(t,x) = \tilde{a}(t,x)$, $b(t,x) = \tilde{b}(t,x)$ for $|x| \leq n$, $\tilde{a}(t,x) = 0$, $\tilde{b}(t,x) = 0$ for $|x| \geq n+3$ for all t,*

$$\int_0^t N_s dV_s < \infty \text{ (a.s.)}, \tag{3.14}$$

and, for all $x, y \in \mathbb{R}^d$,

$$\begin{aligned}|\tilde{a}(t,x)| + \left\|\tilde{\beta}(t,x)\right\|^2 \leq N_t, \quad & 2(x-y)(\tilde{\alpha}(t,x) - (\tilde{\alpha}(t,y)) \\ + \left\|\tilde{\beta}(t,x) - \tilde{\beta}(t,y)\ \right\|^2 \leq \quad & N_t(x-y)^2 \text{ (a.e. } dP \times dV_t),\end{aligned} \tag{3.15}$$

where

$$\tilde{\alpha}(t) = \tilde{a}(t)\frac{dA(t)}{dV_t}, \quad \tilde{\beta}(t) = \tilde{b}(t)C^{1/2}(t).$$

Proof. We fix $n \geq 1$. Let T_t be an orthogonal $d_1 \times d_1$ matrix, and let Λ_t be a diagonal matrix of the same dimension so that $C_t = T_t\Lambda T_t^*$. It is well known that such matrices exist, and they can be chosen to be completely measurable. For every element of the matrix $b(t,x)T_t$, we construct by means of Lemma 3.17 the corresponding truncated elements which we assign to the matrix $b'(t,x)$, and then define $\tilde{b}(t,x) = b'(t,x)T_t^*$. Consider the function $\eta \in C_0^\infty(\mathbb{R}^d)$, $\eta(x) = 1$ for $|x| \leq n+2$, $\eta(x) = 0$ for $|x| \geq n+3$, $0 \leq \eta \leq 1$, and define $\tilde{a}(t,x) = a(t,x)\eta(x)$. We shall prove that there exists a process N_t such that the assertions of the lemma hold for $\tilde{a}$, $\tilde{b}$, N.

It is not hard to see that by Lemmas 3.16 and 3.17 there exists a process $N_t^{(1)}$ satisfying condition (3.14) and such that

$$\begin{aligned}\left\|\tilde{\beta}(t,x) - \tilde{\beta}(t,y)\right\|^2 &= \left\|\tilde{b}(t,x)T_t - \tilde{b}(t,y)T_t\Lambda_t^{1/2}\right\|^2 \\ &\leq N_t^{(1)}(x-y)^2 + \left\|(b(t,x)T_t - b(t,y))T_t\Lambda_t^{1/2}\right\|^2 \\ &= N_t^{(1)}(x-y)^2 + \|\beta(t,x) - \beta(t,y)\|.\end{aligned} \tag{3.16}$$

Moreover, by Lemma 3.16 there exists a process $N_t^{(2)}$ satisfying (3.14) and such that

$$\begin{aligned} 2(x-y)(\tilde{\alpha}(t,x)-(\tilde{\alpha}(t,y)) &= 2(x-y)(\alpha(t,x))(\eta(x)-\eta(y)) \\ &\quad +2(x-y)\eta(y)(\alpha(t,x)-\alpha(t,y)) \qquad (3.17) \\ &\le 2(\eta(x)\wedge\eta(y))(x-y)(\alpha(t,x)-\alpha(t,y))+N_t^{(2)}(x-y)^2. \end{aligned}$$

Further, by Lemma 3.16 there exists a process $N_t^{(3)}$ satisfying (3.14) and such that

$$2\left|\tilde{\alpha}(t,x)\right|+\left\|\tilde{\beta}(t,x)\right\|^2\le N_t^{(3)}. \qquad (3.18)$$

Finally, if $|x|$, $|y|\le n+2$, then $\eta(y)=1$ and the second inequality in (3.15) is satisfied by (3.16), (3.17), and (3.11), with $N=N^{(2)}+N^{(3)}+K(n+3)$. If $|x|$, $|y|\ge n+1$, then $\tilde{\beta}(x)=\tilde{\beta}(y)=0$ and (3.15) is satisfied with $N=N^{(2)}+K(n+3)$. If one of the values of $|x|$, $|y|$ is less than $n+1$ and the other is greater than $n+2$, then $|x-y|\ge 1$, and (3.15) is satisfied by (3.17), (3.18), and (3.11), with $N=N^{(2)}+N^{(3)}+K(n+3)$, since one of the values of $\tilde{\beta}(t,x)$, $\tilde{\beta}(t,y)$ is zero. Thus, inequalities (3.15) are satisfied with $N=N^{(1)}+N^{(2)}+N^{(3)}+K(n+3)$ and the proof of the lemma is complete. □

Lemma 3.19. *Suppose there exists a completely measurable process $N_t\ge 0$ such that $\int_0^T N_s dV_s<\infty$ (a.s.) for all $T>0$ and, for all x, y,*

$$|\alpha(t,x)|+|\beta(t,x)|^2\le N_t \text{ (a.e. } dP\times dV_t),$$
$$2\left|x-y\right|\left|\alpha(t,x)-\alpha(t,y)\right|+\left\|\beta(t,x)-\beta(t,y)\right\|^2\le N_t\left|x-y\right|^2 \text{ (a.e. } dP\times dV_t).$$

Then equation (3.9) has a unique solution.

Proof. We have already proved the uniqueness of the solution of (3.9). Define

$$\psi(t)=\exp\left(-3\int_0^t N_s dV_s-|x_0|\right),$$

$$x_t^0=x_0,\; x^{r+1}=x_0+\int_0^t a(s,x^r(s))dA(s)+\int_0^t b(s,x^r(s))dm(s),\; r>0. \qquad (3.19)$$

Using Itô's formula for $\left(x^{r+1}(t)-x^r(t)\right)^2\psi(t)$, it is easy to show that, for some local martingale $m^r(t)$, $r \geq 1$,

$$\left(x^{r+1}(t) - x^r(t)\right)^2 \psi(t) = \int_0^t \Big\{2\left(x^{r+1}(s) - x^r(s)\right)\left(\alpha(s, x^r(s)) - \alpha(s, x^{r-1}(s))\right)$$

$$+ \left\|\beta(s, x^r(s)) - \beta(s, x^{r-1}(s))\right\|^2 - 3N_s\left(x^{r+1}(s) - x^r(s)\right)^2 \psi_s\Big\}dV_s + m^r(t)$$

$$\leq \int_0^t \Big\{\left|x^{r+1}(s) - x^r(s)\right| N_s \left|x^r(s) - x^{r-1}(s)\right| + N_s\left(x^r(s) - x^{r-1}(s)\right)^2$$

$$- 3N_s\left(x^{r+1}(s) - x^r(s)\right)^2\Big\}\psi_s dV_s + m^r(t)$$

$$\leq \int_0^t \left\{\frac{3}{2}N_s\left(x^r(s) - x^{r-1}(s)\right)^2 - \frac{5}{2}N_s\left(x^{r+1}(s) - x^r(s)\right)^2\right\}\psi_s dV_s + m^r(t)$$

or

$$\left(x^{r+1}(t) - x^r(t)\right)^2 \psi(t) + \int_0^t \frac{5}{2}N_s\left(x^{r+1}(s) - x^r(s)\right)^2 \psi_s dV_s$$
$$\leq \int_0^t \frac{3}{2}N_s\left(x^r(s) - x^{r-1}(s)\right)^2 + m^r(t).$$

In the last inequality, we put $t = \tau^i \wedge \tau$, where τ^i are stopping times such that $\tau^i \uparrow \infty$ and $m^r\left(t \wedge \tau^i\right)$ is a martingale, and the stopping time τ will be specified later. Computing the expectations and letting $\tau_i \uparrow \infty$ we conclude that

$$\mathbb{E}\left[5\int_0^\tau \left(x^{r+1}(s) - x^r(s)\right)^2 N_s\psi_s dV_s + 2\left(x^{r+1}(\tau) - x^r(\tau)\right)^2 \psi_\tau\right]$$
$$\leq 3\mathbb{E}\int_0^\tau \left(x^r(s) - x^{r-1}(s)\right)^2 N_s\psi_s dV_s, \ r \geq 1,$$

where $\left(x^r(\tau+1) - x^r(\tau)\right)^2 \psi_\tau$ is taken equal to zero on the set where $\tau = \infty$. Similarly,

$$\mathbb{E}\left\{\left(x^1(\tau)\right)^2 \psi_\tau + 2\int_0^\tau \left(x^1(s)\right)^2 N_s\psi_s dV_s\right\}$$
$$\leq \mathbb{E}x_0^2 e^{-|x_0|} + \mathbb{E}\int_0^\tau N_s\psi_s dV_s \leq \mathbb{E}x_0^2 e^{-|x_0|} + \frac{1}{3} < \infty,$$

where $\left(x^1(\tau)\right)^2 \psi_\tau$ is also taken equal to zero on the set where $\tau = \infty$. From these inequalities it follow that, for some constant N' and for all r, τ,

$$\mathbb{E}\int_0^\tau \left(x^{r+1}(s) - x^r(s)\right)^2 N_s\psi_s dV_s \leq N'\left(\frac{3}{5}\right)^r,$$

$$\mathbb{E}\left(x^{r+1}(\tau) - x^r(\tau)\right)^2 \psi_\tau \leq N'\left(\frac{3}{5}\right)^r.$$

We now take $\tau = \tau(r) = \inf\left\{t : \left(x^{r+1}(t) - x^r(t)\right)^2 \psi_t \geq r^{-4}\right\}$ and conclude that

$$r^{-4}P\{\tau(r) < \infty\} \leq N'\left(\frac{3}{5}\right)^r,$$

$$P\left\{\sup_{t\geq 0}\left(x^{r+1}(t)-x^r(t)\right)^2\psi_t\geq r^{-4}\right\}\leq N'r^4\left(\frac{3}{5}\right)^r.$$

Hence, by the Borel-Cantelli lemma the series

$$\sum_{r=1}^{\infty}\left|x^{r+1}(t)-x^r(t)^2\right|\psi_t^{1/2}$$

converges uniformly for almost all ω on each finite time segment to some continuous process $x(t)$. Passing to the limit in (3.19), we complete the proof of the lemma.□

Lemma 3.20. *We choose $n>0$, $\tilde{a}$, $\tilde{b}$, N from Lemma 3.18. If in equation* (3.9) *a, b are replaced by $\tilde{a}$, $\tilde{b}$, then the equation has a unique solution.*

Proof. For $r=1,2,\dots$ we define functions $\tilde{a}^r$, $\tilde{b}^r$ by convolution of $\tilde{a}$, $\tilde{b}$ on x with a δ-type sequence of infinitely differentiable, compactly supported, nonnegative functions of the form $\zeta(j_r x)j_r^d$, where $j_r\geq 0$, $j_r\to\infty$ as $r\to\infty$. Since for any r the first derivatives with respect to x and the functions $\tilde{a}^r$, $\tilde{b}^r$ can be bounded in terms of the maxima of $|\tilde{a}|$, $|\tilde{b}|$, by the first inequality in (3.15) the conditions of Lemma 3.19 are satisfied for $\tilde{a}$, $\tilde{b}$. Therefore, for every $r>0$, there exists a unique solution of the equation

$$x^r(t)=x_0+\int_0^t\tilde{a}^r(s,x^r(s))dA(s)+\int_0^t\tilde{b}^r(s,x^r(s))dm(s). \tag{3.20}$$

Since $\tilde{a}(t,x)=0$, $\tilde{b}(t,x)=0$ for $|x|\geq n+3$ and ζ is a compactly supported function, it can be assumed with no loss of generality that $\tilde{a}^r(t,x)=0$, $\tilde{b}^r(t,x)=0$ for $|x|\geq n+4$. In this case if $|x_0|\leq n+4$, then the process $x^r(t)$ never leaves the set $\{x:|x|\leq n+4\}$. If $|x_0|\geq n+4$, then $x^r(t)=x_0$ for all t. This implies that

$$\sup_{t\geq 0}\sup_r|x^r(t)-x_0|\leq 2n+8 \text{ (a.s.)} \tag{3.21}$$

Further, we choose N_t from Lemma 3.18. We note that by (3.15) for $\tilde{\alpha}^r=\tilde{a}^r(dA/dV)$, $\tilde{\beta}^r=\tilde{b}^rC^{1/2}$, we have

$$\left\|\tilde{\beta}^r(t,x^r(t))\right\|^2\leq\sup_x\left\|\tilde{\beta}^r(t,x)\right\|^2\leq\sup_x\left\|\tilde{\beta}(t,x)\right\|^2\leq N_t,\quad|\tilde{\alpha}^r(t,x^r(t))|\leq N_t. \tag{3.22}$$

We define $\psi_t=\exp\left(-\int_0^t N_s dV_s\right)$ and show next that

$$\lim_{r,p\to\infty}\sup_\tau\mathbb{E}\left|x^r(\tau)-x^p(\tau)\right|^2\psi_t=0, \tag{3.23}$$

where the supremum is taken over all stopping times τ and $\left(x^r(\tau)-x^p(\tau)\right)^2\psi_t$ is set equal to zero on when $\tau=\infty$.

By Itô's formula, for some local martingale $m^{r,p}(t)$,

$$\begin{aligned}
d\left|x^r(t)-x^p(t)\right|^2\psi_t &= \Big\{2\left(x^r(t)-x^p(t)\right)\left(\tilde{\alpha}^r(t,x^r(t))-\tilde{\alpha}^r(t,x^p(t))\right)\\
&+\left\|\tilde{\beta}^r(t,x^r(t))-\tilde{\beta}^r(t,x^p(t))\right\|^2-N_t\left(x^r(t)-x^p(t)\right)^2\Big\}\psi_t dV_t\\
&+\Big\{2\left(x^r(t)-x^p(t)\right)\left(\tilde{\alpha}^r(t,x^p(t))-\tilde{\alpha}^p(t,x^p(t))\right)\\
&+\left\|\tilde{\beta}^r(t,x^p(t))-\tilde{\beta}^p(t,x^p(t))\right\|^2\Big\}\psi_t dV_t\\
&+2\left[\tilde{\beta}^r(t,x^p(t))-\tilde{\beta}^p(t,x^p(t)),\tilde{\beta}^r(t,x^r(t))-\tilde{\beta}^r(t,x^p(t))\right]\psi_t dV_t+dm^{r,p}(t),
\end{aligned} \tag{3.24}$$

where, for two $d\times d_1$ matrices, $[\sigma_1,\sigma_2]$ is the sum of all products of the form $\sigma_1^{ij}\sigma_2^{ij}$. Application of the second inequality in (3.15) and the Cauchy-Schwartz inequality shows that

$$\begin{aligned}
&2\left(x^r(t)-x^p(t)\right)\left(\tilde{\alpha}^r(t,x^r(t))-\tilde{\alpha}^r(t,x^p(t))\right)\\
&\quad+\left\|\tilde{\beta}^r(t,x^r(t))-\tilde{\beta}^r(t,x^p(t))\right\|^2-N_t\left(x^r(t)-x^p(t)\right)^2\le 0
\end{aligned}$$

(a.e. $dP\times dV_t$). Moreover, the functions

$$\left(x^r(t)-x^p(t)\right)\left(\tilde{\alpha}^r(t,x^p(t))-\tilde{\alpha}^p(t,x^p(t))\right)\psi_t$$

by (3.21), (3.22) are bounded in absolute value by $4(2n+8)N_t\psi_t$, which is summable with respect to $dP\times dV_t$, and for all$t$, ω. Also, these function tend to zero as r, $p\to\infty$ by continuity in x of the compactly supported $\tilde{\alpha}(t,x)$. For all t, ω, we have $|\tilde{\alpha}^r(t,x)-\tilde{\alpha}(t,x)|$ as $r\to\infty$ uniformly with respect to $x\in\mathbb{R}^d$. Therefore,

$$\lim_{p,r\to\infty}\mathbb{E}\int_0^\infty\left|\left(x^r(s)-x^p(s)\right)\left(\tilde{\alpha}^r(s,x^p(s))-\tilde{\alpha}^p(s,x^p(s))\right)\right|\psi_s dV_s=0.$$

Similarly,

$$\lim_{r,p\to\infty}\mathbb{E}\int_0^\infty\left\|\tilde{\beta}^r(s,x^p(s))-\tilde{\beta}^p(s,x^p(s))\right\|^2\psi_s dV_s=0,$$

$$\lim_{r,p\to\infty}\mathbb{E}\int_0^\infty\left|\left[\tilde{\beta}^r(s,x^p(s))-\tilde{\beta}^p(s,x^p(s)),\tilde{\beta}^r(s,x^r(s))-\tilde{\beta}^r(s,x^p(s))\right]\right|\psi_s dV_s.$$

It is now clear that (3.23) follows from (3.24). From (3.23), in turn, we obtain the existence of a subsequence $r(i)$ such that, for all stopping times τ,

$$\mathbb{E}\left|x^{r(i+1)}(\tau)-x^{r(i)}(\tau)\right|\psi_\tau\le 2^{-i}.$$

Finally, as in Lemma 3.19, from this we obtain the uniform convergence of $x^{r(i)}(t)$ with respect to t on any finite time interval, and by passing to the limit in (3.19) along this subsequence $r(i)$ we complete the proof of the lemma. □

Proof of Theorem 3.14. Denote by a^n, b^n the functions $\tilde{a}$, $\tilde{b}$ constructed in Lemma 3.18 for a given $n > 0$. By Lemma 3.20, for every $n > 0$, there exists a solution of the equation

$$x^n(t) = x_0 + \int_0^t a^n(s, x^n(s))dA(s) + \int_0^t b^n(s, x^n(s))dm(s).$$

Let $\tau^n = \inf\{t : |x^n(t)| \geq n\}$, $\tau^{n,m} = \tau^n \wedge \tau^m$. Since $a^n(t,x) = a(t,x)$ and $b^n(t,x) = b(t,x)$ for $|x| \leq n$, the processes $x^n(t \wedge \tau^{n,m})$, $x^m(t \wedge \tau^{n,m})$ satisfy the same equation

$$dz(t) = \chi_{t<\tau^{n,m}} a(t, z(t))dA(t) + \chi_{t<\tau^{n,m}} b(t, z(t))dm(t).$$

Hence these processes coincide and $x^n(t) = x^m(t)$ on $\{t \leq \tau^{n,m}\}$ (a.s.). Therefore, $\tau^n \leq \tau^m$ (a.s.) for $n \leq m$ and there exists $\tau = \lim_{n\to\infty} \tau^n$ (a.s.). Also, for almost all ω and $t < \tau$, the process $x(t) = \lim_{n\to\infty} x^n(t)$ is defined, and for every n and all t (a.s.)

$$x(t \wedge \tau^n) = x^n(t \wedge \tau^n) = x_0 + \int_0^{t\wedge\tau^n} a(s, x(s))dA(s) + \int_0^{t\wedge\tau^n} b(s, x(s))dm(s).$$

It remains to prove that $\tau = \infty$. Define

$$\psi_t = \exp\left(-\int_0^t K_s dV_s - |x_0|\right).$$

As we have done repeatedly, by means of (3.12) and Itô's formula we find that

$$\mathbb{E}(x^n(\tau^n))^2\psi_{\tau^n}\chi_{\tau^n<\infty} \leq \mathbb{E}x_0^2\psi_0 + \mathbb{E}\int_0^\infty K_{t\psi_t} dV_t < \infty.$$

Therefore

$$n^2\mathbb{E}\psi_{\tau^n}\chi_{\tau^n<\infty} \leq N, \text{ or } \mathbb{E}\psi_{\tau^n}\chi_{\tau^n<\infty} \to 0 \text{ and } \tau^n \to \infty \quad \text{(a.s.)}$$

as $n \to \infty$. The proof of the theorem is complete. □

Remark 3.21. The assumption of complete measurability of $a(t,x)$, $b(t,x)$, $K_t(R)$ ensures that the corresponding stochastic integrals with respect to $dm(t)$ are defined and the integrals with respect to $dA(t)$ and dV_t are $\mathcal{F}_t$-consistent. Theorem 3.14 can hold without this assumption. For example, we can instead assume that $dA(t) \ll dt$, $a(t,x)$ is measurable in (t,ω) and is $\mathcal{F}_t$-consistent, as are those elements of the matrix b which are multiplied by $dm^i(t)$ with $d\langle m^i\rangle_t \ll dt$, while the remaining elements of b are completely measurable, and $K_t(R)$ is progressively measurable.

3.4. *Uniqueness Theorem: A Priori Estimates and Finite-Dimensional Approximations*

In this section we prove Theorems 3.8 and 3.10 and prepare for the proof of Theorem 3.6.

Proof of Theorem 3.8. By Theorem 2.17, for $y^n = u^n - u^0$ we find for all t (a.s.)

$$|y^n(t)|_H^2 = \int_0^t \Big\{2(v^n(s) - v^0(s))\,(A(v^n(s),s) - A(v^0(s),s)) + \big|B(v^n(s),s) - B(v^0(s),s)\big|_Q^2\Big\}ds + m_t^n + \big|u_0^n - u_0^0\big|_H^2,$$

where m_t^n is a local martingale with $m_0^n = 0$. We use this equality and apply Itô's formula to compute $|y^n(t)|_H^2\, e^{-Kt}$. After this, we use the monotonicity of (A, B) (condition (A_2)) and the equality $y = v^n - v^0$ (a.e. in (t,ω)). Then, for some local martingales $\tilde{m}_t^n$, we obtain

$$|y^n(t)|_H^2\, e^{-Kt} \le \big|u_0^n - u_0^0\big|_H^2 + \tilde{m}_t^n \equiv \xi^n(t) \text{ (a.s.)}$$

We see that the local martingales $\xi^n(t)$ are nonnegative. Hence $\xi^n(t)$ is a supermartingale and, by assumption, $\mathbb{E}\xi^n(0) \to 0$. Therefore, for every $\varepsilon > 0$,

$$\lim_{n\to\infty}\left\{\sup_{t\le\tau}\mathbb{E}\xi^n(t) + P\left\{\sup_{t\le\tau}\xi^n(t) \ge \varepsilon\right\}\right\} = 0.$$

Application of the inequality $|y^n(t)|_H^2 \le \xi^n(t)e^{Kt}$ completes the proof of Theorem 3.8. □

Proof of Theorem 3.10. Let v be some solution of equation (3.4) and let u be its continuous modification in H. By Theorem 2.17, for all $t \in [0,T]$ on a single set of full probability,

$$|u(t)|_H^2 = |u_0|_H^2 + 2\int_0^t v(s)A(v(s))ds + 2\int_0^t \widehat{u}(s)\big(B(v(s))dw(s) + dz(s)\big) + \left\langle \int_0^t B(v(s))dw(s) + z\right\rangle_t. \tag{3.25}$$

Let

$$\tau_n = \inf\left\{t : |u(t)|_H^2 \ge n\right\} \wedge T.$$

It is obvious that the stochastic integral on the right side of (3.26) is a local martingale with localizing sequence τ_n. Therefore,

$$\mathbb{E}\,|u(t\wedge\tau_n)|_H^2 = \mathbb{E}\,|u_0|_H^2 + 2\mathbb{E}\int_0^{t\wedge\tau_n} v(s)A(v(s))ds + \mathbb{E}\int_0^{t\wedge\tau_n} |B(v^n(s))|_Q^2 + \mathbb{E}\,|z(t\wedge\tau_n|_H^2 + 2\mathbb{E}\left(\int_0^{t\wedge\tau_n} B(v(s))dw(s), z(t\wedge\tau_n)\right)_H.$$

Hence, by assumption (A_3)

$$\mathbb{E}\,|u(t\wedge\tau_n)|_H^2 \le \mathbb{E}\,|u_0|_H^2 - \alpha\mathbb{E}\int_0^{t\wedge\tau_n} |v(s)|_V^p\, ds + \int_0^{t\wedge\tau_n} (f(s) + K\,|u(s)|_H^2\, ds + \mathbb{E}\,|z(t\wedge\tau_n|_H^2 + 2\mathbb{E}\left(\int_0^{t\wedge\tau_n} B(v(s))dw(s), z(t\wedge\tau_n)\right)_H. \tag{3.26}$$

We now make use of the elementary inequality

$$2ab \le \varepsilon a^2 + \frac{1}{\varepsilon} b^2, \quad \varepsilon > 0. \tag{3.27}$$

Applying this inequality to the last term in (3.26), we obtain

$$\begin{aligned}\mathbb{E}\,|u(t\wedge\tau_n)|_H^2 \le \mathbb{E}\,|u_0|_H^2 - \alpha\mathbb{E}\int_0^{t\wedge\tau_n} |v(s)|_V^p\,ds + \int_0^{t\wedge\tau_n} (f(s) + K\,|u(s)|_H^2\,ds \\ + \left(1+\frac{1}{\varepsilon}\right)\mathbb{E}\langle z\rangle_T + \varepsilon\mathbb{E}\int_0^{t\wedge\tau_n} |B(v(s))|_Q^2\,ds.\end{aligned} \tag{3.28}$$

On the other hand, in view of inequality (3.7),

$$\mathbb{E}\int_0^{t\wedge\tau_n} |B(v(s))|_Q^2\,ds \le c\mathbb{E}\int_0^{t\wedge\tau_n} \left(f(s) + |v(s)|_V^p + |u(s)|_H^2\right) ds. \tag{3.29}$$

Combining (3.28) and (3.29), and choosing ε sufficiently small, for some α_1 we obtain

$$\begin{aligned}\mathbb{E}\,|u(t\wedge\tau_n)|_H^2 \le \mathbb{E}\,|u_0|_H^2 - \alpha_1\mathbb{E}\int_0^{t\wedge\tau_n} |v(s)|_V^p\,ds \\ + c\left(\mathbb{E}\int_0^t f(s)ds + \mathbb{E}\langle z\rangle_T + \int_0^t \mathbb{E}\,|u(s\wedge\tau_n)|_H^2\,ds\right).\end{aligned} \tag{3.30}$$

From this by the Gronwall lemma

$$\sup_{t\le T}\mathbb{E}\,|u(t\wedge\tau_n)|_H^2 \le c\left(\mathbb{E}\,|u_0|_H^2 + \mathbb{E}\int_0^T f(s)ds + \mathbb{E}\langle z\rangle_T\right). \tag{3.31}$$

Thus, from (3.30) and (3.31) it follows that

$$\sup_{t\le T}\mathbb{E}\,|u(t\wedge\tau_n)|_H^2 + \mathbb{E}\int_0^{t\wedge\tau_n} |v(s)|_V^p\,ds \le c\left(\mathbb{E}\,|u_0|_H^2 + \mathbb{E}\int_0^{\tau} f(s)ds + \mathbb{E}\langle z\rangle_T\right). \tag{3.32}$$

Continuity of $|u(t)|_H^2$ in t implies $\lim_{n\to\infty}\tau_n = T$. Therefore, passing to the limit as $n\to\infty$ in (3.32) (by Fatou's lemma and the monotone convergence theorem), we obtain

$$\sup_{t\le T}\mathbb{E}\,|u(t)|_H^2 + \mathbb{E}\int_0^T |v(s)|_V^p\,ds \le c\left(\mathbb{E}\,|u_0|_H^2 + \mathbb{E}\int_0^{\tau} f(s)ds + \mathbb{E}\langle z\rangle_T\right). \tag{3.33}$$

To complete the proof of Theorem 3.10 it remains to show that an analogous estimate also holds for $\mathbb{E}\sup_{t\le\tau}|u(t)|_H^2$. For this we note that from (3.25) because of (A_3) and (3.27) it follows that

$$\begin{aligned}\mathbb{E}\sup_{t\le T}|u(t\wedge\tau_n)|_H^2 \le \mathbb{E}\,|u_0|_H^2 + \mathbb{E}\int_0^T \left(f(s)+K|u(s)|_H^2\right)ds \\ + 2\mathbb{E}\sup_{t\le T}\left|\int_0^{t\wedge\tau_n} \widehat{u}(s)\big(B(v(s))dw(s)+dz(s)\big)\right|^2 \\ + 2\mathbb{E}\left|\int_0^T B(v(s))dw(s)\right|_H^2 + 2\mathbb{E}\langle z\rangle_T.\end{aligned} \tag{3.34}$$

Define

$$M(t)=\int_0^t B(v(s))dw(s)+z(t).$$

Using the Burkholder inequality (Theorem 2.13) and (3.27), we obtain

$$\begin{aligned}\mathbb{E}\sup_{t\le T}\left|\int_0^{t\wedge\tau_n}\widehat{u}(s)\big(B(v(s))dw(s)+dz(s)\big)\right|^2 \le 3\mathbb{E}\left(\int_0^{t\wedge\tau_n}|u(s)|_H^2\,d\langle M\rangle_s\right)^{1/2} \\ \le \frac{3}{2}\varepsilon\,\mathbb{E}\sup_{t\le T}|u(t\wedge\tau_n)|_H^2+\frac{3}{2\varepsilon}\mathbb{E}\,|M(T)|_H^2 \\ \le \frac{3}{2}\varepsilon\,\mathbb{E}\sup_{t\le T}|u(t\wedge\tau_n)|_H^2+\frac{3}{\varepsilon}\mathbb{E}\int_0^T|B(v(s))|_Q^2\,ds+\frac{3}{\varepsilon}\mathbb{E}\langle z\rangle_T.\end{aligned} \tag{3.35}$$

Combining (3.34) and (3.35), choosing ε small, and using (3.29), (3.33), we obtain

$$\mathbb{E}\sup_{t\le T}|u(t\wedge\tau_n)|_H^2\le c\left(\mathbb{E}\,|u_0|_H^2+\mathbb{E}\int_0^T f(s)ds+\mathbb{E}\langle z\rangle_T\right).$$

Because of the continuity of $|u(t)|_H$, the assertion of Theorem 3.10 follows from this on the basis of Fatou's lemma. Theorem 3.10 is proved. □

We will now prepare for the proof of Theorem 3.6. For this we approximate equation (3.4) by equations in finite-dimensional spaces.

Fix orthonormal bases $\{h_i\}$, $\{e_i\}$ in the spaces H and E, respectively, and assume that $h_i\in V$, $i=1,2,\ldots$, while e_i are the eigenvectors of the covariance operator Q of the Wiener process with the corresponding eigenvalues $\lambda_i>0$: $Qe_i=\lambda_i e_1$; this is possible because Q is a completely continuous (compact) operator. Then, for each i, $\lambda_i^{-\frac{1}{2}}(w(t),e_i)_E$ is a one-dimensional standard Wiener process, and for different i these process are independent.

For $n\ge 1$ we consider the following system of stochastic equations for $(u_n^1(t),\ldots,u_n^n(t))\in\mathbb{R}^n$

$$
\begin{aligned}
u_n^i(t) &= (h_i, u_0)_H + \int_0^t h_i A\left(\sum_{j=1}^n u_n^j(s)h_j, s\right) ds \\
&+ \sum_{j=1}^n \int_0^t \left(h_i, B\left(\sum_{k=1}^n u_n^k(s)h_k, s\right) e_j\right)_H d(w(s), e_j)_E \\
&+ (h_i, z(t))_H,\ i = 1, \ldots, n,\ t \in [0, T].
\end{aligned}
\tag{3.36}
$$

Lemma 3.22. *System* (3.36) *has a unique $\mathcal{F}_t$-measurable solution which is continuous in t.*

Proof. It suffices to show that system (3.36) can be written in the form (3.9) and to verify assumptions of Theorem 3.14. We note immediately that the measurability conditions are verified by using Remark 3.21 at the end of Section 3.3, and we leave this verification to the reader.

We choose $d = n$, $d_1 = 2n$ and we set

$$a(t,x) = (a^i(t,x),\, i = 1, \ldots, n),$$

$$b(t,x) = (b^{ij}(t,x),\, i = 1, \ldots, n,\, j = 1, \ldots, 2n),$$

$$m(t) = (m^i(t),\, i = 1, \ldots, 2n),\ A(t) = t,$$

$$m^i(t) = (w(t), e_i)_E,\ m^{n+i}(t) = (z(t), h_i)_H,\ 1 \le i \le n,$$

$$a^i(t,x) = h_i A\left(\sum_{j=1}^n x^j h_j, t\right),\ 1 \le i \le n,$$

$$b^{ij}(t,x) = \begin{cases} \left(h_i, B\left(\sum_{k=1}^n x^k h_k, t\right) e_j\right)_H, & 1 \le i, j \le n, \\ \delta^{n+i,j}, & 1 \le i \le n,\ n+1 \le j \le 2n, \end{cases}$$

where $\delta^{k,j}$ is the Kronecker symbol. With these notations it is obvious that (3.36) can be written in form (3.9). As in Section 3.3, we introduce the functions $c^{ij}(t)$, $\alpha(t,x)$, $\beta(t,x)$, corresponding to $V_t = t + \langle w \rangle_t + \langle z \rangle_t$. It is clear that (3.10) follows from (A_4) and (A_5).

Further, we choose $x, y \in \mathbb{R}^n$ and set $u = \sum_{i=1}^n x^i h_i$, $v = \sum_{i=1}^n y^i h_i$. It is no hard to see that

$$
\begin{aligned}
\|\beta(t,x) - \beta(t,y)\|^2 dV_t &= \sum_{i,j=1}^n \left(h_i, (B(u,t) - B(v,t))e_j\right)_H^2 \lambda_j dt \\
&\le \sum_{j=1}^\infty \lambda_j \left|(B(u,t) - B(v,t))e_j\right|_H^2 dt = |B(u,t) - B(v,t)|_Q^2 dt.
\end{aligned}
\tag{3.37}
$$

Hence, by condition (A_2)

$$\begin{aligned}&\left\{2(x-y)(a(t,x)-a(t,y))+\|\beta(t,x)-\beta(t,y)\|^2\right\}dV_t\\&\qquad\le 2(u-v)big(A(u,t)-A(v,t))dt+|B(u,t)-B(v,t)|_Q^2\,dt\\&\qquad\le K\,|u-v|_H^2\,dt=K\,(u-v)^2\,dt.\end{aligned}$$

Therefore, condition (3.11) is satisfied. Moreover, continuity of $a(t,x)$ in x can be derived from Lemma 3.2, and continuity of $b(t,x)$, from the continuity of $B(u,t)$ which, in turn, follows from (A_2). It remains to check condition (3.12).

Define

$$\xi(t)=\int_0^t b(s,x)dm(s).$$

For every $\varepsilon>0$ we have by (3.7) in analogy to (3.37)

$$\begin{aligned}d\langle\xi\rangle_t&\le(1+\varepsilon)\sum_{i,j=1}^n\lambda_j\,(h_i,B(u,t)\,e_j)_H^2\,dt+\left(1+\frac{1}{\varepsilon}\right)\sum_{i=1}^n d\langle m^{n+i}\rangle_t\\&\le|B(u,t)|_Q^2\,dt+\left(1+\frac{1}{\varepsilon}\right)d\langle z\rangle_t+\varepsilon c\big(f(t)+|u|_H^2+|u|_V^p\big)dt,\end{aligned}\tag{3.38}$$

where c does not depend on ε. We note further that

$$2(x\alpha(t,x)+\|\beta(t,x)\|^2\,dV_t=2uA(u,t)dt+d\langle\xi\rangle_t.\tag{3.39}$$

Combining (3.38) and (3.39) with condition (A_3) and then choosing ε sufficiently small, we conclude that

$$\begin{aligned}\left(2x\alpha(t,x)+\|\beta(t,x)\|^2\right)dV_t&\le c\left(x^2+f(t)\right)dt+cd\langle z\rangle_t\\&\le c(1+x^2)\left(1+f(t)\frac{dt}{dV_t}\right)dV_t,\end{aligned}$$

where c is a constant not depending on t,x,ω. Hence, condition (3.12) is satisfied, and the proof of the lemma is complete. □

Denote by Π_n the projection operator of V^* onto the span of $\{h_1,\dots,h_n\}$, by π_n the projection operator of E onto the span of $\{e_1,\dots,e_n\}$, and set $u_n(t)=\sum_{i=1}^n u_n^i h_i$. It is then obvious that (3.36) is equivalent to the equation

$$u_n(t)=u_0+\int_0^t\Pi_nA(u_n(s),s)ds+\int_0^t\Pi_nB(u_n(s),s)\,\pi_ndw(s)+\Pi_nz(t).\tag{3.40}$$

To complete the preparations, we need the following result.

Theorem 3.23. *1) There exists a constant C_0 such that, for all n at once,*

$$\mathbb{E}\sup_{t\le\tau}|u_n(t)|_H^2+\mathbb{E}\int_0^T|u_n(t)|_V^p\,dt\le C_0\left(\mathbb{E}\,|u_0|_H^2+\mathbb{E}\int_0^T f(t)dt+\mathbb{E}\langle z\rangle_T\right);$$

2) With

$$M_n(t) = \int_0^t \Pi_n B(u_n(s))\pi_n dw(s), \tag{3.41}$$

the function u_n *satisfies*

$$\begin{aligned}\mathbb{E}e^{-ct}\,|u_n(t)|_H^2 &= \mathbb{E}\,|\Pi_n u_0|_H^2 \\ &+ \mathbb{E}\int_0^t e^{-cs}\left\{2u_n(s)A(u_n(s)) + |\Pi_n B(u_n(s))\pi_n|_H^2 - c\,|u_n(s)|_H^2\right\} ds \\ &+ 2\mathbb{E}\int_0^t e^{-cs} d\langle z, M_n\rangle_s + \mathbb{E}\int_0^t e^{-cs} d\langle \Pi_n z\rangle_s\end{aligned}$$

for all $t \le T$ *and for every constant* c.

The proof of this theorem is almost a word-for-word repetition of the proof of Theorem 3.10, and we leave it to the reader. We only point out that

$$u_n \Pi_n A(u_n) = u_n A(u_n), \quad |\Pi_n B(u_n)\pi_n|_Q = |B(u_n)|_Q\,,$$

and that assertion 2) is proved by applying the usual Itô formula, since $u_n(t)$ lies in a finite-dimensional subspace of H.

3.5. *Existence of Solution and the Markov Property: Passing to the Limit by the Method of Monotonicity*

In this section we complete the proof of Theorem 3.6 and prove Theorem 3.11.

We denote by $\mathcal{T}$ the σ-algebra of progressively measurable sets on $[0,T] \times \Omega$ and by $\tilde{\mathcal{T}}$, its completion with respect to the measure $\ell \times P$, where ℓ is Lebesgue measure on $[0,T]$. We define $S = ([0,T] \times \Omega, \tilde{\mathcal{T}}, \ell \times P)$. From the results of the preceding section (Theorem 3.23) it follows that there exists a subsequence $\{\kappa\}$ of the natural numbers along which, for some u, v, $u_\infty(T)$,

$$u_\kappa \to u \text{ weakly in } L_2(S;H), \tag{3.42}$$

$$u_\kappa \to v \text{ weakly in } L_p(S;V), \tag{3.43}$$

$$u_\kappa(T) \to u_\infty(T) \text{ weakly in } L_2(\Omega, \mathcal{F}_T, P; H). \tag{3.44}$$

By Theorem 3.23 and conditions (A_3), (A_4) it may be assumed that

$$Au_\kappa \to A_\infty \text{ weakly in } L_q(S;V^*), \tag{3.45}$$

$$\Pi_\kappa B(u_\kappa)\pi_\kappa \to A_\infty \text{ weakly in } L_2(S;L_Q(E;H)). \tag{3.46}$$

Strictly speaking, u, v, $u_\infty(T)$, A_∞, and B_∞ are all equivalence classes, but in each class u, v, A_∞, B_∞ it is possible to choose a progressive measurable representative, and in the class $u_\infty(T)$, a $\mathcal{F}_T$-measurable representative. We henceforth consider only these representatives and preserve for them the notation of the corresponding classes. We note also that since the embedding of V in H is dense, it follows that $u = v$ for a.a. (t, ω).

Further, it is well known that a strongly continuous linear operator is weakly continuous. Therefore, because of (3.46),

$$\int_0^t \Pi_\kappa B(u_\kappa(s))\pi_\kappa dw(s) \to \int_0^t B_\infty(s)dw(s) \tag{3.47}$$

weakly in $L_2(S; H)$ and in $L_2(\Omega, \mathcal{F}_t, P; H)$ for each t

Let y be a bounded random variable, and let $\psi(t)$ be a bounded function on $[0, T]$. It follows from (3.40) that, for each h_i and $\kappa \geq 1$

$$\begin{aligned}&\mathbb{E}\int_0^T y\psi(t)(u_\kappa(t), h_i)_H dt\\ &= \mathbb{E}\int_0^T y\psi(t)h_i\left(u_0 + \int_0^t A(u_\kappa(s))ds + \int_0^t \Pi_\kappa B(u_\kappa(s))\pi_\kappa\, dw(s) + z(t)\right)dt.\end{aligned}$$

Passing to the limit in this equality and using (3.43), (3.45)–(3.47), we obtain

$$\begin{aligned}&\mathbb{E}\int_0^T y\psi(t)(v(t), h_i)_H dt\\ &= \mathbb{E}\int_0^T y\psi(t)h_i\left(u_0 + \int_0^t A_\infty(s)ds + \int_0^t B_\infty(s)dw(s) + z(t)\right)dt.\end{aligned}$$

From this it follows that for a.a. (t, ω)

$$v(t) = u_0 + \int_0^t A_\infty(s)ds + \int_0^t B_\infty(s)dw(s) + z(t). \tag{3.48}$$

In the same way, using (3.44)–(3.47), we find that (a.s.)

$$u_\infty(T) = u_0 + \int_0^T A_\infty(s)ds + \int_0^T B_\infty(s)dw(s) + z(T). \tag{3.49}$$

By Theorem 2.17, there exists an $\mathcal{F}_t$-consistent function continuous in t with values in H which coincides with $v(t)$ for a.a. (t, ω) and is equal to the right side of (3.48) for all $t \in [0, T]$ and $\omega \in \Omega'$, where $P(\Omega') = 1$. We identify v with u; this is possible, since $u = v$ a.e. in (t, ω). In view of (3.49) we then have

$$u_\infty(T) = u(T) \text{ (a.s.)}. \tag{3.50}$$

By the same Theorem 2.17, for all $(t, \omega) \in [0, T] \times \Omega'$,

$$\begin{aligned}|u(t)|_H^2 = |u_0|_H^2 + 2\int_0^t v(s)A_\infty(s)ds + 2\int_0^t \widehat{u}(s)\big(B_\infty(s)dw(s) + dz(s)\big)\\ + \langle M_\infty + z\rangle_t,\end{aligned} \tag{3.51}$$

where

$$M_\infty(t) = \int_0^t B_\infty(s)dw(s).$$

The remaining argument in the proof of Theorem 3.6 are standard in the theory of monotone operators and do not require special attention of the reader, because, on the first reading, these arguments produce the impression of a collection of trivial computations, mysteriously leading to the required result.

Let $y(t,\omega)$ be an $\mathcal{F}_t$-consistent function, measurable in (t,ω) with values in V and satisfying

$$\mathbb{E}\int_0^T \left(|(y(t))|_V^p + |(y(t))|_H^2\right) dt < \infty. \tag{3.52}$$

Define

$$\begin{aligned}\mathcal{O}_\kappa = \mathbb{E}\int_0^T e^{-ct}\Big\{&2(u_\kappa(t) - y(t))(A(u_\kappa(t) - A(y(t)) - c\,|u_v(t) - y(t)|_H^2 \\ &+ |\Pi_\kappa B(u_\kappa(t))\pi_\kappa - \Pi_\kappa B(y(t))\pi_\kappa|_Q^2\Big\}dt.\end{aligned}$$

From (A_2) it follows that, with a suitable choice of c,

$$\mathcal{O}_\kappa \leq 0. \tag{3.53}$$

Further, we represent $\mathcal{O}_\kappa$ in the form $\mathcal{O}_\kappa = \mathcal{O}_\kappa^1 + \mathcal{O}_\kappa^2$ where

$$\mathcal{O}_\kappa^1 = \mathbb{E}\int_0^T e^{-ct}\left\{2(u_\kappa(t)A(u_\kappa(t)) - c\,|u_\kappa(t)|_H^2 + |\Pi_\kappa B(u_\kappa(t))\pi_\kappa|_Q^2\right\}dt,$$

and $\mathcal{O}_\kappa^2 = \mathcal{O}_\kappa - \mathcal{O}_\kappa^1$. By Theorem 3.23,

$$\mathcal{O}_\kappa^1 = \mathbb{E}e^{-ct}\,|u_\kappa(T)|_H^2 - \mathbb{E}\,|\Pi_\kappa u_0|_H^2 - 2\,\mathbb{E}\int_0^T e^{-ct}d\langle M_\kappa, z\rangle_t - \mathbb{E}\int_0^T e^{-ct}d\langle \Pi_\kappa z\rangle_t,$$

with M_κ defined in (3.41). After integration by parts,

$$\begin{aligned}\mathbb{E}\int_0^T e^{-ct}d\langle M_\kappa, z\rangle_t &= \mathbb{E}\,e^{-cT}\left(\int_0^T \Pi_\kappa B(u_\kappa(s))\pi_\kappa dw(s), z(T)\right)_H \\ &\quad + c\,\mathbb{E}\int_0^T e^{-ct}\left(\int_0^t \Pi_\kappa B(u_\kappa(s))\pi_\kappa dw(s), z(t)\right)_H dt.\end{aligned}$$

From this and (3.47) it follows that

$$\begin{aligned}\limsup_{\kappa\to\infty}\mathcal{O}_\kappa^1 = \mathbb{E}e^{-cT}\,|u(T)|_H^2 - \mathbb{E}\,|u_0|_H^2 - \mathbb{E}\int_0^T e^{-ct}d\langle z\rangle_t \\ - 2\,\mathbb{E}\int_0^T e^{-ct}d\langle M_\infty, z\rangle_t + \delta\, e^{-cT},\end{aligned} \tag{3.54}$$

where $\delta = \limsup\limits_{\kappa\to\infty}\mathbb{E}\,|u_\kappa(T)|_H^2 - \mathbb{E}\,|u(T)|_H^2 \geq 0$ by (3.44). On the other hand, because of (3.51),

$$\mathbb{E}e^{-cT}|u(T)|_H^2 - \mathbb{E}|u_0|_H^2 = \mathbb{E}\int_0^T e^{-ct}\left\{2(v(t)A_\infty(t) - c|u(t)|_H^2 + |B_\infty(t)|_Q^2\right\}dt + 2\mathbb{E}\int_0^T e^{-ct}d\langle M_\infty, z\rangle_t + \mathbb{E}\int_0^T e^{-ct}d\langle z\rangle_t. \tag{3.55}$$

Comparing (3.54) and (3.55) we see that

$$\limsup_{\kappa\to\infty} \mathcal{O}_\kappa^1 = \mathbb{E}\int_0^T e^{-ct}\left\{2(v(t)A_\infty(t) - c|u(t)|_H^2 + |B_\infty(t)|_Q^2\right\}dt + \delta e^{-cT}. \tag{3.56}$$

Further, from (3.42), (3.43), (3.45), and (3.47) it follows that

$$\lim_{\kappa\to\infty} \mathcal{O}_\kappa^2 = \mathbb{E}\int_0^T e^{-ct}\Big(2y(t)A(y(t)) - 2y(t)A_\infty(t) - 2v(t)A(y(t)) + 2c(u(t),y(t))_H - c|y(t)|_H^2 - 2\big(B_\infty(t),B(y(t))\big)_Q + |B(y(t))|_Q^2\Big)dt. \tag{3.57}$$

Combining (3.56) and (3.57) we find, in view of (3.53),

$$\mathbb{E}\int_0^T e^{-ct}\Big\{2(v(t)-y(t))(A_\infty(t) - A(y(t))) - c|u(t)-y(t)|_H^2 + |B_\infty(t) - B(y(t))|_Q^2\Big\}dt + \delta e^{-cT} \le 0. \tag{3.58}$$

Setting $y = v$ in (3.58), we see that $B_\infty(t) = B(v(t))$ (a.e. in (t,ω)), and

$$\delta = \limsup_{\kappa\to\infty} \mathbb{E}|u_\kappa(T)|_H^2 - \mathbb{E}|u(T)|_H^2 = 0. \tag{3.59}$$

On the other hand, it follows from (3.58) that

$$\mathbb{E}\int_0^T e^{-ct}\left\{2(v(t)-y(t))\big(A_\infty(t) - A(y(t))\big) - c|u(t)-y(t)|_H^2\right\}dt \le 0. \tag{3.60}$$

Suppose now that $x(t,\omega)$ is a process with values in V which satisfies an inequality analogous to (3.52), and let $y = v - \lambda x$, $\lambda \in \mathbb{R}_+$. From (3.60) we then obtain the inequality

$$\mathbb{E}\int_0^T e^{-ct}\Big\{x(t)\Big(A_\infty(t) - A\big(v(t)+\lambda x(t)\big)\Big) - c\lambda|x(t)|_H^2\Big\}dt \le 0.$$

Letting λ go to zero, we find from this by (A_4) and the Lebesgue theorem that

$$\mathbb{E}\int_0^T x(t)(A_\infty(t) - A(v(t)))dt \le 0.$$

Since x is arbitrary, the last inequality implies $A_\infty(t) = A(v(t))$ a.e. in (t,ω), which together with (3.48) and an earlier proved equality $B_\infty(t) = B(v(t))$ a.e. in (t,ω) complete the proof of Theorem 3.6. □

Corollary 3.24. *Let v be a solution of* (3.4) *and u, its continuous modification in H, and let u_n be solutions of* (3.40). *Then for any $t \le T$*

$$\lim_{n\to\infty} \mathbb{E}\,|u_n(t) - u(t)|_H^2 = 0.$$

Proof. Note that we proved equality (3.59) in which u is a continuous modification of v in H. On the other hand, from (3.44) we have

$$\mathbb{E}\,|u(T)|_H^2 \le \lim_{\kappa\to\infty} \mathbb{E}\,|u_\kappa(T)|_H^2$$

and so

$$\lim_{\kappa\to\infty} \mathbb{E}\,|u_\kappa(T)|_H^2 = \mathbb{E}\,|u(T)|_H^2 .$$

Returning again to (3.44) and recalling that a weakly convergent sequence converges if the norm of the limit is equal to the limit of the norm, we conclude that

$$\lim_{\kappa\to\infty} \mathbb{E}\,|u_\kappa(T) - u(T)|_H^2 = 0.$$

Here $u(T)$ does not depend on the sequence $\{\kappa\}$ by Theorem 2.3. We have thus proved that any subsequence of the sequence $\{u_n(T),\ n = 1, 2, 3, \ldots\}$ contains a further subsequence converging strongly in $L_2(\Omega, \mathcal{F}_T, P; H)$ to $u(T)$. This implies the convergence of the whole sequence: $u_n(T) \to u(T)$ as $n \to \infty$. It remains to note that, in place of segment $[0, T]$, we can consider any smaller segment. □

Proof of Theorem 3.11. Together with (3.4), we consider the equation

$$v(t) = u_s + \int_s^t A(v(r), r)dr + \int_s^t B(v(r), r)dw(r), \tag{3.61}$$

which we solve for $t \in [s, T]$ and $\mathbb{E}\,|u(s)|_H^2 < \infty$. The definitions of a V-solution and H-solution are made for equation (3.61) in the usual way.

All the assertions proved for equation (3.4) carry over to (3.61) in a natural way; in particular, (3.61) has an H-solution $u(t) = u(t, s, u_s)$. We now assume that $u_s = x \in H$ is non-random and, for $t \in [s, T]$ and a bounded Borel function $f(h)$ on H, we define

$$M_{s,x} f(u(t)) = \mathbb{E} f(u(t, s, x)). \tag{3.62}$$

To prove the theorem it obviously suffices to show that (3.62) defines a Borel function of x, and for the H-solution $u(t)$ of (3.4),

$$\mathbb{E}\left\{f(u(t))|\mathcal{F}_s\right\} = M_{s,u(s)} f(u(t) \text{ (a.s.)} \tag{3.63}$$

The required property of Borel measurability is easily proved by means of a corresponding analog of Theorem 3.8, which shows that for continuous f the function $M_{s,x} f(u(t))$ is also continuous in x. It suffices to prove the relation (3.63) only for continuous f. Moreover, for simplicity, we assume that $s = 0$. This can always be achieved by changing the time variable.

We note, first of all, that $u(t,0,x)$ does not depend on $\mathcal{F}_0$; this follows from Corollary 3.24 and the fact that in Section 3.3 solutions of stochastic equations were constructed in the final analysis by passing to the limit in equations with coefficients that satisfy a Lipschitz condition, and for such equations the independence of a solution with nonrandom initial data from $\mathcal{F}_0$ is known [42, 81]. Further, we approximate $u_0(\omega)$ by step functions $u_0^n(\omega)$ and let Γ_n be the set of values of $u_0^n(\omega)$. It is easy to see that $u(t,0,u_0^n)$ and

$$\sum_{x\in\Gamma_n} \chi_{\{x\}}(u_0^n)u(t,0,x)$$

satisfy the same equation and therefore coincide (a.s.) From this and the independence of $u(t,0,x)$ and $\mathcal{F}_0$, it follows that

$$\mathbb{E}\{f(u(t,0,u_0^n))|\mathcal{F}_0\} = M_{0,u_0^n} f(u(t)) \quad \text{(a.s.)}$$

Letting here $n\to\infty$ and using Theorem 3.8 we obtain (3.63) for continuous f. The proof of Theorem 3.11 is complete. □

4. Itô Stochastic Partial Differential Equations

4.1. *Introduction*

This section is devoted to applications of the results of Section 3 to stochastic partial differential equations. We consider the first boundary-value problem for nonlinear equations of second order. The latter merit special attention, since filtering of diffusion process reduces to the investigation of equations of this type [1, 2, 82].

The results pertaining to nonlinear equations are related to Ref. [62], where equations without stochastic terms are considered, while elements of the theory of linear equations are presented following our work [64].

We start by recalling the definitions and some basic facts from the theory of Sobolev spaces.

Let $\mathbb{R}^d$ be d-dimensional Euclidean space with a fixed basis. Denote by α, β, γ, α_i, β_i, γ_i, $(i=1,2,\ldots)$ the coordinate vectors in this space and also the null vector. If α is the null vector, then D^α is the identity operator, while if α is the $i-th$ basis vector, then $D^\alpha = \partial/\partial x^i$. Suppose further that G is a domain in $\mathbb{R}^d$, Γ is the boundary of G, $m\ge 1$ is an integer, $p\in(1,\infty)$.

Definition 4.1. The Sobolev space $W_p^m(G)$ is the space of real functions defined on G with finite norm

$$\|u\|_{m,p} \equiv \left(\sum_{\alpha_1\ldots\alpha_m} \|D^{\alpha_1}\ldots D^{\alpha_m}u\|_p^p\right)^{1/p},$$

where $D^{\alpha_1},\ldots,D^{\alpha_m}$ are generalized derivatives, and

$$\|g\|_p = \left(\int_G |g(x)|^p\,dx\right)^{1/p}.$$

Definition 4.2. The Sobolev space $\overset{\circ}{W}{}_p^m(G)$ is the closure of $C_0^\infty(G)$ (the space of infinitely differentiable functions with compact supports in G) in the norm $\|\cdot\|_{m,p}$.

Theorem 4.3. *The spaces $\overset{\circ}{W}{}_p^m(G)$ are separable reflexive Banach spaces relative to the norm $\|\cdot\|_{m,p}$.*

Theorem 4.4 (Sobolev Embedding Theorem). *If the domain G is bounded, its boundary Γ is regular, and $2(d - mp) \leq pd$, then $W_p^m(G) \subset L_2(G)$ and this embedding is dense and continuous. In particular, there exists a constant c such that $\|u\|_2 \leq c\,\|u\|_{m,p}$ for all $u \in W_p^m(G)$.*

Proofs of these theorems can be found in Refs. [83–85]; the conditions on the boundary we call regular are also presented there.

The following assertion is obvious.

Theorem 4.5. *The Sobolev space $W_2^m(\mathbb{R}^d)$ is a Hilbert space; it is continuously and densely imbedded in $L_2(\mathbb{R}^d)$.*

We further need below the so-called Friederichs inequality [83, 85] which is given in the following theorem.

Theorem 4.6. *Suppose that the domain G is bounded and its boundary is regular. Then there exists a constant $c > 0$ such that, for every $u \in \overset{\circ}{W}{}_p^m(G)$,*

$$\|u\|_{m,p} \leq c \left(\sum_{|\alpha_1|+\ldots+|\alpha_m|=m} \|D^{\alpha_1} \ldots D^{\alpha_m} u\|_p \right).$$

The next fact is also well know; it follows easily by means of the Fourier transform and the Parseval equality.

Theorem 4.7. *There exists a constant $c > 0$ such that, for every $u \in W_2^m(\mathbb{R}^d)$,*

$$\|u\|_{m,2} \leq c \left(\sum_{|\alpha_1|+\ldots+|\alpha_m|=m} \|D^{\alpha_1} \ldots D^{\alpha_m} u\|_2 + \|u\|_2 \right).$$

We will also need the spaces $W_q^{-m}(G)$ where $q = p/(p-1)$; Refs. [51, 85] present a detailed account of these spaces. It is assumed below that either G is bounded and its boundary is regular, or $G = \mathbb{R}^d$ and $p = 2$.

Definition 4.8. The negative norm of an element $f \in L_q(G)$ is

$$\|f\|_{-m,q} = \sup(f, u)_0,$$

where

$$(f, u)_0 = \int_G f(x) u(x) dx$$

and the supremum is taken over the set of all functions $u \in \overset{\circ}{W}{}_p^m(G)$ for which $\|u\|_{m,p} = 1$.

Definition 4.9. The space with negative norm $W_q^{-m}(G)$ is the completion of $L_q(G)$ in the norm $\|\cdot\|_{-m,p}$.

It is clear that, for $f \in L_q(G)$ and $u \in \overset{\circ}{W}{}_p^m(G)$, we have

$$|(f,u)_0| \leq \|f\|_q \cdot \|u\|_p \leq \|f\|_q \cdot \|u\|_{m,p} \, .$$

Definition 4.9 is therefore correct.

From the above definitions it follows immediately that there are the natural embeddings

$$\overset{\circ}{W}{}_p^m(G) \subset L_p(G) \subset W_p^{-m}(G).$$

The duality between $\overset{\circ}{W}{}_p^m(G)$ and $W_q^{-m}(G)$ is defined by means of the scalar product in $L_2(G)$: if $v \in \overset{\circ}{W}{}_p^m(G)$, $v^* \in W_q^{-m}(G)$, $v_n \in C_0^\infty$, $v_n^* \in L_q(G)$, and $\|v_n - v\|_{m,p} \to 0$, then we define

$$\langle\!\langle v, v^* \rangle\!\rangle = \lim_{n\to\infty} (v_n, v_n^*)_0.$$

It is known that, for every continuous linear functional on $\overset{\circ}{W}{}_p^m(G)$, there exists a $v^* \in W_q^{-m}(G)$ such that this functional is equal to $\langle\!\langle v, v^* \rangle\!\rangle$. Similarly, every continuous linear functional on $W_q^{-m}(G)$ can be written as $\langle\!\langle v, v^* \rangle\!\rangle$ for some $v \in \overset{\circ}{W}{}_p^m(G)$. Moreover, for $v \in \overset{\circ}{W}{}_p^m(G)$ and $v^* \in W_p^{-m}(G)$,

$$\|v\|_{m,p} = \sup_{w^*} \frac{\langle\!\langle v, v^* \rangle\!\rangle}{\|w^*\|_{-m,q}}, \quad \|v^*\| = \sup_{v^*} \frac{\langle\!\langle w, v^* \rangle\!\rangle}{\|w\|_{m,q}}.$$

The relation $\langle\!\langle \cdot, \cdot, \rangle\!\rangle$ makes it possible to identify the dual space of $\overset{\circ}{W}{}_p^m(G)$ with $W_q^{-m}(G)$. We note that under this identification the duality $\langle\!\langle \cdot, \cdot, \rangle\!\rangle$ plays a decisive role. In particular, the space $\overset{\circ}{W}{}_2^m(G)$ is a Hilbert space and, as any other Hilbert space, can be identified in the well-known way with its dual space by means of its own scalar product. At the same time, of course, $\overset{\circ}{W}{}_2^m(G) \neq W_2^{-m}(G)$. Thus, different bilinear forms make it possible to construct dual spaces of $\overset{\circ}{W}{}_2^m(G)$ in different ways. We shall use these considerations in the proof of Theorem 4.20 below.

We identify the dual space of $\overset{\circ}{W}{}_p^m(G)$ with $W_q^{-m}(G)$, so that the result of applying a functional $v^* \in W_q^{-m}(G)$ to an element $v \in \overset{\circ}{W}{}_p^m(G)$ is written as $\langle\!\langle v, v^* \rangle\!\rangle$. This notation differs from the notation vv^* used in Section 2 and 3 because both v and v^* can now be ordinary functions and then vv^* can be misinterpreted as the usual product of two functions.

Theorem 4.10. *Let* $2(d - mp) \le pd$. *Then* $\overset{\circ}{W}{}_p^m(G) \subset L_2(G) \subset W_q^{-m}(G)$, *where each embedding is dense and continuous. If* $v^* \in L_2(G)$ *and* $v \in \overset{\circ}{W}{}_p^m$ *then* $\langle\langle v, v^*\rangle\rangle = (v, v^*)_0$.

4.2. *First Boundary-Value Problem for Nonlinear Stochastic Parabolic Equations*

Let $(\Omega, \mathcal{F}, P)$ be a complete probability space; let $\{\mathcal{F}_t\}$ be an increasing family of complete σ-algebras imbedded in $\mathcal{F}$; let Gbe a bounded domain with a regular boundary or $G = \mathbb{R}^d$ and $p = 2$. We assume that $2(d - mp) \le pd$. Let $z(t)$ be a continuous square-integrable martingale (relative to $\mathcal{F}_t$) with values in $L_2(G)$ and let $w(t)$ be a Wiener process with values in some separable Hilbert space E and with covariance operator Q. By Theorem 4.10 the spaces $V = \overset{\circ}{W}{}_p^m(G)$, $H = L_2(G)$, $V^* = W_q^{-m}(G)$ satisfy assumptions a) – d) of Section 3.

In the cylinder $[0, T] \times G$ for fixed $T > 0$ we consider the problem

$$\begin{aligned} du(t,x,\omega) = &-(-1)^{|\alpha_1|+\ldots+|\alpha_m|} D^{\alpha_1}\ldots D^{\alpha_m} A_{\alpha_1\ldots\alpha_m}\big(D^{\beta_1}\ldots D^{\beta_m}u(t,x,\omega),t,x,\omega\big)dt \\ &+ B\big(D^{\beta_1}\ldots D^{\beta_m}u(t,x,\omega),t,x,\omega\big)dw(t,\omega) + dz(t,x,\omega), \end{aligned} \tag{4.1}$$

$$u(0,x,\omega) = u_0(x,\omega), \ x \in G, \tag{4.2}$$

$$D^{\beta_0}\ldots D^{\beta_{m-1}}u\,|_s = 0 \text{ for all } \beta_0, \ldots, \beta_{m-1} \tag{4.3}$$

such that

$$|\beta_0| + \ldots + |\beta_{m-1}| \le m - 1,$$

where S is the lateral surface of the cylinder $[0, T] \times G$; summations over all the values of the repeated indices α_i is understood; the functions A, B depend on t, x, ω, and all the derivatives of u with respect to x of order no greater than m; A and u are real functions; B is a function with values in E; in the second term in (4.1) the scalar product in E is understood.

We assume that for all collections of real numbers $\xi = (\xi^{\beta_1\ldots\beta_m})$ and any $e \in E$ the functions $A(\xi, t, x, \omega)$, $B(\xi, t, x, \omega)_e = (B(\xi, t, x, \omega), e)_E$ are

(1) measurable in (t, x, ω),
(2) measurable in (x, ω) relative to the product of $\mathcal{F}_t$ and the Borel σ-algebra on $\mathbb{R}^d$ for every fixed $t \in [0, T]$,
(3) continuous in ξ for every fixed t, x, ω.

Suppose further that there exist a constant $K > 0$ and a nonnegative function $f(t, x, \omega)$ possessing the same measurability properties as $A(0, t, x, \omega)$ and such that, for all $t, x, \omega, \xi, \alpha_1, \ldots, \alpha_m$,

$$|A_{\alpha_1,\ldots,\alpha_m}(\xi,t,x,\omega)| \le f^{1/q}(t,x,\omega) + K \sum_{\beta_1,\ldots,\beta_m} \left|\xi^{\beta_1,\ldots,\beta_m}\right|^{p-1}, \tag{4.4}$$

$$|B(\xi,t,x,\omega)|_E^2 \le f(t,x,\omega) + K \sum_{\beta_1,\dots,\beta_m} |\xi^{\beta_1,\dots,\beta_m}|^p + K\,|\xi^{0,\dots,0}|^2. \tag{4.5}$$

Moreover, it is assumed that $u_0(x,\omega)$ is measurable relative to the product of $\mathcal{F}_0$ and the Borel σ-algebra on $\mathbb{R}^d$, and, for all t,ω,

$$\|u_0\|_2 < \infty, \quad \|f(t)\|_1 < \infty, \quad \mathbb{E}\,\|u_0\|_2^2 < \infty, \quad \int_0^T \|f(t)\|_1\,dt < \infty.$$

These assumptions enable us to give a precise meaning to equation (4.1), to the initial condition (4.2), and to the boundary conditions (4.3) in the following manner.[d] We note that by (4.4) and the Hölder inequality

$$\left(A_{\alpha_1\dots\alpha_m}(D^{\beta_1}\dots D^{\beta_m}u), D^{\alpha_1}\dots D^{\alpha_m}v\right)_0 \le c\left(\|f(t)\|_1^{1/q} + \|u\|_{m,p}^{p-1}\right)\|v\|_{m,p}. \tag{4.6}$$

Therefore, for $t\in[0,T]$ the left side of (4.6) is a linear functional on $W_p^m(G)$ and, in particular, on $V = \overset{\circ}{W}{}_p^m(G)$. As we know, there exists a unique element $A(u,t,\omega)\in V^* = W_q^{-m}(G)$ for which for all $v\in V$

$$\langle\!\langle v, A(u,t,\omega)\rangle\!\rangle = -\left(A_{\alpha_1\dots\alpha_m}(D^{\beta_1}\dots D^{\beta_m}u,t,\omega), D^{\alpha_1}\dots D^{\alpha_m}v\right)_0.$$

It is clear that the function $A(u,t,\omega)$ satisfies the measurability conditions and conditions (A_1), (A_4) of Section 3.2. Further, we note that for $g\in H$ by (4.5)

$$\left(\left|B(D^{\beta_1}\dots D^{\beta_m}u,t)\right|_E, g\right)_0 \le \|f(t)\|_1^{1/2}\cdot\|g\|_H + N\left(\|u\|_V^{p/2}\cdot\|g\|_H + \|u\|_H\cdot\|g\|\right).$$

This implies that $B(u,t,\omega)\in\mathcal{L}(E,H)$ is defined for $t\in[0,T]$, $\omega\in\Omega$, $u\in V$ according to the formula

$$(B(u,t,\omega)e,g)_0 = \left((B(D^{\beta_1}\dots D^{\beta_m}u,t,\cdot,\omega),e)_E, g\right)_0,$$

where $e\in E$, $g\in H$. Moreover, if e_i, h_i are orthonormal bases in E, H, respectively, then

$$\begin{aligned}\|B(u,t)\|^2 &= \sum_{i,j}\left(B(u,t)e_i,h_j\right)^2 \le \left\|\,\left|B(D^{\beta_1}\dots D^{\beta_m}u,t)\right|_E\right\|_2^2\\ &\le \|f(t)\|_1 + N\left(\|u\|_V^p + \|u\|_H^2\right).\end{aligned} \tag{4.7}$$

A similar inequality obviously holds for $\left\|B(u,t)Q^{1/2}\right\|^2$. Hence, $B(u,t)$ satisfies inequality (3.7). Moreover, it is clear that if $u=u(t,\omega)$ is a function with values in V which is measurable in (t,ω) and $\mathcal{F}_t$-consistent, then $B(u(t,\omega),t,\omega)$ possesses the same measurability properties as an element of $\mathcal{L}_Q(E,H)$.

Thus, equation (3.4) can be considered for the operators A, B defined above.

Definition 4.11. A V-solution (H-solution) of problem (4.1)–(4.3) is a V-solution (H-solution) of equation (3.4). A continuous modification of problem (4.1)–(4.3) in H is defined similarly.

[d] For notational convenience we will occasionally omit all or some of the arguments in various functions.

Remark 4.12. A function $v(t) \in V$ is a V-solution of problem (4.1)–(4.3) if and only if it is appropriately measurable, inequality (3.5) is satisfied, and, for every $\eta \in V$ and almost all (t, ω),

$$\begin{aligned}(v(t), \eta)_0 = (u_0, \eta)_0 &- \int_0^t \left(A_{\alpha_1 \dots \alpha_m}(D^{\beta_1} \dots D^{\beta_m} v, s), D^{\alpha_1} \dots D^{\alpha_m} \eta\right)_0 \\ &+ \int_0^t \int_{\mathbb{R}^d} B(D^{\beta_1} \dots D^{\beta_m} v, s, x) \eta(x) dx dw(s) + (z(t), \eta)_0.\end{aligned} \tag{4.8}$$

In others words, a V-solution is a solution of problem (4.1)–(4.3) in the sense of the integral identity.

To prove this, it suffices to use Remark 3.3 and the fact that, for $\eta \in V$, the norm $\left|B(D^{\beta_1} \dots D^{\beta_m} v(t, x), t, x)\right|_E$ multiplied by $|\eta(x)|$ is integrable over $\mathbb{R}^d$, and hence for any $e \in E$

$$\begin{aligned}\widehat{\eta} B(v, t) e = (\eta, B(v, t) e)_H &= \int_{\mathbb{R}^d} \left(B(D^{\beta_1} \dots D^{\beta_m} v, t, x), e\right)_E \eta(x) dx \\ &= \left(\int_{\mathbb{R}^d} B(D^{\beta_1} \dots D^{\beta_m} v, t, x) \eta(x) dx, e\right)_E .\end{aligned}$$

A similar remark holds for an H-solution of problem (4.1)–(4.3).

Remark 4.13. Relation (4.8) is obtained if we formally multiply (4.1) by η, integrate over t, x, integrate by parts in x, and use the boundary conditions (4.3). In our interpretation of a solution of (4.1)–(4.3) we start from (4.8) in which there are no boundary conditions. They are accounted in (4.8) only through the membership of v in the space $\overset{\circ}{W}{}_p^m(G)$. In this connection we note that for $p > d$ by one of the Sobolev embedding theorems each function of $\overset{\circ}{W}{}_p^m(G)$ is equal almost everywhere to a function which has derivatives of order less than or equal to $m-1$, and these derivatives are continuous in the closure of G and vanish on the boundary of G.

Definition 4.14. We say that equation (4.1) satisfies the condition of strong parabolicity if the operators A, B satisfy conditions (A_2), (A_3) of Section 3.2.

Conditions which are sufficient for strong parabolicity in terms of the original functions $A_{\alpha_1 \dots \alpha_m}(\xi, t, x, \omega)$, $B(\xi, t, x, \omega)$ will be given below. The next theorem follows automatically from the results of Section 3.2.

Theorem 4.15. *If equation* (4.1) *is strongly parabolic, then the assertions of Theorems 3.6, 3.11 hold.*

Verification of the condition of strong parabolicity in the general case is a problem of colossal difficulty even for $B \equiv 0$. However, by generalizing the results of Ref. [62], in certain cases it is possible to give simple sufficient conditions for strong parabolicity of equation (4.1).

We say that the algebraic condition of strong parabolicity is satisfied if, for every fixed t, x, ω, the functions $A_{\alpha_1\dots\alpha_m}(\xi^{\beta_1\dots\beta_m}, t, x)$, and $B(\xi^{\beta_1\dots\beta_m}, t, x)$ are differentiable once with respect to ξ everywhere except for a set having a finite number of points of intersection with each line in the space of coordinates ξ, the derivatives are locally summable on each line in this space, and there exist constants $\varepsilon > 0$ and $N \in \mathbb{R}$ such that, for all $\xi^{\alpha_1\dots\alpha_m}$, $\eta^{\alpha_1\dots\alpha_m}$, t, x, ω,

$$\begin{aligned} &-2A^{\beta_1\dots\beta_m}_{\alpha_1\dots\alpha_m}(\xi^{\gamma_1\dots\gamma_m}, t, x)\eta^{\alpha_1\dots\alpha_m}\eta^{\beta_1\dots\beta_m} + \left|B^{\beta_1\dots\beta_m}(\xi^{\gamma_1\dots\gamma_m}, t, x)\eta^{\beta_1\dots\beta_m}\right|^2_Q \\ &+\varepsilon \sum_{|\alpha_1|+\dots+|\alpha_m|=m} |\xi^{\alpha_1\dots\alpha_m}|^{p-2}\,|\eta^{\alpha_1\dots\alpha_m}|^2 \le N\left(\eta^{0\dots0}\right)^2, \end{aligned} \tag{4.9}$$

where

$$A^{\beta_1\dots\beta_m}_{\alpha_1\dots\alpha_m} = \frac{\partial A_{\alpha_1\dots\alpha_m}}{\partial \xi^{\beta_1\dots\beta_m}}, \quad B^{\beta_1\dots\beta_m} = \frac{\partial B}{\partial \xi^{\beta_1\dots\beta_m}},$$

and in the second term of the left side of (4.9) summation over all $\beta_1..\beta_m$ is carried out before computing the norm $|\cdot|_Q$.

The next theorem justifies the name of condition (4.9).

Theorem 4.16. *Suppose that the algebraic condition of strong parabolicity is satisfied, together with all other assumptions of this section regarding the functions $A(\xi^{\beta_1\dots\beta_m})$, $B(\xi^{\beta_1\dots\beta_m})$. Then equation (4.1) is strongly parabolic.*

Proof. We use the formula of Hadamard

$$B(\xi) - B(\eta) = \int_0^1 B^{\beta_1\dots\beta_m}(\xi t + (1-t)\eta)(\xi^{\beta_1\dots\beta_m} - \eta^{\beta_1\dots\beta_m})dt,$$

and also the Cauchy-Schwartz inequality and the definition of the operators A, B. We then obtain

$$\begin{aligned} I(v_1, v_2) &\equiv 2\langle\!\langle v_1 - v_2, A(v_1) - A(v_2)\rangle\!\rangle + |B(v_1) - B(v_2)|^2_Q \\ &= 2\big(D^{\alpha_1}\dots D^{\alpha_m}(v_1 - v_2), A_{\alpha_1\dots\alpha_m}(D^{\beta_1}\dots D^{\beta_m}v_1) - A_{\alpha_1\dots\alpha_m}(D^{\beta_1}\dots D^{\beta_m}v_2)\big)_0 \\ &\quad + \left\| \left|B(D^{\beta_1}\dots D^{\beta_m}v_1) - B(D^{\beta_1}\dots D^{\beta_m}v_2)\right|_Q \right\|^2_2 \le N\,\|v_1 - v_2\|^2_H \\ &\quad -\varepsilon \sum_{|\alpha_1|+\dots+|\alpha_m|=m} \int_G \left(\int_0^1 \left|D^{\alpha_1}\dots D^{\alpha_m}(v_2 + t(v_1 - v_2))\right|^{p-2} dt\right) \\ &\quad\quad \times |D^{\alpha_1}\dots D^{\alpha_m}(v_1 - v_2)|^2\,dx. \end{aligned} \tag{4.10}$$

We have thus proved that the condition of monotonicity (A_2) of Section 3.2 is satisfied. In fact, equality (4.10) enables us to verify the coercivity condition (A_3):

From (4.10) and (4.7) we have, with $v_1 = v$, $v_2 = 0$,

$$2\langle\langle v, A(v)\rangle\rangle + |B(v)|_Q^2 \le I(v,0) + 2\langle\langle v, A(0)\rangle\rangle - |B(0)|_Q^2 + 2\,|B(v)|_Q\,|B(0)|_Q$$
$$\le N\left(\frac{1}{\delta} + \delta\right)\|f(t)\|_1 + N\delta\,\|v\|_V^p + 2\delta\,\|v\|_V^p + \frac{2}{\delta}\,\|A(0)\|_{V^*}^q + N\,\|v\|_H^2$$
$$- \frac{\varepsilon}{p-1}\sum_{|\alpha_1|+\ldots+|\alpha_m|=m}\|D^{\alpha_1}\ldots D^{\alpha_m}v\|_p^p,$$

where $\delta > 0$ and N does not depend on δ.

Using Theorems 4.6 and 4.7, noting that the growth condition (A_4) has already been verified, and choosing δ sufficiently small, we conclude that the coercivity condition is satisfied. The proof of the theorem is complete. □

Example 4.17. Let $d = 1$, $E = \mathbb{R}$, and suppose that (4.1) has the form

$$\begin{aligned} du(t,x) = a(t,\omega)(-1)^{m+1}\frac{\partial^m}{\partial x^m}\left(\left|\frac{\partial^m}{\partial x^m}u(t,x)\right|^{p-2}\frac{\partial^m}{\partial x^m}u(t,x)\right)dt \\ + b(t,\omega)\left|\frac{\partial^m}{\partial x^m}u(t,x)\right|^{p/2}dw(t), \end{aligned} \tag{4.11}$$

where a, b are appropriately measurable and also bounded processes. Here the algebraic condition of strong parabolicity becomes

$$-2(p-1)a + \frac{p^2}{4}b^2 \le -\varepsilon,$$

where $\varepsilon > 0$ is a constant. If this condition is satisfied, then by Theorem 4.15 we obtain assertions regarding the existence, uniqueness, stability with respect to the initial data, and the Markov property of solutions of (4.11) with the "boundary" condition $u \in \overset{\circ}{W}{}_p^m(G)$.

Equation (4.11) coincides with (3.2). Equations of the form (3.3) are studied in the next section.

4.3. *Cauchy Problem for Linear Second-Order Equations*

In this section we continue the study of equation (4.1), assuming that $m = 1$, $G = \mathbb{R}^d$, and A and B are linear functions of ξ which are generally not equal to zero for $\xi = 0$. Moreover, all assumptions of Section 4.2 are naturally assumed to be satisfied.

Problem (4.1)–(4.3) becomes

$$\begin{aligned} du(t,x) &= D^\alpha\big(a_{\alpha\beta}(t,x)D^\beta u(t,x) + f_\alpha(t,x)\big)dt \\ &\quad + \big(b_\alpha(t,x)D^\alpha u(t,x) + g(t,x)\big)dw(t) + dz(t,x), \end{aligned} \tag{4.12}$$

$$u(t,\cdot) \in L_2(\mathbb{R}^d), u(0,x) = u_0(x), x \in \mathbb{R}^d, \tag{4.13}$$

where $a_{\alpha\beta}$, f_α are real functions and b_α and g are functions with values in E. Conditions (4.4) and (4.5) are equivalent to the boundedness of $a_{\alpha\beta}$ and $|b_\alpha|_E$, together with the inequality

$$\sum_\alpha \mathbb{E}\int_0^T \|f_\alpha\|_2^2\,dt + \mathbb{E}\int_0^T \|\,|g|_E\,\|_2^2\,dt < \infty.$$

A solution of problem (4.12), (4.13) is understood in the sense of the integral identity (4.8); for a V-solution the identity is satisfied for almost all (t,ω), and for an H-solution, for each t with probability one.

Lemma 4.18. *Suppose that, for all $x, \eta \in \mathbb{R}^d$, $t \in [0,T]$, $\omega \in \Omega$,*

$$2\sum_{i,j=1}^d a_{ij}(t,x)\eta^i\eta^j - \left|\sum_{i=1}^d b_i(t,x)\eta^i\right|_Q^2 \geq \varepsilon\,|\eta|^2. \tag{4.14}$$

where $\varepsilon > 0$ is a constant, $a_{ij} = a_{\alpha\beta}$, $b_i = b_\alpha$ if α is the i-th and β the j-th coordinate vectors. Then the algebraic condition of strong parabolicity is satisfied.

The lemma is easily proved by means of inequalities of the type

$$2a_{0\alpha}\eta^0\eta^i \leq |a_{0\alpha}\,\varepsilon\,|\eta^i|^2 + \varepsilon^{-1}|a_{0,\alpha}\,|\eta^0|^2.$$

The next result is a direct corollary of Lemma 4.18, Theorems 4.1 and 4.2, and also Corollary 3.7.

Theorem 4.19. *Suppose that condition (4.14) is satisfied. Then there exists a function $u(t,\omega)$ defined on $[0,T]\times\Omega$ with values in $L_2(\mathbb{R}^d)$, strongly continuous in t in $L_2(\mathbb{R}^d)$, $\mathcal{F}_t$-consistent and such that*

(a) For almost all (t,ω),

$$u \in W_2^1(\mathbb{R}^d);$$

(b)

$$\mathbb{E}\sup_{t\leq T}\|u(t)\|_2^2 + \mathbb{E}\int_0^T \|u(t)\|_{1,2}^2\,dt < \infty;$$

(c) for each $\eta \in W_2^1(\mathbb{R}^d)$,

$$\begin{aligned}(u(t),\eta)_0 = (u_0,\eta)_0 &+ \int_0^t (-1)^\alpha\big(D^\beta u(s), a_{\alpha\beta}(s)D^\alpha\eta\big)_0 \\ &+ \int_0^t (-1)^\alpha (f_\alpha(s), D^\alpha\eta)_0 ds \\ &+ \int_0^t (b_\alpha(s)D^\alpha u(s) + g(s),\eta)_0 dw(s) + (z(t),\eta)_0\end{aligned} \tag{4.15}$$

for all $t\in[0,T]$ (a.s.)

Theorems 3.8 and 3.11 enable us to establish uniqueness, stability with respect to the initial data, and the Markov property of the function u considered in Theorem 4.19 [u is an H-solution of problem (4.12), (4.13)]. These properties of u are proved simply by referring to Theorems 3.8 and 3.11, and we are not going to discuss them.

We now turn to a more important question, the question of raising the smoothness of a solution of problem (4.12), (4.13). One reason for addressing this question is that filtering of diffusion processes [86] leads to equations analogous to (4.15), but with the inner product in $L_2(\mathbb{R}^d)$ replaced by the inner product in $W_2^m(\mathbb{R}^d)$, i.e., the index 0 in the inner product in (4.15) is replaced by m so that

$$(\varphi, \psi)_m = (D^{\alpha_1} \dots D^{\alpha_m}\varphi, D^{\alpha_1} \dots D^{\alpha_m}\psi)_0.$$

Thus, we denote the modified equation (4.15) by $(4.15)_{\mathrm{m}}$. For $(4.15)_{\mathrm{m}}$ to be meaningful for sufficiently smooth u_0 and coefficients $a_{\alpha\beta}$, f_α, b_α, g, z, it suffices to restrict attention to functions u belonging to $W_2^{m+1}(\mathbb{R}^d)$ [for a.a. (t,ω)]. However, the filtering density is equal to the function $(1-\Delta)^m u$ multiplied by some function of time. Thus, assertions are needed regarding the membership of the solution of equation $(4.15)_{\mathrm{m}}$ not in $W_2^{m+1}(\mathbb{R}^d)$ but in $W_2^{2m}(\mathbb{R}^d)$ [for a.a. (t,ω)]. However, merely an assertion on the existence of a solution of $(4.15)_{\mathrm{m}}$ with values in $W_2^{2m}(\mathbb{R}^d)$ is of little use, since from the filtering theory it is known a priori only that the solutions belongs to $W_2^{m+1}(\mathbb{R}^d)$; if we wish to prove its smoothness, we must have not only a theorem on the existence of a solution of $(4.15)_{\mathrm{m}}$ with values in $W_2^{2m}(\mathbb{R}^d)$ but also a theorem on the uniqueness of a solution with values in $W_2^{m+1}(\mathbb{R}^d)$. Thus, a theorem on raising the smoothness for equation $(4.15)_{\mathrm{m}}$ is required. A proof of the corresponding result is given in Ref. [64]. Not to obscure the exposition with technical details, we here prove the theorem on raising smoothness only for equation (4.15).

Throughout the remainder of the paper we fix an integer $m > 0$ and assume that $z(t)$ is a square-integrable martingale with values in $W_2^m(\mathbb{R}^d)$ which is continuous in t in $W_2^m(\mathbb{R}^d)$ for all t, ω, the functions $a_{\alpha\beta}$ (resp. b_α) have m derivatives (resp. weak derivatives) in x, are continuous (weakly continuous) in x, and these derivatives of $a_{\alpha\beta}$, b_α are bounded (for b_α in the norm of E) uniformly with respect to t, x, ω. Suppose that, for all (t,ω), the functions $f_\alpha \in W_2^m(\mathbb{R}^d)$, $u_0 \in W_2^m(\mathbb{R}^d)$, and

$$\mathbb{E}\,\|u_0\|_{m,2}^2 + \mathbb{E}\int_0^T \|f_\alpha(t)\|_{m,2}^2\,dt < \infty.$$

Let $g \equiv 0$. This condition involves no loss of generality, since the integral of g with respect to $dw(t)$ can be included in $z(t)$. Finally, we assume that condition (4.14) is satisfied.

Theorem 4.20. *There exists a set $\Omega' \subset \Omega$ such that $P(\Omega') = 1$ and for $\omega \in \Omega'$ the function $u(t)$ of Theorem 4.19 belongs to $W_2^m(\mathbb{R}^d)$ and is continuous in t in the* *norm of* $W_2^m(\mathbb{R}^d)$. *Moreover,* $u \in W_2^{m+1}(\mathbb{R}^d)$ *a.e. in* (t,ω), *and*

$$\mathbb{E}\sup_{t\le T}\|u\|_{m,2}^2 + \mathbb{E}\int_0^T \|u(s)\|_{m+1}^2\,ds < \infty.$$

Proof. We set $H = W_2^m(\mathbb{R}^d)$ and identify H with its dual by means of $(\cdot,\cdot)_m$. It is then easily seen from the Parseval equality that W_2^{m-1} is identified with V^*, the space dual to $V = W_2^{m+1}(\mathbb{R}^d)$. Thus, $V \subset H \subset V^*$, and it is obvious that each embedding is dense and continuous. As in the preceding section, it is easy to verify that the formulas

$$\langle\!\langle \eta, A(t)u\rangle\!\rangle = \left(a_{\alpha\beta}(t)D^\beta u + f_\alpha(t), (-1)^\alpha D^\alpha \eta\right)_m,$$

$$\left((B(t)u)\,e, \eta\right)_m = \left((b_\alpha(t), e)_E D^\alpha u, \eta\right)_m$$

define bounded linear operators $A(t) : V \to V^*$, $B(t)u : E \to H$. The reader can also verify without difficulty that the functions $A(t)u$, $B(t)u$ satisfy conditions (A_1)–(A_4) of Section 3.2 and also the measurability conditions of this section (see, e.g. Ref. [64]).

Hence, by Corollary 3.7 in our case equation (4.1) has an H-solution $\tilde{u}(t)$. Application of Remark 4.12 shows that, for each $\eta \in W_2^m(\mathbb{R}^d)$,

$$\begin{aligned}(\tilde{u}(t),\eta)_m = (u_0,\eta)_m &+ \int_0^t \left(a_{\alpha\beta}(s)D^\beta\tilde{u}(s) + f(s), (-1)^{|\alpha|}D^\alpha\eta\right)_m \\ &+ \int_0^t \left(b_\alpha(s)D^\alpha\tilde{u}(s),\eta\right)_m dw(s) + (z(t),\eta)_m\end{aligned} \tag{4.16}$$

for all $t \in [0,T]$ (a.s.)

To avoid misunderstanding we note that (4.16) does not coincide with $(4.15)_{\mathrm{m}}$ if $m > 0$ and $a_{\alpha\beta}$, b_α depend on x.

If in (4.16) in place of η we substitute $(1-\Delta)^{-m}\eta$ and use the fact that by the Parseval equality $(f,g)_m = (f,(1-\Delta)^m g)_0$, where $g \in W_2^{2m}(\mathbb{R}^d)$, then we see that in (4.16) in place of m it is possible to write 0. After this, by Theorem 3.8 we obtain

$$\sup_{t\le T}\|u(t) - \tilde{u}(t)\|_2 = 0 \text{ (a.s.)},$$

and $u(t)$ possesses the same properties as $\tilde{u}(t)$.

The proof of the theorem is complete. □

In conclusion, we discuss the significance of condition (4.14) for the validity of Theorem 4.19. For $d = 1$ and $E = \mathbb{R}$ we consider the equation

$$du(t,x) = \frac{1}{2}\frac{\partial^2 u(t,x)}{\partial x^2}dt + \sigma\frac{\partial u(t,x)}{\partial x}dw(t),$$

where σ is a constant and the initial condition $u_0(x)$ is nonrandom. If Theorem 4.19 is valid for this equation, then from (4.15) and the Parseval equality it follows that

$$\int_{\mathbb{R}} \widehat{u}(t,y)\widehat{\eta}(y)dy = \int_{\mathbb{R}} \widehat{u}_0(y)\widehat{\eta}(y)dy - \frac{1}{2}\int_0^t \left(\int_{\mathbb{R}} y^2\widehat{u}(s,y)\widehat{\eta}(y)dy\right) ds$$
$$- \sqrt{-1}\,\sigma \int_0^t \left(\int_{\mathbb{R}} y\widehat{u}(s,y)\widehat{\eta}(y)dy\right) dw(s),$$

where $\widehat{u}$, $\widehat{\eta}$ are the Fourier transforms of u, η. Here it is easy to interchange the integrals if $\widehat{\eta}(y)$ is a compactly supported function, and we then find that for almost all (t, y, ω)

$$\widehat{u}(t,y) = u_0(y) - \frac{1}{2}y^2 \int_0^t \widehat{u}(s,y)ds - \sqrt{-1}\,\sigma y \int_0^t \widehat{u}(s,y)dw(s). \tag{4.17}$$

We fix a y for which equation (4.17) holds for almost all (t, ω), and we denote the right side of (4.17) by $\varphi(t, y)$. Then $\varphi(t, y)$ satisfies (4.17) for all t (a.s.). The solution of the equation for φ is known:

$$\varphi(t,y) = e^{-\frac{1}{2}(1-\sigma^2)y^2 t - \sqrt{-1}\,\sigma y w(t)}\, \widehat{u}_0(y).$$

Since $\widehat{u}(t, y) = \varphi(t, y)$ a.e. in (t, y, ω), it follows that

$$\mathbb{E}\int_0^T \|u(t)\|_{1,2}^2\, dt = \int_{\mathbb{R}} (1+y^{-2})\frac{e^{-(1-\sigma^2)y^2 T} - 1}{1-\sigma^2}\, \widehat{u}_0^2(y)\, dy. \tag{4.18}$$

This implies that the left side of (4.18) is finite for all $u_0 \in L_2(\mathbb{R})$ if and only if $|\sigma| < 1$. The last condition in the present case is equivalent to (4.14). This example demonstrates the necessity of condition (4.14) for the validity of Theorem 4.19, and also the necessity of the coercivity condition (A_3) of Section 3.2 for the validity of the results in that section.

References

[1] R. S. Liptser and A. N. Shiryaev, *Statistics of Random Processes.* (Nauka, Moscow, 1974). In Russian.

[2] N. V. Krylov and B. L. Rozovskii, On conditional distribuitions of diffusion processes, *Izv. Akad. Nauk SSSR, Ser. Mat.* **42**(2), 356–378, (1978).

[3] N. T. J. Bailey, Stochastic birth, death, and migration processes for spatially distributed populations, *Biometrika.* **55**(1), 189–198, (1968).

[4] J. F. Crow and M. Kimura, *An Introduction to Population Genetics Theory.* (Harper and Row, New York, 1970).

[5] G. Malécot. Identical loci and relationship. In *Proc. 5th Berkeley Symp. Math. Stat. Prob., IV*, pp. 317–332. Calif. Univ. Press, (1967).

[6] D. A. Dawson, Stochastic evolution equations, *Math. Biosci.* **15**(3–4), 287–316, (1972).

[7] W. H. Fleming, *Distributed parameter stochastic systems in population biology*, vol. 107, *Lect. Notes Econ. Math. Syst.*, pp. 179–191. Springer, (1975).

[8] W. Feller. Diffusion processes in genetics. In *Proc. Second Berkeley Symp. 1. Math. Stat. Prob.*, pp. 227–246. Calif. Univ. Press, Berkeley.

[9] I. M. Gel'fand and N. Y. Vilenkin, *Generalized Functions: Applications of Harmonic Analysis.* (Academic Press, 1964).
[10] E. A. Novikov, Functionals and the method of random forces in the theory of turbulence, *Zh. Eksp. Teor. Fiz.* **47**(5), 1919–1926, (1966).
[11] A. S. Monin and A. M. Yaglom, *Statistical Fluid Mechanics.* (MIT Press, 1975).
[12] V. I. Klyatskin, *Stochastic Description of Dynamical Systems with Fluctuating Parameters.* (Nauka, Moscow, 1975). In Russian.
[13] B. Simon, *The* $P(\Phi)_2$ *Model of Euclidean Qauntum Field Theory.* (Mir, Moscow, 1976). Russian translation.
[14] T. Hilda and L. Strett, On quantum theory in terms of white noise, *Nagoya Math. J.* **68**(Dec.), 21–34, (1977).
[15] D. A. Dawson, Stochastic evolution equations and related measure processes, *J. Multivar. An.* **5**(1), 1–52, (1975).
[16] S. Albeverio and R. Hoegh-Krohn, Dirichiet forms and diffusion processes on rigged Hilbert spaces, *Z. Wahr. Verw. Geb.* **40**(1), 1–57, (1977).
[17] Y. L. Daletskii, Infinite-dimensional elliptic operators and parabolic equations related to them, *Usp. Mat. Nauk.* **22**(4), 3–54, (1967).
[18] V. V. Baklan, The existence of solutions of stochastic equations in Hilbert space, *Dopavidi Akad. Nauk Ukr. RSR.* **10**, 1299–1303, (1963).
[19] V. V. Baklan, Equations in variational derivatives and Markov processes, *Dokl. Akad. Nauk SSSR.* **159**(4), 707–710, (1964).
[20] Y. L. Daletskii, Multiplicative operators of diffusion processes and differential equations in sections of vector bundies, *Usp. Mat. Nauk.* **30**(2), 209–210, (1975).
[21] Y. I. Belopol'skaya and Y. L. Daletskii, Diffusion processes in smooth Banach spaces and manifolds, *Tr. Mosk. Mat. Obshch.* **37**, 78–79, (1978).
[22] Y. I. Belopol'skaya and Z. I. Nagolkina, On multiplicative representations of solutions of stochastic equation, *Dopovidi Akad, Nauk Ukr. RSR.* **11**, 977–969, (1977).
[23] V. V. Baklan, The Cauchy problem for equations of parabolic type in infinite-dimensional space, *Mat. Fiz. Resp. Mezhved. Sb.* **7**, 18–25, (1970).
[24] V. V. Baklan. On a class of stochastic partial differential equations. In *The Behavior of Systems in Random Media*, pp. 3–7. Kiev, (1976). In Russian.
[25] A. V. Balakrishnan, *Introduction to Optimization Theory in a Hilbert Space.* (Springer, 1971).
[26] A. V. Balakrishnan, Stochastic optimization theory in Hilbert spaces, I, *Appl. Math. Opt.* **1**(2), 97–120, (1974).
[27] A. V. Balakrishnan, *Stochastic bilinear partial differential equations*, vol. 111, *Lect. Notes Econ. Math. Syst.*, pp. 1–43. Springer, (1975).
[28] R. F. Curtain, Estimation theory for abstract evolution equations excited by general white noise processes, *SIAM J. Cont. Optim.* **14**(6), 1124–1149, (1976).
[29] R. F. Curtain, Stochastic evolution equations with general white noise disturbance, *J. Math. Anal. Appl.* **60**(3), 570–595, (1977).
[30] R. F. Curtain and P. L. Falb, Stochastic differential equations in Hilbert space, *J. Diff. Eqs.* **10**(3), 412–430, (1971).
[31] M. Metivier, Intégrate stochastique par rapport a des processus a veleurs dans un espace de Banach reflexif, *Teor. Veroyatn. Ee Primen.* **19**(4), 787–816, (1974).
[32] M. Mativier and G. Pistone, Une formule d'isometrie pour l'intégrale stochastique Hilbertienne et equations d'évolution linêaires stochastiques, *Z. Wahr. Verw. Geb.* **33**, 1–18, (1975).
[33] A. Friedman, *Partial Differential Equations.* (Krieger, 1976).
[34] K. Ito. On stochastic differential equations. In *Matematika. Periodical Collection of*

Translations of Foreign Articles, vol. 1(1), pp. 78–116, (1957).

[35] B. L. Rozovskii, On stochastic partial differential equations, *Mat. Sb.* **96**(2), 314–341, (1975).

[36] L. G. Margulis and B. L. Rozovskii, Fundamental solutions of stochastic partial differential equations and filtration of diffusion processes, *Usp. Mat. Nauk.* **33**(2), 197, (1978).

[37] A. Shimizu, Construction of a solution of a certain evolution equation, *Nagoya Math. J.* **66**, 23–36, (1977).

[38] R. Markus, Parabolic Ito equations, *Trans. Am. Math. Soc.* **198**, 177–190, (1974).

[39] A. Shimizu. Construction of a solution of a certain evolution equation, II. Preprint.

[40] M. Metivier and J. Pellanmail. A basic course on general stochastic integration. In *Publ. Sém. Math. Inf. Rennes. Inst. Rech. Inf. Syst. Aleatories*, vol. 83, pp. 1–55, (1977).

[41] A. Bensoussan, *Filtrage Optimale des Systemes Linéaires.* (Dunod, Paris, 1971).

[42] J.-L. Lions, *Some Methods of Solving Nonlinear Boundary Value Problems.* (Mir, Moscow, 1972). Russian translation.

[43] A. Bensoussan and R. Temam, Equations aux derives partielles stochastiques non linéaires (1), *Isr. J. Math.* **11**(1), 95–129, (1972).

[44] M. M. Vainberg, *Variational Method and Method of Monotone Operators in the Theory of Nonlinear Equations.* (Halsted Press, 1974).

[45] G. Minty, Monotone (nonlinear) operators in Hilbert spaces, *Duke Math. J.* **29**(3), 341–346, (1962).

[46] F. E. Browder, Nonlinear elliptic boundary-value problems, *Bull. Am. Math. Soc.* **69** (6), 862–974, (1963).

[47] F. E. Browder, Nonlinear equations of evolution and nonlinear accretive operators in Banach spaces, *Bull. Am. Math. Soc.* **73**(6), 867–874, (1967).

[48] H. Brezis, *Operateurs Maximaux Monotones et Semi-Groupes de Contractions dans les Espaces de Hilbert.* North-Holland Mat. Stud., (North-Holland, 1973).

[49] F. E. Browder. Nonlinear operators and nonlinear equations of evolution in Banach spaces. In *Proceedings of Symposia in Pure Mathematics, XVIII, Part 2.* Am. Math. Soc., Providence, RI, (1976).

[50] K. Gaevskii, K. Greger, and K. Zakharias, *Nonlinear Operator Equations and Operator Differential Equations.* (Mir, Moscow, 1978). Russian translation.

[51] K. Yosida, *Functional Analysis.* (Springer, 1974).

[52] E. Pardoux, Sur des equations aux derives partielles stochastiques monotones, *C. R. Acad. Sci.* **275**(2), A101–A103, (1972).

[53] E. Pardoux. Equations aux derives partielles stochastiques non lineaires monotones. Etude de solutions fortes de type Ito, (1975). Thése Doct. Sci. Math. Univ. Paris Sud.

[54] V. A. Lebedev, On the uniqueness of a week solution of a system of stochastic differential equations, *Teor. Veroyatn. Ee Primen.* **23**(1), 153–161, (1978).

[55] A. Bensoussan and R. Temam, Equations stochastiques du type Navier-Stokes, *J. Funct. Anat.* **13**(2), 195–222, (1973).

[56] A. M. Vershik and O. A. Ladyzhenskaya, On the evolution of the measure defined by the Navier-Stokes equations and on the solvability of the Cauchy problem for the statistical Hopf equation, *Dokl. Akad. Nauk SSSR.* **226**(1), 26–29, (1976).

[57] A. M. Vershik and O. A. Ladyzhenskaya, On the evolution of the measure defined by the Navier-Stokes equations and on the solvability of the Cauchy problem for the statistical equation of E. Hopf, *Zap. Nauchn. Sem. Leningr. Otd. Mat. Inst. Akad. Nauk SSSR.* **59**, 3–24, (1976).

[58] A. A. Arsen'ev, Construction of a turbulent measure for the system of Navier – Stokes

equation, *Dokl. Akad. Nauk SSSR.* **225**(1), 18–20, (1975).

[59] M. I. Vishik and A. I. Komech, Infinite-dimensional parabolic equations connected with stochastic partial differential equations, *Dokl. Akad. Nauk SSSR.* **233**(5), 769–772, (1977).

[60] M. I. Vishik and A. I. Komech. On the solvability of the Cauchy problem for the direct Kolmogrov equation corresponding to a stochastic equation of Navier-Stokes type. In *Complex Analysis and Its Applications*, pp. 126–136. Nauka, Moscow, (1978). In Russian.

[61] M. Viot. Solutions faibles d'équations aux derives partielles stochastiques non linéaires, (1976). Thése Doct. Sci. Univ. Pierre Marie Curie.

[62] M. I. Vishik, Quasilinear strongly elliptic systems of differential equations having divergence form, *Tr. Mosk.Mat. Obshch.* **12**, 125–184, (1963).

[63] N. V. Krylov and B. L. Rozovskii. On Cauchy problem for superparabolic stochastic differential equation. In *Proc. Third Soviet-Japanese Symposium on Probability Theory, Tashkent*, pp. 77–79. (1975).

[64] N. V. Krylov and B. L. Rozovskii, On the Cauchy problem for linear stochastic partial differential equations, *Izv. Akad, Nauk SSSR, Ser. Mat.* **41**(6), 1329–1347, (1977).

[65] I. I. Gihman and A. V. Skorokhod, *Stochastic Differential Equations.* (Springer, 1972).

[66] H. Kunita, Stochastic integrals based on martingales taking values in Hilbert space, *Nagoya Math. J.* **38**, 41–52, (1970).

[67] P. A. Meyer. Notes sur les integrals stochastiques, I. Intégrales Hilbertiennes. vol. 581, *Lect. Notes Math.*, pp. 446–463. Springer, (1977).

[68] L. Gross. Abstract wiener space. In *Proc. 5th Berkeley Sympos. Math. Stat. Prob., 1965–1966, Vol. 2, Part 1*, pp. 31–42. (1967).

[69] L. Gross, Potential theory on Hilbert space, *J. Funct. Anal.* **1**(2), 123–181, (1968).

[70] H.-H. Kuo, *Gaussian measures in Banach spaces.* vol. 463, *Lect. Notes Math.*, (Springer, 1975).

[71] B. Gaveau, Intégrate stochastique radonifiante, *C. R. Acad. Sci.* **276**(8), A617–A620, (1973).

[72] D. Lepingle and J. Y. Ouvrard, Martingales browniennes Hilbertinnes, *C. R. Acad. Sci.* **276**(18), A1225–A1228, (1973).

[73] M. Metivier. Integration with respect to process of linear functionals. Preprint.

[74] M. Metivier, *Reelle und Vektorwertige Quasimartingale und die Theorie der Stochastischen Integration.* vol. 607, *Lect. Notes Math.*, (Springer, 1977).

[75] M. Metivier and G. Pistone, Su rune equation d'évolution stochastique, *Bull. Soc. Math. France.* **104**, 65–85, (1976).

[76] P. A. Meyer. Un cours sur les integrals stochastiques. In *Sem. Prob. X*, vol. 511, *Lect. Notes Math.*, pp. 249–400. Springer, (1976).

[77] K. Kuratowski, *Topology, Vol. 1.* (Academic Press, 1966).

[78] N. Kazamaki. Note on a stochastic integral equation. vol. 258, *Lect. Notes Math.*, pp. 105–108. Springer, (1972).

[79] C. Doléans-Dade, On the existence and unicity of solutions of stochastic integral equations, *Z. Wahr. Verw. Geb.* **36**(2), 93–101, (1976).

[80] P. E. Protter, On the existence, uniqueness, convergence and explosions of solutions of systems of stochastic integral equations, *Ann. Probab.* **5**(2), 243–261, (1977).

[81] L. I. Gal'chuk, On the existence and uniqueness of a solution for stochastic equations over a semimartingale, *Teor. Veroyatn. Ee Primen.* **23**(4), 782–795, (1978).

[82] E. Pardoux. Filtrage de diffusions avec conditiones frontiers: caracterisation de la densitéconditionelle. In *Statistique Processus Stochastiques, Proceedings, Grenoble*, vol. 636, *Lect. Notes Math.*, pp. 163–188. Springer, (1977).

[83] S. L. Sobolev, *Applications of Functional Analysis in Mathematical Physics.* (Amer. Math. Soc., 1969).
[84] S. M. Nikol'skii, *Approximation of Functions of Several Variables and Imbedding Theorems.* (Nauka, Moscow, 1969). In Russian.
[85] J.-L. Lions and E. Magenes, *Problèmes aux Limites non Homogenès et Applications, Vol. 2.* (Dunod, Paris, 1968).
[86] S. G. Krein, *Linear Differential Equations in Banach Space* (Amer. Math. Soc., 1972).

Chapter 2

Predictability of the Burgers Dynamics Under Model Uncertainty

Dirk Blömker and Jinqiao Duan*

Institut für Mathematik
RWTH Aachen, 52062 Aachen, Germany

Complex systems may be subject to various uncertainties. A great effort has been concentrated on predicting the dynamics under uncertainty in initial conditions. In the present work, we consider the well-known Burgers equation with random boundary forcing or with random body forcing. Our goal is to attempt to understand the stochastic Burgers dynamics by predicting or estimating the solution processes in various diagnostic metrics, such as mean length scale, correlation function and mean energy. First, for the linearized model, we observe that the important statistical quantities like mean energy or correlation functions are the same for the two types of random forcing, even though the solutions behave very differently. Second, for the full nonlinear model, we estimate the mean energy for various types of random body forcing, highlighting the different impact on the overall dynamics of space-time white noises, trace class white-in-time and colored-in-space noises, point noises, additive noises or multiplicative noises.

Contents

1. Introduction

The Burgers equation has been used as a simplified prototype model for hydrodynamics and infinite dimensional systems. It is often regarded as a one-dimensional Navier-Stokes equation. Our motivation for considering this equation comes from the modeling of the hydrodynamics and thermodynamics of the coupled atmosphere-

*Department of Applied Mathematics, Illinois Institute of Technology, Chicago, IL 60616, USA

ocean system. At the air-sea interface, the atmosphere and ocean interact through heat flux and freshwater flux with a fair amount of uncertainty [1–3]. These translate into random Neumann boundary conditions for temperature or salinity. The Dirichlet boundary condition is also appropriate under other physical situations. The fluctuating wind stress forcing corresponds to a random body forcing for the fluid velocity field. The coupled atmosphere-ocean system is quite complicated and numerical simulation is the usual approach at this time. In this paper, we consider a simplified model for this system, i.e., we consider the Burgers equation with random Neumann boundary conditions and random body forcing. Although the stochastic Burgers equation is widely studied, most work we know are for Dirichlet boundary conditions or periodic boundary conditions [4–7]. The reference [8] studied the control of deterministic Burgers equation with Neumann boundary conditions.

We consider the stochastic Burgers equation with boundary forcing on the interval $[0, L]$

$$\partial_t u + u \cdot \partial_x u = \nu \partial_x^2 u \tag{1.1}$$

$$\partial_x u(\cdot, 0) = \alpha\eta \qquad \partial_x u(\cdot, L) = 0. \tag{1.2}$$

Here $\alpha > 0$ denotes the noise strength and η is white noise, i.e., η is a generalized Gaussian process with $\mathbb{E}\eta(t) = 0$ and $\mathbb{E}\eta(t)\eta(s) = \delta(t-s)$. The restriction to noise on the left boundary is only for simplicity. Analogous results will be true, if forces act on both sides of the domain.

We will see that boundary forcing coincides with point forcing at the boundary. Thus we also look at point forcing. As a simple example for point forcing, we consider

$$\partial_t v + v \cdot \partial_x v = \nu \partial_x^2 v + \alpha\delta_0\eta \tag{1.3}$$

$$u(\cdot, -L) = u(\cdot, L) = 0\ , \tag{1.4}$$

where δ_0 is the Delta-distribution.

We will compare solutions of (1.1) and (1.3) with solutions of the stochastic Burgers equation with body forcing.

$$\partial_t v + v \cdot \partial_x v = \nu \partial_x^2 v + \sigma\xi \tag{1.5}$$

either subject to Dirichlet or Neumann boundary conditions. Here the noise strength is denoted by $\sigma > 0$ and ξ is space-time white noise. I.e., ξ is a generalized Gaussian process with $\mathbb{E}\xi(t,x) = 0$ and $\mathbb{E}\xi(t,x)\xi(s,y) = \delta(t-s)\delta(x-y)$. We will also consider trace class body noise, i.e., noise that is white in time but colored in space.

For the linearized equations, we will compare statistical quantities of both solutions, which are frequently used. One of them is the *mean energy*

$$\frac{1}{L}\int_0^L \mathbb{E}[u(t,x) - \overline{u}(t)]^2 dx, \tag{1.6}$$

where

$$\overline{u}(t) = \frac{1}{L}\int_0^L u(t,x)dx.$$

Another important quantity, which gives information about the characteristic size of pattern, is the *mean correlation function*

$$C(t,r) := \frac{1}{L}\int_0^L \mathbb{E}[u(t,x) - \overline{u}(t)] \cdot [u(t,x+r) - \overline{u}(t)]dx, \tag{1.7}$$

which is usually averaged over all points r with a given distance from 0. We obtain the *averaged mean correlation function*

$$\widehat{C}(t,r) = \frac{1}{2}[C(t,r) + C(t,-r)] \tag{1.8}$$

where we employ the canonical odd and 2L-periodic extension of u in order to define $C(t,r)$ for any $r \in \mathbb{R}$.

For the linearized equation the *main result* states that mean energy and averaged mean correlation function are the same for solutions of (1.1) and (1.5). Nevertheless the solutions behave completely different. Furthermore, we give some qualitative properties like, for instance, the typical pattern size. This should carry over to a transient regime (i.e., small times) for the corresponding nonlinear equations.

For the full nonlinear Burgers model, we estimate the mean energy for various types of random body forcing, highlighting the different impact on the overall dynamics of space-time white noises, trace class white-in-time and colored-in-space noises, point noises, additive noises or multiplicative noises.

In the following, we discuss linear dynamics in §2 and nonlinear dynamics in §3.

2. Linear Theory

Define

$$A = \nu\partial_x^2$$

with

$$D(A) = \{w \in H^2([0,L]) \ : \ \partial_x w(0) = 0, \ \partial_x w(L) = 0\}$$

It is well-known (cf. e.g. [9]) that A has an orthonormal base of eigenfunctions $\{e_k\}_{k\in\mathbb{N}_0}$ in $L^2([0,L])$ with corresponding eigenvalues $\{\lambda_k\}_{k\in\mathbb{N}_0}$. In our situation $e_0(x) = 1/\sqrt{L}$, $e_k(x) = \sqrt{2/L}\cdot\cos(\pi kx/L)$ for $k \in \mathbb{N}$, and $\lambda_k = -\nu(k\pi/L)^2$. Moreover A generates an analytic semigroup $\{e^{tA}\}_{t\geq 0}$. (cf. e.g. [10]).

In fact, $e^{tA}v_0$ is the solution of the following evolution problem

$$\partial_t v = Av, \qquad \partial_x v(\cdot,0) = \partial_x v(\cdot,L) = 0, \qquad v(0,x) = v_0(x). \tag{2.1}$$

The solution is

$$e^{tA}v_0(x) := v(t,x) = \sum_{k=0}^{\infty} < v_0, e_k > e^{\lambda_k t}e_k, \quad t > 0,\ 0 < x < L, \tag{2.2}$$

where $< \cdot, \cdot >$ is the usual scalar product in $L^2(0, L)$.

We now consider the following linearized problems. First

$$\partial_t u = Au, \qquad \partial_x u(\cdot, 0) = \alpha \partial_t \beta, \quad \partial_x u(\cdot, L) = 0. \tag{2.3}$$

Here the white noise η is given by the generalized derivative of a standard Brownian motion (cf. e.g. [11]), and α is the noise intensity.

Secondly,

$$\partial_t v = Av + \sigma \partial_t W, \qquad \partial_x v(\cdot, 0) = 0, \quad \partial_x v(\cdot, L) = 0, \tag{2.4}$$

where the space-time white noise is given by the generalized derivative of an Id-Wiener process. Namely, $W(t) = \sum_{k \in \mathbb{N}_0} \beta_k(t) e_k$, where $\{\beta_k\}_{k \in \mathbb{N}}$ is a family of independent standard Brownian motions, and σ is the noise intensity.

It is known (cf. e.g. [12]) that (2.4) has a unique weak solution given by the stochastic convolution (taking initial condition to be zero)

$$W_A(t) = \sigma \cdot \int_0^t e^{(t-\tau)A} dW(\tau) = \sigma \cdot \sum_{k \in \mathbb{N}_0} \int_0^t e^{(t-\tau)\lambda_k} d\beta_k(\tau) e_k. \tag{2.5}$$

We define the Neumann map $\mathcal{D}$ by

$$(1 - A)\mathcal{D}\gamma = 0, \quad \partial_x \mathcal{D}\gamma(0) = \gamma, \; \partial_x \mathcal{D}\gamma(L) = 0$$

for any $\gamma \in \mathbb{R}$. It is known that $\mathcal{D} : \mathbb{R} \mapsto H^2([0, L])$ is a continuous linear operator. In fact, we have explicit expression for this linear operator

$$\mathcal{D}(\gamma) = \frac{e^x + e^{2L} e^{-x}}{1 - e^{2L}} \gamma. \tag{2.6}$$

From Ref. [13] or Ref. [14] we immediately obtain, that (2.3) has a unique weak solution (taking initial condition to be zero)

$$Z(t) = (1 - A) \int_0^t e^{(t-\tau)A} \mathcal{D} \alpha d\beta(\tau). \tag{2.7}$$

In the next section we derive explicit formulas for Z in term of Fourier series.

2.1. *Mean Energy*

To obtain the Fourier series expansion for Z, consider for $e \in D(A)$ and $\gamma \in \mathbb{R}$

$$\begin{aligned} < \mathcal{D}(\gamma), (1 - A)e >_{L^2([0,L])} &= < \mathcal{D}(\gamma), e > - \int_0^L \mathcal{D}(\gamma) \cdot e_{xx} dx \\ &= < \mathcal{D}(\gamma), e > - \int_0^L \mathcal{D}(\gamma)_{xx} \cdot e dx + \mathcal{D}(\gamma)_x \cdot e |_{x=0}^{x=L} \\ &= -\gamma e(0), \end{aligned} \tag{2.8}$$

by the definition of $\mathcal{D}$. Hence,

$$\begin{aligned} < Z(t), e_k > &= < \int_0^t e^{(t-\tau)A}\mathcal{D}\alpha d\beta(\tau), (1-A)e_k > \\ &= \int_0^t e^{(t-\tau)\lambda_k} < \mathcal{D}\alpha d\beta(\tau), (1-A)e_k > \\ &= \alpha e_k(0) \cdot \int_0^t e^{(t-\tau)\lambda_k} d\beta(\tau). \end{aligned} \tag{2.9}$$

We now obtain

$$Z(t) = \alpha \cdot \sum_{k\in\mathbb{N}_0} e_k(0) \int_0^t e^{(t-\tau)\lambda_k} d\beta(\tau) e_k. \tag{2.10}$$

Finally,

$$Z(t) = \alpha e_1(0) \cdot \sum_{k\in\mathbb{N}} \int_0^t e^{(t-\tau)\lambda_k} d\beta(\tau) e_k + \alpha e_0^2(0)\beta(t) \tag{2.11}$$

and

$$W_A(t) = \sigma \cdot \sum_{k\in\mathbb{N}} \int_0^t e^{(t-\tau)\lambda_k} d\beta_k(\tau) e_k + \sigma\beta_0(t)\ . \tag{2.12}$$

If we now choose $\sigma = \alpha e_1(0)$, we readily obtain that

$$\mathbb{E}\|Z(t) - \overline{Z}(t)\|^2 = \sigma^2 \cdot \sum_{k\in\mathbb{N}} \int_0^t e^{2\tau\lambda_k} d\tau = \mathbb{E}\|W_A(t) - \overline{W_A}(t)\|^2\ ,$$

where $\|\cdot\|$ is the norm in $L^2([0,T])$. Hence, the mean energy in both cases is given by $\sigma^2 L^{-1} \sum_{k\in\mathbb{N}} \int_0^t e^{2\tau\lambda_k} d\tau$.

For the mean energy we can prove the following theorem, which is similar to the results of [15] and [16].

Theorem 2.1. *Fix $\sigma^2 = \alpha^2/L$, then the mean energy $C_Z(t,0) = C_{W_A}(t,0)$ behaves like $C_1(\alpha^2/L)\sqrt{t/\nu}$ for $t \ll L^2/\nu$, and like $C_2\alpha^2/\nu$ for $t \gg L^2/\nu$.*

The main difference to body forcing is the scaling in the length-scale L. The long-time scaling is independent of L, while the transient scaling is.

2.2. *Correlation Function*

To obtain results for the correlation function, we think of Z and W_A to be periodic on $[-L, L]$, and symmetric w.r.t. 0. I.e., we choose the standard $2L$-periodic extension respecting the Neumann boundary conditions on $[0, L]$. To be more precise, we extend Z and W_A in a Fourier series in the basis e_k, which we then consider to be defined on the whole $\mathbb{R}$.

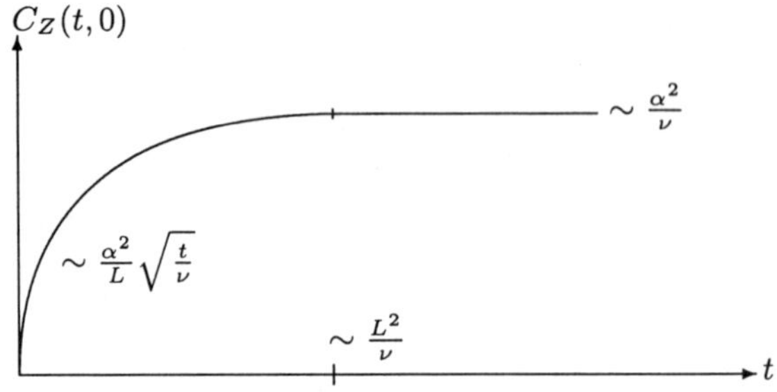

Fig. 2.1. The scaling of the mean energy for boundary forcing.

We consider firstly for $k, l \neq 0$

$$\int_0^L e_k(x)e_l(x+r)dx = \begin{cases} Le_k(0)e_l(0)\frac{l((-1)^{k+l}-1)}{\pi(l^2-k^2)}\sin(\pi lr/L) : k \neq l \\ Le_k(0)^2\cos(\pi kr/L) \qquad : k = l \end{cases}.$$

Now relying on the independence of the Brownian motions, it is straightforward to verify

$$\begin{aligned} C_{W_A}(t,r) &= \frac{1}{L}\mathbb{E} < W_A(t,x) - \overline{W_A}(t), W_A(t,x+r) - \overline{W_A}(t) > \\ &= \frac{\alpha^2 e_1^2(0)}{L} \cdot \sum_{k\in\mathbb{N}} \int_0^t e^{2\tau\lambda_k} d\tau \cos(\pi kr/L) , \end{aligned} \tag{2.13}$$

as $\alpha^2 e_1^2(0) = \sigma^2$. Furthermore,

$$\begin{aligned} C_Z(t,r) &= \frac{1}{L}\mathbb{E} < Z(t,x) - \overline{Z}(t), Z(t,x+r) - \overline{Z}(t) > \qquad (2.14) \\ &= C_{W_A}(t,r) + \frac{e_1^2(0)}{\pi} \sum_{\substack{k,l=1 \\ k\neq l}}^{\infty} \int_0^t e^{\tau(\lambda_k+\lambda_l)} d\tau \frac{l((-1)^{k+l}-1)}{(l^2-k^2)} \sin(\pi lr/L) . \end{aligned}$$

Obviously, C_Z and C_{W_A} do not coincide, but let us now look at the averaged correlation function

$$\widehat{C}(t,r) = \frac{1}{2}[C(t,r) + C(t,-r)] .$$

Then it is obvious that

$$\widehat{C}_{W_A}(t,r) = C_{W_A}(t,r) = \widehat{C}_Z(t,r) \neq C_Z(t,r) . \tag{2.15}$$

Now

Theorem 2.2. *For $\alpha^2 e_1^2(0) = \sigma^2$ the mean energy and the averaged mean correlation functions $\widehat{C}_{W_A}$ and $\widehat{C}_Z$ for Z and W_A coincide for any $t \geq 0$.*

This is somewhat surprising, as realizations of Z and W_A behave completely different, when the condition $\alpha^2 e_1^2(0) = \sigma^2$ is satisfied. See e.g. Figure 2.2 and Figure 2.3.

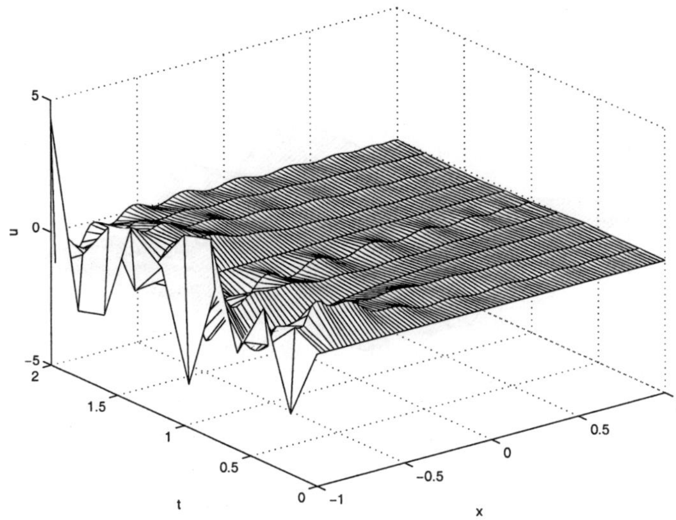

Fig. 2.2. *Random boundary condition*: One realization of the solution of the equation (2.3) for $L = 1$, $\nu = 1$, $\alpha = 1$ and initial condition $u(x, 0) = 0$.

It is even more surprising, as the scaling behavior of quantities like mean energy and mean correlation functions are an important tool in applied science, which for example is used to determine the size of characteristic length scales and the universality class the model belongs to. Here both linear models lie in the same class, although their behavior differs completely.

The scaling behavior with respect to L and t of the mean energy can be described using the results of [15], where the mean surface width for very general models was discussed. Therefore we focus on the scaling properties of the mean correlation function. Here we also want to investigate the dependence on α and ν.

First we consider the scaling properties of the correlation function $\widehat{C}_Z(t, r)$ or C_{W_A}, as given in (2.14). We are especially interested in the smallest zero of the function, which gives information about characteristic length scales or pattern sizes. For this, we use the *normalized correlation function*.

$$\rho_Z(t, r) = \frac{\widehat{C}_Z(t, r)}{\widehat{C}_Z(t, 0)}. \tag{2.16}$$

Note that $C_Z(t, 0)$ is the mean energy and the maximum of $r \mapsto \widehat{C}_Z(t, r)$.

We begin with some technical results. For any continuously differentiable and integrable function $f : \mathbb{R}^+ \to \mathbb{R}$ we obtain using the mean value theorem

$$\left| \int_0^\infty f(x)dx - \sum_{k=1}^\infty f(k) \right| \leq \sum_{k=1}^\infty \sup_{\eta \in (k-1,k)} |f'(\eta)|. \tag{2.17}$$

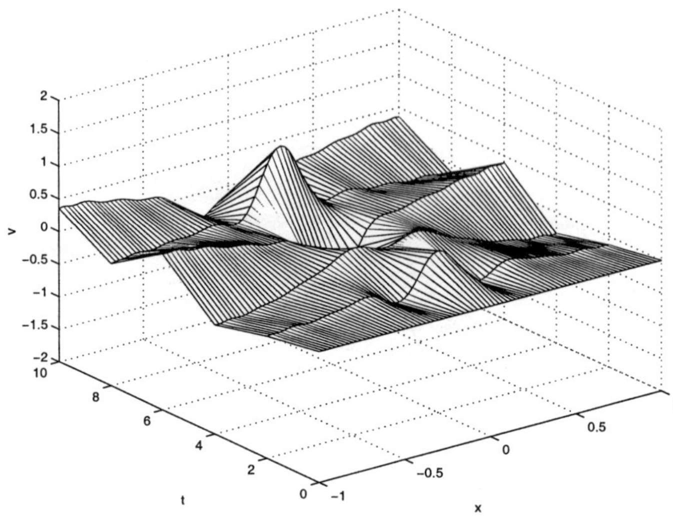

Fig. 2.3. *Random body forcing*: One realization of the solution of the equation (2.4) for $L = 1$, $\nu = 1$, $\sigma = 1$ and initial condition $v(x, 0) = 0$.

For $f(k) := e^{-2\tau\nu k^2\pi^2/L^2}\cos(k\pi r/L)$ it is easy to verify that

$$|f'(k)| \le e^{-\tau\nu k^2\pi^2/L^2}\cdot\frac{\pi}{L}\cdot\left[r + \frac{4\sqrt{\tau\nu}}{\sqrt{2e}}\right],$$

where we used that $x^s e^{-x^2\alpha} \le (2\alpha e)^{-1/2}$ for any $x, \alpha \ge 0$. Hence,

$$\begin{aligned}\sum_{k=1}^{\infty}\sup_{\eta\in(k-1,k)}|f'(\eta)| &\le \sum_{k=1}^{\infty} e^{-\tau\nu(k-1)^2\pi^2/L^2}\cdot\frac{\pi}{L}\cdot[r + \frac{4}{\sqrt{2e}}\sqrt{\tau\nu}]\\ &\le \frac{\pi}{L}[r+\frac{4}{\sqrt{2e}}\sqrt{\tau\nu}](1+\int_0^\infty e^{-\tau\nu k^2\pi^2/L^2}dk)\\ &= \frac{\pi}{L}[r+\frac{4}{\sqrt{2e}}\sqrt{\tau\nu}](1+\frac{L}{\pi\sqrt{\tau\nu}}\cdot\int_0^\infty e^{-k^2}dk) \qquad (2.18)\end{aligned}$$

and

$$\int_0^t\sum_{k=1}^{\infty}\sup_{\eta\in(k-1,k)}|f'(\eta)|d\tau \le C\frac{1}{L}\sqrt{\frac{t}{\nu}}[r+\sqrt{t\nu}][L+\sqrt{t\nu}]. \qquad (2.19)$$

Moreover,

$$\begin{aligned}\frac{1}{L}\int_0^t\int_0^\infty f(k)dkd\tau &= \frac{1}{\pi}\int_0^\infty\frac{1-e^{-2t\nu k^2}}{2\nu k^2}\cos(kr)dk\\ &= \frac{1}{\pi}\sqrt{\frac{t}{\nu}}\cdot G(\frac{r}{\sqrt{\nu t}}), \qquad (2.20)\end{aligned}$$

with $G(x) := \int_0^\infty \frac{1-e^{-2k^2}}{2k^2}\cos(kx)dk$.

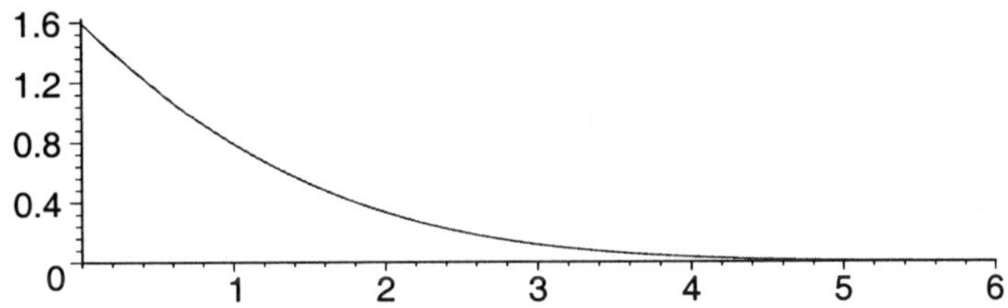

Fig. 2.4. A sketch of G

Using (2.15) we immediately obtain

$$C_Z(t,r) = \frac{\alpha^2}{L}\frac{1}{2\pi}\sqrt{\frac{t}{\nu}}\cdot G(\frac{r}{\sqrt{\nu t}}) + \mathcal{O}\left(\frac{\alpha^2\sqrt{t}}{L^3\sqrt{\nu}}[r+\sqrt{t\nu}][L+\sqrt{t\nu}]\right).$$

Note that the approximation with G is not L-periodic in r, while $C_Z(t,r)$ is. The solution is that the error term is $\mathcal{O}(1)$ for r near L.

For the normalized correlation function we deduce

$$\rho_Z(t,r) := \frac{\widehat{C}_Z(t,r)}{\widehat{C}_Z(t,0)} = \frac{G(\frac{r}{\sqrt{\nu t}}) + \mathcal{O}\left(\frac{1}{L^2}[r+\sqrt{t\nu}][L+\sqrt{t\nu}]\right)}{G(0) + \mathcal{O}\left(\frac{1}{L^2}[\sqrt{t\nu}][L+\sqrt{t\nu}]\right)}.$$

From the properties of G we infer the following:

Theorem 2.3. *Given $\delta \in (0,1)$ and sufficiently small $\epsilon_2 > 0$, there exists some $\epsilon_1 > 0$ and three constants $0 < C_1 < C_2 < C_3$ depending only on δ and ϵ_2 such that for $t < \epsilon_1 L^2/\nu$ the following holds:*

$$\rho_Z(t,r) \geq \delta \quad \textit{for} \quad r \in [0, C_1\sqrt{t\nu}]$$

and

$$|\rho_Z(t,r)| < \epsilon_2 \quad \textit{for} \quad r \in [C_2\sqrt{t\nu}, C_3\sqrt{t\nu}].$$

Note that we did not show that the correlation function has a zero, but it is arbitrary small in a point $r_\epsilon \approx \sqrt{t\nu}$. Therefor the theorem says that the typical length-scale is $\sqrt{t\nu}$, at least for times $t \ll L^2/\nu$.

For $t \to \infty$ we immediately obtain that

$$\widehat{C}_Z(\infty,r) = \frac{\alpha^2}{\pi^2\nu}\sum_{k=1}^{\infty}\frac{1}{k^2}\cos(k\pi r/L) =: \frac{\alpha^2}{\nu}F(r/L)$$

and

$$|\widehat{C}_Z(t,r) - \widehat{C}_Z(\infty,r)| \leq \frac{\alpha^2}{\pi^2\nu}e^{-2t\nu\pi^2/L^2}\sum_{k=1}^{\infty}\frac{1}{k^2}.$$

We can look for the explicit representation of F, which is a 2-periodic function, and compute explicitly the zero, but all we need from F is, that for a given small

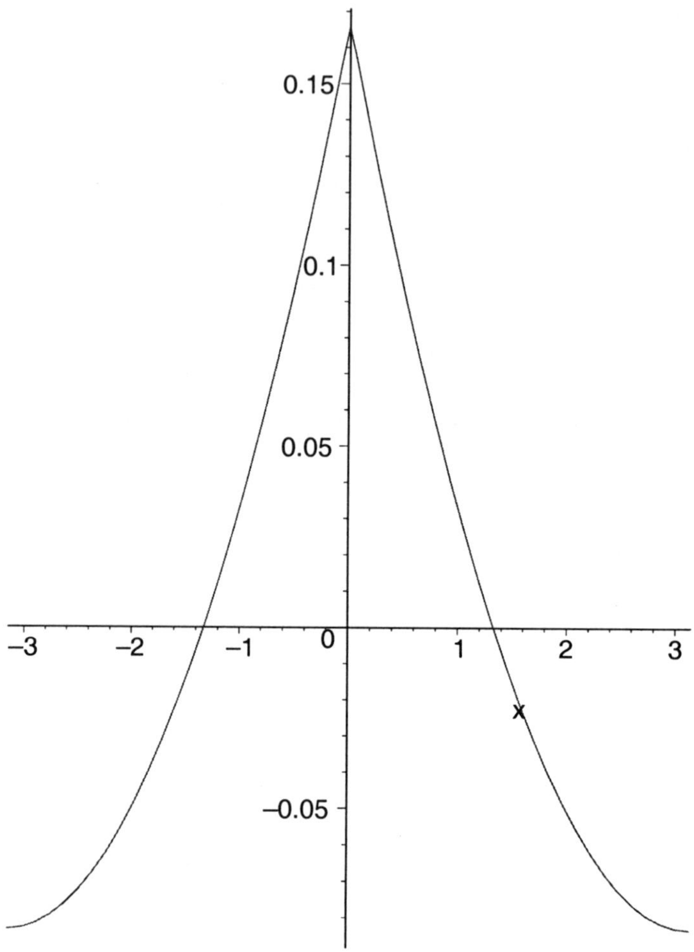

Fig. 2.5. A sketch of F

enough $\delta > 0$ there is a $x_\delta > 0$ such that $F > \delta$ on $[0, x_\delta]$. Moreover, there is some x_0 such that $F(x_0) = 0$.

Consider the normalized correlation function

$$\rho_Z(t,r) = \frac{\widehat{C}_Z(t,r)}{\widehat{C}_Z(t,0)} = \frac{F(r/L) + \mathcal{O}(e^{-2t\nu\pi^2/L^2})}{F(0) + \mathcal{O}(e^{-2t\nu\pi^2/L^2})}.$$

Assume that $t\nu \gg L^2$ (i.e., there is some small $\epsilon > 0$ such that $\epsilon t\nu > L^2$). Now,

$$\rho_Z(t, x_0 L) = \mathcal{O}(e^{-2t\nu\pi^2/L^2})$$

and

$$\rho_Z(t, xL) \geq \frac{\delta}{F(0)} + \mathcal{O}(e^{-2t\nu\pi^2/L^2}) > 0$$

for any $x < x_\delta$.

So for $t \gg L^2/\nu$ the first zero of ρ_Z should be of order L. A more precise formulation is:

Theorem 2.4. *Given $\delta \in (0, 0.8)$ and $\delta \gg \epsilon_2 > 0$, there exists some $\epsilon_1 > 0$, a constant $C > 0$, and a point $x_o > 0$ depending only on δ and ϵ_2 such that for $t > L^2/(\nu\epsilon_1)$ we obtain the following:*

$$\rho_Z(t,r) \geq \delta \quad \textit{for} \quad r \in [0, CL]$$

and

$$|\rho_Z(t, x_0 L)| < \epsilon_2.$$

Thus the theorem tells us that for $t \gg L^2/\nu$, the typical length scale is of order L, which is the size of system. This result is true for both boundary and body forcing.

3. Nonlinear Theory

For the nonlinear results we leave the setting of boundary forcing. Mainly, due to the lack of a-priori estimates. Usually, for Neumann boundary conditions one relies on the maximum principle to bound solutions, but the solution for boundary forcing is quite rough. Therefore, we hardly get useful results. Only, transient bounds for small times are possible to establish. For the next sections, we focus first on body forcing and later on point forcing. We will see later that boundary forcing is actually just a point forcing in a point at the boundary.

The main results of this sections are uniform bounds on the energy and thus on the correlation function C, as $|C(t,r)| \leq C(t,0)$, and $C(t,0)$ is the energy. Furthermore, we show that for $t \to 0$ the linear regime dominates. In Ref. [17] also Hölder-continuity for the mean energy was shown for a quasigeostrophic model. We conclude this section by a qualitative discussion on upper bounds for the energy using additive and multiplicative trace-class noise.

3.1. *Body forcing - Mean energy bounds*

Here we provide bounds on the mean energy for the body forcing case. We consider additive space-time white noise case first, and show that the mean energy and thus the correlation function is uniformly bounded in time. This result is known (cf. Ref. [18]) for Burgers equation using the celebrated Cole-Hopf transformation, but we provide here a simple proof for completeness. Furthermore, our proof is based on energy estimates and it is easily adapted to other types of equations and additional terms in the equation. In contrast to that Cole-Hopf transformation is strictly limited to the standard Burgers equation.

For a long time for space-time white noise only uniform bounds for logarithmic moments were known. See Ref. [13, Lemma 14.4.1] or Ref. [19]. In Ref. [18] the transformation to a stochastic heat equation via the celebrated Cole-Hopf transformation was used to study finiteness of moments. Here we rely on a much simpler tool, which can also be applied to other equations. See for instance [17] for a quasigeostrophic model, where our analysis would apply, too.

Consider

$$\partial_t u + u \cdot \partial_x u = \nu \partial_x^2 u + \sigma \partial_t W \,, \tag{3.1}$$

$$u(\cdot, -L) = u(\cdot, L) = 0, \qquad u(x,0) = u_0(x). \tag{3.2}$$

Here W is a Q-Wiener process with a continuous operator $Q \in \mathcal{L}(L^2)$. Thus W might be cylindrical, and we include the case of space-time white noise.

Using the semigroup $e^t A$ the solution for this system is (see Ref. [13] or [4]):

$$u(t) = e^{tA} u_0 - \int_0^t e^{(t-\tau)A}(\lambda \Phi_\lambda(\tau) + \frac{1}{2}\partial_x u(\tau,x)^2) d\tau + \Phi_\lambda(t) \,, \tag{3.3}$$

where for some $\lambda \geq 0$ fixed later

$$\Phi_\lambda(t) = \sigma\alpha \int_0^t e^{(t-\tau)(A-\lambda)} dW(\tau)$$

solves

$$\partial_t \Phi = \nu \partial_x^2 \Phi - \lambda \Phi + \sigma \partial_t W$$

subject to

$$\Phi(\cdot, -L) = \Phi(\cdot, L) = 0, \qquad \Phi(x,0) = 0 \,.$$

Our main result is now:

Theorem 3.1. *Consider initial conditions u_0 with $E\|u_0\|^2 < \infty$, which are independent of the Wiener process W (e.g. deterministic). Then the mean energy of the solution of (3.3) is uniformly bounded in time. I.e.,*

$$\sup_{t \geq 0} \mathbb{E}\|u(t) - \overline{u}(t)\|^2 < \infty \,.$$

Remark 3.2. Actually, we prove that $\sup_{t\geq 0} \mathbb{E}\|u(t)\|^2 < \infty$. The main problem in the proof is that after applying Gronwall-type estimates we end up with terms $\mathbb{E}\exp\{\int_0^t \|\Phi_\lambda(s)\|_{L^\infty}^2\}$. This might blow up in finite time, as second order exponential moments of the Gaussian may fail to exist, if t is too large. This is why we introduced artificially additional dissipation in the equation for Φ_λ, in order to get exponential moments small.

For the proof of Theorem 3.1 define

$$v(t) = u(t) - \Phi_\lambda(t) \quad \text{for } t \geq 0, \ \lambda \geq 0 \,. \tag{3.4}$$

We see that v is a weak solution of

$$\partial_t v + \frac{1}{2}\partial_x (v + \Phi_\lambda)^2 = \nu \partial_x^2 v + \lambda \Phi_\lambda \tag{3.5}$$

$$v(\cdot,-L) = v(\cdot,L) = 0, \qquad v(x,0) = u_0(x). \tag{3.6}$$

The following calculation is now only formal, but it can easily be made rigorous using for instance spectral Galerkin approximations. Taking the scalar product in (3.5) yields

$$\begin{aligned}\frac{1}{2}\partial_t\|v\|^2 &= -\|v_x\|^2 + \int_{-L}^{L}(v+\Phi_\lambda)^2 v_x\,dx + \int_{-L}^{L}\Phi_\lambda v\,dx \\ &\le -\|v_x\|^2 + \|\Phi_\lambda\|_{H^{-1}}\|v_x\| + |v_x\|\|\Phi_\lambda\|_{L^4}^2 + 2\|v_x\|\|\Phi_\lambda\|_{L^\infty}\|v\| \\ &\le -\frac{1}{2}c_p^2\|v\|^2 + 4\|v\|^2\|\Phi_\lambda\|_\infty^2 + 2\lambda^2\|\Phi_\lambda\|_{L^4}^4 + 2\|\Phi_\lambda\|_{H^{-1}}^2\,,\end{aligned}$$

where we used Young inequality ($ab \le \frac{1}{8}a^2 + 2b^2$), and Poincare inequality $\|v\| \le c_p\|v_x\|$. Now, from Gronwall-type inequalities

$$\begin{aligned}\|v(t)\|^2 &\le e^{-c_p^2 t + 8\int_0^t \|\Phi_\lambda\|_\infty^2 d\tau}\|u(0)\|^2 \\ &\quad + \int_0^t e^{-c_p^2(t-s)+8\int_s^t\|\Phi_\lambda\|_\infty^2 d\tau} 4(\lambda^2\|\Phi_\lambda\|_{L^4}^4 + \|\Phi_\lambda\|_{H^{-1}}^2)ds.\end{aligned} \tag{3.7}$$

Now we use the following lemma, which is easily proved by Fernique's theorem, if we consider Φ_λ as a Gaussian in $L^2([0,t_0],L^\infty)$.

Lemma 3.3. *Fix $K > 0$ and $t_0 > 0$, then there is a λ_0 such that*

$$\sup_{t\in[0,t_0]} \mathbb{E}\exp\{16\int_0^t \|\Phi_\lambda(s)\|_{L^\infty}^2 ds\} \le K^2$$

for all $\lambda \ge \lambda_0$.

Furthermore, we use that all moments of $\|\Phi_\lambda\|_{L^\infty}$ and $\|\Phi_\lambda\|_{H^{-1}}$ are uniformly bounded in time. This is easily proven, using for instance the celebrated factorization method.

Now we first fix $K > 0$, and then t_0 such that $e^{-c_p^2 t}K < \frac{1}{4}$. This yields for $t \in [0,t_0]$ and λ sufficiently large

$$\mathbb{E}\|v(t)\|^2 \le e^{-c_p^2 t}K\mathbb{E}\|u(0)\|^2 + 4K\int_0^t e^{-c_p^2(t-s)}\Big(\mathbb{E}(\lambda^2\|\Phi_\lambda\|_{L^4}^4 + \|\Phi_\lambda\|_{H^{-1}}^2)^2\Big)^{1/2}ds\,,$$

using Hölder, Lemma 3.3, and the independence of $u(0)$ from Φ_λ. We now find a constant C depending on t_0 and K such that

$$\sup_{t\in[0,t_0]} \mathbb{E}\|v(t)\|^2 \le K\mathbb{E}\|u(0)\|^2 + C \qquad \text{and} \qquad \mathbb{E}\|v(t_0)\|^2 \le \frac{1}{4}\mathbb{E}\|u(0)\|^2 + C.$$

Using $\mathbb{E}\|u(t)\|^2 \le 2\mathbb{E}\|v(t)\|^2 + 2\mathbb{E}\|\Phi_\lambda(t)\|^2$ yields for a different constant C

$$\sup_{t\in[0,t_0]} \mathbb{E}\|u(t)\|^2 \le K\mathbb{E}\|u(0)\|^2 + C \qquad \text{and} \qquad \mathbb{E}\|u(t_0)\|^2 \le \frac{1}{2}\mathbb{E}\|u(0)\|^2 + C.$$

Now we repeat the argument for $k \in \mathbb{N}$ by defining $v(t) = u(kt_0 + t) - \tilde{\Phi}_\lambda(t)$, where $\tilde{\Phi}_\lambda(t)$ has the same distribution than $\Phi_\lambda(t)$ due to a time shift of the Brownian motion. Now v solves again (3.5) with initial condition $u(kt_0)$. Note that by construction $u(kt_0)$ is independent of $\tilde{\Phi}_\lambda$.

Repeating the arguments as before yields for $k \in \mathbb{N}_0$

$$\sup_{t\in[0,t_0]} \mathbb{E}\|u(t + kt_0)\|^2 \leq K\mathbb{E}\|u(kt_0)\|^2 + C$$

and

$$\mathbb{E}\|u((k+1)t_0)\|^2 \leq \frac{1}{2}\mathbb{E}\|u(kt_0)\|^2 + C\,.$$

Now the following lemma, which is a trivial statement on discrete dynamical systems, finishes the proof.

Lemma 3.4. *Suppose for $q < 1$ and some $C > 0$ we have $a_{n+1} < qa_n + C$, then a_n is bounded by*

$$a_n < \frac{C}{1-q} + a_0\,.$$

3.2. *Point forcing - Mean energy bounds*

Consider hyperviscous Burgers equation with point-forcing. We would like to proceed exactly the way, we did in the previous section, But we can not, as for point forcing, the solution of the linear equation might fail to be in L^∞. This is why we add additional damping. Hyperviscous Burgers equation has been studied in several occasions. See for example [20–22].

Consider for some $\epsilon > 0$ the operator $A_\epsilon = -\nu(-\Delta)^{1+\epsilon}$, where Δ is the Laplacian subject to Dirichlet boundary conditions. Then the hyperviscous Burgers equation is given by

$$\partial_t u + u\cdot\partial_x u = A_\epsilon u + \alpha\delta_0\dot{\beta} \tag{3.8}$$

$$u(\cdot,-L) = u(\cdot,L) = 0, \qquad u(x,0) = u_0(x). \tag{3.9}$$

Here, β is a standard Brownian motion and δ_0 the Delta-distribution.

Using the semigroup e^{tA_ϵ} the solution for this system is (see Ref. [13] or [4]):

$$u(t) = e^{tA_\epsilon}u_0 - \int_0^t e^{(t-\tau)A_\epsilon}(\lambda\Phi_\lambda(\tau) + \frac{1}{2}\partial_x u(\tau,x)^2)d\tau + \Phi_\lambda(t) \tag{3.10}$$

where for some $\lambda \geq 0$ fixed later

$$\Phi_\lambda(t) = \alpha\int_0^t e^{(t-\tau)(A_\epsilon-\lambda)}\delta_0(x)d\beta(\tau)$$

solves

$$\partial_t\Phi = A_\epsilon\Phi - \lambda\Phi + \alpha\delta_0\dot{\beta}$$

subject to

$$\Phi(\cdot,-L)=\Phi(\cdot,L)=0, \qquad \Phi(x,0)=0\,.$$

Using the standard orthonormal basis $\{e_k\}_{k\in\mathbb{N}}$ of eigenfunctions of A_ϵ given by $e_k(x)=\sqrt{1/L}\cdot\sin(-L+\frac{\pi kx}{2L})$ with corresponding eigenvalues $\lambda_k=-(\pi k/2L)^{2+2\epsilon}$, we see

$$\Phi_\lambda(t)=\alpha\sum_{k=1}^{\infty}\int_0^t e^{(t-\tau)(\lambda_k-\lambda)}d\beta(\tau)e_k(0)e_k\,. \tag{3.11}$$

Note that the Fourier-coefficients of that series are not at all independent. Thus we cannot rely on the better regularity results available for the stochastic convolution of the previous chapter. Especially, for $\epsilon=0$ we cannot show that $\Phi_\lambda(t)\in L^\infty([-L,L])$.

Note that the series expansion of boundary and point forcing is very similar. Thus we can regard boundary forcing at a point forcing at the boundary, when the equation is subject to Neumann boundary conditions.

Our main result is now:

Theorem 3.5. *For all $\epsilon>0$ and all initial conditions u_0 independent of β with $E\|u_0\|^2<\infty$ the solution of (3.10) satisfies that the mean energy is uniformly bounded in time. I.e.,*

$$\sup_{t\geq 0}\mathbb{E}\|u(t)-\overline{u}(t)\|^2<\infty\,.$$

We will proceed exactly as in the previous section. Now $v=u-\Phi_\lambda$ is a weak solution of

$$\partial_t v+\frac{1}{2}\partial_x(v+\Phi_\lambda)^2=\nu\partial_x^2 v+\lambda\Phi_\lambda\,, \tag{3.12}$$

again subject to Dirichlet boundary conditions and initial condition $v(0)=u_0$.

Now consider first the nonlinear term for some small $\delta>0$. Using Hölder, Sobolev embedding of $H^{\frac{1}{2}-\frac{1}{p}}$ into L^p and the bound

$$\|u\|_{H^{2\gamma}}\leq C\|A_\epsilon^{\gamma/(1+\epsilon)}u\|$$

yields

$$\begin{aligned}\int_{-L}^{L}v\Phi_\lambda v_x dx &\leq \|v\|_{L^{2+\delta}}\|\Phi_\lambda\|_{L^{(4+2\delta)/\delta}}\|v_x\|\\ &\leq C\|A_\epsilon^{\frac{1}{4}\frac{1}{1+\epsilon}\frac{\delta}{2+\delta}}v\|\|\Phi_\lambda\|_{L^{(4+2\delta)/\delta}}\|A_\epsilon^{\frac{1}{2}\frac{1}{1+\epsilon}}v\|\,,\end{aligned}$$

Now we can easily find an $\delta>0$ sufficiently small such that there is a $p=p(\epsilon)\in(2,\infty)$ such that (using interpolation inequality)

$$\int_{-L}^{L}v\Phi_\lambda v_x dx\leq C\|v\|_{L^2}\|\Phi_\lambda\|_{L^p}\|A_\epsilon^{\frac{1}{2}}v\|\,.$$

Now we can use the same proof as in the section before. We only need that $\Phi_\lambda(t) \in L^\infty(0,L)$. To be more precise, an easy calculation using the series expansion of (3.11) shows that for any $\epsilon > 0$

$$\sup_{t\geq 0} \mathbb{E}\|\Phi_\lambda(t)\|_{L^\infty}^2 \leq C_\epsilon \sup_{t\geq 0} \mathbb{E}\|\Phi_\lambda(t)\|^2_{H^{\frac{1+\epsilon}{2}}} \to 0 \quad \text{as } \lambda \to \infty .$$

It is now straightforward to prove an analog to Lemma 3.3. The remainder of the proof is analogous to the section before.

Let us remark that we could even simplify that proof a little bit, by avoiding second order exponentials of Φ_λ. In that case we could work with $\lambda = 0$

3.3. *Body forcing - Transient behavior*

Let us focus on Burgers equation with body forcing. The results for hyperviscous Burgers with point-forcing are completely analogous. We will prove:

Theorem 3.6. *Let u be a solution of (3.1) and consider for simplicity $u(0) = 0$. Denote by*

$$E_u(t) = \mathbb{E}\|u(t) - \overline{u}(t)\|^2$$

the mean energy of $u(t)$, then there is some δ_0 such that

$$E_u(t) = E_{\Phi_0}(t) + \mathcal{O}(t^{\frac{1}{2}+\delta_0}) \quad \textit{for } t \to 0 .$$

To be more precise, for some $t_0 > 0$ sufficiently small there is a constant $C > 0$ such that $|E_u(t) - E_{\Phi_0}(t)| \leq Ct^{\frac{1}{2}+\delta_0}$ for all $t \in [0, t_0]$.

As we know from results like Theorem 2.1 that $E_{\Phi_0}(t)$ behaves like $\sqrt{t}$ for small t, we can conclude that the linear regime dominates for small t.

We could explicitly calculate δ_0, but omit this for simplicity of presentation.

For the proof of Theorem 3.6 use

$$|E_u(t) - E_{\Phi_0}(t)| \leq C\mathbb{E}\|v(t)\|^2 ,$$

where we used Cauchy-Schwarz inequality and uniform bounds on $\mathbb{E}\|u(t)\|^2$ and $\mathbb{E}\|\Phi_0(t)\|^2$. Using (3.7) with $\lambda = 0$ and $u(0) = 0$ yields together with Lemma 3.3

$$\mathbb{E}\|v(t)\|^2 \leq C \int_0^t (\mathbb{E}\|\Phi_0(t)\|^4_{H^{-1}})^{1/2} dt .$$

It is now easy to show that $\mathbb{E}\|\Phi_0(t)\|^4_{H^{-1}}$ behaves like $t^{2\delta_0}$ for some $\delta_0 > 0$, which can be explicitly calculated using the methods of Theorem 2.1. Theorem 3.6 is now proved.

A simple corollary using Hölders inequality is:

Corollary 3.7. *Under the assumptions of Theorem 3.6, we know for the mean correlation function*

$$C_u(t,r) = C_{\Phi_0}(t,r) + \mathcal{O}(t^{\frac{1}{2}+\delta_0}) \quad \textit{for } t \to 0 \textit{ and all } r .$$

Notice that this result is only useful for small times and small r, as seen from the qualitative behavior of C_{Φ_0}, which is similar to the results shown in section 2.2.

3.4. *Trace class noise: Additive vs. multiplicative body noises*

Consider again a solution of the following Burgers equation:

$$\partial_t u + u \cdot \partial_x u = \nu \partial_x^2 u + \sigma \dot{W} \tag{3.13}$$

$$u(\cdot, 0) = 0, \qquad u(\cdot, L) = 0, \qquad u(x, 0) = u_0(x), \tag{3.14}$$

where $\{W(t)\}_{t\geq 0}$ is a Brownian motion, with covariance Q, taking values in the Hilbert space $L^2(0, L)$ with the usual scalar product $\langle \cdot, \cdot \rangle$. We assume that the trace $Tr(Q)$ is finite. So $\dot{W}$ is noise colored in space but white in time.

Applying the Itô's formula, we obtain

$$\frac{1}{2} d\|u\|^2 = \langle u, dW\rangle + [\langle u, u_{xx} - uu_x\rangle + \frac{1}{2}\sigma^2 Tr(Q)]dt. \tag{3.15}$$

as before $\langle u, uu_x\rangle = 0$. Thus

$$\frac{d}{dt}\mathbb{E}\|u\|^2 = -2\|u_x\|^2 + \sigma^2 Tr(Q). \tag{3.16}$$

By the Poincare inequality $\|u\|^2 \leq c\|u_x\|^2$ for some positive constant depending only on the length L, we have

$$\frac{d}{dt}\mathbb{E}\|u\|^2 \leq -\frac{2}{c}\|u\|^2 + \sigma^2 Tr(Q). \tag{3.17}$$

Then using the Gronwall inequality, we finally get

$$\mathbb{E}\|u\|^2 \leq \mathbb{E}\|u_0\|^2 e^{-\frac{2}{c}t} + \frac{1}{2}c\sigma^2 Tr(Q)[1 - e^{-\frac{2}{c}t}]. \tag{3.18}$$

Note that the first term in this estimate involves with initial data, and the second term involves with the noise intensity σ as well as the trace of the noise covariance.

We now consider multiplicative body noise forcing.

$$\partial_t u + u \cdot \partial_x u = \nu \partial_x^2 u + \sigma u \dot{w}, \tag{3.19}$$

with the same boundary condition and initial condition as above, where w_t is a scalar Brownian motion. So $\dot{w}$ is noise homogeneous in space but white in time.

By the Itô's formula, we obtain

$$\frac{1}{2} d\|u\|^2 = \langle u, \sigma u dw\rangle + [\langle u, \nu u_{xx} - uu_x\rangle + \frac{1}{2}\sigma^2\|u\|^2]dt. \tag{3.20}$$

Thus

$$\begin{aligned} \frac{d}{dt}\mathbb{E}\|u\|^2 &= -2\nu\|u_x\|^2 + \sigma^2\|u\|^2 \\ &\leq (\sigma^2 - \frac{2\nu}{c})\|u\|^2. \end{aligned} \tag{3.21}$$

Therefore,

$$\mathbb{E}\|u\|^2 \le \mathbb{E}\|u_0\|^2 e^{(\sigma^2 - \frac{2\nu}{c})t}. \tag{3.22}$$

Note here that the multiplicative noise affects the mean energy growth or decay rate, while the additive noise affects the mean energy upper bound.

Acknowledgments

Part of this work was done at the Oberwolfach Mathematical Research Institute, Germany and the Institute of Applied Mathematics, the Chinese Academy of Sciences, Beijing, China. This work was partly supported by the NSF Grants DMS-0209326 & DMS-0542450 and DFG Grant KON 613/2006.

References

[1] T. F. Stocker, D. G. Wright, and L. A. Mysak, A zonally averaged, coupled ocean-atmosphere model for paleoclimate studies, *J. Climate.* **5**, 773–797, (1992).

[2] H. A. Dijkstra, *Nonlinear Physical Oceanography.* (Kluwer Academic Publishers, Boston, 2000).

[3] J.-L. Lions, R. Temam, and S. H. Wang, On the equations of the large-scale ocean, *Nonlinearity.* **5**(5), 1007–1053, (1992). ISSN 0951-7715.

[4] G. Da Prato, A. Debussche, and R. Temam, Stochastic Burgers' equation, *NoDEA Nonlinear Differential Equations Appl.* **1**(4), 389–402, (1994). ISSN 1021-9722.

[5] W. E and E. Vanden Eijnden, Statistical theory for the stochastic Burgers equation in the inviscid limit, *Comm. Pure Appl. Math.* **53**(7), 852–901, (2000). ISSN 0010-3640.

[6] I. D. Chueshov and P.-A. Vuillermot, Long-time behavior of solutions to a class of quasilinear parabolic equations with random coefficients, *Ann. Inst. H. Poincaré Anal. Non Linéaire.* **15**(2), 191–232, (1998). ISSN 0294-1449.

[7] I. D. Chueshov and P. A. Vuillermot, Long-time behavior of solutions to a class of stochastic parabolic equations with homogeneous white noise: Itô's case, *Stochastic Anal. Appl.* **18**(4), 581–615, (2000). ISSN 0736-2994.

[8] H. V. Ly, K. D. Mease, and E. S. Titi, Distributed and boundary control of the viscous Burgers' equation, *Numer. Funct. Anal. Optim.* **18**(1-2), 143–188, (1997). ISSN 0163-0563.

[9] R. Courant and D. Hilbert, *Methods of mathematical physics. Vol. II.* Wiley Classics Library, (John Wiley & Sons Inc., New York, 1989). ISBN 0-471-50439-4. Partial differential equations, Reprint of the 1962 original, A Wiley-Interscience Publication.

[10] A. Lunardi, *Analytic semigroups and optimal regularity in parabolic problems.* Progress in Nonlinear Differential Equations and their Applications, 16, (Birkhäuser Verlag, Basel, 1995). ISBN 3-7643-5172-1.

[11] L. Arnold, *Stochastic differential equations: theory and applications.* (Wiley-Interscience [John Wiley & Sons], New York, 1974). Translated from the German.

[12] G. Da Prato and J. Zabczyk, *Stochastic equations in infinite dimensions.* vol. 44, *Encyclopedia of Mathematics and its Applications*, (Cambridge University Press, Cambridge, 1992). ISBN 0-521-38529-6.

[13] G. Da Prato and J. Zabczyk, *Ergodicity for infinite-dimensional systems.* vol. 229, *London Mathematical Society Lecture Note Series*, (Cambridge University Press, Cambridge, 1996). ISBN 0-521-57900-7.

[14] G. Da Prato and J. Zabczyk, Evolution equations with white-noise boundary conditions, *Stochastics Stochastics Rep.* **42**(3-4), 167–182, (1993). ISSN 1045-1129.
[15] D. Blömker, S. Maier-Paape, and T. Wanner, Roughness in surface growth equations, *Interfaces Free Bound.* **3**(4), 465–484, (2001). ISSN 1463-9963.
[16] D. Blömker, S. Maier-Paape, and T. Wanner, Surface roughness in molecular beam epitaxy, *Stoch. Dyn.* **1**(2), 239–260, (2001). ISSN 0219-4937.
[17] D. Blömker, J. Duan, and T. Wanner, Enstrophy dynamics of stochastically forced large-scale geophysical flows, *J. Math. Phys.* **43**(5), 2616–2626, (2002). ISSN 0022-2488.
[18] J. A. León, D. Nualart, and R. Pettersson, The stochastic Burgers equation: finite moments and smoothness of the density, *Infin. Dimens. Anal. Quantum Probab. Relat. Top.* **3**(3), 363–385, (2000). ISSN 0219-0257.
[19] G. Da Prato and D. Gatarek, Stochastic Burgers equation with correlated noise, *Stochastics Stochastics Rep.* **52**(1-2), 29–41, (1995). ISSN 1045-1129.
[20] J. P. Boyd, Hyperviscous shock layers and diffusion zones: monotonicity, spectral viscosity, and pseudospectral methods for very high order differential equations, *J. Sci. Comput.* **9**(1), 81–106, (1994). ISSN 0885-7474.
[21] C. Gugg, H. Kielhöfer, and M. Niggemann, On the approximation of the stochastic Burgers equation, *Comm. Math. Phys.* **230**(1), 181–199, (2002). ISSN 0010-3616.
[22] L. Machiels and M. O. Deville, Numerical simulation of randomly forced turbulent flows, *J. Comput. Phys.* **145**(1), 246–279, (1998). ISSN 0021-9991.

Chapter 3

Asymptotics for the Space-Time Wigner Transform with Applications to Imaging

Liliana Borcea, George Papanicolaou* and Chrysoula Tsogka†

Computational and Applied Mathematics, Rice University
6100 Main Street, Houston, TX 77005-1892
borcea@caam.rice.edu

We consider the space-time Wigner transform of the solution of the random Schrödinger equation in the white noise limit and for high frequencies. We analyze in particular the strong lateral diversity limit in which the space-time Wigner transform becomes weakly deterministic. We also show how to use these asymptotic results in broadband array imaging in random media.

Contents

1. Introduction

In this paper we analyze the self-averaging property of the space-time Wigner transform for solutions of the random Schrödinger equation, in a particular asymptotic regime. We start with the wave equation in a random medium and then use the parabolic or paraxial approximation, which is valid when waves propagate primarily in a preferred direction and backscattering is negligible. This approximation is

*Department of Mathematics, Stanford University, Stanford, CA 94305. (papanico@math.stanford.edu)

†Department of Mathematics, University of Chicago, Chicago, IL 60637. (tsogka@math.uchicago.edu)

widely used in random wave propagation [1–5] and it is justified in some special cases in Ref. [6], in the regime that we consider here. The parabolic wave field satisfies a random Schrödinger equation, which we consider in the white noise limit. White noise limits for random ordinary differential equations have been analyzed extensively [7–9]. For random partial differential equations, white noise limits are considered in [10] for diffusion equations and, more recently, in Refs. [11, 12], for the random Schrödinger equation.

The resulting Itô-Schrödinger equation for the limit wave field is a stochastic partial differential equation of independent interest that is analyzed in Ref. [5] and in a wider context in Ref. [13, 14]. We consider here the high frequency limit of this equation, using the space-time Wigner transform. This is a slight extension of the high frequency limits analyzed in Refs. [15, 16] and in Ref. [11], using the spatial Wigner transform. The limit process satisfies an Itô-Liouville partial differential equation that arises from a stochastic flow [14, 17].

We analyze this Itô-Liouville equation in the strong lateral diversity limit, where the propagating wave beam is wide with respect to the correlation length of the random inhomogeneities in the transverse direction, orthogonal to the axis of the beam. The importance of this limit in time reversal was pointed out in Ref. [18] and it was analyzed later in Refs. [15, 16, 19, 20], using the spatial Wigner transform. Applications to imaging are considered in Refs. [21–23], especially applications of the space-time Wigner transform, but the strong lateral diversity limit is not analyzed there.

We dedicate this work to Boris Rozovskii on the occasion of his 60th birthday.

2. The parabolic approximation

Let $P(\vec{\mathbf{x}}, t)$ be the solution of the acoustic wave equation

$$\frac{1}{c^2(\vec{\mathbf{x}})}\frac{\partial^2 P}{\partial t^2} - \Delta P = 0, \quad t > 0, \ \vec{\mathbf{x}} \in \mathbb{R}^3, \tag{2.1}$$

with a given excitation source at time $t = 0$ and in a medium with sound speed $c(\vec{\mathbf{x}})$ that is fluctuating about the mean value c_o, taken as constant for simplicity. We model the fluctuations of $c(\vec{\mathbf{x}})$ as a random process

$$c(\vec{\mathbf{x}}) = c_o\left[1 + \sigma_o \mu\left(\frac{\vec{\mathbf{x}}}{\ell}\right)\right]^{-1/2} \tag{2.2}$$

where μ is a normalized, bounded and statistically homogeneous random field, with mean zero and with smooth and rapidly decaying correlation function

$$E\{\mu(\vec{\mathbf{x}} + \vec{\mathbf{x}}')\mu(\vec{\mathbf{x}}')\} = R(\vec{\mathbf{x}}). \tag{2.3}$$

Here the normalization means that

$$R(\vec{0}) = 1, \quad \int d\vec{\mathbf{x}}\, R(\vec{\mathbf{x}}) = 1, \tag{2.4}$$

so that ℓ in (2.2) is the correlation length of the fluctuations.

We consider a regime with weak fluctuations ($\sigma_o \ll 1$) where backscattering of the waves by the medium can be neglected and where we can study $P(\vec{\mathbf{x}}, t)$ with the parabolic approximation [1]. For this, we take the z coordinate in the direction of propagation of the waves and we let $\vec{\mathbf{x}} = (z, \mathbf{x})$, with $\mathbf{x}$ the two dimensional vector of coordinates transverse to the direction of propagation. In the parabolic approximation the wave field is given by

$$P(z, \mathbf{x}, t) = \frac{1}{2\pi} \int \widehat{P}(z, \mathbf{x}, \omega) e^{-i\omega t} d\omega, \quad \widehat{P}(z, \mathbf{x}, \omega) \approx e^{ikz} \psi(z, \mathbf{x}, k), \tag{2.5}$$

where $k = \omega / c_o$ is the wavenumber and ψ is a complex valued amplitude satisfying the Schrödinger equation

$$2ik \frac{\partial \psi}{\partial z} + \Delta_{\mathbf{x}} \psi + k^2 \sigma_o \mu \left(\frac{z}{\ell}, \frac{\mathbf{x}}{\ell} \right) \psi = 0, \quad z > 0, \tag{2.6}$$

with $\Delta_{\mathbf{x}}$ denoting the Laplacian in $\mathbf{x}$. This equation is obtained by substituting $e^{ikz}\psi$ in the reduced wave equation for $\widehat{P}$

$$\Delta \widehat{P} + k^2 n^2(\vec{\mathbf{x}}) \widehat{P} = 0,$$

with index of refraction $n(\vec{\mathbf{x}}) = c_o / c(\vec{\mathbf{x}})$ given by

$$n^2(\vec{\mathbf{x}}) = 1 + \sigma_o \mu \left(\frac{\vec{\mathbf{x}}}{\ell} \right), \tag{2.7}$$

and by neglecting the term $\frac{\partial^2 \psi}{\partial z^2}$ under the hypothesis that ψ is slowly varying in z (i.e., $k \left| \frac{\partial \psi}{\partial z} \right| \gg \left| \frac{\partial^2 \psi}{\partial z^2} \right|$).

We now have an initial value problem for the wave amplitude ψ, governed by equation (2.6) with initial condition

$$\psi(0, \mathbf{x}, k) = \psi_o(\mathbf{x}, k). \tag{2.8}$$

We assume that ψ_o is a compactly supported function with frequency dependence in the positive interval

$$\omega \in \left[\omega_o - \frac{B}{2}, \omega_o + \frac{B}{2} \right], \tag{2.9}$$

centered at ω_o and with bandwidth B. The negative image of this interval is also included if the initial data is real.

3. Scaling and the asymptotic regime

To carry out an asymptotic analysis of the wave field (2.5) we write the Schrödinger equation (2.6) in dimensionless form

$$2ik' \frac{\partial \psi}{\partial z'} + \frac{L_z}{k_o L_x^2} \Delta_{\mathbf{x}'} \psi + k_o L_z \sigma_o (k')^2 \mu \left(\frac{L_z z'}{\ell}, \frac{L_x \mathbf{x}'}{\ell} \right) \psi = 0, \tag{3.1}$$

with scaled variables

$$\mathbf{x} = L_x \mathbf{x}', \quad z = L_z z', \quad \omega = \omega_o \omega', \quad k = k_o k', \quad c = c_o c'. \tag{3.2}$$

Here $k_o = \omega_o / c_o$ is the central wavenumber, L_z quantifies the distance of propagation and L_x is a transversal length scale which we take to be the width of the propagating beam. Note that the scaled sound speed has constant mean $< c' >= 1$. Therefore, since the scaled wavenumber k' is the same as the scaled frequency ω', we shall replace ω' by k' from now on.

To simplify notation we drop the primes on the scaled variables and we introduce three dimensionless parameters depending on the random medium

$$\epsilon = \frac{\ell}{L_z}, \quad \delta = \frac{\ell}{L_x}, \quad \sigma = \sigma_o \delta \epsilon^{-\frac{3}{2}}, \tag{3.3}$$

and the reciprocal of the Fresnel number

$$\theta = \frac{L_z}{k_o L_x^2} = \frac{1}{2\pi} \frac{\left(\frac{\lambda_0 L_z}{L_x}\right)}{L_x}. \tag{3.4}$$

Here λ_o is the central wavelength and the reciprocal of the Fresnel number is written as the ratio of the focusing spot size in time reversal imaging, $\lambda_o L_z / L_x$, and the transversal length scale L_x.

The scaled form of equation (3.1) is

$$2ik\frac{\partial \psi}{\partial z} + \theta \Delta_{\mathbf{x}} \psi + \frac{1}{\epsilon^{1/2}} \mu\left(\frac{z}{\epsilon}, \frac{\mathbf{x}}{\delta}\right) \frac{\sigma k^2 \delta}{\theta} \psi = 0, \quad z > 0 \tag{3.5}$$

and we study it in the asymptotic regime

$$\epsilon \ll \delta \ll 1, \quad \theta \ll 1, \quad \sigma = O(1). \tag{3.6}$$

Thus, we suppose that the waves travel many correlation lengths in the random medium ($\epsilon \ll 1$) and, to be consistent with the parabolic approximation, we take $L_x \ll L_z$ (i.e., $\epsilon \ll \delta$). We also take $\theta \ll 1$, which means that the time reversal imaging spot size is much smaller than L_x

$$\frac{\lambda_o L_z}{L_x} \ll L_x. \tag{3.7}$$

Finally, we scale the strength of the fluctuations in (3.3) and (3.6) so that we can take the white noise limit $\epsilon \to 0$ in (3.5).

The asymptotic regime (3.6) can be realized with several scale orderings. In this paper we assume that

$$\epsilon \ll \theta \ll \delta \ll 1, \tag{3.8}$$

which amounts to taking $\epsilon \to 0$ as the first in a sequence of three limits. This leads to an Itô-Schrödinger equation for the limit ψ. The second limit $\theta/\delta \to 0$ implies that we are in a high frequency regime

$$\frac{\lambda_o}{\ell} \ll \frac{\epsilon}{\delta} \ll 1. \tag{3.9}$$

In imaging, the spot size is small in this limit, when compared with the correlation length

$$\frac{\lambda_o L_z}{L_x} \ll \ell \ll L_x. \tag{3.10}$$

This is a regime in which we can derive an Itô-Liouville equation for the Wigner transform of ψ, under the additional assumption of isotropy of the fluctuations of the sound speed. Finally, we take the strong lateral diversity limit $\delta \ll 1$, which allows us to show that the appropriately smoothed Wigner transform is self-averaging.

Other scale orderings consistent with (3.6) are

$$\theta \ll \epsilon \ll \delta \ll 1 \tag{3.11}$$

and

$$\epsilon \ll \delta \leq \theta \ll 1. \tag{3.12}$$

The ordering (3.11) is considered in Ref. [15], in a study of statistical stability of time reversal in random media. It is a high frequency regime and it gives similar results to those obtained here. The scale ordering (3.12) is consistent with $\lambda_o \sim \ell$ and it is used in numerical simulations in Refs. [21–24], in the context of array imaging of sources and reflectors. In the parabolic approximation this scaling is analyzed in Ref. [16]. The theory is not so well developed when the parabolic approximation does not apply. Nevertheless, it appears from the numerical simulations in Refs. [21–24] that the statistical stability that we have in regimes (3.8) or (3.11) is valid in the case (3.12).

4. The Itô-Liouville equation for the Wigner transform

In this section we give, without details, the Itô-Liouville equation for the Wigner transform of ψ in the limits $\epsilon \to 0$ and $\theta \to 0$. We then state the main result of this paper, which is that in the limit $\delta \to 0$ we have self-averaging for smooth linear functionals of the space-time Wigner transform. The proof is given in section 5. We consider an application of this self-averaging property in section 6, where we look at coherent interferometric imaging in random media, as introduced in Refs. [22, 23].

4.1. *The white noise limit*

Let us emphasize with the notation $\psi^\epsilon(z, \mathbf{x}, k)$ the dependence on ϵ of the wave amplitude satisfying (3.5). This amplitude depends on θ and δ as well, but since these are kept fixed in the first limit we suppress them from the notation. The initial wave field ψ_o is assumed independent of ϵ.

It follows from [11, 12] that as $\epsilon \to 0$, $\psi^\epsilon(z,\mathbf{x},k)$ converges weakly, in law, to the solution $\psi(z,\mathbf{x},k)$ of Itô-Schrödinger equation

$$d\psi = \left[\frac{i\theta}{2k}\Delta_{\mathbf{x}} - \frac{k^2\sigma^2\delta^2}{8\theta^2}R_o(0)\right]\psi dz + \frac{ik\sigma\delta}{2\theta}\psi d\mathcal{B}\left(z,\frac{\mathbf{x}}{\delta}\right), \quad z>0, \qquad (4.1)$$
$$\psi(0,\mathbf{x},k) = \psi_o(\mathbf{x},k), \quad \text{at } z=0.$$

Here $\mathcal{B}(z,\mathbf{x})$ is a Brownian random field that is smooth in the transverse variable $\mathbf{x}$. The mean of $\mathcal{B}$ is zero and its correlation is given by

$$E\{\mathcal{B}(z_1,\mathbf{x}_1)\mathcal{B}(z_2,\mathbf{x}_2)\} = z_1 \wedge z_2 R_o(\mathbf{x}_1-\mathbf{x}_2), \qquad (4.2)$$

where $z_1 \wedge z_2 = \min\{z_1,z_2\}$ and

$$R_o(\mathbf{x}) = \int_{-\infty}^{\infty} R(z,\mathbf{x})dz. \qquad (4.3)$$

Because of our assumptions on R in section 2 we have that R_o is smooth and rapidly decaying. This is used in section 5 to deduce the statistical stability of the smoothed Wigner transform of ψ, in the limit $\theta/\delta \to 0$ and $\delta \to 0$.

4.2. *The high frequency limit and the space-time Wigner transform*

As in section 4.1, we now use the notation $\psi^\theta(z,\mathbf{x},k)$ to emphasize the dependence of the solution of Itô-Schrödinger equation (4.1) on the parameter θ. We study the high frequency limit $\theta \to 0$ with the space-time Wigner transform

$$W^\theta(z,\overline{\mathbf{x}},\overline{k},\mathbf{q},r)$$
$$= \int\frac{d\widetilde{\mathbf{x}}}{(2\pi)^2}\int\frac{d\widetilde{k}}{2\pi}e^{i\mathbf{q}\cdot\widetilde{\mathbf{x}}-i\widetilde{k}r}\psi^\theta\left(z,\overline{\mathbf{x}}+\frac{\theta\widetilde{\mathbf{x}}}{2},\overline{k}+\frac{\theta\widetilde{k}}{2}\right)\overline{\psi^\theta\left(z,\overline{\mathbf{x}}-\frac{\theta\widetilde{\mathbf{x}}}{2},\overline{k}-\frac{\theta\widetilde{k}}{2}\right)}, \qquad (4.4)$$

where the bar on ψ^θ denotes complex conjugate. The r variable in W^θ is dual to $\widetilde{k}$ and it represents the distance traveled by the waves in a medium with constant speed $< c >= 1$, during a travel time $t = r/ < c >$. The argument $\mathbf{q}$ in W^θ is a two dimensional vector that is dual to $\widetilde{\mathbf{x}}$.

For an arbitrary but fixed z, the L^2 norm of the Wigner transform W^θ is determined by the space and frequency L^2 norm of the initial wave function ψ_o^θ

$$\|W^\theta(z,\cdot)\|_{L^2} = \left[\int d\overline{\mathbf{x}}\int d\overline{k}\int d\mathbf{q}\int dr\,\left|W^\theta(z,\overline{\mathbf{x}},\overline{k},\mathbf{q},r)\right|^2\right]^{1/2}$$
$$= \left[\int d\overline{\mathbf{x}}\int d\overline{k}\int\frac{d\widetilde{\mathbf{x}}}{(2\pi)^2}\int\frac{d\widetilde{k}}{2\pi}\left|\psi^\theta\left(z,\overline{\mathbf{x}}+\frac{\theta\widetilde{\mathbf{x}}}{2},\overline{k}+\frac{\theta\widetilde{k}}{2}\right)\right|^2\left|\psi^\theta\left(z,\overline{\mathbf{x}}-\frac{\theta\widetilde{\mathbf{x}}}{2},\overline{k}-\frac{\theta\widetilde{k}}{2}\right)\right|^2\right]^{1/2}$$
$$= \frac{\|\psi^\theta(z,\cdot)\|_{L^2}^2}{(2\pi\theta)^{3/2}} = \frac{\|\psi_o^\theta\|_{L^2}^2}{(2\pi\theta)^{3/2}}, \qquad (4.5)$$

because the Itô-Schrödinger equation (4.1) preserves the space and frequency L^2 norm of its solution [5]. This means that with a proper definition and scaling of the initial wave function ψ_o^θ [25], we can bound the L^2 norm of $W^\theta(z,\cdot)$ uniformly with respect to θ.

We formally obtain an Itô-Liouville equation for the high frequency limit W as follows [12]. We use Itô's formula to get from (4.1) an equation for $\psi_1^\theta\overline{\psi_2^\theta} = \psi^\theta(z,\mathbf{x}_1,k_1)\overline{\psi^\theta(z,\mathbf{x}_2,k_2)}$, with

$$\mathbf{x}_1 = \overline{\mathbf{x}} + \frac{\theta\widetilde{\mathbf{x}}}{2}, \quad \mathbf{x}_2 = \overline{\mathbf{x}} - \frac{\theta\widetilde{\mathbf{x}}}{2}, \tag{4.6}$$

$$k_1 = \overline{k} + \frac{\theta\widetilde{k}}{2}, \quad k_2 = \overline{k} - \frac{\theta\widetilde{k}}{2} \tag{4.7}$$

and then we Fourier transform in $\widetilde{\mathbf{x}}$ and $\widetilde{k}$ and take the limit $\theta/\delta \to 0$. The variables $\overline{\mathbf{x}}$, $\widetilde{\mathbf{x}}$, $\overline{k}$ and $\widetilde{k}$ are independent of the small parameters. We have

$$\begin{aligned} d\left(\psi_1^\theta\overline{\psi_2^\theta}\right) &= \left[\tfrac{i\theta}{2\overline{k}+\theta\widetilde{k}}\left(\tfrac{1}{4}\Delta_{\overline{\mathbf{x}}} + \tfrac{1}{\theta}\nabla_{\overline{\mathbf{x}}}\cdot\nabla_{\widetilde{\mathbf{x}}} + \tfrac{1}{\theta^2}\Delta_{\widetilde{\mathbf{x}}}\right) - \tfrac{i\theta}{2\overline{k}-\theta\widetilde{k}}\left(\tfrac{1}{4}\Delta_{\overline{\mathbf{x}}} - \tfrac{1}{\theta}\nabla_{\overline{\mathbf{x}}}\cdot\nabla_{\widetilde{\mathbf{x}}} + \tfrac{1}{\theta^2}\Delta_{\widetilde{\mathbf{x}}}\right)\right. \\ &\quad \left. + \tfrac{\left(\overline{k}^2 - \frac{\theta^2}{4}\widetilde{k}^2\right)\sigma^2\delta^2}{4\theta^2} R_0\left(\tfrac{\theta|\widetilde{\mathbf{x}}|}{\delta}\right) - \tfrac{\left(\overline{k}^2 + \frac{\theta^2}{4}\widetilde{k}^2\right)\sigma^2\delta^2}{4\theta^2} R_0(0)\right] \psi_1^\theta\overline{\psi_2^\theta}dz \\ &\quad + \tfrac{i\sigma\delta}{2\theta}\psi_1^\theta\overline{\psi_2^\theta}\left[\left(\overline{k} + \tfrac{\theta}{2}\widetilde{k}\right)d\mathcal{B}\left(z, \tfrac{\overline{\mathbf{x}}}{\delta} + \tfrac{\theta\widetilde{\mathbf{x}}}{2\delta}\right) - \left(\overline{k} - \tfrac{\theta}{2}\widetilde{k}\right)d\mathcal{B}\left(z, \tfrac{\overline{\mathbf{x}}}{\delta} - \tfrac{\theta\widetilde{\mathbf{x}}}{2\delta}\right)\right] \end{aligned} \tag{4.8}$$

and, using the smoothness of $\mathcal{B}$ and R_o in the transverse variables, we have further

$$d\mathcal{B}\left(z, \frac{\overline{\mathbf{x}}}{\delta} + \frac{\theta\widetilde{\mathbf{x}}}{2\delta}\right) - d\mathcal{B}\left(z, \frac{\overline{\mathbf{x}}}{\delta} - \frac{\theta\widetilde{\mathbf{x}}}{2\delta}\right) = \frac{\theta\widetilde{\mathbf{x}}}{\delta}\cdot\nabla_{\overline{\mathbf{x}}}d\mathcal{B}(z,\overline{\mathbf{x}}) + O\left(\frac{\theta}{\delta}\right)^2 \tag{4.9}$$

and

$$R_o\left(\frac{\theta\widetilde{\mathbf{x}}}{\delta}\right) = R_o(\mathbf{0}) + \frac{\theta}{\delta}\widetilde{\mathbf{x}}\cdot\nabla R_o(\mathbf{0}) + \frac{\theta^2}{2\delta^2}\sum_{i,j=1}^{2}\partial_{ij}^2 R_o(\mathbf{0})\tilde{x}_i\tilde{x}_j + O\left(\frac{\theta}{\delta}\right)^3. \tag{4.10}$$

The equation for W follows by Fourier transforming (4.8) in $\widetilde{\mathbf{x}}$ and $\widetilde{k}$, using the expansions (4.9) and (4.10) and letting $\theta/\delta \to 0$. We simplify the result by assuming that the fluctuations are isotropic so that $R_o(\mathbf{x}) = R_o(|\mathbf{x}|)$. This gives

$$\nabla R_o(\mathbf{0}) = \mathbf{0} \quad \text{and} \quad \partial_{ij}^2 R_o(\mathbf{0}) = R''(0)\delta_{ij} \tag{4.11}$$

and we obtain for W the Itô-Liouville equation

$$\begin{aligned} dW &= \left[\tfrac{\mathbf{q}}{\overline{k}}\cdot\nabla_{\overline{\mathbf{x}}} - \tfrac{|\mathbf{q}|^2}{2\overline{k}^2}\tfrac{\partial}{\partial r} + \tfrac{\overline{k}^2 D_\kappa}{2}\Delta_{\mathbf{q}} + \tfrac{\delta^2 D_r}{2}\tfrac{\partial^2}{\partial r^2}\right] W dz \\ &\quad + \tfrac{\sigma\overline{k}}{2}\nabla_{\mathbf{q}}W\cdot\nabla_{\overline{\mathbf{x}}}d\mathcal{B}\left(z,\tfrac{\overline{\mathbf{x}}}{\delta}\right) - \tfrac{\sigma\delta}{2}\tfrac{\partial W}{\partial r}d\mathcal{B}\left(z,\tfrac{\overline{\mathbf{x}}}{\delta}\right), \quad z > 0 \end{aligned} \tag{4.12}$$

with initial condition

$$W(z=0,\overline{\mathbf{x}},\overline{k},\mathbf{q},r) = W_o(\overline{\mathbf{x}},\overline{k},\mathbf{q},r) \tag{4.13}$$

and with the positive diffusion coefficients

$$D_\kappa = -\frac{\sigma^2 R_o''(0)}{4} \quad \text{and} \quad D_r = \frac{\sigma^2 R_o(0)}{4}. \tag{4.14}$$

Equation (4.12) was also derived in Ref. [22] and we note that it is a stochastic flow equation [14, 17] that is the starting point of the analysis in this paper. We want to study the limit of the process W as $\delta \to 0$.

4.3. *Statement of the strong lateral diversity limit*

Now that we have the Itô-Liouville equation (4.12), we emphasize the dependence of the process on δ by writing $W^\delta(z, \overline{\mathbf{x}}, \overline{k}, \mathbf{q}, r)$. We assume that the initial condition W_o is independent of δ.

The mean $\mathcal{W}^\delta = E\{W^\delta\}$ is considered in section 4.4 and it follows, as is easily seen from (4.12), that as $\delta \to 0$, $\mathcal{W}^\delta$ converges weakly to the solution $\mathcal{W}$ of the phase space advection-diffusion equation

$$\frac{\partial \mathcal{W}}{\partial z} = \mathcal{L}\mathcal{W} \tag{4.15}$$

with initial conditions

$$\mathcal{W}(0, \overline{\mathbf{x}}, \overline{k}, \mathbf{q}, r) = W_o(\overline{\mathbf{x}}, \overline{k}, \mathbf{q}, r), \tag{4.16}$$

where

$$\mathcal{L} = \frac{\mathbf{q}}{\overline{k}} \cdot \nabla_{\overline{\mathbf{x}}} - \frac{|\mathbf{q}|^2}{2\overline{k}^2} \frac{\partial}{\partial r} + \frac{\overline{k}^2 D_\kappa}{2} \Delta_{\mathbf{q}}. \tag{4.17}$$

This deterministic equation is solved explicitly in Ref. [22].

However, the point-wise variance of W^δ is not zero for any δ and it does not vanish as $\delta \to 0$. This means that W^δ is randomly fluctuating and it does not converge to a deterministic process as $\delta \to 0$ in the strong, point-wise sense. Nevertheless, we do have convergence in a weak sense as follows.

Theorem 4.1. *Suppose that W_o is in L^2 and it does not depend on δ. Then, given any smooth and rapidly decaying test function $\varphi(\overline{\mathbf{x}}, \overline{k}, \mathbf{q}, r)$, we have that*

$$< W^\delta, \varphi > (z) = \int d\overline{\mathbf{x}} \int d\overline{k} \int d\mathbf{q} \int dr\, \varphi(\overline{\mathbf{x}}, \overline{k}, \mathbf{q}, r) W^\delta(z, \overline{\mathbf{x}}, \overline{k}, \mathbf{q}, r) \tag{4.18}$$

converges in probability as $\delta \to 0$ to $< \mathcal{W}, \varphi > (z)$, for any $z > 0$.

Theorem 4.1 is proved in section 5. It states that even though W^δ does not have a deterministic point-wise limit, it is weakly self-averaging. That is, smooth linear functionals of W^δ become deterministic in the limit $\delta \to 0$. This is the property that can be exploited in applications such as imaging in random media, as we explain in section 6.

4.4. *The mean space-time Wigner transform*

Taking expectations in (4.12) we get that $\mathcal{W}^\delta(z,\overline{\mathbf{x}},\mathbf{q},r) = E\{W^\delta(z,\overline{\mathbf{x}},\mathbf{q},r)\}$ satisfies the phase space advection-diffusion equation

$$\frac{\partial \mathcal{W}^\delta}{\partial z} = \mathcal{L}_\delta \mathcal{W}, \tag{4.19}$$

with initial condition $W_o(\overline{\mathbf{x}},\overline{k},\mathbf{q},r)$, where

$$\mathcal{L}_\delta = \mathcal{L} + \frac{\delta^2 D_r}{2}\frac{\partial^2}{\partial r^2}. \tag{4.20}$$

Equivalently, $\mathcal{W}^\delta$ is given as an expectation

$$\mathcal{W}^\delta(z,\overline{\mathbf{x}},\boldsymbol{\kappa},r) = E\left\{W_o\left(\mathbf{X}^\delta(z),\overline{k},\mathbf{Q}^\delta(z),\mathcal{R}^\delta(z)\right)\right\}, \tag{4.21}$$

where $\left\{\mathbf{X}^\delta(z),\mathbf{Q}^\delta(z),\mathcal{R}^\delta(z)\right\}$ is the Itô diffusion process with generator $\mathcal{L}_\delta$ and with initial condition

$$\mathbf{X}^\delta(0) = \overline{\mathbf{x}}, \quad \mathbf{Q}^\delta(0) = \mathbf{q}, \quad \mathcal{R}^\delta(0) = r. \tag{4.22}$$

For $z > 0$, the Itô stochastic differential equations are

$$\begin{aligned} d\mathbf{X}^\delta(z) &= \frac{1}{\overline{k}}\mathbf{Q}^\delta(z)dz, \\ dQ_j{}^\delta(z) &= \overline{k}\sqrt{D_\kappa}\, d\mathcal{B}_j(z), \quad j=1,2, \quad \mathbf{Q}^\delta = \left(Q_1^\delta, Q_2^\delta\right), \\ d\mathcal{R}^\delta(z) &= -\frac{|\mathbf{Q}^\delta(z)|^2}{2\overline{k}^2}dz - \delta\sqrt{D_r}d\mathcal{B}(z), \end{aligned} \tag{4.23}$$

where the driving is with three independent standard Brownian motions $\mathcal{B}(z)$ and $\mathcal{B}_j(z)$, for $j = 1, 2$. The same process $\left\{\mathbf{X}^\delta(z),\mathbf{Q}^\delta(z),\mathcal{R}^\delta(z)\right\}$ also determines expectations of higher powers of W^δ

$$E\left\{\left|W^\delta(z,\overline{\mathbf{x}},\mathbf{q},r)\right|^n\right\} = E\left\{\left|W_o\left(\mathbf{X}^\delta(z),\overline{k},\mathbf{Q}^\delta(z),\mathcal{R}^\delta(z)\right)\right|^n\right\}, \quad n \geq 1. \tag{4.24}$$

As $\delta \to 0$, we see that $\mathcal{W}^\delta$ converges to the solution of (4.15), computed explicitly in Ref. [22]. Actually, all one point moments converge as $\delta \to 0$,

$$E\left\{\left|W^\delta(z,\overline{\mathbf{x}},\mathbf{q},r)\right|^n\right\} \to E\left\{\left|W_o\left(\mathbf{X}(z),\overline{k},\mathbf{Q}(z),\mathcal{R}(z)\right)\right|^n\right\}, \quad n \geq 1, \tag{4.25}$$

where $\{\mathbf{X}(z),\mathbf{Q}(z),\mathcal{R}(z)\}$ is the δ independent Itô diffusion

$$\begin{aligned} d\mathbf{X}(z) &= \frac{1}{\overline{k}}\mathbf{Q}(z)dz, \\ dQ_j(z) &= \overline{k}\sqrt{D_\kappa}\, d\mathcal{B}_j(z), \quad j=1,2, \quad \mathbf{Q} = (Q_1,Q_2), \\ d\mathcal{R}(z) &= -\frac{|\mathbf{Q}(z)|^2}{2\overline{k}^2}dz, \quad z > 0, \end{aligned} \tag{4.26}$$

with initial conditions

$$\mathbf{X}(0) = \overline{\mathbf{x}}, \quad \mathbf{Q}(0) = \mathbf{q} \text{ and } \mathcal{R}(0) = r. \tag{4.27}$$

Clearly, W^δ does not have a point-wise deterministic limit because the limit variance is not zero

$$E\left\{\left|W_o\left(\mathbf{X}(z),\overline{k},\mathbf{Q}(z),\mathcal{R}(z)\right)\right|^2\right\} - \left|E\left\{W_o\left(\mathbf{X}(z),\overline{k},\mathbf{Q}(z),\mathcal{R}(z)\right)\right\}\right|^2 \neq 0. \quad (4.28)$$

5. Self-averaging of the smoothed space-time Wigner transform, in the strong lateral diversity limit

In this section we prove Theorem 4.1. We begin by calculating the form of the infinitesimal generator $\mathcal{A}_\delta$ of Itô-Liouville process $W^\delta(z,\overline{\mathbf{x}},\overline{k},\boldsymbol{\kappa},r)$, considered as a process in the space of continuous functions in z with values in the space $\mathcal{S}'$ of Schwartz distributions w over $\mathbb{R}^2\times\mathbb{R}\times\mathbb{R}^2\times\mathbb{R}$.

Let F be a real valued test function on $\mathbb{R}$ and define for each test function φ in $\mathcal{S}$ over $\mathbb{R}^2\times\mathbb{R}\times\mathbb{R}^2\times\mathbb{R}$ the function $f(w)$ by

$$f(w) = F(< w,\varphi >). \quad (5.1)$$

We have that

$$\begin{aligned}\mathcal{A}_\delta f(w) &= \frac{d}{dz}E\{F(< W^\delta(z),\varphi >)|W^\delta(0)=w\}|_{z=0}\\ &= \mathcal{D}^\delta(w)F'(< w,\varphi >) + \mathcal{M}^\delta(w)F''(< w,\varphi >)\end{aligned} \quad (5.2)$$

where

$$\mathcal{D}^\delta(w) =< w,\mathcal{L}_\delta^\star\varphi >, \quad (5.3)$$

with $\mathcal{L}_\delta^\star$ the adjoint of $\mathcal{L}_\delta$ in (4.20). The $\mathcal{M}^\delta$ can be written as the sum of three terms

$$\begin{aligned}\mathcal{M}_1^\delta(w) = \frac{\sigma^2\delta^2}{8}&\int d\overline{\mathbf{x}}\int d\overline{k}\int d\mathbf{q}\int dr\int d\overline{\mathbf{x}}'\int d\overline{k}'\int d\mathbf{q}'\int dr'\, w(\overline{\mathbf{x}},\overline{k},\mathbf{q},r)w(\overline{\mathbf{x}}',\overline{k}',\mathbf{q}',r')\\ &\times R_o\left(\frac{\overline{\mathbf{x}}-\overline{\mathbf{x}}'}{\delta}\right)\frac{\partial\varphi(\overline{\mathbf{x}},\overline{k},\mathbf{q},r)}{\partial r}\frac{\partial\varphi(\overline{\mathbf{x}}',\overline{k}',\mathbf{q}',r')}{\partial r'},\end{aligned} \quad (5.4)$$

$$\begin{aligned}&\mathcal{M}_2^\delta(w)\\ &= -\frac{\sigma^2\delta}{4}\int d\overline{\mathbf{x}}\int d\overline{k}\int d\mathbf{q}\int dr\int d\overline{\mathbf{x}}'\int d\overline{k}'\int d\mathbf{q}'\int dr'\, w(\overline{\mathbf{x}},\overline{k},\mathbf{q},r)w(\overline{\mathbf{x}}',\overline{k}',\mathbf{q}',r')\overline{k}'\times\\ &\nabla_{\mathbf{x}'}R_o\left(\frac{\overline{\mathbf{x}}-\overline{\mathbf{x}}'}{\delta}\right)\cdot\nabla_{\mathbf{q}'}\varphi(\overline{\mathbf{x}}',\overline{k}',\mathbf{q}',r')\frac{\partial\varphi(\overline{\mathbf{x}},\overline{k},\mathbf{q},r)}{\partial r},\end{aligned} \quad (5.5)$$

$$\begin{aligned}&\mathcal{M}_3^\delta(w)\\ &= -\frac{\sigma^2}{8}\int d\overline{\mathbf{x}}\int d\overline{k}\int d\mathbf{q}\int dr\int d\overline{\mathbf{x}}'\int d\overline{k}'\int d\mathbf{q}'\int dr'\, w(\overline{\mathbf{x}},\overline{k},\mathbf{q},r)w(\overline{\mathbf{x}}',\overline{k}',\mathbf{q}',r')\overline{k}\overline{k}'\times\\ &\sum_{j,l=1}^{2}\partial_{lj}^2R_o\left(\frac{\overline{\mathbf{x}}-\overline{\mathbf{x}}'}{\delta}\right)\frac{\partial\varphi(\overline{\mathbf{x}},\overline{k},\mathbf{q},r)}{\partial q_j}\frac{\partial\varphi(\overline{\mathbf{x}}',\overline{k}',\mathbf{q}',r')}{\partial q_l'}.\end{aligned} \quad (5.6)$$

Now we get from (5.3) that as $\delta \to 0$,

$$\lim_{\delta \to 0} \mathcal{D}^\delta(w) = \mathcal{D}(w) =< w, \mathcal{L}^\star \varphi >, \tag{5.7}$$

uniformly for w bounded in L^2. Here $\mathcal{L}^\star$ is the adjoint of $\mathcal{L}$ defined by (4.17). Furthermore,

$$\lim_{\delta \to 0} \mathcal{M}_j^\delta(w) = 0, \quad \text{for } j = 1, 2, 3,$$

uniformly for w bounded in L^2, as we show next.

From (5.4)-(5.6) we see that it is enough to show that $\mathcal{M}_3^\delta(w) \to 0$. By the Cauchy-Schwartz inequality, we have

$$|\mathcal{M}_3^\delta(w)| \leq \|w\|_{L^2}^2 J^\delta(\varphi), \tag{5.8}$$

where

$$[J^\delta(\varphi)]^2 = \left(\frac{\sigma^2}{8}\right)^2 \int d\overline{\mathbf{x}} \int d\overline{k} \int d\mathbf{q} \int dr \int d\overline{\mathbf{x}}' \int d\overline{k}' \int d\mathbf{q}' \int dr'$$
$$\left[\overline{k}\overline{k}' \sum_{j,l=1}^{2} \partial_{lj}^2 R_o \left(\frac{\overline{\mathbf{x}} - \overline{\mathbf{x}}'}{\delta}\right) \frac{\partial \varphi(\overline{\mathbf{x}}, \overline{k}, \mathbf{q}, r)}{\partial q_j} \frac{\partial \varphi(\overline{\mathbf{x}}', \overline{k}', \mathbf{q}', r')}{\partial q_l'}\right]^2 .$$

Since R_o is rapidly decaying at infinity, we see that for any fixed test function φ, $J^\delta(\varphi)$ tends to zero as $\delta \to 0$.

We have shown therefore that for functions $f(w)$ of the form (5.1), with φ in $\mathcal{S}$ fixed and uniformly for w bounded in L^2,

$$\mathcal{A}^\delta f(w) \to \mathcal{A} f(w) = \mathcal{D}(w) F'(< w, \varphi >), \tag{5.9}$$

where

$$\mathcal{D}(w) =< w, \mathcal{L}^\star \varphi > . \tag{5.10}$$

The operator $\mathcal{A}$ is the generator of the deterministic process $\mathcal{W}(z, \overline{\mathbf{x}}, \overline{k}, \mathbf{q}, r)$, that is the solution of (4.15). Since the limit process is deterministic, it follows that convergence in law implies convergence in probability, weakly in $\mathcal{S}'$. It also follows that functions of the form (5.1) are sufficient again since the limit is deterministic [14]. The moment condition needed for tightness for processes in $C([0, Z], \mathcal{S}')$ or in $D([0, Z], \mathcal{S}')$ [26] is easily obtained as in Ref. [20], and we omit it here.

6. Application to imaging

In this section we consider applications to imaging a source in a random medium from measurements of the wave field P at an array of transducers. The setup is shown in Figure 3.1, where we introduce a new coordinate system, with scaled range $\zeta = L - z$ measured from the array and with cross range (transverse) coordinates $\mathbf{x}$ defined with respect to the center $\vec{\mathbf{y}}^\star = (L, \mathbf{0})$ of the source, which can be small or

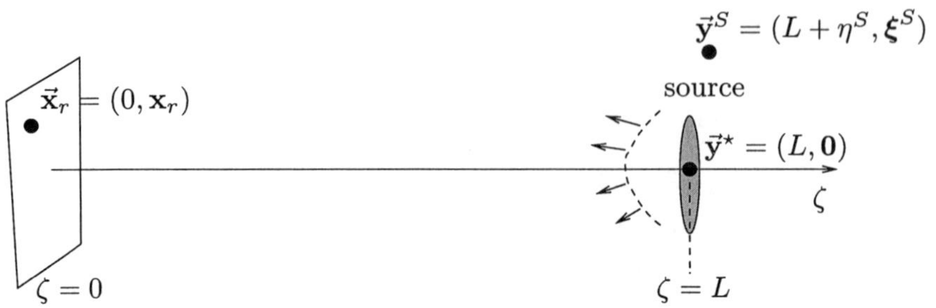

Fig. 3.1. Setup for imaging a distributed source with a planar array of transducers

spatially distributed. The array is in the plane $\zeta = 0$ and it consists of receivers at N discrete locations $\vec{\mathbf{x}}_r = (0, \mathbf{x}_r)$, where we record the traces $P(\vec{\mathbf{x}}_r, t)$ over a time window $t \in [0, T]$ that we suppose is long enough for

$$P(\vec{\mathbf{x}}_r, t) \approx 0 \quad \text{for } t > T$$

to hold. This allows us to simplify the analysis by neglecting the effect of a finite measurement time window.

The goal in imaging is to estimate the support of the source from the traces at the array and this is done very efficiently with Kirchhoff migration [27, 28], if there are no fluctuations of the sound speed. However, in random media Kirchhoff migration gives noisy and unpredictable results, in the sense that they lack statistical stability. As an alternative to Kirchhoff migration we introduced in Refs. [22, 23] a new, coherent interferometric (CINT) imaging functional, which is a statistically smoothed migration. The resolution analysis of CINT is given in Ref. [22]. Here we use the result of section 4.3 to show that it is statistically stable in the asymptotic regime (3.8).

6.1. *Migration*

In this and the following sections all variables are scaled as in section 3. The wave field $\widehat{P}$ at receiver location $\vec{\mathbf{x}}_r = (0, \mathbf{x}_r)$ is

$$\widehat{P}(\mathbf{x}_r, \omega) \approx e^{i(\delta/\epsilon)^2 k/\theta L} \psi(L, \mathbf{x}_r, k), \tag{6.1}$$

where we used that $k_o L_z = \delta^2/(\theta\epsilon^2)$ and we dropped the range coordinate in the argument of $\widehat{P}$, as it is always $\zeta = 0$ at the array. We also kept the definition $\psi = \psi(z, \mathbf{x}, k)$, with $z = L - \zeta$, which means that at the array the range coordinate in ψ is $z = L$. The parabolic amplitude ψ solves equation (3.5) with initial data $\psi_o(\mathbf{x}, k)$ and it depends on the three small parameters ϵ, θ and δ. In the previous sections we emphasized this dependence using superscripts, before taking limits. Here we don't use the superscripts to simplify notation and we keep ϵ, θ and δ fixed until the very end where we apply the theoretical results of section 4.3.

Classic Kirchhoff migration imaging [27, 28] estimates the support of the source by migrating (back propagating) the traces $P(\mathbf{x}_r, t)$ to search points $\vec{\mathbf{y}}^S = (L + \eta^S, \boldsymbol{\xi}^S)$, in a fictitious, homogeneous medium, with scaled sound speed $< c >= 1$ and by summing over the receivers. The scaled distance from $\vec{\mathbf{x}}_r$ to $\vec{\mathbf{y}}^S$ is

$$\left[(L+\eta^S)^2 + \left(\frac{L_x}{L_z}\right)^2 \left|\boldsymbol{\xi}^S - \mathbf{x}_r\right|^2\right]^{\frac{1}{2}} \approx L + \eta^S + \frac{\epsilon^2}{2\delta^2} \frac{\left|\boldsymbol{\xi}^S - \mathbf{x}_r\right|^2}{(L+\eta^S)} \tag{6.2}$$

and it equals the scaled travel time $\tau(\mathbf{x}_r, \vec{\mathbf{y}}^S)$, since the scaled mean speed is $< c >= 1$. This gives the migration phase

$$(k_o L_z)\omega\tau(\vec{\mathbf{x}}_r, \vec{\mathbf{y}}^S) \approx \frac{\delta^2}{\epsilon^2\theta} k(L+\eta^S) + \frac{k}{2\theta} \frac{\left|\boldsymbol{\xi}^S - \mathbf{x}_r\right|^2}{(L+\eta^S)} \tag{6.3}$$

and the migrated wave field to $\vec{\mathbf{y}}^S$

$$\widehat{P}(\mathbf{x}_r, \omega) e^{-i(k_o L_z)\omega\tau(\vec{\mathbf{x}}_r, \vec{\mathbf{y}}^S)} \approx \psi(L, \mathbf{x}_r, k) \exp\left\{-i\frac{\delta^2}{\epsilon^2}\frac{k}{\theta}\eta^S - i\frac{k}{2\theta}\frac{\left|\boldsymbol{\xi}^S - \mathbf{x}_r\right|^2}{(L+\eta^S)}\right\}. \tag{6.4}$$

The Kirchhoff migration image is given by

$$\mathcal{I}^{\mathrm{KM}}(\vec{\mathbf{y}}^S) = \sum_{r=1}^{N} \int d\omega\, \widehat{P}(\mathbf{x}_r, \omega) e^{-i(k_o L_z)\omega\tau(\vec{\mathbf{x}}_r, \vec{\mathbf{y}}^S)} \tag{6.5}$$

and, as shown in Refs. [21–23], it lacks statistical stability with respect to the realizations of the random medium and it gives noisy results that are difficult to interpret.

We consider next coherent interferometric imaging, which is a statistically smoothed version of migration [22, 23]. Before describing this method, let us make the assumption that the array receivers are placed on a square mesh, in a square aperture of area a^2. The scaled mesh size is h and we suppose that it is small, so we can write

$$\sum_{r=1}^{N} \approx \frac{1}{h^2}\int d\mathbf{x} \sim \int d\mathbf{x}, \tag{6.6}$$

with $\mathbf{x}$ varying continuously in the array aperture and with symbol $\sim$ denoting approximate, up to a multiplicative constant.

6.2. *Coherent interferometric imaging*

The coherent interferometric imaging technique was introduced in Refs. [22, 23] and it uses the coherence in the data traces $P(\mathbf{x}, t)$ to obtain reliable images in random media. There are two characteristic coherent parameters in the data:

- The decoherence frequency Ω_d, which is the difference in the frequencies ω_1 and ω_2 over which $\psi(z, \mathbf{x}, k_1)$ and $\psi(z, \mathbf{x}, k_2)$ become uncorrelated.
- The decoherence length X_d, which is the distance $|\mathbf{x}_1 - \mathbf{x}_2|$ over which $\psi(z, \mathbf{x}_1, k)$ and $\psi(z, \mathbf{x}_2, k)$ become uncorrelated.

These decoherence parameters depend on the statistics of the random medium and the range z and they are described in detail in the next section, for the asymptotic regime (3.8).

Coherent interferometry (CINT) is a migration technique that works with cross correlations of the traces, instead of the traces themselves. These cross correlations are computed locally over space-time windows of size $X_d \times \Omega_d$ and they are called coherent interferograms. We give in section 6.2.2 the mathematical expression of the CINT imaging function and then we study its statistical stability. The CINT functional and its resolution properties are motivated and analyzed in Refs. [22, 23].

6.2.1. *The decoherence length and frequency*

The decoherence length and frequency can be determined from the decay over $\widetilde{\mathbf{x}} = (\mathbf{x}_1 - \mathbf{x}_2)/\theta$ and $\widetilde{k} = (k_1 - k_2)/\theta$ of the expectation

$$\left\langle \psi\left(z, \overline{\mathbf{x}} + \frac{\theta\widetilde{\mathbf{x}}}{2}, \overline{k} + \frac{\theta\widetilde{k}}{2}\right) \overline{\psi\left(z, \overline{\mathbf{x}} - \frac{\theta\widetilde{\mathbf{x}}}{2}, \overline{k} - \frac{\theta\widetilde{k}}{2}\right)} \right\rangle,$$

which we calculated explicitly in Ref. [22], by solving equation (4.19). The moment formula is given by

$$\begin{aligned}
&\left\langle \psi\left(z, \overline{\mathbf{x}} + \frac{\theta\widetilde{\mathbf{x}}}{2}, \overline{k} + \frac{\theta\widetilde{k}}{2}\right) \overline{\psi\left(z, \overline{\mathbf{x}} - \frac{\theta\widetilde{\mathbf{x}}}{2}, \overline{k} - \frac{\theta\widetilde{k}}{2}\right)} \right\rangle \\
&\approx \frac{-\overline{k}^2 \varphi_1(z, \widetilde{k})}{4\pi^2 z^2} \exp\left\{ -\frac{\widetilde{k}^2 \delta^2 D_r z}{2} - \frac{\overline{k}^2 D_\kappa \varphi_2(z, \widetilde{k}) z |\widetilde{\mathbf{x}}|^2}{6} \right\} \\
&\int d\overline{\xi} \int d\tilde{\xi} \exp\left\{ \frac{i\widetilde{k}|\overline{\mathbf{x}} - \overline{\xi}|^2}{2z} + \frac{i\overline{k}}{z}(\overline{\mathbf{x}} - \overline{\xi}) \cdot (\widetilde{\mathbf{x}} - \tilde{\xi}) + \overline{k}^2 \varphi_3(z, \widetilde{k}) \widetilde{\mathbf{x}} \cdot \tilde{\xi} - \overline{k}^2 \varphi_4(z, \widetilde{k}) |\tilde{\xi}|^2 \right. \\
&\qquad \left. - \frac{\overline{k}^2 D_\kappa \varphi_2(z, \widetilde{k}) \varphi_5(z, \widetilde{k}) z}{6} \left[\widetilde{\mathbf{x}} \cdot \tilde{\xi} + \varphi_5(z, \widetilde{k}) |\tilde{\xi}|^2 \right] \right\} \\
&\qquad \psi_o\left(\overline{\xi} + \frac{\theta\tilde{\xi}}{2}, \overline{k} + \frac{\theta\widetilde{k}}{2}\right) \overline{\psi_o\left(\overline{\xi} - \frac{\theta\tilde{\xi}}{2}, \overline{k} - \frac{\theta\widetilde{k}}{2}\right)}.
\end{aligned} \tag{6.7}$$

This result is obtained in the white noise limit and the approximation involves a simplification of the exact formula for small θ. The coefficients in this moment formula (6.7) are given by

$$\varphi_1(z,\widetilde{k}) = \frac{z\sqrt{-i\widetilde{k}D_\kappa}}{\sinh^{1/2}\left(z\sqrt{-i\widetilde{k}D_\kappa}\right)}\coth^{1/2}\left(z\sqrt{-i\widetilde{k}D_\kappa}\right), \tag{6.8}$$

$$\varphi_2(z,\widetilde{k}) = \frac{3i}{\widetilde{k}zD_\kappa}\left(\frac{\sqrt{-i\widetilde{k}D_\kappa}}{\tanh(z\sqrt{-i\widetilde{k}D_\kappa})} - \frac{1}{z}\right), \tag{6.9}$$

$$\varphi_3(z,\widetilde{k}) = \frac{i}{2\widetilde{k}z}\left(\frac{3z\sqrt{-i\widetilde{k}D_\kappa}}{\sinh(z\sqrt{-i\widetilde{k}D_\kappa})} - \frac{1}{\cosh(z\sqrt{-i\widetilde{k}D_\kappa})} - 2\right), \tag{6.10}$$

$$\varphi_4(z,\widetilde{k}) = \frac{D_\kappa\tanh(z\sqrt{-i\widetilde{k}D_\kappa})}{2\sqrt{-i\widetilde{k}D_\kappa}}\left(1 - \frac{\tanh(z\sqrt{-i\widetilde{k}D_\kappa})}{z\sqrt{-i\widetilde{k}D_\kappa}}\right) \tag{6.11}$$

$$\varphi_5(z,\widetilde{k}) = \frac{1}{\cosh(z\sqrt{-i\widetilde{k}D_\kappa})}. \tag{6.12}$$

To determine the decoherence length we let $\widetilde{k}\to 0$ and then study the decay over $|\widetilde{\mathbf{x}}|$ of the right hand side in (6.7).

In the limit $\widetilde{k}\to 0$, we get from (6.8)-(6.12) that

$$\varphi_j(z,\widetilde{k}) = \begin{cases} 1 + O(\widetilde{k}), & j = 1,2,5 \\ O(\widetilde{k}^{1/2}), & j = 3,4 \end{cases} \tag{6.13}$$

and equation (6.7) simplifies to

$$\begin{aligned}&\left\langle \psi\left(z,\overline{\mathbf{x}}+\frac{\theta\widetilde{\mathbf{x}}}{2},\overline{k}\right)\overline{\psi\left(z,\overline{\mathbf{x}}-\frac{\theta\widetilde{\mathbf{x}}}{2},\overline{k}\right)}\right\rangle \approx \frac{-\overline{k}^2}{4\pi^2z^2}\exp\left\{-\frac{\overline{k}^2D_\kappa z|\widetilde{\mathbf{x}}|^2}{6}\right\}\\ &\int d\overline{\boldsymbol{\xi}}\int d\tilde{\boldsymbol{\xi}}\psi_o\left(\overline{\boldsymbol{\xi}}+\frac{\theta\tilde{\boldsymbol{\xi}}}{2},\overline{k}\right)\overline{\psi_o\left(\overline{\boldsymbol{\xi}}-\frac{\theta\tilde{\boldsymbol{\xi}}}{2},\overline{k}\right)}\\ &\exp\left\{\frac{i\overline{k}}{z}(\overline{\mathbf{x}}-\overline{\boldsymbol{\xi}})\cdot(\widetilde{\mathbf{x}}-\tilde{\boldsymbol{\xi}}) - \frac{\overline{k}^2D_\kappa z}{6}\left[\widetilde{\mathbf{x}}\cdot\tilde{\boldsymbol{\xi}}+|\tilde{\boldsymbol{\xi}}|^2\right]\right\},\end{aligned} \tag{6.14}$$

with the explicit integration depending on the spatial support of the wave source function ψ_o. For example, in the case of a spatially distributed source, where

$$\psi_o(\overline{\boldsymbol{\xi}}\pm\theta\tilde{\boldsymbol{\xi}}/2,\overline{k}) \approx \psi_o(\overline{\boldsymbol{\xi}},\overline{k}),$$

the integration over $\tilde{\boldsymbol{\xi}}$ gives

$$\left\langle \psi\left(z,\overline{\mathbf{x}}+\frac{\theta\widetilde{\mathbf{x}}}{2},\overline{k}\right)\overline{\psi\left(z,\overline{\mathbf{x}}-\frac{\theta\widetilde{\mathbf{x}}}{2},\overline{k}\right)}\right\rangle \approx \frac{-3}{2\pi z^3 D_\kappa}\exp\left\{-\frac{\overline{k}^2 D_\kappa z|\widetilde{\mathbf{x}}|^2}{2}\right\}\int d\overline{\boldsymbol{\xi}}\,|\psi_o(\overline{\boldsymbol{\xi}},\overline{k})|^2$$
$$\exp\left\{-\frac{3}{2z^3 D_\kappa}\left|\overline{\mathbf{x}}-\overline{\boldsymbol{\xi}}-\frac{i\overline{k}z^2 D_k}{2}\widetilde{\mathbf{x}}\right|^2\right\}.$$

In the case of a small source with support of $O(\theta)$, where

$$\psi_o \rightsquigarrow \theta^{-2}\psi_o\left(\frac{\overline{\boldsymbol{\xi}}}{\theta},\overline{k}\right),$$

we can let $\overline{\boldsymbol{\xi}} \rightsquigarrow \theta\overline{\boldsymbol{\xi}}$ in (6.14) and obtain

$$\left\langle \psi\left(z,\overline{\mathbf{x}}+\frac{\theta\widetilde{\mathbf{x}}}{2},\overline{k}\right)\overline{\psi\left(z,\overline{\mathbf{x}}-\frac{\theta\widetilde{\mathbf{x}}}{2},\overline{k}\right)}\right\rangle \sim \frac{-\overline{k}^2}{4\pi^2 z^2}\exp\left\{-\frac{\overline{k}^2 D_\kappa z|\widetilde{\mathbf{x}}|^2}{8}+\frac{i\overline{k}}{z}\overline{\mathbf{x}}\cdot\widetilde{\mathbf{x}}\right\}$$
$$\int d\overline{\boldsymbol{\xi}}\int d\tilde{\boldsymbol{\xi}}\,\psi_o\left(\overline{\boldsymbol{\xi}}+\frac{\tilde{\boldsymbol{\xi}}}{2},\overline{k}\right)\overline{\psi_o\left(\overline{\boldsymbol{\xi}}-\frac{\tilde{\boldsymbol{\xi}}}{2},\overline{k}\right)}\exp\left\{-\frac{\overline{k}^2 D_\kappa z|\widetilde{\mathbf{x}}|^2}{6}\left|\tilde{\boldsymbol{\xi}}+\frac{\widetilde{\mathbf{x}}}{2}\right|^2-\frac{i\overline{k}}{z}\tilde{\boldsymbol{\xi}}\cdot\overline{\mathbf{x}}\right\}.$$

In either case, the decay in $\widetilde{\mathbf{x}}$ occurs as a Gaussian function, with standard deviation of $O\left(\frac{1}{\overline{k}\sqrt{zD_\kappa}}\right)$. This means that the scaled decoherence length is

$$X_d(\overline{k}) \sim \frac{\theta}{\overline{k}\sqrt{zD_\kappa}} \tag{6.15}$$

and it corresponds to the scaled expected time reversal spot size derived in Refs. [15, 22]. From the analysis in Refs. [15, 18, 22] we know that the effective aperture is given, in scaled variables, by

$$a_e = \sqrt{D_\kappa z^3} \tag{6.16}$$

We can now write

$$X_d(\overline{k}) \sim \frac{\theta}{\overline{k}\kappa_d} \tag{6.17}$$

where

$$\kappa_d = \frac{a_e}{z} = \sqrt{D_\kappa z}. \tag{6.18}$$

The uncertainty in the direction of arrival of the waves in the random medium [23] is κ_d/θ.

Next, we estimate the decoherence frequency by setting $\widetilde{\mathbf{x}} \to 0$ in (6.7)

$$\left\langle \psi\left(z,\overline{\mathbf{x}},\overline{k}+\frac{\theta\tilde{k}}{2}\right)\overline{\psi\left(z,\overline{\mathbf{x}},\overline{k}-\frac{\theta\tilde{k}}{2}\right)}\right\rangle = \frac{-\overline{k}^2\varphi_1(z,\tilde{k})}{4\pi^2 z^2}\exp\left\{-\frac{\tilde{k}^2\delta^2 D_r z}{2}\right\}\int d\overline{\boldsymbol{\xi}}\int d\tilde{\boldsymbol{\xi}}$$
$$\exp\left\{\frac{i\tilde{k}|\overline{\mathbf{x}}-\overline{\boldsymbol{\xi}}|^2}{2z}-\frac{i\overline{k}}{z}(\overline{\mathbf{x}}-\overline{\boldsymbol{\xi}})\cdot\tilde{\boldsymbol{\xi}}-\overline{k}^2\varphi_4(z,\tilde{k})|\tilde{\boldsymbol{\xi}}|^2-\frac{\overline{k}^2 D_\kappa\varphi_2(z,\tilde{k})\varphi_5^2(z,\tilde{k})z|\tilde{\boldsymbol{\xi}}|^2}{6}\right\}$$
$$\psi_o\left(\overline{\boldsymbol{\xi}}+\frac{\theta\tilde{\boldsymbol{\xi}}}{2},\overline{k}+\frac{\theta\tilde{k}}{2}\right)\overline{\psi_o\left(\overline{\boldsymbol{\xi}}+\frac{\theta\tilde{\boldsymbol{\xi}}}{2},\overline{k}-\frac{\theta\tilde{k}}{2}\right)} \tag{6.19}$$

and by taking the large $\widetilde{k}$ approximation in (6.19). We obtain from (6.8)-(6.12) that

$$\varphi_1(z,\widetilde{k}) \approx z\sqrt{-2i\widetilde{k}D_\kappa} e^{-\frac{z}{2}\sqrt{\frac{\widetilde{k}D_\kappa}{2}}(1-i)}, \tag{6.20}$$

$$\varphi_j(z,\widetilde{k}) = O(\widetilde{k}^{-1/2}), \quad j = 2,4 \tag{6.21}$$

$$\varphi_5(z,\widetilde{k}) \approx e^{-z\sqrt{\frac{\widetilde{k}D_\kappa}{2}}(1-i)}, \tag{6.22}$$

which means that as $\widetilde{k}$ increases, the decay in (6.19) is determined by the factor

$$\exp\left\{-\frac{z}{2}\sqrt{\frac{\widetilde{k}D_\kappa}{2}} - \frac{\widetilde{k}^2\delta^2 D_r z}{2}\right\}$$

and therefore, that the scaled decoherence frequency is

$$\Omega_d \sim \min\left\{\frac{\theta}{z^2 D_\kappa}, \frac{\theta}{\delta\sqrt{D_r z}}\right\} \approx \frac{\theta}{z^2 D_\kappa} \quad \text{as } \delta \to 0. \tag{6.23}$$

In conclusion, both X_d and Ω_d are small, of order θ in our scaling, which means that we can cover many decoherence lengths with an array aperture of $O(1)$ and we can fit many frequency intervals of width Ω_d in a broad bandwidth $B/\omega_o = O(1)$. This is a key point for achieving the self-averaging property of the CINT imaging function discussed below.

6.2.2. *The coherent interferometric imaging function as a smoothed space-time Wigner transform*

Consider a smooth window $\chi(r;\rho)$ of length $O(\rho)$, with Fourier transform

$$\widehat{\chi}(\widetilde{k};\rho^{-1}) = \int \chi(r;\rho) e^{i\widetilde{k}r} dr, \tag{6.24}$$

supported in the wavenumber interval $|\widetilde{k}| \leq \rho^{-1}$, where

$$\rho \sim \frac{\theta}{\Omega_d} = O(1). \tag{6.25}$$

Let also $\Phi(\boldsymbol{\kappa};\kappa_d)$ be a smooth function of two dimensional vectors $\boldsymbol{\kappa}$, with support in a disk of radius $O(\kappa_d)$, with κ_d quantifying the uncertainty in the direction of arrival of the waves in the random medium, as explained in section 6.2.1. The Fourier transform of Φ is denoted by

$$\widehat{\Phi}(k\widetilde{\mathbf{x}};\kappa_d^{-1}) = \int \Phi(\boldsymbol{\kappa};\kappa_d) e^{-ik\boldsymbol{\kappa}\cdot\widetilde{\mathbf{x}}} d\boldsymbol{\kappa} \tag{6.26}$$

and it is supported in the disk $k|\widetilde{\mathbf{x}}| \leq \kappa_d^{-1}$.

Using the windows (6.24), (6.26) and the migrated wave field (6.4), we define the coherent interferometric imaging function [23]

$$\mathcal{I}^{\mathrm{CINT}}(\vec{\mathbf{y}}^S;\rho,\kappa_d) \sim \int d\overline{k} \int d\overline{\mathbf{x}} \int d\widetilde{k}\, \widehat{\chi}(\widetilde{k};\rho^{-1})\, e^{-i(\delta/\epsilon)^2 \widetilde{k}\eta^S} \int d\widetilde{\mathbf{x}}\, \widehat{\Phi}(\overline{k}\widetilde{\mathbf{x}};\kappa_d^{-1}) \psi\left(L,\overline{\mathbf{x}}+\frac{\theta\widetilde{\mathbf{x}}}{2},\overline{k}+\frac{\theta\widetilde{k}}{2}\right)\overline{\psi\left(L,\overline{\mathbf{x}}-\frac{\theta\widetilde{\mathbf{x}}}{2},\overline{k}-\frac{\theta\widetilde{k}}{2}\right)} \exp\left\{-i\frac{(\overline{k}+\frac{\theta\widetilde{k}}{2})}{2\theta}\frac{\left|\boldsymbol{\xi}^S-\overline{\mathbf{x}}-\frac{\theta\widetilde{\mathbf{x}}}{2}\right|^2}{(L+\eta^S)}+i\frac{(\overline{k}-\frac{\theta\widetilde{k}}{2})}{2\theta}\frac{\left|\boldsymbol{\xi}^S-\overline{\mathbf{x}}+\frac{\theta\widetilde{\mathbf{x}}}{2}\right|^2}{(L+\eta^S)}\right\} \tag{6.27}$$

that becomes after simplifying the exponent,

$$\mathcal{I}^{\mathrm{CINT}}(\vec{\mathbf{y}}^S;\rho,\kappa_d) \sim \int d\overline{k} \int d\overline{\mathbf{x}} \int d\widetilde{k}\, \widehat{\chi}(\widetilde{k};\rho^{-1})\, e^{-i(\delta/\epsilon)^2 \widetilde{k}\eta^S} \int d\widetilde{\mathbf{x}}\, \widehat{\Phi}(\overline{k}\widetilde{\mathbf{x}};\kappa_d^{-1}) \psi\left(L,\overline{\mathbf{x}}+\frac{\theta\widetilde{\mathbf{x}}}{2},\overline{k}+\frac{\theta\widetilde{k}}{2}\right)\overline{\psi\left(L,\overline{\mathbf{x}}-\frac{\theta\widetilde{\mathbf{x}}}{2},\overline{k}-\frac{\theta\widetilde{k}}{2}\right)} \exp\left\{-i\overline{k}\widetilde{\mathbf{x}}\cdot\frac{\left(\overline{\mathbf{x}}-\boldsymbol{\xi}^S\right)}{(L+\eta^S)}-i\widetilde{k}\frac{\left|\boldsymbol{\xi}^S-\overline{\mathbf{x}}\right|^2}{2(L+\eta^S)}\right\}. \tag{6.28}$$

Now note that because the support of $\chi(r;\rho)$ is $O(1)$, the imaging function is nonzero if the range offset satisfies

$$\eta^S \le O\left(\epsilon^2/\delta^2\right) \ll 1.$$

We set then $\eta^S \rightsquigarrow \frac{\epsilon^2}{\delta^2}\eta^S$ and write approximately for $\vec{\mathbf{y}}^S = \left(L+\frac{\epsilon^2}{\delta^2}\eta^S,\boldsymbol{\xi}^S\right)$,

$$\mathcal{I}^{\mathrm{CINT}}(\vec{\mathbf{y}}^S;\rho,\kappa_d) \sim \int d\overline{k} \int d\overline{\mathbf{x}} \int d\widetilde{k}\, \widehat{\chi}(\widetilde{k};\rho^{-1}) \int d\widetilde{\mathbf{x}}\, \widehat{\Phi}(\overline{k}\widetilde{\mathbf{x}};\kappa_d^{-1}) \psi\left(L,\overline{\mathbf{x}}+\frac{\theta\widetilde{\mathbf{x}}}{2},\overline{k}+\frac{\theta\widetilde{k}}{2}\right)\overline{\psi\left(L,\overline{\mathbf{x}}-\frac{\theta\widetilde{\mathbf{x}}}{2},\overline{k}-\frac{\theta\widetilde{k}}{2}\right)} \exp\left\{i\overline{k}\widetilde{\mathbf{x}}\cdot\frac{\left(\boldsymbol{\xi}^S-\overline{\mathbf{x}}\right)}{L}-i\widetilde{k}\left(\eta^S+\frac{\left|\boldsymbol{\xi}^S-\overline{\mathbf{x}}\right|^2}{2L}\right)\right\}. \tag{6.29}$$

Next, we use the definition (4.4) of the Wigner transform in (6.29) and obtain

$$\mathcal{I}^{\mathrm{CINT}}(\vec{\mathbf{y}}^S;\rho,\kappa_d) \sim \int d\overline{k} \int d\overline{\mathbf{x}} \int d\mathbf{q} \int dr\, W(L,\overline{\mathbf{x}},\overline{k},\mathbf{q},r)$$
$$\int d\widetilde{k}\, \widehat{\chi}(\widetilde{k};\rho^{-1}) \exp\left[-i\widetilde{k}\left(\eta^S + \frac{\left|\boldsymbol{\xi}^S - \overline{\mathbf{x}}\right|^2}{2L} - r\right)\right] \tag{6.30}$$
$$\int d\widetilde{\mathbf{x}}\, \widehat{\Phi}(\overline{k}\widetilde{\mathbf{x}};\kappa_d^{-1}) \exp\left[i\overline{k}\widetilde{\mathbf{x}}\cdot\frac{\boldsymbol{\xi}^S - \overline{\mathbf{x}}}{L} - i\mathbf{q}\cdot\widetilde{\mathbf{x}}\right].$$

Finally, changing variables $\mathbf{q} = \overline{k}\boldsymbol{\kappa}$, we get

$$\mathcal{I}^{\mathrm{CINT}}(\vec{\mathbf{y}}^S;\rho,\kappa_d) \sim \int d\overline{\mathbf{x}} \int d\boldsymbol{\kappa} \int dr\, \chi\left(\eta^S + \frac{\left|\boldsymbol{\xi}^S - \overline{\mathbf{x}}\right|^2}{2L} - r;\rho\right)$$
$$\Phi\left(\frac{\boldsymbol{\xi}^S - \overline{\mathbf{x}}}{L} - \boldsymbol{\kappa};\kappa_d\right) \int d\overline{k}\, W(L,\overline{\mathbf{x}},\overline{k},\overline{k}\boldsymbol{\kappa},r). \tag{6.31}$$

In conclusion, the coherent imaging function is given by the Wigner transform, smoothed by convolution over directions $\boldsymbol{\kappa}$ and range r and by integration over the array locations $\overline{\mathbf{x}}$ and wavenumbers $\overline{k}$. The self-averaging of $\mathcal{I}^{\mathrm{CINT}}$ follows from Theorem 4.1.

Acknowledgments

The work of L. Borcea was partially supported by the Office of Naval Research, under grant N00014-02-1-0088 and by the National Science Foundation, grants DMS-0604008, DMS-0305056, DMS-0354658. It was also supported by INRIA in the group POEMS of P. Joly. The work of G. Papanicolaou was supported by grants ONR N00014-02-1-0088, 02-SC-ARO-1067-MOD 1 and NSF DMS-0354674-001. The work of C. Tsogka was partially supported by the Office of Naval Research, under grant N00014-02-1-0088 and by 02-SC-ARO-1067-MOD 1.

References

[1] F. Tappert. The parabolic approximation method. In *Wave Propagation and Underwater Acoustics*, vol. 70, *Lecture Notes in Physics*, pp. 224–287. Springer-Verlag, Berlin, (1977).

[2] V. I. Tatarskii, A. Ishimaru, and V. U. Zavorotny, *Wave Propagation in Random Media (Scintillation)*. (SPIE and IOP, 1993).

[3] B. J. Uscinski, *Wave Propagation and Scattering*. (Oxford Clarendon Press, 1986).

[4] J. Fouque, G. Papanicolaou, and Y. Samuelides, Forward and markov approximation: The strong intensity fluctuations regime revisited, *Waves in Random Media*. **8**, 303–314, (1998).

[5] D. A. Dawson and G. Papanicolaou, A random wave process, *Appl. Math. Optim.* **12**, 97–114, (1984).

[6] F. Bailly, J. Clouet, and J. Fouque, Parabolic and white noise approximation for waves in random media, *SIAM Jour. Appl. Math.* **56**, 1445–1470, (1996).

[7] R. Z. Khasminskii, A limit theorem for solutions of differential equations with random right hand side, *Theory Probab. Appl.* **11**, 390–406, (1966).

[8] G. Blankenship and G. C. Papanicolaou, Stability and control of stochastic systems with wide-band noise disturbances, *SIAM J. Appl. Math.* **34**, 437–476, (1978).

[9] H. Kushner, *Approximation and Weak Convergence Methods for Random Processes, with Applications to Stochastic Systems Theory.* (MIT Press, 1984).

[10] R. Bouc and E. Pardoux, Asymptotic analysis of pdes with wide-band noise disturbances and expansion of the moments, *Stochastic Analysis and Applications.* **2**, 369–422, (1984).

[11] A. Fannjiang, White-noise and geometrical optics limits of wigner-moyal equation for wave beams in turbulent media, *Comm. Math. Phys.* **254**, 289–322, (2005).

[12] A. Fannjiang, White-noise and geometrical optics limits of wigner-moyal equation for beam waves in turbulent media. ii. two-frequency formulation, *J. Stat. Phys.* **120**, 543–586, (2005).

[13] D. A. Dawson. Measure-valued markov processes. In ed. P. L. Hennequin, *Ecole d' Ete de Probabilite de Saint-Flour XXI—1991. [Saint-Flour Summer School on Probability Theory XXI—1991]*, Lecture Notes in Mathematics, 1541. Springer-Verlag, (1993).

[14] B. L. Rozovkii, *Stochastic Evolution Systems: Linear Theory and Applications to Non-linear Filtering.* (Kluwer Academic Press, 1990).

[15] G. Papanicolaou, L. Ryzhik, and K. Solna, Statistical stability in time reversal, *SIAM J. Applied Mathematics.* **64**(4), 1133–1155, (2004).

[16] G. Papanicolaou, L. Ryzhik, and K. Solna, Self-averaging from lateral diversity in the Itô-Schrödinger equation, *SIAM J. Multiscale Modeling and Simulation.* (2007). To appear.

[17] H. Kunita, *Stochastic Flows and Stochastic Differential Equations.* vol. 24, *Cambridge Stud. Adv. Math.*, (Cambridge University Press, UK, 1997).

[18] P. Blomgren, G. Papanicolaou, and H. Zhao, Super-resolution in time-reversal acoustics, *Journal of the Acoustical Society of America.* **111**, 238–248, (2002).

[19] G. Bal, On the self-averaging of wave energy in random media, *Multiscale Model. Simul.* **2**, 398–420, (2004).

[20] G.Bal, G. Papanicolaou, and L. Ryzhik, Self-averaging in time reversal for the parabolic wave equation, *Stoch. Dyn.* **2**, 507–531, (2002).

[21] L. Borcea, G. Papanicolaou, and C. Tsogka, Theory and applications of time reversal and interferometric imaging, *Inverse Problems.* **19**, S134–164, (2003).

[22] L. Borcea, G. Papanicolaou, and C. Tsogka, Interferometric array imaging in clutter, *Inverse Problems.* **21**(4), 1419–1460, (2005).

[23] L. Borcea, G. Papanicolaou, and C. Tsogka, Adaptive interferometric imaging in clutter and optimal illumination, *Inverse Problems.* **22**(4), 1405–1436, (2006).

[24] L. Borcea, G. C. Papanicolaou, C. Tsogka, and J. Berryman, Imaging and time reversal in random media, *Inverse Problems.* **18**(5), 1247–1279, (2002).

[25] G.Bal and L. Ryzhik, Time reversal and refocusing in random media, *SIAM Jour. Appl. Math.* **63**, 1475–1498, (2003).

[26] J. Fouque, La convergence en loi pour les processus a valeurs dans un espace nucleaire, *Ann. Inst. H. Poincare Probab. Statist.* **20**, 225–245, (1984).

[27] J. F. Claerbout and S. M. Doherty, Downward continuation of moveout-corrected seismograms, *Geophysics.* **37**(5), 741–768, (1972).

[28] G. Beylkin and R. Burridge, Linearized inverse scattering problems in acoustics and elasticity, *Wave Motion.* **12**(1), 15–52, (1990).

Chapter 4

The Korteweg-de Vries Equation with Multiplicative Homogeneous Noise

Anne de Bouard and Arnaud Debussche*
CNRS et Université Paris-Sud,
UMR 8628, Bât. 425, 91405 ORSAY CEDEX, FRANCE
Anne.deBouard@math.u-psud.fr.

We prove the global existence and uniqueness of solutions both in the energy space and in the space of square integrable functions for a Korteweg-de Vries equation with noise. The noise is multiplicative, white in time, and is the multiplication by the solution of a homogeneous noise in the space variable.

Contents

1. Introduction and statement of the results

The aim of the paper is to prove the global existence and uniqueness of strong solutions for a Korteweg-de Vries equation with noise, which may be written in Itô form as

$$du + (\partial_x^3 u + \frac{1}{2}\partial_x(u^2))dt = u\varphi dW \tag{1.1}$$

where u is a random process defined on $(t, x) \in \mathbb{R}^+ \times \mathbb{R}$, W is a cylindrical Wiener process on $L^2(\mathbb{R})$ and φ is a convolution operator on $L^2(\mathbb{R})$ defined by

$$\varphi f(x) = \int_{\mathbb{R}} k(x-y)f(y)dy, \text{ for } f \in L^2(\mathbb{R})$$

where the convolution kernel k is an $H^1(\mathbb{R}) \cap L^1(\mathbb{R})$ function of $x \in \mathbb{R}$. Here $H^1(\mathbb{R})$ is the usual Sobolev space of square integrable functions of the space variable x, having their first order derivative in $L^2(\mathbb{R})$. Considering a complete orthonormal

*ENS de Cachan, Antenne de Bretagne, Campus de Ker Lann, Av. R. Schuman, 35170 BRUZ, FRANCE, arnaud.debussche@bretagne.ens-cachan.fr

system $(e_i)_{i\in\mathbb{N}}$ in $L^2(\mathbb{R})$, we may alternatively write W as

$$W(t,x) = \sum_{i\in\mathbb{N}} \beta_i(t)\varphi e_i(x), \tag{1.2}$$

$(\beta_i)_{i\in\mathbb{N}}$ being an independent family of real valued Brownian motions. Hence, the correlation function of the process φW is

$$\mathbb{E}(\varphi W(t,x)\varphi W(s,y)) = c(x-y)(s\wedge t), \quad x,y\in\mathbb{R}, \quad s,t>0,$$

where

$$c(z) = \int_{\mathbb{R}} k(z+u)k(u)du.$$

The existence and uniqueness of solutions for stochastic KdV equations of the type (1.1) but with an additive noise have been studied in Refs. [1], [2], [3]. Here we extend those results to equation (1.1), that is the multiplicative case with homogeneous noise.

Note that an equation of this form, but with an additional weak dissipation has been considered in Ref. [4]. Indeed, in this latter case where a dissipative term is added, such a noise may be viewed as a perturbation of the dissipation. Although our existence and uniqueness results would easily extend to the case where weak dissipation is added, the dissipative term is of no help in the existence proof, so we prefer stating the result for equation (1.1).

Assuming $k \in H^1(\mathbb{R}) \cap L^1(\mathbb{R})$ will allow us to prove the global existence and uniqueness of solutions to equation (1.1) in the energy space $H^1(\mathbb{R})$, that is in the space where both quantities

$$m(u) = \frac{1}{2}\int_{\mathbb{R}} u^2(x)dx \tag{1.3}$$

and

$$H(u) = \frac{1}{2}\int_{\mathbb{R}} (\partial_x u)^2 dx - \frac{1}{6}\int_{\mathbb{R}} u^3 dx \tag{1.4}$$

are well defined. Note that m and H are conserved quantities for the equation without noise, that is

$$\partial_t u + \partial_x^3 u + \frac{1}{2}\partial_x(u^2) = 0. \tag{1.5}$$

It is important to solve equation (1.1) in the energy space, indeed most of the studies on the qualitative behavior of the solutions are done in this space. One of our aim in the future is to analyse the qualitative influence of a noise on a soliton solution of the deterministic equation, as we did in the additive case in Ref. [5], and this requires the use of the hamiltonian (1.4). However, our method of construction of solutions easily extends to treat the case of a kernel $k \in L^2(\mathbb{R}) \cap L^1(\mathbb{R})$, obtaining global existence and uniqueness in $L^2(\mathbb{R})$. It seems difficult to get a result with less regularity.

It may be noted also that the use of a noise of the form given in (1.1) naturally brings some localization in the noisy part of the equation, at least in the limit where the amplitude of the noise goes to zero, and when the initial state is a solitary wave – or soliton – solution of the deterministic equation, that is a well localized solution which propagates with a constant shape and velocity. This localization in the noise was a missing ingredient in the study of the influence of an additive noise on the propagation of a soliton (see Ref. [5]).

The precise existence result is the following, and the method we use to prove it closely follows the method in Ref. [2].

Theorem 1.1. *Assume that the kernel k of the noise satisfies $k \in H^s(\mathbb{R}) \cap L^1(\mathbb{R})$, $s = 0$ or 1. Then for any u_0 in $H^s(\mathbb{R})$, there is a unique adapted solution u with paths almost surely in $C(\mathbb{R}^+; H^s(\mathbb{R}))$ of equation (1.1). Moreover, $u \in L^2(\Omega; C(\mathbb{R}^+; L^2(\mathbb{R})))$.*

As in Refs. [2, 3], we use the functional framework introduced by Bourgain to study dispersive equations. Following [6], [7], [8], for $s, b \in \mathbb{R}$, $X_{b,s}$ denotes the space of tempered distributions $f \in \mathcal{S}'(\mathbb{R}^2)$ for which the norm

$$\|f\|_{X_{b,s}} = \left(\int\int_{\mathbb{R}^2} (1+|\xi|)^{2s}(1+|\tau-\xi^3|)^{2b}|\widehat{f}(\tau,\xi)|^2 d\tau d\xi\right)^{1/2}$$

is finite, where $\widehat{f}(\tau,\xi)$ stands for the space-time Fourier transform of $f(t,x)$. In the same way, we set for $b, s_1, s_2 \in \mathbb{R}$,

$$\|f\|_{\widetilde{X}_{b,s_1,s_2}} = \left(\int\int_{\mathbb{R}^2} |\xi|^{2s_2}(1+|\xi|)^{2s_1}(1+|\tau-\xi^3|)^{2b}|\widehat{f}(\tau,\xi)|^2 d\tau d\xi\right)^{1/2}$$

and

$$\widetilde{X}_{b,s_1,s_2} = \{f \in \mathcal{S}'(\mathbb{R}^2),\ \|f\|_{\widetilde{X}_{b,s_1,s_2}} < +\infty\}.$$

Note that the use of the space $\widetilde{X}_{b,s_1,s_2}$ is necessary here. Indeed, since we work with stochastic equations driven by white in time noises, we cannot require too much time regularity, and we have to choose $0 < b < 1/2$. But then the bilinear estimate which allows to treat the KdV equation is not true in the space $X_{b,s}$ as was already mentioned for the additive case in Ref. [2].

For $T \geq 0$, we also introduce the spaces $X^T_{b,s}$ and $\widetilde{X}^T_{b,s_1,s_2}$ of restrictions to $[0,T]$ of functions in $X_{b,s}$ and $\widetilde{X}_{b,s_1,s_2}$. They are endowed with

$$\|f\|_{X^T_{b,s}} = \inf\{\|\widetilde{f}\|_{X_{b,s}},\ \widetilde{f} \in X_{b,s} \text{ and } f|_{[0,T]} = \widetilde{f}|_{[0,T]}\}$$

$$\|f\|_{\widetilde{X}^T_{b,s_1,s_2}} = \inf\{\|\widetilde{f}\|_{\widetilde{X}_{b,s_1,s_2}},\ \widetilde{f} \in \widetilde{X}_{b,s_1,s_2} \text{ and } f|_{[0,T]} = \widetilde{f}|_{[0,T]}\}.$$

Because equation (1.1) is a multiplicative equation with a nonlinear deterministic part, we have to consider first a cut-off version of this equation (see Section 2). As we make use of the functional framework defined above, the cut-off will arise as a function of the norm of the solution of the type $\|\cdot\|_{X^t_{b,s}}$. Moreover, this function of the norm must be a regular function, in order to allow us to use a fixed point argument (i.e. in order that our mapping is a contraction mapping, see Section 2). The fact that the functional spaces we consider are nonlocal spaces in the time variable then brings a lot of technical difficulties, concerning points that would be obvious if we were dealing with more classical function spaces (see e.g. the proof of Lemma 2.1).

The paper is organized as follows: Section 2 is devoted to the proof of several preliminary lemmas and propositions, which once brought together lead quite easily to the proof of global existence and uniqueness for the cut-off version of the equation – or to the local existence and uniqueness for equation (1.1). In Section 3 we prove that the solutions of equation (1.1) are global in time, by using estimates on the moments of the L^2-norm of the solution. Again, due to the spaces we consider for the local existence, the globalization argument is not obvious.

2. Preliminaries and existence for a truncated equation

As is usual, we introduce the mild form of the stochastic Korteweg-de Vries equation (1.1). We denote by $U(t) = e^{-t\partial_x^3}$ the unitary group on $L^2(\mathbb{R})$ generated by the linear equation

$$\frac{\partial u}{\partial t} + \frac{\partial^3 u}{\partial x^3} = 0.$$

Using Fourier transform, we have $\mathcal{F}(U(t)v)(\xi) = e^{it\xi^3}\mathcal{F}(v)(\xi)$. We then rewrite (1.1) as follows

$$u(t) = U(t)u_0 - \frac{1}{2}\int_0^t U(t-r)\partial_x(u^2(r))dr + \int_0^t U(t-r)(u(r)\varphi dW(r)),\ t \geq 0. \quad (2.1)$$

The $X_{b,s}$ and $\widetilde{X}_{b,s_1,s_2}$ norms defined in the introduction have the nice property that they are increasing with T. However, it is more convenient to work with other norms, given by the multiplication by the function $\mathbb{1}_{[0,T]}$. In the case we consider here, that is $0 \leq b < 1/2$, we can prove the following result, stating that the two norms are equivalent.

Lemma 2.1. *Let $s \geq 0$ and $0 \leq b < 1/2$, then there exist two constants C_1, C_2 depending on b but not on T such that for any $f \in X_{b,s}$*

$$C_1\|f\|_{X^T_{b,s}} \leq \|\mathbb{1}_{[0,T]}(t)f\|_{X_{b,s}} \leq C_2\|f\|_{X^T_{b,s}}.$$

Proof. The first inequality is clear and in fact we may choose $C_1 = 1$. For the other inequality, let us set $g(t) = \mathbb{1}_{[0,T]}(t)U(-t)f(t)$ so that

$$\|\mathbb{1}_{[0,T]}(t)f\|^2_{X_{b,s}} = \iint_{\mathbb{R}^2} (1+|\xi|)^{2s}(1+|\tau|)^{2b}|\widehat{g}(\tau,\xi)|^2 d\tau d\xi$$

$$= \iint_{\mathbb{R}^2} (1+|\xi|)^{2s}\|(\mathcal{F}_x g)(\cdot,\xi)\|^2_{H^b_t} d\xi.$$

The result follows from the following inequality

$$\|\mathbb{1}_{[0,T]}h\|_{H^b(\mathbb{R})} \leq C\|h\|_{H^b(\mathbb{R})},\ h \in H^b(\mathbb{R}),$$

which holds for a constant $C \geq 0$ depending on $0 < b < 1/2$. To prove this, we use the following equivalent norm on $H^b(\mathbb{R})$ (see for instance [9]):

$$\|h\|^2_{H^b(\mathbb{R})} = \iint_{\mathbb{R}^2} \frac{|h(t)-h(r)|^2}{|t-r|^{1+2b}} dtdr + \|h\|^2_{L^2(\mathbb{R})}.$$

Clearly, $\|\mathbb{1}_{[0,T]}h\|^2_{L^2(\mathbb{R})} \leq \|h\|^2_{L^2(\mathbb{R})}$. Moreover

$$\begin{aligned}
&\iint_{\mathbb{R}^2} \frac{\left|\mathbb{1}_{[0,T]}(t)h(t) - \mathbb{1}_{[0,T]}(r)h(r)\right|^2}{|t-r|^{1+2b}} drdt \\
&= 2\iint_{r<t} \frac{\left|\mathbb{1}_{[0,T]}(t)h(t) - \mathbb{1}_{[0,T]}(r)h(r)\right|^2}{|t-r|^{1+2b}} drdt \\
&= 2\int_0^T\int_0^t \frac{|h(t)-h(r)|^2}{|t-r|^{1+2b}} drdt + 2\int_0^T\int_{-\infty}^0 \frac{|h(t)|^2}{|t-r|^{1+2b}} drdt \\
&+ 2\int_T^\infty\int_0^T \frac{|h(r)|^2}{|t-r|^{1+2b}} drdt = I + II + III.
\end{aligned}$$

The first term I is less than $\|h\|_{H^b(\mathbb{R})}$. The second and third terms are equal to

$$II = \frac{1}{b}\int_0^T |t|^{-2b}|h(t)|^2 dt,\ III = \frac{1}{b}\int_0^T |T-r|^{-2b}|h(r)|^2 dr.$$

Both are bounded by $C\|h\|_{H^b(0,T)}$. To see this, we note that it is obvious for $b=0$ and results from Hardy inequality for $b=1$ when $H^b(0,T)$ is replaced by $H^1_0(0,T)$. The result follows by interpolation since, for $0 \leq b < 1/2$, $H^b(0,T) = H^b_0(0,T)$. □

We now define $Y^T_{b,0} = X^T_{b,0} \cap \widetilde{X}^T_{b,0,-3/8}$ endowed with the norm

$$\|f\|_{Y^T_{b,0}} = \max\{\|f\|_{X^T_{b,0}}, \|f\|_{\widetilde{X}^T_{b,0,-3/8}}\}.$$

We also use the space $Y^T_{b,1} = X^T_{b,1} \cap \widetilde{X}^T_{b,1,-3/8}$ with a similar definition of its norm. From now on and thanks to Lemma 2.1, we will use the definition $\|v\|^T_{Y_{b,s}} = \|\mathbb{1}_{[0,T]}v\|_{Y_{b,s}}$ each time we take a norm in $Y^T_{b,s}$ with $0 \leq b < 1/2$.

For $u_0 \in L^2(\Omega; H^1(\mathbb{R}))$, we set

$$z(t) = U(t)u_0,\ \text{ and } v(t) = u(t) - z(t).$$

Then (2.1) is rewritten as

$$
\begin{aligned}
v(t) = &-\tfrac{1}{2}\int_0^t U(t-r)\left[\partial_x(v^2(r)) + \partial_x(z^2(r)) + 2\partial_x(z(r)v(r))\right] dr \\
&+ \int_0^t U(t-r)((z(r)+v(r))\varphi dW(r)),\ t \geq 0.
\end{aligned} \tag{2.2}
$$

Let θ be a cut-off function – $\theta(x) = 0$ for $x \geq 2$, $\theta(x) = 1$ for $0 \leq x \leq 1$, with $\theta \in C_0^\infty(\mathbb{R}^+)$ – and let $\theta_R = \theta(\frac{\cdot}{R})$; we consider the cut-off version of (2.2) written for $R > 0$ as:

$$
\begin{aligned}
v_R(t) = &-\frac{1}{2}\int_0^t U(t-r)\left[\theta_R^2\left(\|v_R\|_{Y_{b,0}^r}\right)\partial_x(v_R^2(r))\right] dr \\
&- \int_0^t U(t-r)\left[\theta_R\left(\|v_R\|_{Y_{b,0}^r}\right)\partial_x(z(r)v_R(r))\right] dr \\
&-\frac{1}{2}\int_0^t U(t-r)\left[\partial_x(z^2(r))\right] dr \\
&+ \int_0^t U(t-r)((z(r)+v_R(r))\varphi dW(r)),\ t \geq 0.
\end{aligned} \tag{2.3}
$$

We find v_R as a fixed point of the mapping $\mathcal{T}_R$, $\mathcal{T}_R v_R$ being defined by the right hand side above. Note that the cut-off is made in the L^2 in space norm, even for the H^1 result. We will choose $0 < b < 1/2$ and $1/2 < c$.

We use the following Lemma.

Lemma 2.2. *For any $0 \leq b < 1/2$, $R > 0$, $v \in Y_{b,1}^T$, there exists $C(R)$ such that*

$$
\left\|\theta_R\left(\|v\|_{Y_{b,0}^t}\right) v(t)\right\|_{Y_{b,0}^T} \leq C(R)
$$

and, for $s = 0$ or 1, there is a positive constant C, independent of R, such that

$$
\left\|\theta_R\left(\|v\|_{Y_{b,0}^t}\right) v(t)\right\|_{Y_{b,s}^T} \leq C\,\|v(t)\|_{Y_{b,s}^T}.
$$

Proof. We use arguments similar to the proof of Lemma 2.1. Let $w(t) = U(-t)v(t)$ then, using the same norm on $H^b(\mathbb{R})$ as in Lemma 2.1,

$$
\left\|\theta_R\left(\|v\|_{Y_{b,0}^t}\right) v(t)\right\|_{X_{b,0}^T}^2 \leq C\int_{\mathbb{R}}\left\|\theta_R\left(\|v\|_{Y_{b,0}^t}\right)(\mathcal{F}_x w)(t,\xi)\right\|_{H_t^b([0,T])}^2 d\xi.
$$

The L^2 part of the H^b norm above is easily estimated, while the other part is bounded above by

$$C\int_{\mathbb{R}}\int_0^T\int_0^t \theta_R^2\left(\|v\|_{Y_{b,0}^t}\right)\frac{|(\mathcal{F}_x w)(t,\xi)-(\mathcal{F}_x w)(r,\xi)|^2}{|t-r|^{1+2b}}drdtd\xi$$

$$+C\int_{\mathbb{R}}\int_0^T\int_0^t \left(\theta_R\left(\|v\|_{Y_{b,0}^t}\right)-\theta_R\left(\|v\|_{Y_{b,0}^r}\right)\right)^2\frac{|(\mathcal{F}_x w)(r,\xi)|^2}{|t-r|^{1+2b}}drdtd\xi$$

$$=I+II.$$

Next, we define $\tau_R=\inf\{t\geq 0,\ \|v\|_{Y_{b,0}^t}\geq 2R\}$; then $\theta_R\left(\|v\|_{Y_{b,0}^t}\right)=0$ for $t\geq\tau_R$ and

$$I\leq C\int_{\mathbb{R}}\int_0^{\tau_R}\int_0^t\frac{|(\mathcal{F}_x w)(t,\xi)-(\mathcal{F}_x w)(r,\xi)|^2}{|t-r|^{1+2b}}drdtd\xi$$

$$\leq C\int_{\mathbb{R}}\|(\mathcal{F}_x w)(\cdot,\xi)\|_{H^b(0,\tau_R)}^2 d\xi\leq C\|v\|_{X_{b,0}^{\tau_R}}\leq 2CR.$$

In order to estimate II, we use the fact that for $r<t$,

$$\left(\theta_R\left(\|v\|_{Y_{b,0}^t}\right)-\theta_R\left(\|v\|_{Y_{b,0}^r}\right)\right)^2\leq\frac{C}{R^2}|\theta'|_{L^\infty}^2|\|v\|_{Y_{b,0}^t}-\|v\|_{Y_{b,0}^r}|^2$$

$$\leq\frac{C}{R^2}|\|\mathbb{1}_{[0,t]}v\|_{Y_{b,0}}-\|\mathbb{1}_{[0,r]}v\|_{Y_{b,0}}|^2\leq\frac{C}{R^2}\|\mathbb{1}_{[r,t]}v\|_{Y_{b,0}}^2$$

$$\leq\frac{C}{R^2}\int_{\mathbb{R}}(1+|\eta|^{-3/4})\|\mathbb{1}_{[r,t]}(\mathcal{F}_x w)(.,\eta)\|_{H^b}^2 d\eta.$$

We leave to the reader the estimate of the contribution to II of the L^2 part of the H^b norm above; indeed, it follows the same line as the estimate of the remaining contribution, which is bounded above by

$$\frac{C}{R^2}\int_{\mathbb{R}}\int_{\mathbb{R}}(1+|\eta|^{-3/4})\int_0^{\tau_R}\int_0^t\int_r^t\int_r^{\sigma_2}\frac{|(\mathcal{F}_x w)(\sigma_2,\eta)-(\mathcal{F}_x w)(\sigma_1,\eta)|^2}{|\sigma_2-\sigma_1|^{1+2b}}d\sigma_1 d\sigma_2$$

$$\times\frac{|(\mathcal{F}_x w)(r,\xi)|^2}{|t-r|^{1+2b}}drdtd\eta d\xi$$

$$\leq\frac{C}{R^2}\int_{\mathbb{R}}\int_{\mathbb{R}}(1+|\eta|^{-3/4})\int_0^{\tau_R}\int_0^{\sigma_2}\int_0^{\sigma_1}\left(\int_{\sigma_2}^{\tau_R}\frac{dt}{|t-r|^{1+2b}}\right)|(\mathcal{F}_x w)(r,\xi)|^2 dr$$

$$\times\frac{|(\mathcal{F}_x w)(\sigma_2,\eta)-(\mathcal{F}_x w)(\sigma_1,\eta)|^2}{|\sigma_2-\sigma_1|^{1+2b}}d\sigma_1 d\sigma_2 d\eta d\xi$$

where we have inverted the integrals in the time variables; this last term is in turn bounded above by

$$\frac{C}{R^2}\int_{\mathbb{R}}\int_{\mathbb{R}}(1+|\eta|^{-3/4})\int_0^{\tau_R}\int_0^{\sigma_2}\left(\int_0^{\sigma_1}|\sigma_1-r|^{-2b}|(\mathcal{F}_x w)(r,\xi)|^2dr\right)$$
$$\times\frac{|(\mathcal{F}_x w)(\sigma_2,\eta)-(\mathcal{F}_x w)(\sigma_1,\eta)|^2}{|\sigma_1-\sigma_2|^{1+2b}}d\sigma_1 d\sigma_2 d\eta d_\xi,$$

by the fact that $|\tau_R-r|^{-2b}\le|\sigma_2-r|^{-2b}\le|\sigma_1-r|^{-2b}$ for $0\le r\le\sigma_1\le\sigma_2\le t\le\tau_R$.

Using then the same arguments as in the proof of Lemma 2.1, we finally get

$$II\le\frac{C}{R^2}\int_{\mathbb{R}}\|(\mathcal{F}_x w)(.,\xi)\|^2_{H^b(0,\tau_R)}d\xi\int_{\mathbb{R}}(1+|\eta|^{-3/4})\|(\mathcal{F}_x w)(.,\eta)\|^2_{H^b(0,\tau_R)}d\eta$$
$$\le\frac{C}{R^2}\|v\|^2_{Y^{\tau_R}_{b,0}}\|v\|^2_{X^{\tau_R}_{b,0}}.$$

This, together with the estimate of I implies the first inequality of the Lemma for the $X_{b,0}$ part of the $Y_{b,0}$ norm; the $\widetilde{X}_{b,0,-3/8}$ part, and the second inequality of the Lemma are proved in the same way. □

Next results state the estimates on the bilinear term appearing in (2.3).

Proposition 2.3. *Let $a>0$, $0<b<1/2<c<1$, with $b+c>1$ and a,b,c sufficiently close to $1/2$, then for any $v\in Y^T_{b,s}$, $z\in X^T_{c,s}$, $s=0$ or 1, we have*

$$\|\partial_x(v^2)\|_{Y^T_{-a,s}}\le C\|v\|_{Y^T_{b,0}}\|v\|_{Y^T_{b,s}},$$

$$\|\partial_x(vz)\|_{Y^T_{-a,s}}\le C\left(\|v\|_{Y^T_{b,0}}\|z\|_{X^T_{c,s}}+\|v\|_{Y^T_{b,s}}\|z\|_{X^T_{c,0}}\right),$$

and

$$\|\partial_x(z^2)\|_{Y^T_{-a,s}}\le C\|z\|_{X^T_{c,s}}\|z\|_{X^T_{c,0}}.$$

Proof. These estimates are proved in Ref. [2], Proposition 2.2 and 2.3, for $s=0$. These are easily extended to $s=1$. It suffices to add a factor $1+|\xi|$ in the expression

$$|\langle f,\partial_x(gh)\rangle|$$
$$=\left|\int_\tau\int_\xi \xi\widehat{f}(\tau,\xi)\int_{\tau_1}\int_{\xi_1}\overline{\widehat{g}(\tau-\tau_1,\xi-\xi_1)\widehat{h}(\tau_1,\xi_1)}d\tau_1 d\xi_1\, d\tau d\xi\right|,$$

and to use the fact that $1+|\xi|\le(1+|\xi-\xi_1|)+(1+|\xi_1|)$. □

Remark 2.4. It does not seem possible to get rid of the homogeneous Sobolev space, i.e. of $\widetilde{X}_{b,0,-3/8}$, to get the result of Proposition 2.3, when $b<1/2$, even in the case $s=1$; indeed, a careful reading of the proof of Proposition 2.2 in Ref. [2] shows that the additional factor $|\xi|^{-3/4}$ induced by the use of $\widetilde{X}_{b,0,-3/8}$ is necessary in a region of the integral where $|\xi_1|\ll|\xi|$, so that $|\xi-\xi_1|\sim|\xi|$, and with moreover $|\xi|\le 1$; hence the supplementary factor $(1+|\xi-\xi_1|)(1+|\xi_1|)/(1+|\xi|)$ is of no help there.

It remains to derive the estimates on the stochastic integrals in (2.3). In order to be able to globalize the solutions in Section 3, we will need estimates on all the moments of the stochastic integrals.

Proposition 2.5. *Let $m \in \mathbb{N}$, $s = 0$ or 1, and $v \in L^{2m}(\Omega, X_{b,s}^T)$; then for any $0 \le b \le 1/2$,*

$$\mathbb{E}\left(\left\|\int_0^t U(t-r)\left[v(r)\varphi dW(r)\right]\right\|_{X_{b,s}^T}^{2m}\right) \le C\|k\|_{H^s(\mathbb{R})}^{2m}\mathbb{E}\left(\|v\|_{X_{0,s}^T}^{2m}\right)$$
$$\le CT^{bm}\|k\|_{H^s(\mathbb{R})}^{2m}\mathbb{E}\left(\|v\|_{X_{b,s}^T}^{2m}\right).$$

Moreover

$$\mathbb{E}\left(\left\|\int_0^t U(t-r)\left[v(r)\varphi dW(r)\right]\right\|_{\tilde{X}_{b,s-3/8}^T}^{2m}\right) \le C\left(\|k\|_{H^s(\mathbb{R})}^{2m} + \|k\|_{L^1(\mathbb{R})}^{2m}\right)\mathbb{E}\left(\|v\|_{X_{0,s}^T}^{2m}\right)$$
$$\le CT^{bm}\left(\|k\|_{H^s(\mathbb{R})}^{2m} + \|k\|_{L^1(\mathbb{R})}^{2m}\right)\mathbb{E}\left(\|v\|_{X_{b,s}^T}^{2m}\right).$$

Proof. We prove the result for $s = 1$, the proof is exactly the same for $s = 0$. We set $w(t) = \mathbb{1}_{[0,T]}(t)\displaystyle\int_0^t U(t-r)[v(r)\varphi dW(r)]$. Let

$$g(t) = \mathbb{1}_{[0,T]}(t)\int_0^t U(-r)\left[v(r)\varphi dW(r)\right],$$

then $w(t) = U(t)g(t)$, $t \ge 0$. We have

$$\mathbb{E}\left(\|w\|_{X_{b,1}}^{2m}\right) = \mathbb{E}\left(\left(\iint_{\mathbb{R}^2}(1+|\xi|)^2(1+|\tau|)^{2b}|\widehat{g}(\tau,\xi)|^2 d\tau d\xi\right)^m\right)$$

Choosing Brownian motions $(\beta_k)_{k\in\mathbb{N}}$, defined on $\mathbb{R}$, we have

$$(\mathcal{F}_x g)(t,\xi) = \sum_{k=0}^{\infty}\mathbb{1}_{[0,T]}(t)\int_0^t e^{ir\xi^3}\mathcal{F}_x\left(v(r)\varphi e_k\right)(\xi)d\beta_k(r)$$
$$= \sum_{k=0}^{\infty}\mathbb{1}_{[0,T]}(t)\int_{-\infty}^t \mathbb{1}_{[0,T]}(r)e^{ir\xi^3}\mathcal{F}_x\left(v(r)\varphi e_k\right)(\xi)d\beta_k(r).$$

It follows

$$\widehat{g}(\tau,\xi) = \sum_{k=0}^{\infty}\int_{\mathbb{R}}\mathbb{1}_{[0,T]}(t)\int_{-\infty}^t \mathbb{1}_{[0,T]}(r)e^{ir\xi^3}\mathcal{F}_x\left(v(r)\varphi e_k\right)(\xi)d\beta_k(r)e^{-i\tau t}dt$$
$$= \sum_{k=0}^{\infty}\int_{\mathbb{R}}\mathbb{1}_{[0,T]}(r)e^{ir\xi^3}\mathcal{F}_x\left(v(r)\varphi e_k\right)(\xi)\left(\int_r^{\infty}\mathbb{1}_{[0,T]}(t)e^{-i\tau t}dt\right)d\beta_k(r).$$

By Burkholder inequality, we deduce:

$$\mathbb{E}\left(\left(\iint_{\mathbb{R}^2}(1+|\xi|)^2(1+|\tau|)^{2b}|\widehat{g}(\tau,\xi)|^2 d\tau d\xi\right)^m\right)$$
$$\leq C_m\mathbb{E}\Bigg(\Bigg(\sum_{k=0}^{\infty}\iiint_{\mathbb{R}^3}(1+|\xi|)^2(1+|\tau|)^{2b}\mathbb{1}_{[0,T]}(r)\,|\mathcal{F}_x\left(v(r)\varphi e_k\right)(\xi)|^2$$
$$\times\left|\int_r^{\infty}\mathbb{1}_{[0,T]}(t)e^{-i\tau t}dt\right|^2 dr d\xi d\tau\Bigg)^m\Bigg).$$

It is easy to see that

$$\left|\int_r^{\infty}\mathbb{1}_{[0,T]}(t)e^{-i\tau t}dt\right|^2\leq\min\{T^2,\ 2\tau^{-2}\}.$$

Therefore, using Lemma 2.6 below,

$$\mathbb{E}\left(\left(\iint_{\mathbb{R}^2}(1+|\xi|)^2(1+|\tau|)^{2b}|\widehat{g}(\tau,\xi)|^2 d\tau d\xi\right)^m\right)$$
$$\leq C\mathbb{E}\Bigg(\Bigg(\iiiint_{\mathbb{R}^4}(1+|\xi|)^2(1+|\tau|)^{2b}\min\{T^2,\ 2\tau^{-2}\}\mathbb{1}_{[0,T]}(r)\,|(\mathcal{F}_x v(r))(\xi+\eta)|^2$$
$$\times|\widehat{k}(\eta)|^2 d\eta dr d\xi d\tau\Bigg)^m\Bigg)$$

which in turn we bound from above, using the unitarity of $U(t)$ in L^2 and in H^1, by

$$C\mathbb{E}\Bigg(\Bigg(\iiint_{\mathbb{R}^3}\left[(1+|\xi+\eta|)^2+(1+|\eta|^2)\right]\mathbb{1}_{[0,T]}(r)\,|(\mathcal{F}_x v(r))(\xi+\eta)|^2$$
$$\times|\widehat{k}(\eta)|^2 d\eta dr d\xi\Bigg)^m\Bigg)$$
$$\leq C\left[\|k\|^{2m}_{L^2(\mathbb{R})}\mathbb{E}\left(\|\mathbb{1}_{[0,T]}v\|^{2m}_{X_{0,1}}\right)+\|k\|^{2m}_{H^1(\mathbb{R})}\mathbb{E}\left(\|\mathbb{1}_{[0,T]}v\|^{2m}_{X_{0,0}}\right)\right]$$
$$\leq C\left[\|k\|^{2m}_{L^2(\mathbb{R})}\mathbb{E}\left(\|v\|^{2m}_{X^T_{0,1}}\right)+\|k\|^{2m}_{H^1(\mathbb{R})}\mathbb{E}\left(\|v\|^{2m}_{X^T_{0,0}}\right)\right].$$

For the second statement, we proceed similarly. However, the extra $|\xi|^{-3/4}$ implies that a special treatment of the integral for $|\xi|\leq 1$. On the region $|\xi|\geq 1$, we simply use $|\xi|^{-3/4} \leq 1$. The following estimate is thus sufficient to conclude.

$$\mathbb{E}\Big(\Big(\int_{|\xi|\leq 1}\iint_{\mathbb{R}^2}(1+|\xi|)^2|\xi|^{-3/4}\mathbb{1}_{[0,T]}(r)\,|(\mathcal{F}_x v(r))(\xi+\eta)|^2\,|\widehat{k}(\eta)|^2 d\eta dr d\xi\Big)^m\Big)$$

$$\leq C\mathbb{E}\Big(\Big(\iint_{\mathbb{R}^2}\Big(\int_{|\xi|\leq 1}|\xi|^{-3/4}|\widehat{k}(\eta-\xi)|^2 d\xi\Big)\mathbb{1}_{[0,T]}(r)\,|\mathcal{F}_x(v(r))(\eta)|^2\,d\eta dr\Big)^m\Big)$$

$$\leq C\|\widehat{k}\|_{L^\infty(\mathbb{R})}\mathbb{E}\Big(\|v\|^{2m}_{X^T_{0,0}}\Big)$$

$$\leq C\|k\|_{L^1(\mathbb{R})}\mathbb{E}\Big(\|v\|^{2m}_{X^T_{0,0}}\Big).$$

□

We now give the Lemma used in the above proof.

Lemma 2.6. *Let $v \in X_{0,0}$, then for any complete orthonormal system $(e_k)_{k\in\mathbb{N}}$ of $L^2(\mathbb{R})$, we have*

$$\sum_{k=0}^{\infty}|\mathcal{F}_x\left(v(r)\varphi e_k\right)(\xi)|^2 = \int_{\mathbb{R}}|\mathcal{F}_x(v(r))(\xi+\eta)|^2\,|\widehat{k}(\eta)|^2 d\eta.$$

Proof. We have $\mathcal{F}_x\left(v(r)\varphi e_k\right)(\xi) = \mathcal{F}_x\Big(v(r,x)\langle k(x-y),e_k(y)\rangle_{L^2_y}\Big)(\xi) = \langle\mathcal{F}_x\left(v(r,x)k(x-y)\right)(\xi),e_k(y)\rangle_{L^2_y}$. Therefore, by Parseval identity,

$$\sum_{k=0}^{\infty}|\mathcal{F}_x\left(v(r)\varphi e_k\right)(\xi)|^2 = \|\mathcal{F}_x\left(v(r,x)k(x-y)\right)(\xi)\|^2_{L^2_y},$$

and by Plancherel theorem and an easy computation

$$\sum_{k=0}^{\infty}|\mathcal{F}_x\left(v(r)\varphi e_k\right)(\xi)|^2 = \|\mathcal{F}_{x,y}\left(v(r,x)k(x-y)\right)(\xi,\eta)\|^2_{L^2_\eta} = \|\mathcal{F}_x(v(r))(\xi+.)\bar{\widehat{k}}\|^2_{L^2},$$

which gives the conclusion. □

The proof of the next proposition is left to the reader. It only makes use of classical arguments and ideas similar to those at the end of the proof of Proposition 2.5.

Proposition 2.7. *Let $s=0$ or 1. For any $T_0>0$, any stopping time τ and any predictable process $v \in L^2(\Omega;C([0,T_0\wedge\tau];H^s(\mathbb{R})))$, $\int_0^{\cdot} U(\cdot-r)[\varphi v(r)dW(r)]$ has continuous paths with values in $H^s(\mathbb{R})\cap\dot{H}^{-3/8}(\mathbb{R})$ and for any integer m, there is a constant C_m with*

$$\mathbb{E}\left(\sup_{t\leq T_0\wedge\tau}\left\|\int_0^t U(t-r)[\varphi v(r)dW(r)]\right\|^{2m}_{H^s\cap\dot{H}^{-3/8}}\right) \leq C_m\mathbb{E}\left(\sup_{t\leq T_0\wedge\tau}\|v(t)\|^{2m}_{H^s}\right).$$

We are now able to prove the following existence theorem for the truncated equation.

Theorem 2.8. *Let $s = 0$ or 1 and assume that the convolution kernel of the operator φ satisfies $k \in H^s(\mathbb{R}) \cap L^1(\mathbb{R})$; then for any u_0 in $H^s(\mathbb{R})$, equation (2.3) with $z(t) = U(t)u_0$ has a unique solution $v_R \in Y_{b,s}^{T_0}$, for any b with $0 < b < 1/2$, and any $T_0 \geq 0$. Moreover $v_R \in L^2(\Omega; C([0, T_0]; H^s(\mathbb{R})))$.*

Proof. We use a fixed point argument on equation (2.3). The following lemma, whose first and third estimates were proved in Ref. [2], while the second one can be proved in the same way, is useful.

Lemma 2.9.

- *Let $u_0 \in H^s(\mathbb{R})$, $s = 0$ or 1. For any $T > 0$ and $c > 1/2$, $z = U(\cdot)u_0 \in X_{c,s}^T$ and*

$$\|z\|_{X_{c,s}^T} \leq C(T)\|u_0\|_{H^s(\mathbb{R})}.$$

- *For any $u_0 \in H^s(\mathbb{R}) \cap \dot{H}^{-3/8}(\mathbb{R})$, and any b with $0 \leq b < 1/2$, $z = U(\cdot)u_0 \in Y_{b,s}^T$ and*

$$\|z\|_{Y_{b,s}^T} \leq C(T)(\|u_0\|_{H^s(\mathbb{R})} + \|u_0\|_{\dot{H}^{-3/8}(\mathbb{R})}).$$

- *For any $a, b \in (0, 1)$ with $a + b \leq 1$, and any $f \in Y_{-a,s}^T$, $\int_0^{\cdot} U(\cdot - r)f(r)dr \in Y_{b,s}^T$ and*

$$\left\| \int_0^{\cdot} U(\cdot - r)f(r)dr \right\|_{Y_{b,s}^T} \leq CT^{1-(a+b)}\|f\|_{Y_{-a,s}^T}.$$

We first assume that the hypothesis of Theorem 2.8 hold with $s = 0$. Let us fix a, b, c as in Proposition 2.3, with $a + b < 1$. We fix T_0 and take $T \leq T_0$. Let $v_1, v_2 \in Y_{b,0}^T$, T being also fixed. We set $\tilde{v}_i(t) = \theta_R\left(|v_i|_{Y_{b,0}^t}\right) v_i(t)$, $i = 1, 2$. Then, recalling that $\mathcal{T}_R v_R$ is defined by the right hand side of (2.3), we have

$$\mathbb{E}\left(\|\mathcal{T}_R v_1 - \mathcal{T}_R v_2\|_{Y_{b,0}^T}^2\right)$$

$$\leq C\mathbb{E}\left(\left\| \int_0^t U(t - r)\partial_x \left[(\tilde{v}_1(r))^2 - (\tilde{v}_2(r))^2\right] dr \right\|_{Y_{b,0}^T}^2\right)$$

$$+C\mathbb{E}\left(\left\| \int_0^t U(t - r)\partial_x \left[(\tilde{v}_1(r) - \tilde{v}_2(r))\, z(r)\right] dr \right\|_{Y_{b,0}^T}^2\right)$$

$$+C\mathbb{E}\left(\left\| \int_0^t U(t - r)\left[(v_1(r) - v_2(r))\, \varphi dW(r)\right] \right\|_{Y_{b,0}^T}^2\right)$$

which, by Lemma 2.9 and Proposition 2.5, applied with $m = 1$, is bounded above by

$$CT^{2(1-(a+b))}\mathbb{E}\left(\left\|\partial_x\left((\tilde{v}_1)^2-(\tilde{v}_2)^2\right)\right\|^2_{Y^T_{-a,0}}\right)$$
$$+\,CT^{2(1-(a+b))}\mathbb{E}\left(\left\|\partial_x\left((\tilde{v}_1-\tilde{v}_2)\,z\right)\right\|^2_{Y^T_{-a,0}}\right)+CT^b\mathbb{E}\left(\|v_1-v_2\|^2_{X^T_{b,0}}\right).$$

By Proposition 2.3, it follows

$$\mathbb{E}\left(\|\mathcal{T}_Rv_1-\mathcal{T}_Rv_2\|^2_{Y^T_{b,0}}\right)$$
$$\leq CT^{2(1-(a+b))}\left\{\mathbb{E}\left(\|\tilde{v}_1-\tilde{v}_2\|^2_{Y^T_{b,0}}\|\tilde{v}_1+\tilde{v}_2\|^2_{Y^T_{b,0}}\right)+\mathbb{E}\left(\|\tilde{v}_1-\tilde{v}_2\|^2_{Y^T_{b,0}}\right)\|z\|^2_{X^T_{c,0}}\right\}$$
$$+CT^b\mathbb{E}\left(\|v_1-v_2\|^2_{X^T_{b,0}}\right).$$

By Lemma 2.2,

$$\|\tilde{v}_1+\tilde{v}_2\|^2_{Y^T_{b,0}}\leq C(R).$$

Moreover, it is not difficult to use the arguments of the proof of Lemma 2.2 and prove

$$\|\tilde{v}_1-\tilde{v}_2\|^2_{Y^T_{b,0}}\leq C(R)\,\|v_1-v_2\|^2_{Y^T_{b,0}}.$$

We deduce that for some $\alpha>0$,

$$\mathbb{E}\left(\|\mathcal{T}_Rv_1-\mathcal{T}_Rv_2\|^2_{Y^T_{b,0}}\right)\leq C(R,T_0,\|u_0\|_{L^2(\mathbb{R})})T^\alpha\mathbb{E}\,\|v_1-v_2\|^2_{Y^T_{b,0}}.$$

Thus, $\mathcal{T}_R$ has a unique fixed point $v_R\in L^2(\Omega;Y^T_{b,0})$ for $T\leq T_*$ where T_* is chosen such that

$$C(R,T_0,\|u_0\|_{L^2(\mathbb{R})})T_*^\alpha\leq 1/2.$$

Moreover, using arguments similar to the proof of Proposition 2.5, it can be seen that $\int_0^t U(-r)\left[(z(r)+v_R(r))\,\varphi dW(r)\right]$ is a square integrable martingale in $L^2(\mathbb{R})\cap\dot{H}^{-3/8}(\mathbb{R})$. Since $(U(t))_{t\in\mathbb{R}}$ is strongly continuous on $L^2(\mathbb{R})\cap\dot{H}^{-3/8}(\mathbb{R})$, we deduce that

$$\int_0^t U(t-r)\left[(z(r)+v_R(r))\,\varphi dW(r)\right]\in L^2(\Omega;C([0,T_*];L^2(\mathbb{R})\cap\dot{H}^{-3/8}(\mathbb{R}))).$$

Using then Lemma 2.9 with $b>1/2$ and similar estimates as above, we deduce that v_R is also in $L^2(\Omega;C([0,T_*];L^2(\mathbb{R})\cap\dot{H}^{-3/8}(\mathbb{R})))$.

Then, we construct a solution in $[T_*, 2T_*]$. First, we write equation (2.3) with $t \geq T_*$ in the form

$$v_R(T_*+t) = U(t)v_R(T_*) - \frac{1}{2}\int_0^t U(t-r)\left[\theta_R^2\left(\|v_R\|_{Y_{b,0}^{T_*+r}}\right)\partial_x(v_R^2(T_*+r))\right]dr$$
$$-\int_0^t U(t-r)\left[\theta_R\left(\|v_R\|_{Y_{b,0}^{T_*+r}}\right)\partial_x(z(T_*+r)v_R(T_*+r))\right]dr$$
$$-\frac{1}{2}\int_0^t U(t-r)\left[\partial_x(z^2(T_*+r))\right]dr$$
$$+\int_0^t U(t-r)((z(T_*+r)+v_R(T_*+r))\varphi dW(r)),\ t\geq 0. \tag{2.4}$$

Since $v_R(T_*) \in L^2(\Omega; L^2(\mathbb{R}) \cap \dot{H}^{-3/8}(\mathbb{R}))$, the first term is in $L^2(\Omega; Y_{b,0}^{T_*})$. It is then easily seen that v_R can be found on $[T_*, 2T_*]$ as a fixed point in $L^2(\Omega; Y_{b,0}^{T_*})$ in the same way as on the interval $[0, T_*]$.

Iterating this, we get a solution on $[0, T_0]$ which is in fact in $L^2(\Omega; Y_{b,0}^{T_0})$ and also in $L^2(\Omega; C([0,T_0]; L^2(\mathbb{R}) \cap \dot{H}^{-3/8}(\mathbb{R})))$. This proves the result for $s=0$.

Now suppose that the assumptions hold with $s=1$. Let $v \in L^2(\Omega; Y_{b,1}^T)$; we have, setting $\tilde{v}(t) = \theta_R(\|v\|_{Y_{b,0}^t})v(t)$, and using Lemma 2.9,

$$\|\mathcal{T}_R v\|_{Y_{b,1}^T} \leq CT^{1-(a+b)}\left[\|\tilde{v}\partial_x\tilde{v}\|_{Y_{-a,1}^T} + \|\partial_x(\tilde{v}z)\|_{Y_{-a,1}^T} + \|\partial_x(z^2)\|_{Y_{-a,1}^T}\right]$$
$$+\left\|\int_0^t U(t-r)\left[z(r)\varphi dW(r)\right]\right\|_{Y_{b,1}^T} + \left\|\int_0^t U(t-r)\left[v(r)\varphi dW(r)\right]\right\|_{Y_{b,1}^T},$$

so that by Proposition 2.3,

$$\|\mathcal{T}_R v\|_{Y_{b,1}^T} \leq CT^{1-(a+b)}\left[\|\tilde{v}\|_{Y_{b,0}^T}\|\tilde{v}\|_{Y_{b,1}^T} + \|\tilde{v}\|_{Y_{b,0}^T}\|z\|_{X_{c,1}^T} + \|\tilde{v}\|_{Y_{b,1}^T}\|z\|_{X_{c,0}^T}\right.$$
$$\left.+\|z\|_{X_{c,0}^T}\|z\|_{X_{c,1}^T}\right] + \left\|\int_0^t U(t-r)\left[z(r)\varphi dW(r)\right]\right\|_{Y_{b,1}^T}$$
$$+\left\|\int_0^t U(t-r)\left[v(r)\varphi dW(r)\right]\right\|_{Y_{b,1}^T}.$$

We then make use of Lemma 2.2 and Lemma 2.9 to get

$$\|\mathcal{T}_R v\|_{Y^T_{b,1}} \le CT^{1-(a+b)}\left(\|v\|_{Y^T_{b,1}} + \|z\|_{X^{T_0}_{c,1}}\right)\left(C(R)+\|z\|_{X^{T_0}_{c,0}}\right)$$
$$+\left\|\int_0^t U(t-r)\left[z(r)\varphi dW(r)\right]\right\|_{Y^T_{b,1}} + \left\|\int_0^t U(t-r)\left[v(r)\varphi dW(r)\right]\right\|_{Y^T_{b,1}}$$
$$\le C(R,T_0,\|u_0\|_{L^2(\mathbb{R})})T^{1-(a+b)}\left(\|v\|_{Y^T_{b,1}} + \|u_0\|_{H^1(\mathbb{R})}\right)$$
$$+\left\|\int_0^t U(t-r)\left[z(r)\varphi dW(r)\right]\right\|_{Y^T_{b,1}} + \left\|\int_0^t U(t-r)\left[v(r)\varphi dW(r)\right]\right\|_{Y^T_{b,1}}.$$

Thus, by Proposition 2.5 and Lemma 2.9,

$$\mathbb{E}\left(\|\mathcal{T}_R v\|^2_{Y^T_{b,1}}\right) \le \left[C(R,T_0,\|u_0\|_{L^2(\mathbb{R})})T^{2(1-(a+b))} + CT^b\right]\left(\mathbb{E}\left(\|v\|^2_{Y^T_{b,1}}\right) + \|u_0\|^2_{H^1(\mathbb{R})}\right).$$

This shows that $\mathcal{T}_R$ maps $L^2(\Omega; Y^T_{b,1})$ into itself. Moreover, the ball in $L^2(\Omega; Y^T_{b,1})$ of radius R_0 is invariant by $\mathcal{T}_R$ if $T \le T_{**}$ such that $C(R,T_0,\|u_0\|_{L^2(\mathbb{R})})T_{**}^{2(1-(a+b))} + CT^b_{**} \le 1/2$ and $R_0 \ge \|u_0\|_{H^1(\mathbb{R})}$.

Choosing $T_* \le T_{**}$ in the construction of the solution v_R of (2.3) in L^2, it follows that the solution v_R is in $L^2(\Omega; Y^{T_*}_{b,1})$. We then use similar arguments as above to prove that $v_R \in L^2(\Omega; C([0,T_*]; H^1(\mathbb{R}))$ and

$$\mathbb{E}\left(\sup_{[0,T_*]}\|v_R\|^2_{H^1(\mathbb{R})}\right)$$
$$\le CT_*^{2(1-(a+\tilde{b}))}\left(\mathbb{E}\Big(\|v_R\|^2_{Y^{T_*}_{b,1}}\Big) + \|u_0\|^2_{H^1(\mathbb{R})}\right)\left(C(R)+\|u_0\|_{L^2(\mathbb{R})}\right)^2$$
$$+\mathbb{E}\left(\sup_{[0,T_*]}\left\|\int_0^t U(t-r)\left[(z(r)+v_R(r))\varphi dW(r)\right]\right\|^2_{H^1(\mathbb{R})}\right)$$
$$\le CT_*^{2(1-(a+\tilde{b}))}\left(\mathbb{E}\Big(\|v_R\|^2_{Y^{T_*}_{b,1}}\Big) + \|u_0\|^2_{H^1(\mathbb{R})}\right)\left(C(R)+\|u_0\|_{L^2(\mathbb{R})}\right)^2$$
$$+CT_*\mathbb{E}\left(\sup_{[0,T_*]}\|v_R\|^2_{H^1(\mathbb{R})}\right) + CT_*\|u_0\|^2_{H^1(\mathbb{R})},$$

with $\tilde{b} > 1/2$ and $a+\tilde{b} < 1$. Choosing a smaller T_* if necessary, we deduce

$$\mathbb{E}\left(\sup_{[0,T_*]}\|v_R\|^2_{H^1(\mathbb{R})}\right) \le R_0^2,$$

if $R_0 \ge \|u_0\|_{H^1(\mathbb{R})}$. On $[T_*, 2T_*]$, we use equation (2.4) and obtain by similar arguments

$$\mathbb{E}\left(\|v_R(T_*+\cdot)\|^2_{Y^{T_*}_{b,1}}\right) \le C(T_*)\mathbb{E}\left(\|v_R(T_*)\|^2_{H^1(\mathbb{R})\cap\dot{H}^{-3/8}(\mathbb{R})}\right) + \|u_0\|^2_{H^1(\mathbb{R})},$$

and

$$\mathbb{E}\left(\sup_{[T_*,2T_*]}\|v_R\|^2_{H^1(\mathbb{R})}\right)\leq C\mathbb{E}\left(\|v_R(T_*)\|^2_{H^1(\mathbb{R})\cap\dot{H}^{-3/8}(\mathbb{R})}\right)+R_0^2.$$

We know from the L^2 construction that $v_R\in L^2(\Omega;C([0,T_0];\dot{H}^{-3/8}(\mathbb{R})))$, therefore

$$\mathbb{E}\left(\sup_{[T_*,2T_*]}\|v_R\|^2_{H^1(\mathbb{R})}\right)\leq C\mathbb{E}\left(\|v_R(T_*)\|^2_{H^1(\mathbb{R})}\right)+R_0^2+\mathbb{E}\left(\sup_{[0,T_0]}\|v_R\|^2_{\dot{H}^{-3/8}(\mathbb{R})}\right).$$

It is now easy to iterate this argument and deduce that the solution v_R is in $L^2(\Omega;Y^{T_0}_{b,1})$ and also in $L^2(\Omega;C([0,T_0];H^1(\mathbb{R})))$. □

Theorem 2.8 gives the following local in time existence result for the non truncated equation.

Corollary 2.10. *Let $s=0$ or 1 and assume that $k\in H^s(\mathbb{R})\cap L^1(\mathbb{R})$; then for any $u_0\in H^s(\mathbb{R})$, there is a stopping time $\tau^*(u_0,\omega)$ a.s. positive, such that (2.1) has an adapted solution u, defined a.s. on $[0,\tau^*(u_0)[$, unique in some class, and with paths a.s. in $C([0,\tau^*(u_0)[;H^s(\mathbb{R}))$. If $s=1$, the uniqueness holds among solutions with paths in $C([0,\tau^*(u_0)[;H^1(\mathbb{R}))$ a.s; moreover, the stopping time $\tau^*(u_0)$ satisfies*

$$\tau^*(u_0)=+\infty\quad\text{or}\quad\limsup_{t\nearrow\tau^*(u_0)}\|u-U(\cdot)u_0\|_{Y^t_{b,0}}=+\infty,\ a.s.$$

Remark 2.11. Let us explain what we mean by a solution on the random interval $[0,\tau^*(u_0)[$. This means that u is defined on $[0,\tau^*(u_0)[$ and is an adapted process such that for any stopping time $\tau<\tau^*(u_0)$ the following holds on $[0,\tau]$:

$$u(t)=U(t)u_0-\frac{1}{2}\int_0^t U(t-r)\partial_x(u^2(r))dr+\int_0^{t\wedge\tau}U(t-r)(u(r)\varphi dW(r)).$$

Proof. Let $z(t)=U(t)u_0$ and let $v_R\in Y^{T_0}_{b,s}$ for any $b<1/2$ and any $T_0>0$ be the solution of (2.3) given by Theorem 2.8. We then set $\tau_R=\inf\{t\geq 0,\ \|v_R\|_{Y^t_{b,0}}\geq R\}$; for $t\in[0,\tau_R]$, we have $\theta_R(\|v_R\|_{Y^t_{b,0}})=1$, hence v_R is a solution of (2.2) on $[0,\tau_R]$. It is not difficult to see that τ_R is non decreasing in R and that $v_{R+1}=v_R$ on $[0,\tau_R]$. Hence we may define u on $[0,\tau^*(u_0)[$ with $\tau^*(u_0)=\lim_{R\to\infty}\tau_R$ by setting $u(t)=v_R(t)+z(t)$ for $t\in[0,\tau_R]$ and u is then a solution of (2.1) on $[0,\tau^*(u_0)[$. The uniqueness for u holds in the class $z+Y^{\tau_R}_{b,0}$ for any R and it is not difficult to see that any solution u with paths in $C([0,\tau^*(u_0)[;H^1(\mathbb{R}))$ is in this class. The last property of the lemma is an immediate consequence of the definition of $\tau^*(u_0)$. □

3. Global existence

As already seen, Theorem 2.8 gives a local in time existence result for the equation without cut-off. In the present section, we end the proof of Theorem 1.1 by showing that those solutions are globally defined in time. To that aim we need an estimate on $\|v\|_{Y^T_{b,0}}$. We will use the following result.

Proposition 3.1. *Assume that $k \in L^2(\mathbb{R})$. Let $u \in C([0,\tau]); L^2(\mathbb{R}))$ be a solution of equation* (2.1) *with $u_0 \in L^2(\mathbb{R})$, where τ is a stopping time. Then, for any $m \geq 1$, $u \in L^{2m}(\Omega; C([0,\tau]; L^2(\mathbb{R})))$ and for any $T > 0$*

$$\mathbb{E}\left(\sup_{t\in[0,\tau\wedge T]} |u(t)|^{2m}_{L^2(\mathbb{R})}\right) \leq C(T, \|u_0\|_{L^2(\mathbb{R})}, m).$$

Proof. The result is a straightforward consequence of Itô formula. We prove it for $m = 2$. For $m \leq 2$ it then follows from Hölder inequality. For $m \geq 2$, the proof is similar.

We apply Itô formula to $M(u) = \|u\|^2_{L^2(\mathbb{R})}$ and obtain after a regularization argument and easy computations (see Ref. [1] for more details in the case of an additive noise or Ref. [10] for the case of the stochastic Schrödinger equation):

$$\begin{aligned} M(u(\tau\wedge r)) &= M(u_0) + 2\sum_{k=0}^{\infty}\int_0^{\tau\wedge r}\int_{\mathbb{R}} u^2(\sigma,x)\varphi e_k(x)dxd\beta_k(\sigma) \\ &+ \|k\|^2_{L^2(\mathbb{R})}\int_0^{\tau\wedge r} M(u(\sigma))d\sigma. \end{aligned}$$

We take the square of this identity and deduce:

$$\begin{aligned} &\mathbb{E}\left(\sup_{r\in[0,\tau\wedge T]} M^2(u(r))\right) \\ &\leq 2M^2(u_0) + 4\mathbb{E}\left(\sup_{r\in[0,\tau\wedge T]}\left|\sum_{k=0}^{\infty}\int_0^{r\wedge T}\int_{\mathbb{R}} u^2(\sigma,x)\varphi e_k(x)dxd\beta_k(\sigma)\right|^2\right) \\ &+2\|k\|^4_{L^2(\mathbb{R})}\mathbb{E}\left(\left(\int_0^{\tau\wedge T} M(u(\sigma))d\sigma\right)^2\right) \\ &\leq 2M^2(u_0) + \left(4\|k\|^2_{L^2(\mathbb{R})} + 2T\|k\|^4_{L^2(\mathbb{R})}\right)\mathbb{E}\left(\int_0^{\tau\wedge T} M^2(u(r))dr\right), \end{aligned}$$

thanks to Burkholder and Hölder inequalities, and to Lemma 2.6. The result follows from Gronwall Lemma. □

Let v_R be the solution given by Theorem 2.8, let $T_0 > 0$ be fixed, and let $\tau_R = \inf\{t \in [0, T_0],\ \|v_R\|_{Y^t_{b,0}} \geq R\}$; then on $[0,\tau_R]$, $v_R + z$ is a solution to (2.1) which is a.s. in $C([0,\tau_R]; L^2(\mathbb{R}))$ and Proposition 3.1 applies:

$$\mathbb{E}\left(\sup_{t\in[0,\tau_R\wedge T_0]} |v_R(t) + z(t)|^{2m}_{L^2(\mathbb{R})}\right) \leq C(T_0, \|u_0\|_{L^2(\mathbb{R})}, m). \tag{3.1}$$

We now show that this implies an estimate on the $Y^T_{b,0}$ norm of v_R.

Lemma 3.2. *Let v_R be the solution of the truncated equation* (2.3), *then there exists a constant $C(T_0, \|u_0\|_{L^2(\mathbb{R})})$ independent of R such that*

$$\mathbb{E}\left(\|v_R\|_{Y^{\tau_R}_{b,0}}\right) \leq C(T_0, \|u_0\|_{L^2(\mathbb{R})}).$$

Proof.
Step 1: Using similar arguments as in the beginning of the proof of Theorem 2.8, taking into account the fact that v_R satisfies equation (2.3), we prove using Lemma 2.2 that for $T_0 \geq T \geq 0$,

$$\|v_R\|_{Y_{b,0}^{T\wedge\tau_R}} \leq CT^{1-(a+b)}\left[\|v_R\|^2_{Y_{b,0}^{T\wedge\tau_R}} + \|z\|^2_{X_{c,0}^{T_0}}\right]$$

$$+\left\|\int_0^t U(t-r)\left[u_R(r)\varphi dW(r)\right]\right\|_{Y_{b,0}^{T_0}}$$

$$\leq CT^{1-(a+b)}\left[\|v_R\|^2_{Y_{b,0}^{T\wedge\tau_R}} + C(T_0)\left\|u_0\right\|^2_{L^2(\mathbb{R})}\right]$$

$$+\left\|\int_0^t U(t-r)\left[u_R(r)\varphi dW(r)\right]\right\|_{Y_{b,0}^{T_0}},$$

with $u_R(t) = v_R(t) + U(t)u_0$. We set

$$K_1 = K_1(\omega) = CT_0^{1-(a+b)}C(T_0)\left\|u_R\right\|^2_{C([0,T_0];L^2(\mathbb{R}))} + \left\|\int_0^t U(t-r)\left[u_R\varphi dW(r)\right]\right\|_{Y_{b,0}^{T_0}},$$

then

$$CT^{1-(a+b)}\left\|v_R\right\|^2_{Y_{b,0}^{T\wedge\tau_R}} - \left\|v_R\right\|_{Y_{b,0}^{T\wedge\tau_R}} + K_1 \geq 0.$$

Therefore, if we choose $T = T(\omega)$ such that $T^{1-(a+b)} = \dfrac{3}{16CK_1}$, we have

$$\|v_R\|_{Y_{b,0}^{T\wedge\tau_R}} \leq 2K_1.$$

Note indeed that $v_R(0) = 0$ and that $\|v_R\||Y^t_{b,0}$ is a continuous function of t. Similarly, for any $k \geq 0$, we define

$$v_R^k(t) = u_R(t) - U(t-kT)u_R(kT), \quad t \in [kT, (k+1)T],$$

with $T = T(\omega)$ chosen above. Then the same argument shows that

$$\left\|v_R^k\right\|_{Y_{b,0}^{[kT\wedge\tau_R,(k+1)T\wedge\tau_R]}} \leq 2K_1,$$

where we use the space $Y_{b,0}^{[T_1,T_2]}$ whose definition is exactly the same as $Y_{b,0}^T$ but $[0,T]$ is replaced by $[T_1,T_2]$.
Step 2: Since u_R is a solution of (2.1) on $[0,\tau_R]$, we may write for any $t \in [0,\tau_R]$, $u_R(t)$ as

$$\begin{aligned} u_R(t) &= U(t)u_0 - \tfrac{1}{2}\int_0^t U(t-r)\partial_x(u_R^2(r))dr + \int_0^t U(t-r)\left[u_R(r)\varphi dW(r)\right] \\ &= U(t)u_0 - \frac{1}{2}\sum_{k=0}^{k_t}\int_{kT}^{(k+1)T\wedge t} U(t-r)\partial_x\left[\left(v_R^k(r) + U(r-kT)u_R(kT)\right)^2\right]dr \\ &+ \int_0^t U(t-r)\left[u_R(r)\varphi dW(r)\right], \end{aligned}$$

where k_t is the integer part of t/T. Using this decomposition and the unitarity of $U(\sigma)$ in $\dot{H}^{-3/8}$ for any σ, we deduce that for any $t \in [0, \tau_R]$,

$$\begin{aligned}
&\|u_R(t) - U(t)u_0\|_{\dot{H}^{-3/8}(\mathbb{R})} \\
&\leq \frac{1}{2}\sum_{k=0}^{k_t} \left\| \int_{kT}^{(k+1)T\wedge t} U(t-r)\partial_x \left[\left(v_R^k(r) + U(r-kT)u_R(kT) \right)^2 \right] dr \right\|_{\dot{H}^{-3/8}(\mathbb{R})} \\
&+ \left\| \int_0^t U(t-r)\left[u_R(r)\varphi dW(r) \right] \right\|_{\dot{H}^{-3/8}(\mathbb{R})} \\
&\leq \frac{1}{2}\sum_{k=0}^{k_t} \left\| \int_{kT}^{(k+1)T\wedge t} U((k+1)T\wedge t - r)\partial_x \left[\left(v_R^k(r) + U(r-kT)u_R(kT) \right)^2 \right] dr \right\|_{\dot{H}^{-3/8}(\mathbb{R})} \\
&+ \left\| \int_0^t U(t-r)\left[u_R(r)\varphi dW(r) \right] \right\|_{\dot{H}^{-3/8}(\mathbb{R})}.
\end{aligned}$$

Now, suppose that a is fixed with $0 < a < 1/2$ in such a way that Proposition 2.3 holds, and set $\tilde{b} = 1 - a$, so that $\tilde{b} > 1/2$. Then, using the fact that $Y_{\tilde{b},0}^{[T_1,T_2]} \subset C([[T_1,T_2]; L^2(\mathbb{R}) \cap \dot{H}^{-3/8}(\mathbb{R}))$ for any positive T_1, T_2, we have for $t \in [0, \tau_R]$ and $k = 0, \ldots, k_t$,

$$\begin{aligned}
&\left\| \int_{kT}^{(k+1)T\wedge t} U((k+1)T\wedge t - r)\partial_x \left[\left(v_R^k(r) + U(r-kT)u_R(kT) \right)^2 \right] dr \right\|_{\dot{H}^{-3/8}(\mathbb{R})} \\
&\leq \left\| \int_{kT}^{\cdot} U(\cdot - r)\partial_x \left[\left(v_R^k(r) + U(r-kT)u_R(kT) \right)^2 \right] dr \right\|_{C([kT\wedge\tau_R,(k+1)T\wedge\tau_R];\dot{H}^{-3/8}(\mathbb{R}))} \\
&\leq C \left\| \int_{kT}^{\cdot} U(\cdot - r)\partial_x \left[\left(v_R^k(r) + U(r-kT)u_R(kT) \right)^2 \right] dr \right\|_{Y_{\tilde{b},0}^{[kT\wedge\tau_R,(k+1)T\wedge\tau_R]}}.
\end{aligned}$$

By Lemma 2.9, the above term is majorized for each $k \in \{0, \cdots, k_t\}$ by

$$\begin{aligned}
&CT^{1-(a+\tilde{b})} \left\| \partial_x \left[\left(v_R^k + U(\cdot - kT)u_R(kT) \right)^2 \right] \right\|_{Y_{-a,0}^{[kT\wedge\tau_R,(k+1)T\wedge\tau_R]}} \\
&\leq C \left\{ \|v_R^k\|^2_{Y_{\tilde{b},0}^{[kT\wedge\tau_R,(k+1)T\wedge\tau_R]}} + \|u_R(kT)\|^2_{L^2(\mathbb{R})} \right\},
\end{aligned}$$

by Proposition 2.3 and Lemma 2.9 again, since $a + \tilde{b} = 1$. By the result of step 1, we obtain

$$\|u_R(t) - U(t)u_0\|_{\dot{H}^{-3/8}(\mathbb{R})} \leq K_2, \quad t \in [0, \tau_R],$$

with

$$\begin{aligned}
K_2 = &\frac{1}{2}CT_0T^{-1}\left[4K_1^2 + \|u_R\|^2_{C([0,T_0];L^2(\mathbb{R}))} \right] \\
&+ \left\| \int_0^{\cdot} U(\cdot - r)\left[u_R(r)\varphi dW(r) \right] \right\|_{C([0,T_0];\dot{H}^{-3/8}(\mathbb{R}))}.
\end{aligned}$$

Step 3: By Lemma 2.9, and the unitarity of $U(kT)$ on $L^2(\mathbb{R}) \cap \dot{H}^{-3/8}(\mathbb{R})$ we have

$$\begin{aligned}\|U(t-kT)u_R(kT) - U(t)u_0\|_{Y_{b,0}^{[kT\wedge\tau_R,(k+1)T\wedge\tau_R]}} \\ \le C\,\|U(-kT)u_R(kT) - u_0\|_{L^2(\mathbb{R})\cap\dot{H}^{-3/8}(\mathbb{R})} \\ \le C\,\|u_R(kT) - U(kT)u_0\|_{L^2(\mathbb{R})\cap\dot{H}^{-3/8}(\mathbb{R})}\,.\end{aligned}$$

Therefore, using step 2,

$$\begin{aligned}\|U(\cdot - kT)u_R(kT) - U(\cdot)u_0\|_{Y_{b,0}^{[kT\wedge\tau_R,(k+1)T\wedge\tau_R]}} \le K_3 \\ = C\left(K_2 + 2\,\|u_R\|_{C([0,T_0];L^2(\mathbb{R}))}\right).\end{aligned}$$

Finally, for $t \in [kT\wedge\tau_R, (k+1)T\wedge\tau_R]$, we have

$$v_R(t) = v_R^k(t) + U(t-kT)u_R(kT) - U(t)u_0,$$

and we may write, k_0 being the integer part of T_0/T,

$$\begin{aligned}\|v_R\|_{Y_{b,0}^{\tau_R}} &\le \sum_{k=0}^{k_0} \|v_R\|_{Y_{b,0}^{[kT\wedge\tau_R,(k+1)T\wedge\tau_R]}} \\ &\le \sum_{k=0}^{k_0} \|v_R^k\|_{Y_{b,0}^{[kT\wedge\tau_R,(k+1)T\wedge\tau_R]}} + \|U(\cdot - kT)u(kT) - U(\cdot)u_0\|_{Y_{b,0}^{[kT\wedge\tau_R,(k+1)T\wedge\tau_R]}} \\ &\le \left(\frac{T_0}{T}+1\right)(2K_1 + K_3)\,.\end{aligned}$$

Note that T^{-1} is proportional to $K_1^{1/(1-(a+b))}$; by Proposition 3.1 and Proposition 2.7, K_1 and K_3 have all moments finite, and it follows

$$\mathbb{E}\left(\|v_R\|_{Y_{b,0}^{\tau_R}}\right) \le c(T_0, \|u_0\|_{L^2(\mathbb{R})})$$

which concludes the proof of Lemma 3.2. □

It is now straightforward to achieve the proof of Theorem 1.1. Indeed, due to Corollary 2.10, it suffices to see that $\limsup_{R\to\infty} \tau_R \wedge T_0 = T_0$ in probability as R goes to infinity. But this is an easy consequence of Markov inequality and Lemma 3.2, since

$$\mathbb{P}\left(\tau_R < T_0\right) = \mathbb{P}\left(\|v_R\|_{Y_{b,0}^{T_0\wedge\tau_R}} \ge R\right).$$

References

[1] A. de Bouard and A. Debussche, On the stochastic Korteweg-de Vries equation, *J. Funct. Anal.* **154**(1), 215–251, (1998). ISSN 0022-1236.

[2] A. de Bouard, A. Debussche, and Y. Tsutsumi, White noise driven Korteweg-de Vries equation, *J. Funct. Anal.* **169**(2), 532–558, (1999). ISSN 0022-1236.

[3] A. De Bouard, A. Debussche, and Y. Tsutsumi, Periodic solutions of the Korteweg-de Vries equation driven by white noise, *SIAM J. Math. Anal.* **36**(3), 815–855 (electronic), (2004/05). ISSN 0036-1410.

[4] R. L. Herman, The stochastic, damped KdV equation, *J. Phys. A.* **23**(7), 1063–1084, (1990). ISSN 0305-4470.

[5] A. de Bouard and A. Debussche, Random modulation of solitons for the stochastic Korteweg-de Vries equation, *Ann. Inst. H. Poincaré Anal. Non Linéaire.* **24**, (2007). ISSN 0294-1449. To appear.

[6] J. Bourgain, Fourier transform restriction phenomena for certain lattice subsets and applications to nonlinear evolution equations. II. The KdV-equation, *Geom. Funct. Anal.* **3**(3), 209–262, (1993). ISSN 1016-443X.

[7] C. E. Kenig, G. Ponce, and L. Vega, The Cauchy problem for the Korteweg-de Vries equation in Sobolev spaces of negative indices, *Duke Math. J.* **71**(1), 1–21, (1993). ISSN 0012-7094.

[8] C. E. Kenig, G. Ponce, and L. Vega, A bilinear estimate with applications to the KdV equation, *J. Amer. Math. Soc.* **9**(2), 573–603, (1996). ISSN 0894-0347.

[9] R. A. Adams, *Sobolev spaces.* (Academic Press [A subsidiary of Harcourt Brace Jovanovich, Publishers], New York-London, 1975). Pure and Applied Mathematics, Vol. 65.

[10] A. de Bouard and A. Debussche, The stochastic nonlinear Schrödinger equation in H^1, *Stochastic Anal. Appl.* **21**(1), 97–126, (2003). ISSN 0736-2994.

Chapter 5

On Stochastic Burgers Equation Driven by a Fractional Laplacian and Space-Time White Noise

Zdzisław Brzeźniak and Latifa Debbi*

Department of Mathematics, University of York
Heslington, York YO10 5DD, UK
zb500@york.ac.uk

We prove existence and uniqueness of a mild global solution to the Cauchy problem for the stochastic fractional Burgers equation on an interval with multiplicative space time white noise and periodic boundary conditions when the power of the Laplace operator is between 3/4 and 1.

Contents

1. Introduction

It is widely accepted that the classical Burgers equation can be used to model several physical phenomena and because of this it has been extensively studied. Recently its random generalizations have been introduced either by considering noisy force, see Refs. [1, 2] or by assuming randomness of the initial data, see a recent paper [3]. Moreover other types of generalizations have been studied. For example by considering an arbitrary algebraic nonlinearity instead of the quadratic, see Ref. [4] or by replacing the Laplace operator by a non local operator, see Ref. [5]). The first kind is called stochastic Burgers equation and the two latter ones are called generalized Burgers equation. Generalized Burgers equations with fractional differential operators are used to model some anomalous diffusions such as the far

*Institut Elie Cartan, Nancy 1 B.P 239, 54506 Vandoeuvre-Les-Nancy cedex, France & Department of Mathematics, Faculty of Sciences, University Ferhat Abbas, El-Maabouda Sétif 19000, Algeria, debbi@iecn.u-nancy.fr

field (or long time) behavior of the acoustic waves propagating in a gas-filled tube and the wave propagation in viscoelastic medium (see Refs. [6–8] for the generalized Burgers equation and Refs. [9, 10] and the references therein for more details on anomalous diffusions and fractional differential equations). Generalized Burgers equations can also be found in works on continuum mechanics and hydrodynamics (shallow water, bubbly liquid) and in molecular biology.

The equation dealt with in this paper is a generalization of Burgers equation which is characterized by the coexistence of the stochastic noise and the fractional power of the Laplacian. We are motivated by the tri-interaction between the wave steepening given by the nonlinearity and by the small dissipation given by the fractional power of the Laplacian and by the irregular random perturbation given by the cylindrical noise. In particular, the critical power for the Laplacian is the same as the one obtained in the deterministic fractional Burgers' equation [5]. Some other stochastic fractional partial differential equations have been studied in several papers, see Refs. [11–15]. In this work, we are interested in the following Cauchy problem for a Burgers equation driven by a fractional Laplacian and space-time white noise:

$$\begin{cases} du(t) = [-A_\alpha u(t) + Bu^2(t)]\,dt + g(u(t))\,dW(t),\ t > 0, \\ u(0) = u_0, \end{cases} \tag{1.1}$$

where $\frac{3}{2} < \alpha < 2$, $A_\alpha = A^{\frac{\alpha}{2}}$, $A = -\Delta$, $D(A) = H^{2,2}(0,1) \cap H_0^{1,2}(0,1)$. Here by $H^{k,p}(0,1)$, for $k \in \mathbb{N}$, $p \in [1,\infty)$ we denote the Banach space of all $f \in L^p(0,1)$ for which $D^j f \in L^p(0,1)$, $j = 0, 1, \dots, k$. The norm in $H^{k,p}(0,1)$ is given by

$$\| f \|_{H^{k,p}(0,1)} = \left(\sum_{j=0}^{k} |D^j f|_{L^p(0,1)} \right)^{\frac{1}{p}}. \tag{1.2}$$

We define the fractional order Sobolev space $H^{\beta,p}(0,1)$, $\beta \in \mathbb{R}^+ \setminus \mathbb{N}$ by the complex interpolation method, i.e.

$$H^{\beta,p}(0,1) = [H^{k,p}(0,1), H^{m,p}(0,1)]_\vartheta, \tag{1.3}$$

where $k, m \in \mathbb{N}, \vartheta \in (0,1)$, $k < m$, are chosen to satisfy

$$\beta = (1-\vartheta)k + \vartheta m. \tag{1.4}$$

One should bear in mind that the space on the LHS of formula (1.3) does not depend on k, m, ϑ provided they satisfy condition (1.4).

In what follows by $H_0^{s,p}(0,1)$, $s \geq 0$, $p \in (1,\infty)$ we will denote the closure of $C_0^\infty(0,1)$ in the Banach space $H^{s,p}(0,1)$. It is well known, see Theorem 11.1 in Ref. [16] and Theorem 1.4.3.2 on p.317 in Ref. [17] and Theorem 3.40 in Ref. [18] that $H_0^{s,p}(0,1) = H^{s,p}(0,1)$ iff $s \leq \frac{1}{p}$.

Since A is a self-adjoint operator in the Hilbert space $H = L^2(0,1)$, the fractional power, see Ref. [19], $A_\alpha = A^{\frac{\alpha}{2}}$ is well defined and it follows from Theorem 1.15.3 on p. 103 in Ref. [17] that $D(A_\alpha) = [H, D(A)]_{\frac{\alpha}{2}}$, where $[H, D(A)]_{\frac{\alpha}{2}}$ is the complex interpolation space of order $\frac{\alpha}{2}$, see Refs. [16], [17] and Theorem 4.2 in Ref. [20]. Moreover, by Seeley [21], see also Theorem in section 4.4.3 in the monograph [17], we have that

$$D(A_\alpha) = \begin{cases} H^{\alpha,2}(0,1) \cap H_0^{1,2}(0,1), & \text{if } 1 < \alpha \le 2, \\ H_0^{\alpha,2}(0,1), & \text{if } \frac{1}{2} < \alpha \le 1, \\ \tilde{H}^{\alpha,2}(0,1), & \text{if } \alpha = \frac{1}{2}, \\ H^{\alpha,2}(0,1), & \text{if } \alpha < \frac{1}{2}, \end{cases} \tag{1.5}$$

where $\tilde{H}^{\alpha,2}(0,1)$ is the space of all $f \in H^{\alpha,2}(\mathbb{R})$ such that $\text{supp}(f) \subset [0,1]$.

Since the operator A is self-adjoint, A^{-1} exists and is compact, there exists an ONB $\{e_j\}_{j\in\mathbb{N}}$ and a sequence $\{\lambda_j\}_{j\in\mathbb{N}}$ such that $\lambda_j \to \infty$ and $Ae_j = \lambda_j e_j$, $j \in \mathbb{N}$. In fact, in our special case, $e_k = \sqrt{2}\sin k\pi\cdot$ and $\lambda_k = k^2\pi^2$, $k \in \mathbb{N}$. Then it is well known that $D(A_\alpha) = D(A^{\alpha/2}) = \{v \in L^2(0,1) : \sum_{k=1}^\infty \lambda_k^\alpha v_k^2 < \infty\}$, where $v_k = \langle v, e_k\rangle = \sqrt{2}\int_0^1 v(x)\sin k\pi x\,dx$.

Notice that B is the first order differential operator defined by

$$Bu = \frac{\partial u}{\partial x}$$

and $g : \mathbb{R} \to \mathbb{R}$, is a bounded Lipschitz continuous function on $\mathbb{R}$. Let us note here that the Lipschitz condition is not restrictive. In fact, it will be easily seen from our proof that our method works when g is locally Lipschitz . $\{W(t), t \ge 0\}$ is a cylindrical Wiener process on the probability space $(\Omega, \mathcal{F}, \{\mathcal{F}\}_{t\ge 0}, \mathbb{P})$. The initial condition u_0 is a $L^2(0,1)$-valued $\mathcal{F}_0$-measurable function. Let us denote by $\{S_\alpha(t), t \ge 0\}$ the semigroup generated by $-A_\alpha$. The following is now widely accepted definition, see Ref. [22].

Definition 1.1. Suppose that $\frac{3}{2} < \alpha < 2$. An $\mathcal{F}_t$-adapted $L^2(0,1)$-valued continuous process $u = \{u(t), t \ge 0\}$ is called a mild solution of equation (1.1) iff for some $p > \frac{2\alpha}{\alpha - 1}$

$$\mathbb{E} \sup_{t\in[0,T]} |u(t)|_{L^2}^p < \infty, \quad T > 0 \tag{1.6}$$

and for all $t \ge 0$, a.s. the following identity holds

$$u(t) = S_\alpha(t)u_0 + \int_0^t S_\alpha(t-s)Bu^2(s)\,ds + \int_0^t S_\alpha(t-s)g(u(s))\,dW(s). \tag{1.7}$$

In the representation of the stochastic integral in (1.7), with a slight abuse of notation, g is regarded as a nonlinear operator from $H = L^2(0,1)$ to $\mathcal{L}(H)$, the set of bounded linear operators on H, defined by[a] $g(u)(h) = \{(0,1) \ni x \mapsto g(u(x))h(x) \in \mathbb{R}\}$. In other words, the nonlinear operator g is the Nemytski map associated with function g. For $v \in H$, $g(v)$ is given as a multiplicative operator. From the hypothesis that g is bounded we have $\|g(v)\| \le b_0$, where $b_0 = \sup_{\mathbb{R}} |g(x)|$.

[a]Note that because g is bounded, the natural domain of this nonlinear operator is the whole space H.

Remark 1.2. We will see later, in Lemmas 2.9 and 2.13 , that if an $\mathcal{F}_t$-adapted $L^2(0,1)$-valued continuous process $u = \{u(t), t \geq 0\}$ satisfies condition (1.6) then all terms on the RHS of (1.7) make sense.

The main result of our paper is the following

Theorem 1.3. *Let u_0 be a $L^2(0,1)$-values $\mathcal{F}_0$-measurable function such that*

$$\mathbb{E}|u_0|^p_{L^2} < \infty$$

for some $p > \frac{2\alpha}{\alpha-1}$ and let $T > 0$. Then there exists a unique mild solution of equation (1.1).

The paper is organized in the following way. In section 2 we prove existence and uniqueness of a global solutions to the approximate versions of problem (1.1). In section 3 we construct a local maximal solution to problem (1.1) and then prove that it is in fact a global solution.

2. Existence of global solutions to approximating equations

We begin with the following definition used throughout the whole paper.

Definition 2.1. Let $T > 0$ and $p \in [1, \infty]$ be fixed and X is a separable Banach space. By $Z_{T,p}(X)$ we denote the space of all X-valued continuous and $\mathcal{F}_t$-adapted processes $u = \{u(t), t \in [0,T]\}$ such that

$$\|u\|^p_{T,X,p} := \mathbb{E} \sup_{t\in[0,T]} |u(t)|^p_X < \infty. \tag{2.1}$$

When $X = L^2(0,1)$ we will usually denote the space $Z_{T,p}(L^2(0,1))$ by $Z_{T,p}$ and the norm $\|\cdot\|_{T,L^2(0,1),p}$ by $\|\cdot\|_{T,p}$.

We have the following simple but useful observation.

Proposition 2.2. *In the framework of Definition 2.1 the family spaces $Z_{T,p}(X)$ is decreasing with respect to parameter p. To be precise, if $1 \leq p_1 \leq p_2 \leq \infty$, then $Z_{T,p_2}(X) \subset Z_{T,p_1}(X)$ and*

$$\|u\|_{T,X,p_1} \leq \|u\|_{T,X,p_2}, \text{ for all } u \in Z_{T,X,p_2}(X).$$

Proof. Follows immediately from Jensen's Inequality. □

We fix, for the time being, $T > 0$ and $p > 2$.

Let π_n be the projection from H onto the ball $B(0,n) \subset L^2(0,1)$ defined by

$$\pi_n(v) = \begin{cases} v, & \text{if } |v|_{L^2} \leq n, \\ \frac{n}{|v|_{L^2}} v, & \text{if } |v|_{L^2} > n. \end{cases}$$

The following result is well known fact, see Refs. [1, 23, 24].

Lemma 2.3. *The map $\pi_n : L^2(0,1)\to L^2(0,1)$ is globally Lipschitz with Lipschitz constant* 1. *Moreover,*

$$\begin{aligned} |(\pi_n u)^2 - (\pi_n v)^2|_{L^1} &\le 2n|u-v|_{L^2}, \quad \text{for all } u,v \in L^2, \\ |(\pi_n u)^2|_{L^1} &\le n|u|_{L^2}, \quad \text{for all } u \in L^2, \\ |(\pi_n u)^2|_{L^1} &\le |u|^2_{L^2}, \quad \text{for all } u \in L^2. \end{aligned} \tag{2.2}$$

We fix $n \in \mathbb{N}$ and consider operators $H_n, G : Z_{T,p} \to Z_{T,p}$ defined by

$$H_n u(t) = \int_0^t S_\alpha(t-s)B(\pi_n u(s))^2\, ds \tag{2.3}$$

and

$$Gu(t) = \int_0^t S_\alpha(t-s)g(u(s))\, dW(s). \tag{2.4}$$

Let us recall that here g is regarded as a multiplicative operator. The above constitute a special part of the following result.

Lemma 2.4. *Suppose that $q \in [2,\infty]$ and $z \in L^q(0,1)$. Let Z denotes the multiplication operator by z. Then,*

$$\|S_\alpha(t)Z\|^2_{HS} \le 2^{\frac{2}{q}}|z|^2_{L^q}\sum_k e^{-2\pi^\alpha k^\alpha t} = 2^{\frac{2}{q}}|z|^2_{L^q}\|S_\alpha(t)\|^2_{HS},\ t>0,$$

where by $\|.\|_{HS}$ we denote the Hilbert-Schmidt norm.

Proof. Let $e_k = \sqrt{2}\sin k\pi\cdot$ and $\mu_k = \lambda_k^{\alpha/2}$, where $\lambda_k = k^2\pi^2$, $k \in \mathbb{N}$. Then, as explained earlier, $\{e_k\}$ is an ONB of H and $A_\alpha e_k = \mu_k e_k$, $k \in \mathbb{N}$. Then, because both $S_\alpha(t)$ and Z are self-adjoint on H it follows from [25], section V.2.4,

$$\|S_\alpha(t)Z\|^2_{HS} = \sum_k |ZS_\alpha(t)e_k|^2_{L^2} = \sum_k |ze^{-\mu_k t}e_k|^2_{L^2} = \sum_k e^{-2\mu_k t}|ze_k|^2_{L^2}.$$

Let us observe that by the Hölder inequality, $|ze_k|_{L^2} \le |z|_{L^q}|e_k|_{L^r}$, where $\frac{1}{r}+\frac{1}{q} = \frac{1}{2}$. Moreover, since $|e_k|_{L^2} = 1$ and $|e_k|_{L^\infty} = 2^{1/2}$ it follows by applying the Hölder inequality that $|e_k|_{L^r} \le 2^{1/q}$. Therefore,

$$\|S_\alpha(t)Z\|^2_{HS} \le 2^{\frac{2}{q}}|z|^2_{L^q}\sum_k e^{-2\mu_k t},$$

what proves the result. □

The above Lemma implies the following result.

Corollary 2.5. *Suppose that $z \in L^q(0,1)$ for $q \in [2,\infty]$. Let Z denotes the multiplication operator by z. Then, for $\beta < 1 - \frac{1}{\alpha}$, one has*

$$\begin{aligned} \int_0^t s^{-\beta}\|S_\alpha(s)Z\|^2_{HS}\, ds &\le \int_0^\infty s^{-\beta}\|S_\alpha(s)Z\|^2_{HS}\, ds \\ &\le 2^{\frac{2}{q}}|z|^2_{L^q}(2\pi^\alpha)^{\beta-1}\Gamma(1-\beta)\sum_{k=1}^{+\infty} k^{-\alpha(1-\beta)} < \infty, \quad t>0. \end{aligned} \tag{2.5}$$

Proof. From the Fubini Theorem we have

$$\begin{aligned}\int_0^\infty s^{-\beta}\|S_\alpha(s)Z\|_{HS}^2\,ds &\le 2^{\frac{2}{q}}|z|_{L^q}^2\int_0^\infty s^{-\beta}\sum_{k=1}^{+\infty}e^{-2s(k\pi)^\alpha}\,ds\\ &= 2^{\beta-1}2^{\frac{2}{q}}|z|_{L^q}^2\sum_{k=1}^{+\infty}(k\pi)^{\alpha\beta-\alpha}\int_0^\infty \tau^{-\beta}e^{-\tau}d\tau\\ &\le 2^{\frac{2}{q}}|z|_{L^q}^2(2\pi^\alpha)^{\beta-1}\Gamma(1-\beta)\sum_{k=1}^{+\infty}k^{-\alpha(1-\beta)}.\end{aligned}$$

Since by our assumptions $\alpha(1-\beta)>1$ the series on the RHS above is convergent and the result follows. □

Remark 2.6. Let us observe that the constant $(2\pi^\alpha)^{\beta-1}\Gamma(1-\beta)\sum_{k=1}^{+\infty}k^{-\alpha(1-\beta)}$ is equal to $\int_0^\infty s^{-\beta}\|S_\alpha(s)\|_{HS}^2\,ds$.

Let us recall that for $0<\delta\le 1$ the fractional power A_α^δ of A_α, see Ref. [19], is defined as the inverse of operator $A_\alpha^{-\delta}$ defined by

$$A_\alpha^{-\delta}=\frac{1}{\Gamma(\delta)}\int_0^\infty t^{\delta-1}S_\alpha(t)dt.$$

Since, see Lemma 2.6.6 in Ref. [19], the operator $A_\alpha^{-\delta}$ is injective, $A_\alpha^\delta := \left(A_\alpha^{-\delta}\right)^{-1}$ is well defined. Furthermore, it is known that A_α^δ is a closed densely defined operator and that $D(A_\alpha^\delta)=R(A_\alpha^{-\delta})$, where $R(\Lambda)$ denotes the range of an operator Λ.

The following result is just Lemma 3.3 from [26], see also [22], [27] and [28]. We state and prove it here for the convenience of the reader.

Lemma 2.7. *Provided that $\gamma>p^{-1}+\delta$ the operator $R_\gamma : L^p(0,T;L^2(0,1)) \to C([0,T];D(A_\alpha^\delta))$ given by*

$$R_\gamma h(t)=\int_0^t (t-s)^{\gamma-1}S_\alpha(t-s)h(s)\,ds,\ h\in L^p(0,T;L^2(0,1)),$$

is well defined, linear and bounded. Moreover, there exists a constant $C_{p,\gamma-\delta}>0$ such that for all $h\in L^p(0,T;L^2(0,1))$

$$|R_\gamma h|_{C([0,T];D(A_\alpha^\delta))}\le C_{p,\gamma-\delta}T^{\gamma-\delta-\frac{1}{p}}|h|_{L^p(0,T;L^2(0,1))}. \tag{2.6}$$

Proof. Let us fix $h\in L^p(0,T;L^2(0,1))$. Then for $t\in(0,T)$ we have

$$
\begin{aligned}
|R_\gamma h(t)|_{D(A_\alpha^\delta)} &= |A_\alpha^\delta \int_0^t (t-s)^{\gamma-1} S_\alpha(t-s) h(s)\, ds|_{L^2} \\
&\le \int_0^t (t-s)^{\gamma-1} |A_\alpha^\delta S_\alpha(t-s) h(s)|_{L^2}\, ds \\
&\le \int_0^t (t-s)^{\gamma-1} \|A_\alpha^\delta S_\alpha(t-s)\| |h(s)|_{L^2}\, ds.
\end{aligned}
$$

Since A_α is self-adjoint it follows by Lemma 2.10 from [29] that $\|A_\alpha^\delta S_\alpha(s)\| \le s^{-\delta}$, $s > 0$. Hence by applying the Hölder inequality we get

$$
\begin{aligned}
|R_\gamma h(t)|_{D(A_\alpha^\delta)} &\le \int_0^t (t-s)^{\gamma-\delta-1} |h(s)|_{L^2}\, ds \\
&\le \Big(\int_0^T s^{(\gamma-\delta-1)\frac{p}{p-1}}\, ds \Big)^{\frac{p-1}{p}} \Big(\int_0^T |h(s)|_{L^2}^p\, ds \Big)^{\frac{1}{p}}.
\end{aligned}
$$

The integral $\int_0^T s^{(\gamma-\delta-1)\frac{p}{p-1}}\, ds < \infty$ when $\gamma > p^{-1} + \delta$. Hence for all $t \ge 0$

$$
|R_\gamma h(t)|_{D(A_\alpha^\delta)} \le \Big(\frac{1-1/p}{\gamma-\delta-1/p} \Big)^{1-\frac{1}{p}} T^{\gamma-\delta-\frac{1}{p}} |h|_{L^p(0,T;L^2(0,1))}.
$$

□

Remark 2.8. When $\delta = 0$ then $R_\gamma : L^p(0,T;L^2(0,1)) \to C([0,T];L^2(0,1))$ is a bounded linear operator and because the semigroup $\{S_\alpha(t)\}_{t\ge 0}$ is contractive, $C_{p,1} = 1$.

Lemma 2.9. *If $g : \mathbb{R}\to\mathbb{R}$ is linear growth, i.e. for some $b_0, b_1 \ge 0$*

$$
|g(x)| \le b_0 + b_1|x|, \quad x \in \mathbb{R}, \tag{2.7}
$$

$\alpha > 1$ and $p > \frac{2\alpha}{\alpha-1}$, then the nonlinear operator $G : Z_{T,p} \to Z_{T,p}$ is well defined and for $u \in Z_{T,p}$ one has

$$
\|Gu\|_{T,p} \le [C_{p,\gamma} b_0 + \sqrt{2} C_{p,\gamma} b_1 \|u\|_{T,p}] \widehat{C}_{p,\gamma} T^\gamma, \tag{2.8}
$$

where a positive number γ is such that $\frac{1}{p} < \gamma < \frac{1}{2}(1-\frac{1}{\alpha})$ and $C_{p,\gamma}$ is the constant from inequality (2.6) and $\widehat{C}_{p,\gamma}^p := C_p\big((2\pi^\alpha)^{2\gamma-1}\Gamma(1-2\gamma)\sum_{k\ge 1} k^{-\alpha(1-2\gamma)}\big)^{\frac{p}{2}}$.

Moreover, if g is a Lipschitz function, i.e. for some $b_2 \ge 0$

$$
|g(x_2) - g(x_1)| \le b_2 |x_2 - x_1|, \quad x_1, x_2 \in \mathbb{R}, \tag{2.9}
$$

then $G : Z_{T,p} \to Z_{T,p}$ is Lipschitz continuous and for all $u_1, u_2 \in Z_{T,p}$ the following inequality holds

$$
\|Gu_2 - Gu_1\|_{T,p} \le \sqrt{2} C_{p,\gamma} \widehat{C}_{p,\gamma} b_2 T^\gamma \|u_2 - u_1\|_{T,p}. \tag{2.10}
$$

Proof. We begin with an observation that a number γ such that $\frac{1}{p} < \gamma < \frac{1}{2}(1-\frac{1}{\alpha})$ exists thanks to one of our assumptions that $p > \frac{2\alpha}{\alpha-1}$. We choose it and fix for the remaining parts of the proof.

We first consider the case when g is bounded, i.e. g satisfies the condition (2.7) with $b_1 = 0$. Let $u \in Z_{T,p}$. By Lemma 2.4 we have

$$\int_0^t \|S_\alpha(t-s)g(u(s))\|_{HS}^2\,ds \le \int_0^t \|S_\alpha(t-s)\|_{HS}^2 \|g(u(s))\|_{L^\infty}^2\,ds \\ \le b_0^2 \int_0^t \|S_\alpha(s)\|_{HS}^2\,ds. \tag{2.11}$$

Since $\alpha > 1$ and so $0 < 1 - \frac{1}{\alpha}$ by Corollary 2.5 we infer that for all $t > 0$, the integral on the RHS of the inequality (2.11) is convergent. Hence the Itô stochastic integral $\int_0^t S_\alpha(t-s)g(u(s))\,dW(s)$ exists and thus the process $G(u)$ is well defined. We shall prove that in fact $Gu \in Z_{T,p}$.

Employing the factorization method (see Refs. [1, 22, 27]), we can rewrite $Gu(t)$ in the following form

$$Gu(t) = R_\gamma Y(t) = \int_0^t (t-s)^{\gamma-1} S_\alpha(t-s)Y(s)\,ds,\ t \in (0,T], \tag{2.12}$$

where

$$Y(s) = \frac{\sin \pi\gamma}{\pi} \int_0^s (s-r)^{-\gamma} S_\alpha(s-r)g(u(r))\,dW(r),\ s \in [0,T]. \tag{2.13}$$

Arguing as above, we see that because $2\gamma < 1 - \frac{1}{\alpha}$, the RHS of (2.13) is a well defined $L^2(0,1)$-valued stochastic Itô integral. Furthermore, since $\gamma > p^{-1}$, by Lemma 2.7, a process $Gu(t) = [R_\gamma Y](t)$, $t \in [0,T]$ is continuous H-valued provided we can show that a.s. $Y \in L^p(0,T;H)$. In fact, we will show below that a stronger fact holds, namely that $Y \in L^p(\Omega; L^p(0,T;H))$. Indeed by the Burkholder's inequality and Corollary 2.5 there exists a constant $C_p > 0$ such that

$$\int_0^T \mathbb{E}|Y(s)|_{L^2}^p\,ds \le C_p \int_0^T \mathbb{E}\Big(\int_0^s (s-r)^{-2\gamma}\|S_\alpha(s-r)g(u(r))\|_{HS}^2 dr\Big)^{\frac{p}{2}}\,ds \\ \le C_p b_0^p \int_0^T \Big(\int_0^s r^{-2\gamma}\|S_\alpha(r)\|_{HS}^2 dr\Big)^{\frac{p}{2}}\,ds \\ \le C_p b_0^p T\Big((2\pi^\alpha)^{2\gamma-1}\Gamma(1-2\gamma)\sum_{k=1}^{+\infty} k^{-\alpha(1-2\gamma)}\Big)^{\frac{p}{2}} = \widehat{C}_{p,\gamma}^p b_0^p T. \tag{2.14}$$

Since $0 < \gamma < \frac{1}{2} - \frac{1}{2\alpha}$ what implies that $\alpha(1-2\gamma) > 1$ we infer that the last term is finite. Furthermore, by inequalities (2.6) and (2.14), we get

$$\begin{aligned}\mathbb{E}\big(\sup_{t\in[0,T]} |Gu(t)|^p_{L^2}\big) &= \mathbb{E}\big(\sup_{t\in[0,T]} |R_\gamma Y(t)|^p_{L^2}\big) = \mathbb{E}\big(|R_\gamma Y|^p_{C(0,T;L^2(0,1))}\big), \\ &\le C^p_{p,\gamma} T^{\gamma p-1} \mathbb{E}|Y|^p_{L^p(0,T;L^2(0,1))} \\ &\le C^p_{p,\gamma} \widehat{C}^p_{p,\gamma} b_0^p T^{\gamma p} < \infty.\end{aligned}$$

This concludes the proof of inequality (2.8) in the special case considered. Next, we should consider the case $b_0 = 0$. However, the proof in this case is in fact a special case of the proof of the Lipschitz property of the map G given below and hence will be omitted. Moreover, the general case of inequality (2.8) is a consequence of the just proved special case and the proof of the inequality (2.10).

Finally, we shall deal with the Lipschitz property of the map G. Take $u_2, u_1 \in Z_{T,p}$. From the 1st part of the proof we know that $G(u_2)$ and $G(u_1)$ both belong to $Z_{T,p}$. However, in order to estimate the norm in of $G(u_2) - G(u_1)$ we need to use Lemma 2.4 with $q = 2$ while in the 1st part of the proof we used it with $q = \infty$.

We choose γ such that $p^{-1} < \gamma < \frac{1}{2} - \frac{1}{2\alpha}$ and introduce an auxiliary process h by

$$h(s) = \frac{\sin \pi\gamma}{\pi} \int_0^s (s-r)^{-\gamma} S_\alpha(s-r)(g(u_2(r)) - g(u_1(r))\, dW(r),\ t \ge 0.$$

Then we have $Gu_2(t) - Gu_1(t) = R_\gamma h(t)$, where R_γ is the operator defined in Lemma 2.7.

By the Burkholder inequality in view of Lemma 2.4 we have the following sequence of inequalities.

$$\begin{aligned}&\int_0^T \mathbb{E}\big|h(s)\big|^p_{L^2}\, ds \qquad (2.15)\\ &\le \pi^{-p} \int_0^T \mathbb{E}\Big|\int_0^s (s-r)^{-\gamma} S_\alpha(s-r)(g(u_2(r)) - g(u_1(r))\, dW(r)\Big|^p_{L^2} ds \\ &\le C_p \int_0^T \mathbb{E}\Big(\int_0^s (s-r)^{-2\gamma} \|S_\alpha(s-r)(g(u_2(r)) - g(u_1(r))\|^2_{HS} dr\Big)^{\frac{p}{2}} ds \\ &\le C_p 2^{\frac{p}{2}} \int_0^T \mathbb{E}\Big(\int_0^s (s-r)^{-2\gamma} \|S_\alpha(s-r)\|^2_{HS} |(g(u_2(r)) - g(u_1(r))|^2_{L^2} dr\Big)^{\frac{p}{2}} ds \\ &\le C_p 2^{\frac{p}{2}} b_2^p \int_0^T \mathbb{E}\Big(\int_0^s (s-r)^{-2\gamma} \|S_\alpha(s-r)\|^2_{HS} |u_2(r) - u_1(r)|^2_{L^2}\, dr\Big)^{\frac{p}{2}} ds \\ &\le C_p 2^{\frac{p}{2}} b_2^p \mathbb{E}[\sup_{[0,T]} |u_2(r) - u_1(r)|^p_{L^2}] \int_0^T \Big(\int_0^s r^{-2\gamma} \|S_\alpha(r)\|^2_{HS}\, dr\Big)^{\frac{p}{2}} ds \\ &\le \widehat{C}^p_{p,\gamma} 2^{\frac{p}{2}} b_2^p T \|u-v\|^p_{T,p} < \infty.\end{aligned}$$

Therefore, by Lemma 2.7 we have that

$$\begin{aligned}\|Gu - Gv\|^p_{T,p} &= \mathbb{E}|R_\gamma h|^p_{C(0,T;L^2)} \le C^p_{p,\gamma} T^{\gamma p-1}\mathbb{E}|h|^p_{L^p(0,T;L^2)}\\ &\le 2^{\frac{p}{2}} C^p_{p,\gamma}\widehat{C}^p_{p,\gamma} b_2^p T^{\gamma p}\|u-v\|^p_{T,p}.\end{aligned} \tag{2.16}$$

□

Remark 2.10. In the proof of the previous Lemma we used Lemma 2.7 with $\delta = 0$. If instead we used it with $\delta \ge 0$ such that $\delta + \frac{1}{p} < \frac{1}{2}(1-\frac{1}{\alpha})$ then by choosing γ such that

$$\delta + \frac{1}{p} < \gamma < \frac{1}{2}(1 - \frac{1}{\alpha}) \tag{2.17}$$

we would prove that G maps the space $Z_{T,p}(L^2)$ into the space $Z_{T,p}(D(A^\delta_\alpha))$. Moreover, under some appropriate assumptions the inequalities (2.8) and (2.10) take the following forms,

$$\begin{aligned}\|Gu\|_{Z_{T,p}(D(A^\delta_\alpha))} &\le [C_{p,\gamma-\delta}b_0 + \sqrt{2}C_{p,\gamma-\delta}b_1\|u\|_{T,p}]\widehat{C_{p,\gamma}}T^{\gamma-\delta},\\ \|Gu_2 - Gu_1\|_{Z_{T,p}(D(A^\delta_\alpha))} &\le \sqrt{2}C_{p,\gamma-\delta}\widehat{C}_{p,\gamma}b_2T^{\gamma-\delta}\|u_2-u_1\|_{T,p}.\end{aligned}$$

Lemma 2.11. *For each $\alpha > \frac{3}{2}$ there exists a constant $C_\alpha > 0$ such that for all $t > 0$ and for any bounded and strongly-measurable function $v : (0,t)\to L^1(0,1)$ the following inequality holds*

$$\int_0^t |S_\alpha(t-s)\frac{\partial v}{\partial x}(s)|_{L^2}\, ds \le C_\alpha t^{1-\frac{3}{2\alpha}} \sup_{s\le t}|v(s)|_{L^1}. \tag{2.18}$$

Proof. We begin by showing that if $v \in H^{1,2}(0,1)$ then

$$|S_\alpha(s)\frac{\partial v}{\partial x}|_{L^2} \le \sqrt{2}\pi\Big(\sum_{k=1}^{+\infty} 2\pi k^2 e^{-2s(k\pi)^\alpha}\Big)^{\frac{1}{2}}|v|_{L^1},\ s > 0. \tag{2.19}$$

By the Parseval Identity, the fact that $|\cos(k\pi\cdot)|_{L^\infty} \le 1$ and by integration by parts formula we have

$$\begin{aligned}|S_\alpha(s)\frac{\partial v}{\partial x}|^2_{L^2} &= \sum_{k=1}^{+\infty} e^{-2s(k\pi)^\alpha}\langle\frac{\partial}{\partial x}v, e_k\rangle^2\\ &\le \sum_{k=1}^{+\infty} e^{-2s(k\pi)^\alpha}\Big(\sqrt{2}\pi\int_0^1 \frac{\partial v}{\partial x}(x)\sin(k\pi x)\,dx\Big)^2\\ &\le \sum_{k=1}^{+\infty} e^{-2s(k\pi)^\alpha}\Big(k\sqrt{2}\pi\int_0^1 v(x)\cos k\pi x\,dx\Big)^2\\ &\le 2\pi^2\sum_{k=1}^{+\infty} k^2 e^{-2s(k\pi)^\alpha}|v|^2_{L^1}.\end{aligned}$$

This proves (2.19). Hence, if $v \in L^\infty(0,t;L^1(0,1))$, then

$$\int_0^t |S_\alpha(t-s)\frac{\partial v}{\partial x}(s)|_{L^2}\,ds \le \sqrt{2}\pi \int_0^t \Big(\sum_{k=1}^{+\infty} k^2 e^{-2(t-s)(k\pi)^\alpha}\Big)^{\frac{1}{2}} |v(s)|_{L^1}\,ds$$

$$\le \sqrt{2}\pi \sup_{s\le t} |v(s)|_{L^1} \int_0^t \Big(\sum_{k=1}^{+\infty} k^2 e^{-2(t-s)(k\pi)^\alpha}\Big)^{\frac{1}{2}} ds. \quad (2.20)$$

We need the following auxiliary result.

Lemma 2.12. *Let us define a function $h:(0,\infty)\to\mathbb{R}$ by formula*

$$h(s) = \sum_{k=1}^{\infty} k^2 e^{-2s(k\pi)^\alpha},\ s\in(0,\infty).$$

Then there exists $C>0$ such that

$$h(s) \le C s^{-3/\alpha}, \quad s>0.$$

Proof. [Proof of Lemma 2.12] The function h is the Laplace transform of a purely atomic measure $\mu = \sum_{k=1}^\infty k^2 \delta_{2(k\pi)^\alpha}$. Let $U(\sigma) := \mu((0,\sigma]) = \sum_{k\le \frac{1}{\pi}(\frac{\sigma}{2})^{\frac{1}{\alpha}}} k^2$. Then, by Ref. [30], we have

$$\mu((0,\sigma]) = \sum_{k\le [\frac{1}{\pi}(\frac{\sigma}{2})^{\frac{1}{\alpha}}]} k^2 = \frac{1}{6}[\frac{1}{\pi}(\frac{\sigma}{2})^{\frac{1}{\alpha}}]([\frac{1}{\pi}(\frac{\sigma}{2})^{\frac{1}{\alpha}}]+1)(2[\frac{1}{\pi}(\frac{\sigma}{2})^{\frac{1}{\alpha}}]+1)$$

$$U(\sigma) \sim \frac{1}{3}[\pi^{-1}(\frac{\sigma}{2})^{\frac{1}{\alpha}}]^3 = \frac{2^{-\frac{3}{\alpha}}\sigma^{\frac{3}{\alpha}}}{3\pi^3}, \quad \text{as } \sigma\to\infty,$$

where $[\cdot]$ denotes the integer part, we infer that for any $y>0$, $\lim_{\sigma\to\infty}\frac{U(\sigma y)}{U(\sigma)} = y^{\frac{3}{\alpha}}$. Therefore the function L defined by $L(x) = \frac{U(x)}{x^{\frac{3}{\alpha}}}$, $x>0$, varies slowly at infinity and hence by the Abelian theorem, see Theorem XIII.5.2 in Ref. [31] we infer that

$$h(s) \sim \frac{1}{\Gamma(\frac{3}{\alpha}+1)} s^{-\frac{3}{\alpha}} L(s^{-1}), \text{ as } s\to 0.$$

In particular, since $\lim_{x\to\infty} L(x) = \frac{2^{-\frac{3}{\alpha}}}{3\pi^3}$, there exists $t_0>0$ and $C_0>0$ such that $h(s)\le C_0 s^{-\frac{3}{\alpha}}$, for $s\in(0,t_0]$. Since the function $h(s)s^{\frac{3}{\alpha}}$, $s\in(0,\infty)$ is continuous and the $\lim_{s\to 0} h(s)s^{\frac{3}{\alpha}}$ and $\lim_{s\to\infty} h(s)s^{\frac{3}{\alpha}} = \lim_{s\to\infty} s^{\frac{3}{\alpha}} e^{-2s}\sum_{k=1}^\infty k^2 e^{-2s((k\pi)^\alpha-1)}$ exist, then it is bounded on $(0,\infty)$, hence it is on $[0,T]$. □

Combining Lemma 2.12 with inequality (2.20) we infer that Lemma 2.11 is valid for $v\in H^{1,2}(0,1)$. Hence the result follows by standard density argument. □

Proof. [An alternative proof of Lemma 2.12.]

$$h(s) = \sum_{k=1}^{\infty} k^2 e^{-2s(k\pi)^\alpha},\ s\in(0,\infty).$$

Since for each $N > 0$ there exists $C_N > 0$ such that

$$e^{-x} \le C_N x^{-N}, \; x > 0$$

we have, for some $N > 0$,

$$h(s) \le \sum_{k=1}^{\infty} k^2 (2s(k\pi)^\alpha)^{-N} = \frac{\widehat{C}_N}{s^N} \sum_{k=1}^{\infty} k^{2-N\alpha}$$

Taking $N > \frac{3}{\alpha}$ (so that the series $\sum_{k=1}^{\infty} k^{2-N\alpha}$ is convergent) we infer that there exists $\tilde{C}_N > 0$ such that

$$h(s) \le \frac{\tilde{C}_N}{s^N}, \qquad s \in (0,\infty).$$

Inserting the last inequality into (2.20) yields

$$\begin{aligned}\int_0^t |S_\alpha(t-s)\frac{\partial v}{\partial x}(s)|_{L^2}\, ds &\le C \sup_{s\le t} |v(s)|_{L^1} \int_0^t s^{-\frac{3}{2\alpha}}\, ds. \\ &\le C\frac{1}{1-\frac{3}{2\alpha}} \sup_{s\le t} |v(s)|_{L^1} t^{1-\frac{3}{2\alpha}}.\end{aligned}$$

□

Lemma 2.13. *Assume that $\alpha > \frac{3}{2}$ and $T > 0$. Then the nonlinear operator $H_n : Z_{T,p} \to Z_{T,p}$ is well defined and globally Lipschitz. Moreover, for all $n \in \mathbb{N}$ and $u, v \in Z_{T,p}$*

$$\begin{aligned}\mathbb{E} \sup_{t\in[0,T]} \Big| \int_0^t S_\alpha(t-s) B(\pi_n u(s))^2\, ds \Big|_{L^2}^p &\le n^p C_\alpha^p T^{p(1-\frac{3}{2\alpha})} \mathbb{E} \sup_{t\in[0,T]} |u(t)|_{L^2}^p, \\ \mathbb{E} \sup_{t\in[0,T]} \Big| \int_0^t S_\alpha(t-s) B(\pi_n u(s))^2\, ds - \int_0^t S_\alpha(t-s) B(\pi_n v(s))^2\, ds \Big|_{L^2}^p & \\ \le (2n)^p C_\alpha^p T^{p(1-\frac{3}{2\alpha})} \mathbb{E} \sup_{t\in[0,T]} |u(t)-v(t)|_{L^2}^p&, \\ \mathbb{E} \sup_{t\in[0,T]} \Big| \int_0^t S_\alpha(t-s) B(u(s))^2\, ds \Big|_{L^2}^{p/2} &\le C_\alpha^{p/2} T^{\frac{p}{2}(1-\frac{3}{2\alpha})} \mathbb{E} \sup_{t\in[0,T]} |u(t)|_{L^2}^p,\end{aligned} \tag{2.21}$$

where C_α is the constant appearing in the inequality (2.18).

Proof. Take $u, v \in Z_{T,p}$. It is enough to assume that u, v are deterministic functions and $u, v \in C([0,T], L^2)$. Then, by Lemma 2.11 and the second inequality in Lemma 2.3 we have

$$\begin{aligned}\sup_{t\in[0,T]} \Big|\int_0^t S_\alpha(t-s)B(\pi_n u(s))^2\,ds\Big|^p_{L^2} &\le \sup_{t\in[0,T]}\Big(\int_0^t \big|S_\alpha(t-s)B(\pi_n u(s))^2\big|_{L^2}\,ds\Big)^p\\ &\le C_\alpha^p T^{p(1-\frac{3}{2\alpha})}\sup_{s\in[0,T]}|(\pi_n u(s))^2|^p_{L^1},\\ &\le C_\alpha^p T^{p(1-\frac{3}{2\alpha})} n^p \sup_{t\in[0,T]}|u(t)|^p_{L^2},\end{aligned}$$

hence, we get the first inequality and that the operator H_n is well defined. To prove the second inequality, we use again Lemma 2.11 and the first inequality in Lemma 2.3 and arguing as above we get

$$\begin{aligned}\sup_{t\in[0,T]} \Big|\int_0^t S_\alpha(t-s)B\big((\pi_n u(s))^2-(\pi_n v(s))^2\big)\,ds\Big|^p_{L^2}&\\ \le C_\alpha^p T^{p(1-\frac{3}{2\alpha})}&\sup_{s\in[0,T]}|(\pi_n u(s))^2-(\pi_n v(s))^2|^p_{L^1},\\ \le C_\alpha^p T^{p(1-\frac{3}{2\alpha})}&(2n)^p\sup_{t\in[0,T]}|u(t)-v(t)|^p_{L^2}.\end{aligned}$$

Hence we conclude that H_n is globally Lipschitz. We finish with the proof of the third inequality. As before it is enough to assume that $u\in C([0,T],L^2)$ is a deterministic function. Then, by Lemma 2.11 we have

$$\begin{aligned}\sup_{t\in[0,T]} \Big|\int_0^t S_\alpha(t-s)B(u(s))^2\,ds\Big|^{p/2}_{L^2} &\le \sup_{t\in[0,T]}\Big(\int_0^t \big|S_\alpha(t-s)B(u(s))^2\big|_{L^2}\,ds\Big)^{p/2}\\ &\le C_\alpha^{p/2} T^{\frac{p}{2}(1-\frac{3}{2\alpha})}\sup_{s\in[0,T]}|(u(s))^2|^{p/2}_{L^1},\\ &\le C_\alpha^{p/2} T^{\frac{p}{2}(1-\frac{3}{2\alpha})}\sup_{t\in[0,T]}|u(t)|^p_{L^2}.\end{aligned}$$

□

It follows from Lemmas 2.9 and 2.13 that a map

$$\Phi_n : Z_{T,p}\ni u = S_\alpha(\cdot)u_0+[H_n u](\cdot)+[Gu](\cdot)\in Z_{T,p} \tag{2.22}$$

is well defined.

We are now ready to formulate the main result of this section.

Theorem 2.14. *Assume that* $\alpha>\frac{3}{2}$, $p>\frac{2\alpha}{\alpha-1}$ *and that the function* $g:\mathbb{R}\to\mathbb{R}$ *is Lipschitz. Then the problem*

$$\begin{cases} du(t)=\big\{-A_\alpha u(t)+B[(\pi_n u(t))^2]\big\}\,dt+g(u(t))\,dW(t),\ t>0,\\ u(0)=u_0,\end{cases} \tag{2.23}$$

has a unique global mild solution $u_n=\{u_n(t),t\ge0\}$ *such that*

$$\mathbb{E}\sup_{t\in[0,T]}|u_n(t)|^p_{L^2}<\infty,\ \textit{for each } T>0.$$

Proof. Let us fix $n \in \mathbb{N}$. We will prove that the operator Φ_n is a strict contraction provided $T > 0$ is small enough. Then by the Banach Fixed Point Theorem there exists a unique $u_n \in Z_{T,p}$ such that $\Phi_n(u_n) = u_n$. It is standard to show that then u_n is a mild solution to problem 2.23 on the time interval $[0, T]$. It is also standard, see Ref. [26] that this solution can be extended to the whole half-line $[0, \infty)$ and that the extension, denoted also by u_n is a fixed point of Φ_n for each $T > 0$ and in consequence is a mild solution to problem 2.23 on the time interval $[0, \infty)$.

Let us take and fix $u, v \in Z_{T,p}$. Then by Lemma 2.13 we have

$$\begin{aligned} \|H_n u - H_n v\|_{T,p}^p &= \mathbb{E}\big[\sup_{t\in[0,T]} \big|\int_0^t S_\alpha(t-s)(B(\pi_n u(s))^2 - B(\pi_n v(s))^2)\, ds\big|_{L^2}^p\big] \\ &\le 2^p C_\alpha^p n^p T^{p(1-\frac{3}{2\alpha})} \|u - v\|_{T,p}^p. \end{aligned} \tag{2.24}$$

Taking $T \le (6C_\alpha n)^{\frac{2\alpha}{3-2\alpha}}$, we get

$$\|H_n u - H_n v\|_{T,p} \le \frac{1}{3}\|u - v\|_{T,p}.$$

Next, from inequality (2.10) in Lemma 2.9 we infer that G is Lipschitz continuous with Lipschitz constant $\sqrt{2}C_{p,\gamma}\widehat{C}_{p,\gamma}b_2T^\gamma$, where b_2 is the Lipschitz constant of the function g.

Taking $T \le (3\sqrt{2}C_{p,\gamma}\widehat{C}_{p,\gamma}b_2)^{-\frac{1}{\gamma}}$, we infer that $\|Gu - Gv\|_T \le \frac{1}{3}\|u - v\|_T$. From (2.24) and (2.16) and for $T \le \min\{(3\sqrt{2}C_{p,\gamma}\widehat{C}_{p,\gamma}b_2)^{-\frac{1}{\gamma}}, (6nC_\alpha)^{\frac{2\alpha}{3-2\alpha}}\}$, we get $\|\Phi_n u - \Phi_n v\|_T \le \frac{2}{3}\|u - v\|_T$. Thus Φ_n is a contraction on $Z_{T,p}$ and so there exists a unique fixed point $u_n \in Z_{T,p}$ of Φ_n. □

Concerning regularity of solutions to problem (2.23) we have the following result, a standard proof of which will be omitted.

Proposition 2.15. *The solution $\{u_n(t), t \ge 0\}$ of equation (2.23) belongs to the space $C((0,T]; D(A_\alpha^\delta)) \cap C([0,T]; L^2(0,1))$ provided $\delta + \frac{3}{2\alpha} < 1$ and $\delta + p^{-1} < \frac{1}{2}(1 - \frac{1}{\alpha})$.*

3. Global solutions to Burgers equations

In this section we assume that the function g is bounded and Lipschitz, i.e. condition (2.7) with $b_1 = 0$ and condition (2.9) are satisfied.

Let us define a stopping time τ_n by $\tau_n = \inf\{t > 0, |u_n(t)|_{L^2} \ge n\}$. Let $n \ge m$, for $t \le \tau_m$, $\mathbb{P}$ a.s. $\{u_m(s), s \le t\}$ is solution of Equation (2.23). The same argument as in Ref. [26], compare with Lemma 4.11 and the proof of Theorem 4.10, based on the uniqueness part of Theorem 2.14, we get[b] $\mathbb{P} \otimes ds$ a.s, for all $0 \le t \le \tau_n$, $u_n(t) = u_m(t)$. Hence the sequence $\{\tau_n\}_{n>0}$ is increasing. Let $\tau_\infty := \lim_{n\to\infty} \tau_n$. We define the function $u(t)$ for $t \le \tau_\infty$ by $u(t) = u_n(t), t \le \tau_n$. Then we also have that

$$\tau_n = \inf\{t > 0, |u(t)|_{L^2} \ge n\},\ n \in \mathbb{N}. \tag{3.1}$$

[b]The equality is in $L^2(0,1)$.

Then, see Theorem 1.5 in Ref. [32], Theorem 4.10 in Ref. [26] , [33] or [34], $u(t)$, $t \le \tau_\infty$ is a unique maximal local (mild) solution to (1.1) with lifespan τ_∞, i.e. τ_∞ is an accessible, strictly positive $\mathbb{P}$-almost surely stopping time, $(u(t),\ t < \tau_\infty)$ is a progressively measurable process satisfying

$$\limsup_{t \nearrow \tau_\infty} |u(t)|_{L^2} = +\infty \quad \mathbb{P}\text{-almost surely on } \{\tau_\infty < \infty\},$$

and

$$u(t \wedge \tau_n) = S_\alpha(t \wedge \tau_n)u(0) + \int_0^{t\wedge\tau_n} S_\alpha(t \wedge \tau_n - r)B(u(r)^2)\, dr + I_{\tau_n}(g(u))(t \wedge \tau_n),\ t \ge 0 \tag{3.2}$$

for all $n \in \mathbb{N}$, where we define

$$I_{\tau_n}(g(u))(t) = \int_0^t \mathbf{1}_{[0,\tau_n)}(r) S_\alpha(t-r) g(u(r\wedge\tau_n))\, dW(r), \quad t \ge 0.$$

For the choice of the process $I_{\tau_n}(g(u))$ see Proposition C.1.1. Note that the process u has continuous paths, $u \in \mathcal{C}([0,\tau_\infty) ; L^2)$ $\mathbb{P}$-almost surely.

To prove that it is the global solution it is sufficient to prove that $\tau_\infty = \infty$.

Let us consider, for $n \in \mathbb{N}$, a process z_n defined by[c]

$$\begin{aligned} z_n(t) &= \int_0^t 1_{[0,\tau_n)}(s) S_\alpha(t-s) g(u_n(s))\, dW(s),\ t \ge 0, \\ &= \int_0^t 1_{[0,\tau_n)}(s) S_\alpha(t-s) g(u(s))\, dW(s),\ t \ge 0. \end{aligned} \tag{3.3}$$

Let v_n be a process defined pathwise as a solution to the following initial value problem

$$\begin{cases} \frac{dv_n(t)}{dt} = -A_\alpha v_n(t) + B(v_n(t) + z_n(t))^2, & t > 0, \\ v(0) = u_0. \end{cases} \tag{3.4}$$

It follows from Corollary C.1.9 that the processes $z_n(t)$, $t \ge 0$ defined by (3.3) satisfy the following. If $q \in [2,\infty)$, $p \in [1,\infty)$ and

$$\frac{1}{p} < \beta < \frac{1}{2}(1 - \frac{1}{\alpha}) - \delta, \tag{3.5}$$

then there exists $C > 0$ such that for all $T > 0$ and all $n \in \mathbb{N}$,

$$\mathbb{E} \sup_{t\in[0,T]} |z_n(t)|^p_{H^{\alpha\delta,q}(0,1)} \le CT^{p\beta}. \tag{3.6}$$

The following result follows directly from the definitions of the local solution and of the process z_n.

[c]The second equality below follows as $\mathbb{P}$-a.s. $u_n(s) = u(s)$ for $s \in [0,\tau_n)$.

Lemma 3.1. *If $n \in \mathbb{N}$, then*

$$u(t \wedge \tau_n) = z_n(t \wedge \tau_n) + v_n(t \wedge \tau_n),\ t \geq 0. \tag{3.7}$$

Let us also quote the following consequence of a much more general result from a monograph by Runst & Sickel [35], see Theorem 2 in section 4.4.4, Theorem 7.48 in Ref. [36] and section 2.3.5 and Remark 4.4.2/2 p. 324 in Ref. [37].

Proposition 3.2. *Assume that $0 < s_2 < \frac{1}{2}$ and $s_1 > \max\{s_2, \frac{1}{q_1}\}, q_1 \in [2, \infty)$. Then the pointwise multiplication between spaces $H^{s_1,q_1}(0,1)$ and $H^{s_2,2}(0,1)$ is well defined and moreover, there exists $C > 0$ such that*

$$|uv|_{H^{s_2,2}} \leq C|u|_{H^{s_1,q_1}}|v|_{H^{s_2,2}},\ u \in H^{s_1,q_1}(0,1),\ v \in H^{s_2,2}(0,1). \tag{3.8}$$

We will also need the following results.

Proposition 3.3. *Assume that $\beta \in (\frac{1}{2}, 1)$. Then there exists a constant $C > 0$ such that for each $u \in H_0^{\beta,2}(0,1)$ and each $v \in H^{1-\beta,2}(0,1)$ the following inequality is satisfied*

$$\left| \int_0^1 u(x)Dv(x)\,dx \right| \leq C|u|_{H_0^{\beta,2}}|v|_{H^{1-\beta,2}}. \tag{3.9}$$

Lemma 3.4. *Suppose that $T > 0$, a function $z : [0,T] \to H^{s,q}(0,1)$, with s and q such that $\frac{1}{q} < s < \frac{1}{2}$, is continuous, and that a function $v \in C(0,T; H_0^{\frac{\alpha}{2},2})$ is a solution to*

$$\begin{cases} \frac{dv(t)}{dt} = -A_\alpha v(t) + B(v(t)+z(t))^2,\ t > 0, \\ v(0) = u_0. \end{cases} \tag{3.10}$$

Then, there exists a generic constant $C > 0$ such that

$$\begin{aligned} \ln^+ |v(t)|_{L^2} \leq \ln^+ |v(0)|_{L^2} + C(1 + \ln^+ t) + \ln^+ \sup_{0 \leq \tau \leq t} |z(\tau)|_{H^{s,q}}^{\frac{\alpha}{\alpha-1}} \\ + \ln^+ \sup_{0 \leq \tau \leq t} \left[|z(\tau)|_{H^{s,q}}^4 + |z(\tau)|_{H^{1-\alpha/2,2}}^4 \right] + \sup_{0 \leq \tau \leq t} |z(\tau)|_{H^{s,q}}^{\frac{\alpha}{\alpha-1}}, \end{aligned} \tag{3.11}$$

where $\ln^+ x = \max\{0, \ln x\}$.

Proof. [Proof of Proposition 3.3.]

Let $\{e_n\}$ be the ONB of $L^2(0,1)$ as before and let $\{f_n, n \geq 0\}$ be an ONB in $L^2(0,1)$, given by $f_n(x) = \sqrt{2}\cos n\pi x$ for $n \geq 1$ and $f_0(x) = 1$. Assume that $u \in H_0^{\beta,2}(0,1)$ and $v \in C_0^1(0,1)$, where $C_0^1(0,1)$ is the set of differentiable functions on $(0,1)$ with compact support. So that $\langle e_n, Dv \rangle = -n\pi \langle f_n, v \rangle$.

Let us denote by $\widehat{A}$ the -Laplacian with the Neumann boundary conditions, i.e. a linear self-adjoint operator defined by the following

$$\begin{aligned} D(\widehat{A}) &= H^{2,2}(0,1) \\ \widehat{A}(u) &= -\Delta u,\ u \in H^{2,2}(0,1). \end{aligned}$$

It is known that $\widehat{A}(f_k) = (k\pi)^2 f_k$, $k \in \mathbb{N}$ and that $D(\widehat{A}^{\frac{1}{2}-\beta/2}) = H^{1-\beta,2}$. Further, from formula (1.5) we have $D(A^{\frac{\beta}{2}}) = H_0^{\beta,2}(0,1)$. Then, by the Plancherel Theorem we have the following sequence of inequalities

$$\begin{aligned}
|\int_0^1 u(x)Dv(x)\,dx| &= |\sum_{k=1}^{\infty}\langle u, e_k\rangle\langle e_k, Dv\rangle| = |\sum_{k=1}^{\infty} -k\pi\langle u, e_k\rangle\langle v, f_k\rangle| \\
&= |\sum_{k=1}^{\infty}(k\pi)^{\beta}\langle u, e_k\rangle(k\pi)^{1-\beta}\langle v, f_k\rangle| \\
&\le (\sum_{k=1}^{\infty}(k\pi)^{2\beta}\langle u, e_k\rangle^2)^{\frac{1}{2}}(\sum_{k=1}^{\infty}(k\pi)^{2-2\beta}\langle v, f_k\rangle^2)^{\frac{1}{2}} \\
&\le C|u|_{H_0^{\beta,2}}|v|_{H^{1-\beta,2}}(0,1).
\end{aligned}$$

Using the density of $C_0^1(0,1)$ in $H_0^{1-\beta,2} = H^{1-\beta,2}$, we get the result. □

Proof. [Proof of Lemma 3.4.] We begin with an observation that in view of Proposition 3.3 $(B(v^2), v)_{L^2} = 0$ for all $v \in H_0^{\frac{\alpha}{2},2}$. Indeed this is true for $v \in C_0^1(0,1)$, the space $C_0^1(0,1)$ is dense in $H_0^{\frac{\alpha}{2},2}$ and, see Ref. [38], the space $H_0^{\frac{\alpha}{2},2}$ is an algebra with pointwise multiplication.

Next, applying Lemma III.1.2 from [39] the solution v of problem (3.10) we get

$$\frac{1}{2}\frac{d}{dt}|v(t)|_{L^2}^2 = -\langle A_\alpha v(t), v(t)\rangle + 2\langle B(z(t)v(t)), v(t)\rangle + \langle Bz^2(t), v(t)\rangle. \tag{3.12}$$

Since $A_\alpha = A^{\alpha/2}$, where A is positive self-adjoint we infer that

$$\langle A_\alpha v, v\rangle = \langle A^{\alpha/2}v, v\rangle = |A^{\alpha/4}v|_{L^2}^2,\ v \in D(A_\alpha).$$

Since $\alpha \in (\frac{3}{2}, 2)$ so that $\frac{\alpha}{4} \in (\frac{3}{8}, \frac{1}{2})$ in view of equality (1.5) we infer that there exists $C > 0$ such that $|A^{\alpha/4}v|_{L^2}^2 \ge C|v|^2_{H_0^{\frac{\alpha}{2},2}}$, for $v \in D(A_{\alpha/2})$. Therefore[d],

$$\langle A_\alpha v, v\rangle \ge \frac{\nu_1}{2}|v|^2_{H_0^{\frac{\alpha}{2},2}},\ v \in D(A_\alpha). \tag{3.13}$$

Applying Proposition 3.3 with $\beta = \alpha/2$ we infer that

$$|\langle B(zv), v\rangle| \le C|v|_{H_0^{\alpha/2,2}}|zv|_{H^{1-\alpha/2,2}}$$

Let us choose positive numbers s and q such that $\frac{1}{q} < s < \frac{1}{2}$. Since $\alpha \in (\frac{3}{2}, 2)$, we infer that $1 - \frac{\alpha}{2} < \frac{1}{2}$ and therefore by Proposition 3.2, formula (3.8) we infer that that there exists a constant $C > 0$ such that

$$4|\langle B(zv), v\rangle| \le C|v|_{H_0^{\alpha/2,2}}|z|_{H^{s,q}}|v|_{H^{1-\alpha/2,2}}$$

---

[d]One can choose an equivalent norm on $H_0^{\frac{\alpha}{2},2}$ such that the constant $\nu_1 > 0$ in inequality (3.13) is equal to 1.

Moreover, since $0 < 1 - \frac{\alpha}{2} < \frac{\alpha}{2}$ and $H^{0,2}(0,1) = L^2(0,1)$, by the interpolation and definition (1.3) we infer that

$$|v|_{H^{1-\frac{\alpha}{2},2}} \leq |v|_{L^2}^{2-\frac{2}{\alpha}} |v|_{H_0^{\alpha/2,2}}^{\frac{2}{\alpha}-1}, \; v \in H_0^{\frac{\alpha}{2},2}(0,1).$$

Since $\alpha > 1$ we infer by applying the classical Young inequality that for some generic constant $C > 0$

$$4|\langle B(zv), v\rangle| \leq \frac{\nu_1}{4}|v|^2_{H_0^{\alpha/2,2}} + C|z|_{H^{s,q}}^{\frac{\alpha}{\alpha-1}}|v|^2_{L^2}. \tag{3.14}$$

Applying again Proposition 3.3 with $\beta = \alpha/2$ we infer that

$$|\langle B(z^2), v\rangle| \leq C|v|_{H_0^{\alpha/2,2}}|z^2|_{H^{1-\alpha/2,2}}.$$

Arguing as before and with the same choice of constants we have

$$\begin{aligned} 2|\langle B(z^2), v\rangle| &\leq C|v|_{H_0^{\alpha/2,2}}|z|_{H^{s,q}}|z|_{H^{1-\alpha/2,2}} \\ &\leq \frac{\nu_1}{4}|v|^2_{H_0^{\alpha/2,2}} + C|z|^4_{H^{s,q}} + C|z|^4_{H^{1-\alpha/2,2}}. \end{aligned} \tag{3.15}$$

Combining inequalities (3.12) and (3.13) with inequalities (3.14) and (3.15) we infer that

$$\begin{aligned} \frac{d}{dt}|v(t)|^2_{L^2} &\leq -\frac{\nu_1}{2}|v(t)|^2_{H^{\alpha/2,2}} \\ &\quad + C|z(t)|_{H^{s,q}}^{\frac{\alpha}{\alpha-1}}|v(t)|^2_{L^2} + C|z(t)|^4_{H^{s,q}} + C|z(t)|^4_{H^{1-\alpha/2,2}}, \; t \geq 0. \end{aligned} \tag{3.16}$$

Since there exists a constant $\nu_2 > 0$ such that $|v|^2_{H_0^{\alpha/2,2}} \geq \nu_2|v|^2_{L^2}$, $u \in H_0^{\alpha/2,2}(0,1)$, we infer that with $\nu := \frac{\nu_1\nu_2}{2}$,

$$\begin{aligned} \frac{d}{dt}|v(t)|^2_{L^2} &\leq (-\nu + C|z(t)|_{H^{s,q}}^{\frac{\alpha}{\alpha-1}})|v(t)|^2_{L^2} \\ &\quad + C|z(t)|^4_{H^{s,q}} + C|z(t)|^4_{H^{1-\alpha/2,2}}, \; t \geq 0. \end{aligned} \tag{3.17}$$

Applying finally the Gronwall Lemma we obtain that for $t \geq 0$,

$$\begin{aligned} |v(t)|^2_{L^2} &\leq |v(0)|^2_{L^2} e^{\int_0^t (-\nu + C|z(\tau)|_{H^{s,q}}^{\frac{\alpha}{\alpha-1}})\, d\tau} \\ &\quad + C\int_0^t \left[|z(\sigma)|^4_{H^{s,q}} + C|z(\sigma)|^4_{H^{1-\alpha/2,2}}\right] e^{\int_\sigma^t (-\nu + C|z(\tau)|_{H^{s,q}}^{\frac{\alpha}{\alpha-1}})\, d\tau}\, d\sigma, \\ &\leq |v(0)|^2_{L^2} e^{-\nu t + Ct\sup_{0\leq\tau\leq t}|z(\tau)|_{H^{s,q}}^{\frac{\alpha}{\alpha-1}}} \\ &\quad + Ct\sup_{0\leq\tau\leq t}\left[|z(\tau)|^4_{H^{s,q}} + |z(\tau)|^4_{H^{1-\alpha/2,2}}\right] e^{Ct\sup_{0\leq\tau\leq t}|z(\tau)|_{H^{s,q}}^{\frac{\alpha}{\alpha-1}}}. \end{aligned} \tag{3.18}$$

Applying the increasing function $\ln^+ x = \max\{0, \ln x\}$ to both sides of the above inequality and then using the classical inequalities $\ln^+(a+b) \leq \ln^+(a) + \ln^+(b) + \ln 2$ and $\ln^+(ab) \leq \ln^+(a) + \ln^+(b)$, we get the result. □

Proof. [Completion of the proof of Theorem 1.3.] To finish the proof we will employ the Hasminski's method as in Ref. [34], see Ref. [40], Theorem III.4.1, for the finite-dimensional case. Let us recall that

$$\tau_k = \{t \geq 0;\ |u(t)|_{L^2} \geq k\}, \quad k \in \mathbb{N}.$$

In order to prove that $\tau_\infty = +\infty$ $\mathbb{P}$-almost surely it suffices to find a Lyapunov function $V : L^2(0,1) \longrightarrow \mathbb{R}$ and a function $C : [0,\infty) \rightarrow (0,\infty)$ satisfying

$$V \geq 0 \text{ on } L^2(0,1), \tag{3.19}$$

$$q_R := \inf_{|u|_{L^2} \geq R} V(u) \rightarrow \infty \text{ as } R \rightarrow \infty, \tag{3.20}$$

$$\mathbb{E}V(u(0)) < \infty \tag{3.21}$$

and

$$\mathbb{E}V(u(t\wedge\tau_k)) \leq C(t)\big(1 + \mathbb{E}V(u(0))\big) \text{ for all } t \geq 0, k \in \mathbb{N}. \tag{3.22}$$

Inequality (3.22) implies easily that

$$\begin{aligned}\mathbb{P}\{\tau_k < t\} &\leq \frac{1}{q_k}\mathbb{E}\mathbf{1}_{\{\tau_k<t\}}V(u(t\wedge\tau_k)) \\ &\leq \frac{1}{q_k}C(t)\big(1 + \mathbb{E}V(u(0))\big),\end{aligned}$$

so that

$$\lim_{k\rightarrow\infty} \mathbb{P}\{\tau_k < t\} = 0$$

for each fixed $t \geq 0$, and $\mathbb{P}\{\tau_\infty < t\} = 0$ follows.

Set

$$V(u) = \ln^+ |u|_{L^2}, \quad u \in L^2(0,1).$$

Obviously V is uniformly continuous on bounded sets and satisfies (3.19), (3.20). Moreover, (3.22) follows from (3.18) and (3.21) is obviously satisfied for initial condition u_0 such that $\mathbb{E}\ln^+|u_0|_{L^2} < \infty$. □

4. Proof of uniqueness

Suppose that two processes $u^1(t)$, $t \geq 0$ and $u^2(t)$, $t \geq 0$ are solutions to problem (1.1) with the same initial data u_0. We may suppose that both satisfy condition (1.6) with the same constant $p > \frac{2\alpha}{\alpha-1}$. Let us define a stopping time $\tau_R = \tau_R^1 \wedge \tau_R^2$, where for $i = 1,2$ and $R > 0$,

$$\tau_R^i = \inf\{t > 0, |u^i(t)|_{L^2} \geq R\},\ n \in \mathbb{N}. \tag{4.1}$$

Then for all $R > 0$ each u^i satisfies the stopped version of equation (1.7), i.e.

$$\begin{aligned} u^i(t\wedge\tau_n) &= S_\alpha(t\wedge\tau_R)u_0 + \int_0^{t\wedge\tau_R} S_\alpha(t\wedge\tau_R - r)B(u^i(r))^2\,dr \\ &\quad + I_{\tau_R}(g(u^i))(t\wedge\tau_R), \quad \text{a.s.}, t\geq 0 \qquad (4.2)\\ I_{\tau_R}(g(u^i))(t) &= \int_0^t \mathbf{1}_{[0,\tau_R)}(r)S_\alpha(t-r)g(u^i(r\wedge\tau_R))\,dW(r), \quad \text{a.s.}, t\geq 0. \end{aligned}$$

Subtracting the equations (4.2) for $i = 2, 1$ and denoting by u the difference between u_2 and u_1 we infer that for $t \geq 0$, a.s.

$$\begin{aligned} u(t\wedge\tau_R) &= \int_0^{t\wedge\tau_R} S_\alpha(t\wedge\tau_R - r)B\left[(u^2(r))^2 - (u^1(r))^2\right]\,dr \\ &\quad + J_{\tau_R}(t\wedge\tau_R), \\ J_{\tau_R}(t) &:= I_{\tau_R}(g(u^2)-g(u^1))(t) \\ &= \int_0^t \mathbf{1}_{[0,\tau_R)}(r)S_\alpha(t-r)g(u^2(r\wedge\tau_R)) - g(u^1(r\wedge\tau_R))\,dW(r). \end{aligned}$$

Because $|u(t)|_{L^2} \leq R$ for $t \in [0,\tau_R]$, the proof can be easily concluded by applying Gronwall Lemma in conjunction with Lemmata 2.11 and 2.13. Details of this idea are presented below. We shall prove that $\mathbb{E}|u(t\wedge\tau_R)|^p_{L^2} = 0$. i.e $u(t) = 0$ a.s on $\{t \leq \tau_R\}$. In fact, using Lemma A.1 from [34] and the semigroup property, we get, for some $C_p > 0$

$$\begin{aligned} \mathbb{E}|u(t\wedge\tau_R)|^p_{L^2} &\leq C_p\mathbb{E}|\textstyle\int_0^t \mathbf{1}_{(0,\tau_R)}S_\alpha(t-r)B\left[(u^2(r\wedge\tau_R))^2 - (u^1(r\wedge\tau_R))^2\right]dr|^p_{L^2} \\ &\quad + C_p\mathbb{E}|\textstyle\int_0^t \mathbf{1}_{[0,\tau_R)}(r)S_\alpha(t-r)g(u^2(r\wedge\tau_R)) - g(u^1(r\wedge\tau_R))\,dW(r)|^p_{L^2}. \qquad (4.3) \end{aligned}$$

Put the following locally integrable functions:

$$\begin{aligned} \varphi_1(s) := h^{\frac{1}{2}}(s) &= \Big(\sum_{k\geq 1} k^2 e^{-2s(k\pi)^\alpha}\Big)^{\frac{1}{2}}, \; s > 0, \\ \varphi_2(r) &:= \sum_{k\geq 1}\int_0^r \xi^{-2\gamma}e^{-2\xi(k\pi)^\alpha}\,d\xi, \; r > 0 \\ \varphi &:= \max\{\varphi_1, \varphi_2\}. \end{aligned}$$

Arguing as above, using Lemmata 2.6, 2.7 and 2.8 and thanks to the assumptions above, we get the following sequence of inequalities

$$
\begin{aligned}
&\mathbb{E}\Big|\int_0^t \mathbf{1}_{[0,\tau_R)} S_\alpha(t-r) B\left[(u^2(r\wedge\tau_R))^2 - (u^1(r\wedge\tau_R))^2\right]\, dr\Big|_{L^2}^p \\
&\quad \le \mathbb{E}\Big(\int_0^t |\mathbf{1}_{[0,\tau_R)} S_\alpha(t-r) B\left[(u^2(r\wedge\tau_R))^2 - (u^1(r\wedge\tau_R))^2\right]|_{L^2}\, dr\Big)^p \\
&\quad \le 2^{\frac{p}{2}}\pi^p \mathbb{E}\Big(\int_0^t \Big(\sum_{k\ge 1} k^2 e^{-2(t-r)(k\pi)^\alpha}\Big)^{\frac{1}{2}} \left[\mathbf{1}_{[0,\tau_R)} |u^2(r\wedge\tau_R)) - u^1(r\wedge\tau_R)|_{L^2}\right. \\
&\quad\quad \left.|u^2(r\wedge\tau_R)) + u^1(r\wedge\tau_R)|_{L^2}\right] dr\Big)^p \qquad (4.4)\\
&\quad \le 2^{\frac{3p}{2}}\pi^p R^p \mathbb{E}\Big(\int_0^t \varphi_1(t-r) |u^2(r\wedge\tau_R)) - u^1(r\wedge\tau_R)|_{L^2} dr\Big)^p \\
&\quad \le 2^{\frac{3p}{2}}\pi^p R^p \Big(\int_0^t \varphi_1(r) dr\Big)^{p-1} \int_0^t \varphi_1(t-r) \mathbb{E}|u(r\wedge\tau_R)|_{L^2}^p dr \\
&\quad \le C 2^{\frac{3p}{2}}\pi^p R^p t^{(1-\frac{3}{2\alpha})(p-1)} \int_0^t \varphi_1(t-r) \mathbb{E}|u(r\wedge\tau_R)|_{L^2}^p dr,
\end{aligned}
$$

and

$$
\begin{aligned}
\mathbb{E}\Big|\int_0^t & \mathbf{1}_{[0,\tau_R)} S_\alpha(t-r) [g(u^2(r)) - g(u^1(r))]\, dW(r)\Big|_{L^2}^p \\
&\le \mathbb{E}\Big(\int_0^t (t-r)^{\gamma-1} |Y_1(r)|_{L^2}\, dr\Big)^p, \\
&\le \Big(\int_0^t r^{(\gamma-1)(\frac{p}{p-1})}\Big)^{p-1} \int_0^t \mathbb{E}|Y_1(r)|_{L^2}^p\, dr,
\end{aligned}
$$

where

$$
Y_1(r) = \frac{\sin\pi\gamma}{\pi}\int_0^r (r-s)^{-\gamma} S_\alpha(r-s) \mathbf{1}_{[0,\tau_R)}(s)\big(g(u^2(s\wedge\tau_R)) - g(u^1(s\wedge\tau_R))\big)\, dW(s).
$$

Furthermore,

$$
\begin{aligned}
&\mathbb{E}|Y_1(r)|_{L^2}^p \\
&\quad \le C_p b_2^p \mathbb{E}\Big(\int_0^r (r-s)^{-2\gamma} \sum_{k\ge 1} e^{-2(r-s)(k\pi)^\alpha} |u^2(s\wedge\tau_R) - u^1(s\wedge\tau_R)|_{L^2}\Big)^{\frac{p}{2}} ds \\
&\quad\quad \le C_p b_2^p \Big(\sum_{k\ge 1}\int_0^\infty s^{-2\gamma} e^{-2s(k\pi)^\alpha}\, ds\Big)^{\frac{p}{2q}} \\
&\quad\quad\quad \Big(\int_0^r (r-s)^{-2\gamma} \sum_{k\ge 1} e^{-2(r-s)(k\pi)^\alpha} \mathbb{E}|u(s\wedge\tau_R)|_{L^2}^p\, ds\Big).
\end{aligned}
$$

Hence

$$\begin{aligned}
&\mathbb{E}\Big|\int_0^t \mathbf{1}_{[0,\tau_R)} S_\alpha(t-r)\big(g(u^2(r)) - g(u^1(r))\big)\, dW(r)\Big|_{L^2}^p \\
&\le \tilde{C}\int_0^t\int_0^r (r-s)^{-2\gamma}\sum_{k\ge 1} e^{-2(r-s)(k\pi)^\alpha}\mathbb{E}|u(s\wedge\tau_R)|_{L^2}^p\, ds\, dr \\
&\le \tilde{C}\int_0^t\Big(\int_0^{t-s} \xi^{-2\gamma}\sum_{k\ge 1} e^{-2\xi(k\pi)^\alpha}\, d\xi\Big)\mathbb{E}|u(s\wedge\tau_R)|_{L^2}^p\, ds \qquad (4.5)\\
&\le \tilde{C}\int_0^t \varphi_2(t-s)\mathbb{E}|u(s\wedge\tau_R)|_{L^2}^p\, ds.
\end{aligned}$$

Replacing (4.5) and (4.4) in (4.3) we get

$$\mathbb{E}|u(t\wedge\tau_R)|_{L^2}^p \le C_{R,p}\int_0^t \varphi(t-r)\mathbb{E}|u(r\wedge\tau_R)|_{L^2}^p\, dr \quad \text{for all } t\ge 0.$$

Hence the result follows from the application of the Gronwall Lemma and Jensen inequality. This method can work for any exponent $q \ge p$.

A.1. Proof of Lemma 2.3

Lemma A.1.1. *The map $\pi_n : L^2 \to L^2$ is globally Lipschitz with Lipschitz constant 1. Moreover,*

$$\begin{aligned}
|(\pi_n u)^2 - (\pi_n v)^2|_{L^1} &\le 2n|u-v|_{L^2}, \quad \textit{for all } u,v\in L^2(0,1), \qquad (A.1)\\
|(\pi_n u)^2|_{L^1} &\le n|u|_{L^2}, \quad \textit{for all } u\in L^2(0,1),\\
|(\pi_n u)^2|_{L^1} &\le |u|_{L^2}^2, \quad \textit{for all } u\in L^2(0,1).
\end{aligned}$$

Proof. Let us first consider the case when $u, v \in L^2(0,1)$ are such that $|u|_{L^2} > n$, $|v|_{L^2} > n$. Then we have

$$\begin{aligned}
|\pi_n u - \pi_n v|_{L^2}^2 &= |\pi_n u|_{L^2}^2 + |\pi_n v|_{L^2}^2 - 2\langle \pi_n u, \pi_n v\rangle \\
&= n^2 + n^2 - 2n^2\langle \frac{u}{|u|}, \frac{v}{|v|}\rangle = n^2\frac{2|u|\,|v| - 2\langle u,v\rangle}{|u|\,|v|} \\
&\le \frac{n^2}{|u|_{L^2}|v|_{L^2}}\Big(|u|_{L^2}^2 + |v|_{L^2}^2 - 2\langle u,v\rangle\Big) \le |u-v|_{L^2}^2.
\end{aligned}$$

In the case when $|u|_{L^2} > n$ and $|v|_{L^2} \le n$, we have

$$\begin{aligned}
|\pi_n u - \pi_n v|_{L^2}^2 &= \frac{n}{|u|_{L^2}}\Big(n|u|_{L^2} + \frac{|v|_{L^2}}{n}|v|_{L^2}|u|_{L^2} - 2\langle u,v\rangle\Big) \\
&\le \frac{n}{|u|_{L^2}}\Big(|u|_{L^2}^2 + |v|_{L^2}^2 - 2\langle u,v\rangle\Big) \le |u-v|_{L^2}^2.
\end{aligned}$$

The case $|u|_{L^2} < n, |v|_{L^2} < n$ is trivial.

The inequality (A.1) then easily follows. Indeed, by using the trivial identity $a^2 - b^2 = (a+b)(a-b)$, the Hölder inequality and the fact that the projection π_n is both Lipschitz continuous and bounded we have, for $u, v \in L^2(0,1)$, the following inequalities,

$$\begin{aligned} |(\pi_n u)^2 - (\pi_n v)^2|_{L^1} &\le |(\pi_n u) + (\pi_n v)|_{L^2} |(\pi_n u) - (\pi_n v)|_{L^2} \\ &\le 2n|(\pi_n u) - (\pi_n v)|_{L^2}. \end{aligned}$$

In a similar way we can prove the second and the third inequalities. □

B.1. Gronwall Lemma

Lemma B.1.1. *Assume that $\varphi, \psi \in L^1(a,b)$ and $y : [a,b] \to \mathbb{R}$ is an absolutely continuous function. Suppose that these functions satisfy*

$$\frac{dy(t)}{dt} \le \varphi(t)y(t) + \psi(t) \qquad \text{for a.a. } t \in (a,b). \tag{B.1}$$

Then we have

$$y(t) \le y(a)e^{\int_a^t \varphi(\tau)\,d\tau} + \int_a^t \psi(s)e^{\int_s^t \varphi(\tau)\,d\tau} ds, \qquad \text{for all } t \in [a,b]. \tag{B.2}$$

C.1. Some estimates on stopped stochastic convolutions

The main technical tool is the following result stated in Ref. [41] and proven in Ref. [33], Theorem 4.2.7. We formulate the result in the Hilbert space framework as we do not require in this paper Banach space valued Itô integrals. For a notion of γ-radonifying operator see Refs. [26] and [41] or Remark 6.1 in a recent paper [42]. The operator ideal of all gamma-radonifying operators from a separable Hilbert space H into a separable Banach space X will be denoted by $R(H,X)$. Let us recall that with a naturally defined norm, $R(H,X)$ is a separable Banach space, see Ref. [43].

Proposition C.1.1. *Assume that H is a separable Hilbert space. Assume that X is an 2-smoothable Banach. Assume that a linear operator $-A$ is an infinitesimal generator of an analytic semigroup $(S(t))_{t \ge 0}$ on X. Assume finally that ξ is a progressively measurable operator-valued stochastic process such that $S(t-r)\xi(r) \in R(H,X)$ for $t > r \ge 0$, where $R(H,X)$ is the Banach space of all γ-radonifying operators from H to X and there exists $p > 2$ and $\beta_0 \in (\frac{1}{p}, \frac{1}{2})$ such that for some $T \in (0,\infty)$,*

$$|||\xi|||^p_{\beta,p,T} := \int_0^T \left(\int_0^s (s-r)^{-2\beta} \|S(s-r)\xi(r)\|^2_{R(H,X)}\, dr \right)^{p/2} ds < \infty, \ \beta < \beta_0.$$

Assume that $\delta \in [0, \beta_0 - \frac{1}{p})$. Then there exists an $D(A^\delta)$-valued continuous stochastic process $v(t)$, $t \in [0,T]$, such that

$$x(t) = \int_0^t S(t-s)\,\xi(s)\,dW(s), \text{ a.s. for each } t \in [0,T]. \tag{C.1}$$

Moreover, if $\delta + \frac{1}{p} < \beta < \beta_0$ then there exists a constant $C > 0$ independent of ξ and T such that for any accessible stopping time σ,

$$\mathbb{E} \sup_{0\le t\le T} |x(t\wedge\sigma)|^p_{D(A^\delta)} \le C^p T^{p(\beta-\delta)-1} \,\|\!|1_{[0,\sigma)}\xi\|\!|^p_{\beta,p,T} \tag{C.2}$$

$$= C^p T^{p(\beta-\delta)-1} \int_0^T \left(\int_0^{s\wedge\sigma} (s-r)^{-2\beta} \|S(s-r)\xi(r)\|^2_{R(H,X)}\,dr \right)^{p/2} ds.$$

Corollary C.1.2. *Assume that a progressively measurable $\mathcal{L}(H)$-valued stochastic process ξ satisfies, for some $M > 0$ and $T \in (0,\infty)$,*

$$|\xi(t)|_{\mathcal{L}(H)} \le M,\ t \in [0,T], \mathbb{P} - a.s.,$$

where $\mathcal{L}(H)$ is the space of all bounded linear operators in H. Assume that $S(t) \in R(H,X)$ for $t > 0$ and there exists $p > 2$ and $\beta_0 \in (\frac{1}{p}, \frac{1}{2})$ such that for $T > 0$ and $\beta < \beta_0$,

$$\int_0^T \left(\int_0^s r^{-2\beta} \|S(r)\|^2_{R(H,X)}\,dr \right)^{p/2} ds < \infty.$$

Then the assumptions of Proposition C.1.1 are satisfied and therefore the process $x(t)$, $t \ge 0$, satisfies the following inequality

$$\mathbb{E} \sup_{0\le t\le T} |x(t\wedge\sigma)|^p_{D(A^\delta)} \tag{C.3}$$

$$\le C^p M^p T^{p(\beta-\delta)-1} \int_0^T \left(\int_0^{s\wedge\sigma} (s-r)^{-2\beta} \|S(s-r)\|^2_{R(H,X)}\,dr \right)^{p/2} ds.$$

Proof. [Proof of Corollary C.1.2.] It is enough to use the fact, see Ref. [44], that $R(H,X)$ an operator ideal and so for a constant C independent of ξ and $\{S(t)\}_{t\ge 0}$ one has for all $0 \le r < s < \infty$

$$\|S(s-r)\xi(r)\|_{R(H,X)} \le C\|S(s-r)\|_{R(H,X)}|\xi(r)|_{\mathcal{L}(H)} \le CM\|S(s-r)\|_{R(H,X)}. \qquad \square$$

Proof. [Proof of Proposition C.1.1.] Let us choose $\beta \in (\frac{1}{p}+\delta, \beta_0)$. First we define a process y by

$$y(t) = \frac{1}{\Gamma(1-\beta)} \int_0^t (t-s)^{-\beta} S(t-s)\,\xi(s)\,dW(s), \quad t \in [0,T].$$

Then $\mathbb{E}\int_0^T |y(t)|^p_X\,dt < \infty$. Indeed, by the Fubini Theorem and the Burkholder and Young inequalities

$$\mathbb{E}\int_0^T |y(s)|_X^p\,ds = \int_0^T \mathbb{E}|y(s)|_X^p\,ds$$
$$\le C_p\int_0^T \mathbb{E}\Big(\int_0^s (s-r)^{-2\beta}\|S(s-r)\,\xi(r)\|^2_{R(H,X)}\,dr\Big)^{\frac{p}{2}}\,ds \quad (C.4)$$
$$\le C_p\|\!|\xi|\!\|^p_{\beta,p,T} < \infty.$$

Hence, there exists a set of full measure $\tilde{\Omega}$ such that $y(\cdot,\omega)\in L^p(0,T;X)$. For $\omega\in\tilde{\Omega}$ we set

$$x(t,\omega) = \frac{1}{\Gamma(\beta)}\int_0^t (t-s)^{\beta-1}S(t-s)\,y(s,\omega)\,ds, \quad t\in[0,T].$$

Let $\omega\in\tilde{\Omega}$ be fixed. Let us recall without proof the following classical result.

Lemma C.1.3. *Assume that $\delta\ge 0$, $p\in(1,\infty)$ and $1>\beta>\frac{1}{p}+\delta$. Then the map $R_\beta : L^p(0,T;X)\to C([0,T];D(A^\delta))$ given by*

$$R_\beta h(t) = \int_0^t (t-s)^{\beta-1}S(t-s)h(s)\,ds,\ h\in L^p(0,T;X),$$

is well defined, linear and bounded. Moreover, there exists a constant $C_{p,\beta-\delta}>0$ such that the norm of $\|R_\beta\|\le C_{p,\beta-\delta}T^{\beta-\delta-\frac{1}{p}}$.

Since $\beta>\frac{1}{p}+\delta$. By the above Lemma there exists a constant C_T (independent of y (and hence ω)) such that

$$|\,x(t,\omega)\,|^p_{D(A^\delta)}\le C_T\int_0^t |\,y(s,\omega)\,|_X^p\,ds, \quad t\in[0,T].$$

In particular,

$$|\,x(t,\omega)\,|^p_{D(A^\delta)}\le C_T\int_0^{T\wedge\sigma(\omega)} |\,y(s,\omega)\,|_X^p\,ds, \quad t\in[0,T\wedge\sigma(\omega)].$$

It follows that

$$\sup_{0\le t\le T\wedge\sigma(\omega)} |\,x(t,\omega)\,|^p_{D(A^\delta)}\le C_T\int_0^{T\wedge\sigma(\omega)} |\,y(s,\omega)\,|_X^p\,ds. \quad (C.5)$$

Noting that

$$\sup_{0\le t\le T} |\,x(t\wedge\sigma)\,|^p_{D(A^\delta)} = \sup_{0\le t\le T\wedge\sigma} |\,x(t)\,|^p_X,$$

we infer that

$$\sup_{0\le t\le T} |\,x(t\wedge\sigma)\,|^p_{D(A^\delta)}\le C_T\int_0^{\sigma\wedge T} |\,y(s)\,|_X^p\,ds \quad \text{a.s..}$$

Taking expectations gives

$$\mathbb{E}\sup_{0\le t\le T} \mid x(t\wedge\sigma)\mid^p_{D(A^\delta)} \le C_T\mathbb{E}\int_0^{\sigma\wedge T} \mid y(s)\mid^p_X ds. \tag{C.6}$$

Note that

$$\begin{aligned}\mathbb{E}\int_0^{\sigma\wedge T} \mid y(s)\mid^p_X ds &= \mathbb{E}\int_0^T \mid 1_{[0,\sigma)}(s)y(s)\mid^p_X ds \\ &= \int_0^T \mathbb{E}\mid 1_{[0,\sigma)}(s)y(s)\mid^p_X ds.\end{aligned} \tag{C.7}$$

Define next a process

$$\tilde{y}(s) = \frac{1}{\Gamma(1-\beta)}\int_0^s (s-r)^{-\beta}S(s-r)1_{[0,\sigma)}(r)\xi(r)\,dW(r).$$

$\tilde{y}$ is well defined and for $s\in[0,T]$

$$1_{[0,\sigma)}(s)y(s) = \begin{cases}\tilde{y}(s) & \text{if } 1_{[0,\sigma)}(s)=1,\\ 0 & \text{if } 1_{[0,\sigma)}(s)=0.\end{cases}$$

Therefore,

$$\begin{aligned}\mathbb{E}\mid 1_{[0,\sigma)}(s)y(s)\mid^p_X &= \int_{\{1_{[0,\sigma)}(s)=1\}} \mid 1_{[0,\sigma)}(s,\omega)y(s,\omega)\mid^p_X d\mathbb{P}(\omega)\\ &\le \int_{\{1_{[0,\sigma)}(s)=1\}} \mid 1_{[0,\sigma)}(s,\omega)y(s,\omega)\mid^p_X d\mathbb{P}(\omega)\\ &\quad+\int_{\{1_{[0,\sigma)}(s)=0\}} \mid \tilde{y}(s,\omega)\mid^p_X d\mathbb{P}(\omega)\\ &= \int_{\{1_{[0,\sigma)}(s)=1\}} \mid \tilde{y}(s,\omega)\mid^p_X d\mathbb{P}(\omega)\\ &\quad+\int_{\{1_{[0,\sigma)}(s)=0\}} \mid \tilde{y}(s,\omega)\mid^p_X d\mathbb{P}(\omega) = \mathbb{E}\mid \tilde{y}(s)\mid^p_X .\end{aligned}$$

Thus, integrating from 0 to T and then applying as earlier the Fubini Theorem and the Burkholder and Young inequalities we infer that

$$\begin{aligned}&\mathbb{E}\int_0^T \mid 1_{[0,\sigma)}(s)y(s)\mid^p_X ds \le \mathbb{E}\int_0^T \mid \tilde{y}(s)\mid^p_X ds\\ &\le C_p\int_0^T \mathbb{E}\Big(\int_0^s (s-r)^{-2\beta}1_{[0,\sigma)}(r)\|S(s-r)\,\xi(r)\|^2_{R(H,X)}\,dr\Big)^{\frac{p}{2}} ds\\ &\qquad\qquad \le C_p|\!|\!|1_{[0,\sigma)}\xi|\!|\!|^p_{\beta,p,T} < \infty.\end{aligned} \tag{C.8}$$

□

Corollary C.1.4. *Assume that the assumptions of Proposition C.1.1 are satisfied. Let* $0 < T < \infty$ *and* $\tau : \Omega \to [0, T]$ *be a stopping time. Furthermore assume that* $\xi(t)$*,* $t < \tau$ *is an admissible* $R(H, X)$*-valued process with*

$$|\!|\!| 1_{[0,\tau)}\xi |\!|\!|_{\beta,p,T} < \infty, \ \text{if } \beta < \beta_0.$$

Set

$$z(t) = \int_0^t S(t-s)\ 1_{[0,\tau)}(s)\xi(s)\, dW(s).$$

Then, there exists a modification $\tilde{z}$ *of* z *such that*

$$\tilde{z} \in L^p(\Omega; C(0, T; X)).$$

Moreover, there exists $C(T) > 0$*, independent of* ξ *and* τ*, such that for each* $t \in [0, T]$*,*

$$\mathbb{E} \sup_{0 \le s \le t} |\ \tilde{z}(s \wedge \tau)\ |^p_{D(A^\delta)} \le C(T) |\!|\!| 1_{[0,\tau)}\xi |\!|\!|^p_{\beta,p,t}. \tag{C.9}$$

Proof. Define the process $\eta : [0, T] \times \Omega \to R(H, X)$ by

$$\eta(s, \omega) = 1_{[0,\tau(\omega))}(s)\xi(s, \omega).$$

As ξ is admissible and $1_{[0,\tau)}$ is right continuous and adapted, then η right continuous and adapted. In particular, η has a progressively measurable modification. Moreover, as

$$|\!|\!| \eta |\!|\!|_{\beta,p,T} = |\!|\!| \xi |\!|\!|_{\beta,p,T} < \infty,$$

it follows, by Proposition C.1.1 that the process z is well defined and has a continuous modification $\tilde{z}$ with $\tilde{z} \in L^p(\Omega; C(0, T; X))$. Furthermore, we can find $C(T) > 0$, independent of ξ and τ, such that

$$\mathbb{E} \sup_{0 \le s \le t} |\ \tilde{z}(s \wedge \tau)\ |^p_{D(A^\delta)} \le C(T) |\!|\!| 1_{[0,\tau)}(s)\xi |\!|\!|^p_{\beta,p,t}, \quad t \in [0, T]. \tag{C.10}$$

This completes the proof of Corollary C.1.4. □

Finally, we shall prove

Corollary C.1.5. *Assume that the assumptions of Proposition C.1.1 are satisfied. Let* $x(t)$*,* $t \in [0, T]$*, be an* $D(A^\delta)$*-valued continuous stochastic process such that*

$$x(t) = \int_0^t S(t-s)\, \xi(s)\, dW(s), \ \text{a.s. for each } t \in [0, T]. \tag{C.11}$$

Let σ *be an accessible* $[0, T]$*-valued stopping time.*

Let $y(t) = y_\sigma(t)$, $t \in [0,T]$, be an X-valued continuous stochastic process such that

$$y(t) = \int_0^t S(t-s)\, 1_{[0,\sigma)}(r)\xi(r \wedge \sigma)\, dW(s), \text{ a.s. for each } t \in [0,T] \tag{C.12}$$

Then, for each $t \in [0,T]$,

$$x(t \wedge \sigma) = y(t \wedge \sigma), \text{a.s.}.$$

Proof. If $\eta(r) := 1_{[0,\sigma)}(r)\xi(r \wedge \sigma)$, $r \in [0,T]$, then y depends on η in the same way as x depends on ξ and $y - x$ on $\eta - \xi$. Therefore, by Proposition C.1.1

$$\mathbb{E} \sup_{0 \le t \le T} |x(t \wedge \sigma) - y(t \wedge \sigma)|^p \le C(T)\, \| 1_{[0,\sigma)}(\xi - \eta) \|^p_{\beta,p,T} \tag{C.13}$$

Hence, for each $t \in [0,T]$, $\mathbb{E}|x(t \wedge \sigma) - y(t \wedge \sigma)|^p = 0$. In particular, $x(t \wedge \sigma) - y(t \wedge \sigma) = 0$ a.s. as claimed. □

We finish this section with describing applications of the results obtained here to the stochastic evolution equation driven by the fractional Laplace operator or, as called by some authors, stochastic fractional heat equation.

The following result is stated Remark 6.1 in a recent paper [42]. It can be proved by applying Theorem 2.3 from Ref. [45].

Proposition C.1.6. *Assume that K is a linear self-adjoint compact operator in $L^2(0,1)$. Let $\{e_j\}_{j=1}^\infty$ and $\{\lambda_j\}_{j=1}^\infty$ be the corresponding sequence of eigenvectors*[e] *and eigenvalues of K. Let $q \in (1,\infty)$. Then $K : L^2(0,1) \to L^q(0,1)$ is γ-radonifying if and only if*

$$\int_0^1 \left[\sum_{j=1}^\infty \lambda_j^2 |e_j(x)|^2 \right]^{q/2} dx < \infty.$$

Moreover, the γ-radonifying norm of K, denoted by $\|K\|_{R(L^2,L^q)}$, is equivalent to $\left[\int_0^1 [\sum_j \lambda_j^2 |e_j(x)|^2]^{q/2}\, dx \right]^{1/q}$.
In particular, if $\sup_j |e_j|_{L^\infty} < \infty$, then there exists a constant M_q independent of K, such that

$$\|K\|_{R(L^2,L^q)} \le M_q \|K\|_{R(L^2,L^2)} = M_q \|K\|_{HS(L^2,L^2)},$$

where as before $\|K\|_{HS(L^2,L^2)}$ denotes the Hilbert-Schmidt norm of the operator K.

Example C.1.7. Assume that $q \in (1,\infty)$ and let $A_{q,\alpha}$ be the fractional power of order $\frac{\alpha}{2}$ of the operator A_q defined by

$$\begin{cases} D(A_q) = H^{2,q}(0,1) \cap H_0^{1,q}(0,1), \\ Au \quad\; = -\Delta u,\; u \in D(A_q). \end{cases} \tag{C.14}$$

[e] Let us here recall that $\{e_j\}_{j=1}^\infty$ is an ONB of $L^2(0,1)$ and $Ke_j = \lambda_j e_j$, $j \in \mathbb{N}^*$.

Briefly, $A_{q,\alpha} = A_q^{\frac{\alpha}{2}}$. Let us denote the semigroups generated by the operators $A_{q,\alpha}$ by one symbol $\{S_\alpha(t)\}_{t\geq 0}$.

Since $|e_j|_{L^2} = 1$ By Proposition C.1.6 we infer that there exists a constant $C > 0$ such that

$$\|S_\alpha(t)\|^2_{R(L^2,L^q)} \leq C\|S_\alpha(t)\|^2_{HS(L^2,L^2)},\ t > 0. \tag{C.15}$$

Assume that $\beta < \frac{1}{2} - \frac{1}{2\alpha} =: \beta_0(\alpha)$. Then, because $\|S_\alpha(t)\|^2_{HS(L^2,L^2)} = \sum_k e^{-2\pi^\alpha k^\alpha t}$ we infer that for $s > 0$

$$\begin{aligned}
&\int_0^s r^{-2\beta}\|S_\alpha(r)\|^2_{R(L^2,L^q)}\,dr \leq C^2\int_0^\infty r^{-2\beta}\|S_\alpha(r)\|^2_{HS(L^2,L^2)}\,dr\\
&= C^2\int_0^\infty r^{-2\beta}\sum_{k=1}^\infty e^{-2\pi^\alpha k^\alpha r}\,dr = C^2\sum_{k=1}^\infty\int_0^\infty r^{-2\beta}e^{-2\pi^\alpha k^\alpha r}\,dr\\
&= C^2\Gamma(1-2\beta)(2\pi^\alpha)^{2\beta-1}\sum_{k=1}^\infty k^{\alpha(2\beta-1)} =: C^2_{\alpha,\beta} < \infty.
\end{aligned}$$

Therefore

$$\int_0^T\left(\int_0^s r^{-2\beta}\|S_\alpha(r)\|^2_{R(L^2,L^q)}\,dr\right)^{p/2} ds \leq C^p_{\alpha,\beta}T < \infty$$

and so the semigroup $\{S_\alpha(t)\}_{t\geq 0}$ satisfies the assumptions of Corollary C.1.2 with $\beta_0 = \beta_0(\alpha) = \frac{1}{2} - \frac{1}{2\alpha}$.

Hence we proved the following result.

Proposition C.1.8. *Consider the framework of Example C.1.7 with $\alpha \in (1,2]$ and $q \in [2,\infty)$. Assume that a progressively measurable $\mathcal{L}(L^2)$-valued stochastic process ξ satisfies, for some $M > 0$,*

$$|\xi(t)|_{\mathcal{L}(L^2)} \leq M,\ t \in [0,\infty), \mathbb{P} - a.s.. \tag{C.16}$$

Assume that $p \in (2,\infty)$ and $\delta \geq 0$ satisfy the following inequality

$$\delta + \frac{1}{p} < \beta < \frac{1}{2} - \frac{1}{2\alpha} =: \beta_0(\alpha). \tag{C.17}$$

Then there exists an $D(A^\delta_{q,\alpha})$-valued continuous stochastic process $x(t)$, $t \in [0,T]$ such that for $t \geq 0$

$$x(t) = \int_0^t S(t-s)\,\xi(s)\,dW(s),\ a.s.\ for\ each\ t \in [0,T]. \tag{C.18}$$

and there exists a constant $C > 0$ independent of ξ and T such that for any accessible stopping time σ,

$$\mathbb{E}\sup_{0\leq t\leq T}|x(t\wedge\sigma)|^p_{D(A^\delta_{q,\alpha})} \leq C^pM^pT^{p(\beta-\delta)} \tag{C.19}$$

Since $D(A^{\delta}_{\alpha,q}) = D(A_q^{\alpha\delta/2}) \subset H^{\alpha\delta,q}(0,1)$ we infer the following corollaryllary.

Corollary C.1.9. *Consider the framework of Example C.1.7 with $\alpha \in (1,2]$ and $q \in [2,\infty)$. Assume that $p \in (2,\infty)$ and $\delta \geq 0$ satisfy condition (C.17). Then there exists a constant $C > 0$ such that for any $T > 0$, any accessible stopping time σ, any progressively measurable $\mathcal{L}(L^2)$-valued stochastic process ξ satisfying (C.16), the process $x(t)$, $t \geq 0$ from Proposition C.1.8 satisfies the following*

$$\mathbb{E} \sup_{0\leq t\leq T} |x(t\wedge\sigma)|^p_{H^{\alpha\delta,q}} \leq C^p M^p T^{p(\beta-\delta)}. \tag{C.20}$$

Let us finish this section with the following two observations. The first is that given positive numbers α and δ satisfying inequality (C.17) we can always find q such that $\alpha\delta > \frac{1}{q}$. The second is that because of Jensen's inequality, inequalities (C.19) and (C.20) is also valid for all $p_2 \leq p$.

D.1. Pointwise multiplication in Sobolev spaces

Since our proof relies on Theorems 1 and 2 from section 4.4.4 of Ref. [35] and the formulations of these theorems contain some errors we have decided to present them here for the convenience of the reader.

Theorem 1. Assume that

$$0 < s_1 \leq s_2 \tag{1}$$

$$\frac{1}{p} \leq \frac{1}{p_1} + \frac{1}{p_2}, \tag{2}$$

$$\frac{n}{p} - s_1 > \begin{cases} (\frac{n}{p_1} - s_1)_+ + (\frac{n}{p_2} - s_2)_+, & \text{if } \max_i(\frac{n}{p_i} - s_i) > 0, \\ \max_i(\frac{n}{p_i} - s_i) > 0, & \text{otherwise} \end{cases} \tag{3}$$

$$s_1 + s_2 > \frac{n}{p_1} + \frac{n}{p_2} - n, \tag{4}$$

and either

$$q \geq q_1 \text{ if } s_1 < s_2 \tag{5}$$

or

$$q \geq \max(q_1, q_2) \text{ if } s_1 = s_2. \tag{6}$$

(i) Then

$$F^{s_1}_{p_1,q_1} \cdot F^{s_2}_{p_2,q_2} \hookrightarrow F^{s_1}_{p,q} \tag{7}$$

(ii) Also we have

$$B^{s_1}_{p_1,q_1} \cdot B^{s_2}_{p_2,q_2} \hookrightarrow B^{s_1}_{p,q} \tag{8}$$

Theorem 2. Assume that the conditions (1), (2), (4) and (5) are satisfied. Assume also that[f]

$$\frac{n}{p} - s_1 = \begin{cases} (\frac{n}{p_1} - s_1)_+ + (\frac{n}{p_2} - s_2)_+, & \text{if } \max_i(\frac{n}{p_i} - s_i) > 0, \\ \max_i(\frac{n}{p_i} - s_i) > 0, & \text{otherwise} \end{cases} \tag{16}$$

(i) If

$$\{i \in \{1,2\} : s_i = \frac{n}{p_i} \text{ and } p_i > 1\} = \emptyset, \tag{17}$$

then (7) remains true.

(ii) Suppose in addition that[g]

$$q \geq q_2, \text{ if } s_1 - \frac{n}{p_1} = s_2 - \frac{n}{p_2}. \tag{18}$$

If

$$\{i \in \{1,2\} : s_i = \frac{n}{p_i} \text{ and } q_i > 1\}$$

$$\cup \{i \in \{1,2\} : s_i < \frac{n}{p_i} \text{ and } q_i > \frac{n}{\frac{n}{p_i} - s_i}\} = \emptyset, \tag{19}$$

then (8) remains true.

References

[1] G. Da Prato and D. Gatarek, Stochastic Burgers equation with correlated noise, *Stochastic and Stochastic Reports.* **52**, 29–41, (1995).

[2] W. E, K. Khanin, A. Mazel, and Y. Sinai, Invariant measures for Burgers equation with stochastic forcing, *Ann. of Math. (2).* **151**(3), 877–960, (2000). ISSN 0003-486X.

[3] J. Bertoin, C. Giraud, and Y. Isozaki, Statistics of a flux in Burgers turbulence with one-sided Brownian initial data, *Comm. Math. Phys.* **224**(2), 551–564, (2001). ISSN 0010-3616.

[4] J. D. Avrin, The generalized Burgers' equation and the Navier-Stokes equation in $\mathbb{R}^n$ with singular initial data, *Proceedings of the American Mathematical Society.* **101**(1), 29–40, (1987).

[5] P. Biler, T. Funaki, and A. W. Wojbor, Fractal Burgers' equations, *J. Differential Equations.* **148**, 9–46, (1998).

[6] N. Sugimoto, Generalized Burgers equations and fractional calculus, *Nonlinear Wave Motion.* **(A. Jeffrey, Ed)**, 162–179, (1989).

[7] N. Sugimoto, Burgers equations with a fractional derivative; hereditary effects on nonlinear acoustic waves, *J. Fluid Mech.* **225**, 631–653, (1991).

[8] N. Sugimoto and T. Kukatani, Generalized Burgers equations for nonlinear viscoelastic waves, *Wave Motion.* **7**, 447–458, (1985).

[f] In the book on the LHS of (16) one has $\frac{n}{p} - s$. But there is no s in the assertion and instead of s we need to have s_1, i.e. $\frac{n}{p} - s_1$. Another error of this sort can be seen twice on the RHS of formula (16), where it is written: $\max_i(\frac{n}{p_i} - s_1)$. The correct form should be $\max_i(\frac{n}{p_i} - s_i)$.

[g] In the book instead of $s_1 - \frac{n}{p_1}$ below in (18) one has $s_1 - \frac{n}{p}$.

[9] A. Le Mehaute, T. Machado, J. Trigeassou, and J. Sabatier, *Fractional Differentiation and its Applications, FDA'04.* vol. 2004-1, *Procceedings of the first IFAC Workshop,* (International Federation of Automatic Control, ENSEIRB, Bordeaux, France, July 19-21, 2004).

[10] V. Uchaikin and V. M. Zolotarev, *Chance and Stability, Stable Distributions and their Applications.* Modern Probability and Statistics, (VSP., 1999).

[11] J. Angulo, M. Ruiz-Medina, V. Anh, and W. Grecksch, Fractional diffusion and fractional heat equation, *Adv.Appl.Prob.* **32**, 1077–1099, (2000).

[12] R. Dalang and C. Mueller, Some non-linear s.p.d.e.'s that are second order in time, *Electronic Journal of Probab.* **8 no. 1**, 1–21, (2003).

[13] R. Dalang and M. Sanz-Solé, Regularity of the sample paths of a class of second-order spde's., *J. Funct. Anal.* **227, no. 2**, 304–337, (2005).

[14] L. Debbi and M. Dozzi, On the solution of non linear stochastic fractional partial differential equations, *Stochastic Processes and Their Applications.* **115**, 1764–1781, (2005).

[15] C. Mueller, The heat equation with Lévy noise, *Stoch. Proc. Appl.* **74**, 67–82, (1998).

[16] J.-L. Lions and E. Magenes, *Non-homogeneous boundary value problems and applications. Vol. I.* (Springer-Verlag, New York, 1972). Translated from the French by P. Kenneth, Die Grundlehren der mathematischen Wissenschaften, Band 181.

[17] H. Triebel, *Interpolation theory, function spaces, differential operators.* (Johann Ambrosius Barth, Heidelberg, 1995), second edition. ISBN 3-335-00420-5.

[18] W. McLean, *Strongly elliptic systems and boundary integral equations.* (Cambridge University Press, Cambridge, 2000). ISBN 0-521-66332-6; 0-521-66375-X.

[19] A. Pazy, *Semigroups of Linear Operators and Applications to Partial Differential Equations.* (Springer-Verlag, New York, 1983).

[20] M. E. Taylor, *Pseudodifferential Operators.* Princeton University Press, (Princeton, New Jersey, 1981). ISBN 0-691-08282-0.

[21] R. Seeley, Interpolation in L^p with boundary conditions, *Studia Math.* **44**, 47–60, (1972). ISSN 0039-3223. Collection of articles honoring the completion by Antoni Zygmund of 50 years of scientific activity, I.

[22] G. Da Prato and J. Zabczyk, *Stochastic Equations in Infinte Dimensions.* (Springer, Combridge university press, 1992). ISBN 0-521-38529-6.

[23] G. Da Prato, A. Debussche, and R. Temam, Stochastic Burgers' equation, *NoDEA Nonlinear Differential Equations Appl.* **1**(4), 389–402, (1994). ISSN 1021-9722.

[24] K. Twardowska and J. Zabczyk, A note on stochastic Burgers' system of equations, *Stochastic Analyis and Applications.* **22 no. 6**, 1641–1670, (2004).

[25] T. Kato, *Perturbation theory for linear operators.* Die Grundlehren der mathematischen Wissenschaften, Band 132, (Springer-Verlag New York, Inc., New York, 1966).

[26] Z. Brzeźniak, On stochastic convolution in Banach spaces and applications, *Stochastics Stochastics Rep.* **61**(3-4), 245–295, (1997). ISSN 1045-1129.

[27] G. Da Prato, S. Kwapien, and J. Zabczyk, Regularity of solutions of linear stochastic equations in Hilbert space, *Stochastic.* **23**, 1–23, (1987).

[28] D. Gatarek, A note on nonlinear stochastic equations in hilbert spaces, *Statistics & Probability letters.* **17**, 387–394.

[29] H. Fujita and T. Kato, On the Navier-Stokes initial value problem. I, *Arch. Rational Mech. Anal.* **16**, 269–315, (1964). ISSN 0003-9527.

[30] I. S. Gradshteyn, M. Ryzhik, A. Jeffrey, Yu. L. Geronimus, and M. Yu. Tseytlin, *Table of integrals, series, and products.* (Academic Press, Orlando, Florida/San Diego, California/New York/London, 1980).

[31] W. Feller, *An Introduction to Probability Theory and Its Applications : Vol. 2.* (John

Wiley and sons, New York, 1971).

[32] J. Seidler, Da Prato-Zabczyk's maximal inequality revisited. I, *Math. Bohem.* **118** (1), 67–106, (1993). ISSN 0862-7959.

[33] A. Carroll, *The stochastic nonlinear heat equation.* (PhD thesis; The University of Hull, Hull, 1999).

[34] Z. Brzeźniak, B. Maslowski, and J. Seidler, Stochastic nonlinear beam equations, *Probab. Theory Related Fields.* **132**(1), 119–149, (2005). ISSN 0178-8051.

[35] T. Runst and W. Sickel, *Sobolev spaces of fractional order, Nemytskij operators, and nonlinear partial differential equations.* vol. 3, *de Gruyter Series in Nonlinear Analysis and Applications*, (Walter de Gruyter & Co., Berlin, 1996). ISBN 3-11-015113-8.

[36] R. A. Adams, *Sobolev spaces.* (Academic Press, New York/San Francisco/London, 1975). ISBN 0-12-044150-0.

[37] H. Triebel, *Theory of function spaces.* vol. 78, *Monographs in Mathematics*, (Birkhäuser Verlag, Basel, 1983). ISBN 3-7643-1381-1.

[38] H. Amann. Multiplication in Sobolev and Besov spaces. In *Nonlinear analysis*, Sc. Norm. Super. di Pisa Quaderni, pp. 27–50. Scuola Norm. Sup., Pisa, (1991).

[39] T. Roger, *Navier-Stokes equations.* (AMS Chelsea Publishing, Providence, RI, 2001). ISBN 0-8218-2737-5. Theory and numerical analysis, Reprint of the 1984 edition.

[40] R. Z. Has′minskiĭ, *Stochastic stability of differential equations.* vol. 7, *Monographs and Textbooks on Mechanics of Solids and Fluids: Mechanics and Analysis*, (Sijthoff & Noordhoff, Alphen aan den Rijn, 1980). ISBN 90-286-0100-7. Translated from the Russian by D. Louvish.

[41] Z. Brzeźniak and D. Gątarek, Martingale solutions and invariant measures for stochastic evolution equations in Banach spaces, *Stochastic Process. Appl.* **84**(2), 187–225, (1999). ISSN 0304-4149.

[42] Z. Brzeźniak and Y. Li, Asymptotic compactness and absorbing sets for 2d stochastic navier-stokes equations on some unbounded domains, *Trans. Amer. Math. Soc.* **to appear in**. ISSN 0002-9947.

[43] A. Neidhardt, *Stochastic Integrals in 2-Uniformly Smooth Banach Spaces.* (PhD thesis; University of Wisconsin, 1978).

[44] P. Baxendale, Gaussian measures on function spaces, *Amer. J. Math.* **98**(4), 891–952, (1976). ISSN 0002-9327.

[45] Z. Brzeźniak and J. van Neerven, Space-time regularity for linear stochastic evolution equations driven by spatially homogeneous noise, *J. Math. Kyoto Univ.* **43**(2), 261–303, (2003). ISSN 0023-608X.

Chapter 6

Stochastic Control Methods for the Problem of Optimal Compensation of Executives

Abel Cadenillas, Jaksa Cvitanić* and Fernando Zapatero†

Department of Mathematical and Statistical Sciences, University of Alberta Edmonton, Alberta T6G 2G1, Canada.

acadenil@math.ualberta.ca.

We consider the problem of an executive who receives call options as compensation. He can influence the mean return of the stock with his effort, and the level of volatility through his choice of projects. The executive wants to select the optimal effort and choice of projects to maximize the expected utility from the call option minus the disutility associated with the effort. We model this as a stochastic control problem, and present an analytical solution. In this framework, we also introduce the problem of the company that simultaneously wants to minimize overtime volatility and maximize final expected value of the price of the stock. We characterize (and compute numerically for the logarithmic case) the optimal strike price the company should choose. When the executive can affect the mean return of the stock, we find that options should be granted out-of-the-money. In a context of perfect information, the executive would not be able to affect the mean return of the stock independently of the volatility, and it would be optimal to grant options in-the-money. The mathematical techniques are those of stochastic control, convex analysis, and martingale theory.

Contents

*Humanities and Social Sciences, Caltech, M/C 228-77, 1200 E. California Blvd. Pasadena, CA 91125, cvitanic@hss.caltech.edu.

†FBE, Marshall School of Business, University of Southern California, Los Angeles, CA 90089-1427, fzapatero@marshall.usc.edu.

1. Introduction

The optimal compensation of executives is currently one of the most active areas of research in mainstream finance, although it has not received much attention yet in the mathematical finance literature. The problem of the company consists in selecting the optimal compensation package to give the executive, and the problem of the executive consists in selecting the optimal strategy to take advantage of the compensation package.

A call option gives its owner the right to buy one share of an underlying stock at a predetermined price. Options are similar to a leveraged portfolio invested in the underlying security. For that reason, options have become the main choice for companies to compensate executives in order to align their interests with those of the company stockholders. Hall and Leibman [1] and Murphy [2] report statistics about the number of companies that use stock options as the main component of the compensation of their executives. Carpenter [3] and Hall and Murphy [4] consider the problem of valuation of executive stock options, Detemple and Sundaresan [5] develop a pricing model for options that can be applied to executive options, and Hall and Murphy [6] study the problem of choosing the strike price of the options. Stoughton and Wong [7] consider the optimality of stock versus options in a context of industry competition. Aseff and Santos [8] address the broader problem of the optimality of stock options versus other contracts. Kadan and Swinkels [9, 10] compare options vs. stock when there is a possibility of bankruptcy.

We develop a dynamic model in order to address some of the problems mentioned above. In particular, the existent literature considers the incentives of options in a static framework, which does not seem appropriate to address some of the questions considered. One of the potential problems of stock options compensation (see, for instance, Johnson and Tian [11, 12]) is the incentive to the executive to increase volatility. However, volatility is a dynamic concept and it is appropriate to consider problems like these in a dynamic setting. Another element that is missing in the current literature is the computation of the effort of the executive and the tradeoff between the amount of that effort and the disutility resulting from it.

We assume that the executive receives call options as compensation and we study the optimal strike price of the options. The strike price of zero means that the compensation is in stock. In our setting, the effort of the executive affects the mean return of the stock. Besides, the executive can choose among a menu of projects with a tradeoff between expected return and volatility: projects with higher volatility offer a higher expected return (see Cadenillas, Cvitanić and Zapatero [13] for a similar model in a different application).

The executive will choose effort and volatility in order to maximize expected utility coming from compensation minus the effort disutility. The company cares both about the final value of the stock and the overtime volatility. The company chooses the strike price of the option so as to induce the executive to achieve an optimal mix of effort and volatility. Carpenter [14] studies, also in a dynamic setting, the optimal choice of volatility when the effort of the agent cannot affect the stock dynamics and considers several exogenously given strike prices and more general benchmark functions. Bolton and Harris [15] also study a model where the agent has control of the dynamics of the asset, for a different purpose. In Cadenillas Cvitanić and Zapatero [16] we extend the model of this paper to the situation in which the type of the executive δ is unknown by the company, which has a prior distribution for it.

For the executive that maximizes logarithmic utility from the options, minus the quadratic disutility from the effort, we compute the optimal level of effort and volatility in closed form. Based on that, we can compute the optimal strike price to be set by the company. We find that the strike price should be higher, the higher the type (quality) of the executive, and the more the company is interested in maximizing the expected value of the price of the stock (as opposed to minimizing overtime volatility). The relationship between the optimal strike price and the quality of the projects (defined as the additional expected return resulting from accepting an extra unit of risk) is not monotonic.

We also compute the no-arbitrage (complete markets) price of the option. This would be the price of the option for the company and, therefore, would represent the cost the company incurs by granting options. The executive, on the other hand, faces incomplete markets since he cannot trade in the underlying. We compute the value he assigns to the option as the certainty equivalent associated with his utility. The price of the options is lower for the executive (facing incomplete markets) than for the company. We show that options seem to be more efficient when the executive is of a high type relative to the quality of the projects.

Finally, we also compute the optimal effort of the executive when cash is a part of the compensation package, but we are not able to get explicit expressions for the optimal strike price in this case.

The paper is structured as follows. In section 2 we present and solve explicitly the problem of the executive, and in section 3 we present and solve the problem of the company. In section 4 we perform some numerical computations and comparative statics. In Section 5 we compute the option prices from the perspectives of both the company and the executive. Section 6 considers the case of additional compensation in cash. We close the paper with some conclusions.

2. The Executive

2.1. *Stock Dynamics*

Consider a probability space $(\Omega, \mathcal{F}, P)$ endowed with a filtration $(\mathcal{F}_t)$, which is the P-augmentation of the filtration generated by a one-dimensional Brownian motion W. Our benchmark stock has a price that follows a geometric Brownian motion process,

$$dS_t = \mu S_t dt + \sigma S_t dW_t$$

with starting value S_0. Here, $\mu \in [0, \infty)$ and $\sigma \in (0, \infty)$ are exogenous constants.

Definition 2.1. The level of effort is an $(\mathcal{F}_t)$-adapted stochastic process u that satisfies

$$E\left[\int_0^T |u_t|^2 dt\right] < \infty.$$

The value u_t is the amount of effort that the executive puts in the management of the company at time t. The choice of projects is an $(\mathcal{F}_t)$-adapted stochastic process v that satisfies

$$E\left[\int_0^T |v_t S_t|^2 dt\right] < \infty.$$

We assume that at each moment in time, the executive can choose different projects (or strategies) v_t, that are characterized by a level of risk and expected return. We will denote by $\mathcal{A}$ the pairs (u, v) which satisfy the above technical conditions, and say that $\mathcal{A}$ is the class of admissible pairs of effort and choice of projects.

When the company is managed by the executive, the dynamics of the stock price S is given by

$$dS_t = \mu S_t dt + \delta u_t dt + \alpha v_t S_t dt + v_t S_t dW_t. \tag{2.1}$$

Here, $\delta \in [0, \infty)$ and $\alpha \in (0, \infty)$. The higher the value of u, the higher the expected value of the stock. On the other hand, the choice of projects v is equivalent to the choice of the volatility of the stock, although it also has an impact on the expected value. We assume that the executive can choose different projects or strategies that are characterized by a level of risk and expected return. Since $\alpha > 0$, the higher the risk of a project, the higher its expected return. The parameter α is a characteristic of the type of the company. One possible interpretation of α (and, potentially, a way to estimate it empirically) would be the slope of the equivalent of the "Capital Market Line" resulting from all the projects available to the company (more about the distinction between firm-specific and market risk below). On the other hand, δ is a measure of the impact of the effort of the executive on the value of the company. It can be interpreted as an indicator of the quality of the executive.

We emphasize that this is a partial equilibrium setting and we do not compare the dynamics of the stock of this company with the dynamics of other stocks. In the simplest setting, with complete markets, the coexistence of different stocks with different drifts (and maybe same standard deviations) would be consistent with incomplete information about the actions of the agent and its effects on the dynamics of the stock. In theory, in that setting, complete information would be consistent with $\delta = 0$, which is a special case of the model that we consider in this paper.

Without loss of generality, we will assume that $\mu = 0$. We then rewrite equation (2.1) as

$$dS_t = \delta u_t dt + \alpha v_t S_t dt + v_t S_t dW_t. \tag{2.2}$$

2.2. *The Problem of the Executive*

In our model, the executive is risk-averse and experiences disutility as a result of the effort.

Problem 2.2. *The objective of the executive is to select $(\widehat{u}, \widehat{v}) \in \mathcal{A}$ that maximizes the functional J defined by*

$$J(u, v) := E\left[\log\{n(S_T - K)^+\} - \frac{1}{2}\int_0^T u_t^2 dt\right]. \tag{2.3}$$

Here, n is the number of call options the executive receives as a part of his compensation package. The second term of the objective function of the executive represents the disutility from effort. That disutility might be, for example, the result of spending more time working for the company. The other control, v, only involves the choice of projects the company will undertake and has no effect on the disutility of the executive, since it does not require any effort: the executive has a menu of projects and decides the level of risk to undertake.

With this parameterization, the number of options n becomes irrelevant for the executive's problem, because $\log\{n(S_T - K)^+\} = \log n + \log(S_T - K)^+$. We will then assume that $n = 1$. It is obvious that a more general parameterization (such as power utility) would not have this property.

Problem 2.1 of the executive of this paper is similar to the problem of the manager considered by Cadenillas, Cvitanić and Zapatero [13]. In that paper, the manager controls the value of unlevered assets.

2.3. *Optimal Effort and Choice of Projects*

First we introduce the auxiliary exponential martingale Z, defined by

$$Z_t := \exp\left\{-\frac{1}{2}\alpha^2 t - \alpha W_t\right\}. \tag{2.4}$$

Also, consider the function of time $\bar{T}$, defined by

$$\bar{T}_t := \frac{e^{\alpha^2(T-t)} - 1}{\alpha^2}. \tag{2.5}$$

Using the previous notation and given the following quadratic equation in z,

$$\delta^2 \bar{T}_0 z^2 + (S_0 - K)z - 1 = 0, \tag{2.6}$$

where δ is the parameter that measures the type of the executive and K is the strike price of the options, we denote by $\check{z}$ the positive solution of (2.6):

$$\check{z} = \frac{1}{2\delta^2\bar{T}_0}\left((K-S_0)+\sqrt{(K-S_0)^2+4\delta^2\bar{T}_0}\right) \tag{2.7}$$

We now present the optimal controls of the executive:

Theorem 2.3. *Consider the problem of the executive described in Section 2.2. Consider also the exponential martingale of (2.4), the positive number $\check{z}$ given by (2.7), and $\bar{T}$, the time function of (2.5). The optimal effort $\widehat{u}$ of the executive is*

$$\widehat{u}_t = \delta\check{z}Z_t. \tag{2.8}$$

The optimal choice of projects $\widehat{v}$ is given by

$$\widehat{v}_t S_t = \frac{\alpha}{\check{z}Z_t} + \alpha\check{z}\delta^2 Z_t\bar{T}_t. \tag{2.9}$$

The optimal effort and volatility determine that the price of the stock of the firm be given by

$$S_t = \frac{1}{\check{z}Z_t} + K - \check{z}\delta^2 Z_t\bar{T}_t. \tag{2.10}$$

If $S_0 > K$ the above formulas remain valid even for $\delta = 0$, but with

$$\check{z} = \frac{1}{S_0 - K}\ .$$

If $S_0 \leq K$ and $\delta = 0$, then the utility of the executive is minus infinity, since it is impossible in this case to guarantee $S_T > K$ with probability one.

Proof. This theorem can be proved applying the duality method to solve stochastic control problems. We omit the proof, because it is a special case of the proof of Theorem 6.1. □

We observe that the optimal effort and volatility can also be written as functions of the price of the stock. Indeed, from equations (2.8)-(2.10), we obtain that when $\delta > 0$,

$$\widehat{u}_t = \frac{1}{2\delta\bar{T}_t}\left\{(K-S_t)+\sqrt{(K-S_t)^2+4\delta^2\bar{T}_t}\right\} \tag{2.11}$$

and

$$\begin{aligned}\widehat{v}_t S_t &= \frac{\alpha\delta}{\widehat{u}_t} + \alpha\delta\widehat{u}_t\bar{T}_t \\ &= \frac{2\alpha\delta^2\bar{T}_t}{(K-S_t)+\sqrt{(K-S_t)^2+4\delta^2\bar{T}_t}} \\ &\quad + \frac{\alpha}{2}\left[(K-S_t)+\sqrt{(K-S_t)^2+4\delta^2\bar{T}_t}\right].\end{aligned} \tag{2.12}$$

With respect to the optimal effort, as expected, $\widehat{u}$ is increasing in the strike price K: as K goes to infinity, the effort goes to infinity, as well. Besides, we note that $\check{z}$ is decreasing in T, the time to maturity of the option ($\bar{T}_0$ is increasing in T and $\check{z}$ is decreasing in $\bar{T}_0$). Therefore, the larger the maturity of the option, the lower the effort of the executive. The intuition is clear: a larger T has a similar effect on the executive as a reduction of the strike price. The effect of δ (the "type" of executive) depends on whether the option is in-, out-, or at-the-money. When the option is at-the-money, the optimal effort is independent of δ, as we can see by substituting (2.7) in (2.8). We can also check that when the option is in-the-money the effort is increasing in δ, and when the option is out-of-the-money the optimal effort decreases with δ.

Since Z is a martingale, the initial expected value of the effort at any point in time is

$$E[\widehat{u}_t] \;=\; \delta\check{z}. \tag{2.13}$$

With respect to the effect of α, we note that $\bar{T}_0$ is increasing in α and, therefore, $\check{z}$ is decreasing in α. Expected effort is, then, decreasing in α (everything else constant): the better the menu of projects the executive can choose, the lower the expected effort of the executive.

The analysis of the volatility is more complicated. Since $\bar{T}_.$ is a decreasing function of t and $\bar{T}_T = 0$, the second term of (2.9) decreases in expected value as we approach maturity, and will tend to be negligible relative to the first term. Therefore, for short maturities, optimal volatility will tend to decrease with higher strike price. For large maturities, the relation will tend to be the opposite. We also see that the volatility is increasing in the type δ of executive ($\check{z}$ is decreasing in δ, and $\check{z}\delta^2$ is increasing in δ). The economic intuition is straightforward: a high-type executive can afford more volatility because his effort will be more effective to counteract drops in the value of the stock. It is straightforward to see that the expected value of the volatility at a future date t is

$$E[\widehat{v}_t S_t] \;=\; \frac{\alpha}{\check{z}}e^{\alpha^2 t} + \alpha\check{z}\delta^2\bar{T}_t. \tag{2.14}$$

Since $\bar{T}_t$ is increasing in α, the expected volatility is increasing in α. In other words, the higher the expected return-risk tradeoff, the higher the risk the executive will be willing to undertake.

Some of these comparative statics are illustrated in the results of Table 6.1 (that we analyze in detail in section 4). However, in that table the strike price is always the optimal one, and some of the previous conclusions hold for changes in a given parameter with constant strike price.

It is also interesting to study the correlation between optimal effort and optimal choice of projects. By Itô's lemma, and equation (2.9), the dynamics of the optimal volatility are

$$d(\widehat{v}_t S_t) \;=\; (\ldots)dt + \alpha^2\left(\frac{1}{\check{z}Z_t} - \delta^2\check{z}Z_t\bar{T}_t\right)dW_t. \tag{2.15}$$

Equation (2.8) yields

$$d\widehat{u}_t = (\ldots)dt - \alpha\widehat{u}_t dW_t. \tag{2.16}$$

It is clear that their correlation can be either positive or negative. Considering only the instantaneous correlation and ignoring the drift terms, we see that for a short maturity of the option they tend to be negatively correlated, and increases in optimal effort will be typically associated with decreases of the optimal level of volatility.

3. The Company

3.1. *The Problem of the Company*

The company is interested in two things: the expected value and the volatility of the stock price. Specifically, we suppose that the objective of the company is to select the strike price $\widehat{K}$ that solves the following problem:

$$\max_{K\in[0,\infty)} \left\{ \lambda E[S_T] - \int_0^T Var[S_t]dt \right\}. \tag{3.1}$$

Here, T is the maturity of the European call options and λ is an exogenous constant that represents the tradeoff in the preferences of the company between the expected final value and the variance of the stock. The objective of the company has some similarity with quadratic preferences (in which the agent only cares about expected return and variance), but in our model the company cares about overtime variance. The objective function of the company is justified as a way to incorporate risk aversion in a tractable way. Another possible criterion is to select the number of options and the strike price to maximize the expected value of the price of the stock at maturity minus the price of the compensation package. In that case, the number of options and the strike price must be restricted so that the expected utility of the executive be greater or equal than a positive number. A similar problem has been studied by Cadenillas, Cvitanić and Zapatero [13].

3.2. *Optimal Strike Price*

We assume that the company has full information about the parameters that characterize the dynamics of the stock, as well as the preferences of the executive. The objective of the company is given by (3.1). In order to characterize the optimal choice of the company we introduce the following auxiliary notation:

$$A := \frac{e^{3\alpha^2 T} - 1}{3\alpha^2} - \frac{e^{2\alpha^2 T} - 1}{2\alpha^2}, \tag{3.2}$$

$$B := \frac{\delta^4}{\alpha^4}\left\{ \frac{1}{2\alpha^2}e^{2\alpha^2 T} + \left(\frac{2}{\alpha^2} - 2T\right)e^{\alpha^2 T} - \frac{5}{2\alpha^2} - T \right\}, \tag{3.3}$$

and the function h defined by

$$h(z) := \lambda \frac{1}{z}(e^{\alpha^2 T} - 1) - \frac{1}{z^2} A - z^2 B + \lambda z \delta^2 \bar{T}_0, \tag{3.4}$$

where λ is the parameter of (3.1). We also define

$$\eta := \begin{cases} \frac{-S_0 + \sqrt{S_0^2 + 4\delta \bar{T}_0}}{2\delta \bar{T}_0} & \text{if } \delta > 0 \\ \frac{1}{S_0} & \text{if } \delta = 0. \end{cases}$$

We will be interested in maximizing the function h in the interval $[\eta, \infty)$. We now present the optimal strike price for the company.

Theorem 3.1. *Consider the company whose objective is given by (3.1). Let $\bar{z} \in [\eta, \infty)$ be the quantity that maximizes (3.4), Then the optimal strike price is given by*

$$\widehat{K} = S_0 - \frac{1}{\bar{z}} + \bar{z}\delta^2 \bar{T}_0. \tag{3.5}$$

Proof. Our first objective is to compute the variance and the expectation of S_t. Some preliminary computations give

$$E[Z^2(t)] = e^{\alpha^2 t}, \quad E[Z(t)] = 1, \quad E[Z^{-2}(t)] = e^{3\alpha^2 t}, \quad E[Z^{-1}(t)] = e^{\alpha^2 t}.$$

Thus, according to equation (2.10),

$$E[S_t] = \frac{1}{\check{z}} e^{\alpha^2 t} + K - \check{z}\delta^2 \bar{T}_t$$

and

$$\begin{aligned} Var[S_t] &= Var\left[\frac{1}{\check{z} Z_t} - \check{z}\delta^2 Z_t \bar{T}_t\right] \\ &= E\left[\left(\frac{1}{\check{z} Z_t} - \check{z}\delta^2 Z_t \bar{T}_t\right)^2\right] - \left(E\left[\frac{1}{\check{z} Z_t} - \check{z}\delta^2 Z_t \bar{T}_t\right]\right)^2 \\ &= \frac{1}{\check{z}^2}(e^{3\alpha^2 t} - e^{2\alpha^2 t}) + 2\delta^2 \bar{T}_t (e^{\alpha^2 t} - 1) + \check{z}^2 \delta^4 \bar{T}_t^2 (e^{\alpha^2 t} - 1). \end{aligned}$$

Noting that $\bar{T}_T = 0$, $K = S_0 - 1/\check{z} + \check{z}\delta^2 \bar{T}_0$ from (2.6)-(2.7), we see that

$$\begin{aligned} \lambda E[S_T] &= \lambda \left[\frac{1}{\check{z}} e^{\alpha^2 T} + K\right] \\ &= \lambda \left[\frac{1}{\check{z}} e^{\alpha^2 T} + S_0 - 1/\check{z} + \check{z}\delta^2 \bar{T}_0\right] \\ &= \lambda \left[\frac{1}{\check{z}} e^{\alpha^2 T} - 1/\check{z} + \check{z}\delta^2 \bar{T}_0\right] \\ &\quad + \text{ an expression that involves neither } \check{z} \text{ nor } K. \end{aligned}$$

Furthermore,

$$\int_0^T \mathrm{Var}[S_t]dt = \frac{1}{\check{z}^2}\int_0^T (e^{3\alpha^2 t} - e^{2\alpha^2 t})dt + 2\delta^2 \int_0^T \bar{T}_t(e^{\alpha^2 t} - 1)dt$$
$$+ \check{z}^2\delta^4 \int_0^T \bar{T}_t^2(e^{\alpha^2 t} - 1)dt = \frac{1}{\check{z}^2}\left\{\frac{e^{3\alpha^2 T} - 1}{3\alpha^2} - \frac{e^{2\alpha^2 T} - 1}{2\alpha^2}\right\}$$
$$+\check{z}^2\frac{\delta^4}{\alpha^4}\left\{\frac{1}{2\alpha^2}e^{2\alpha^2 T} + \left(\frac{2}{\alpha^2} - 2T\right)e^{\alpha^2 T} - \frac{5}{2\alpha^2} - T\right\}$$
$$+ \text{ an expression that involves neither } \check{z} \text{ nor } K.$$

Disregarding the terms that do not depend on $\check{z}$ or K, we see that (3.1) is equivalent to finding $\bar{z} \in (0, \infty)$ that maximizes the function h defined in (3.4). Here, A and B are given in (3.2) and (3.3). The maximization has to be performed under the constraint that the corresponding K is non-negative. We can find K from (2.6) as

$$K = S_0 - 1/\check{z} + \check{z}\delta^2\bar{T}_0,$$

the equation given in (3.5). We observe that $\bar{z} \in (0, \infty)$ satisfies the restriction

$$S_0 - 1/\bar{z} + \bar{z}\delta^2\bar{T}_0 \geq 0$$

if and only if

$$\delta\bar{T}_0\bar{z}^2 + S_0\bar{z} - 1 \geq 0.$$

That is equivalent to

$$\bar{z} \geq \begin{cases} \frac{-S_0+\sqrt{S_0^2+4\delta\bar{T}_0}}{2\delta\bar{T}_0} & \text{if } \delta > 0 \\ \frac{1}{S_0} & \text{if } \delta = 0 \end{cases} = \eta.$$

□

We do not have a closed-form solution for the constant $\bar{z}$. However, it can be found easily by applying the following necessary condition of optimality:

$$2B\bar{z}^4 - \lambda\delta^2\bar{T}_0\bar{z}^3 + \lambda(e^{\alpha^2 T} - 1)\bar{z} - 2A = 0 \text{ if } \bar{z} > \eta \tag{3.6}$$
$$2B\bar{z}^4 - \lambda\delta^2\bar{T}_0\bar{z}^3 + \lambda(e^{\alpha^2 T} - 1)\bar{z} - 2A \leq 0 \text{ if } \bar{z} = \eta, \tag{3.7}$$

which is the result of differentiating (3.4) and the constraint $\bar{z} \in [\eta, \infty)$ (which is equivalent to $\widehat{K} \in [0, \infty)$). We will perform some numerical exercises in the next section. Next, we consider the case in which the company is only interested in variance minimization.

Corollary 3.2. *Consider a company that chooses the strike price that minimizes overtime variance, i.e.,*

$$\min_{K \geq 0} \int_0^T \mathrm{Var}[S_t]dt. \tag{3.8}$$

Suppose also that $\delta = 0$, i.e., that the executive cannot influence the drift independently of volatility. Then it is optimal to issue at the money options, i.e.,

$$\widehat{K} = S_0. \tag{3.9}$$

Proof. This corresponds to the case $\lambda = 0$. Since also $\delta = 0$, then $B = 0$, and (3.4) implies we have to find $\bar{z}$ that minimizes $\frac{1}{\bar{z}^2}A$. Hence optimal $\bar{z} = \infty$, and (3.5) gives $\widehat{K} = S_0$. □

If the company cares only about the variance and the executive can affect the drift of the stock only through the choice of the level of volatility, the corollary says that options should be optimally issued at-the-money (which is the customary practice). We will see later that in general we get a large range of possible optimal values for the strike price .

Corollary 3.3. *Consider a company that chooses the strike price that maximizes the expected value of the stock, i.e.,*

$$\max_{K \geq 0} E[S_T]. \tag{3.10}$$

Suppose also that $\delta = 0$. Then, the optimal strike price is $\widehat{K} = 0$.

Proof. We note that in the case $\delta = 0$, equation (2.10) gives

$$E[S_T] \;=\; \frac{e^{\alpha^2 T}}{\tilde{z}} + K.$$

Thus, according to (3.5), the company wants to find the value of K that maximizes

$$E[S_T] \;=\; K(1 - e^{\alpha^2 T}) + S_0 e^{\alpha^2 T}.$$

Since $\alpha > 0$, the optimal $\widehat{K}$ is zero. □

When the company only cares about the expected final price of the stock and the executive can influence the drift of the stock only through the choice of projects, this corollary says that it is optimal for the company to give all the compensation in stock.

4. Numerical Computations of the Strike Price

In the previous section we derived the optimal exercise price for a company that cares about both the expected final value of the stock and the overtime volatility. It is expressed in equation (3.5). It depends on the solution to equations (3.6)-(3.7), which do not have an explicit solution. In this section we perform some numerical exercises in order to derive some properties of the optimal strike price. The results are included in Table 6.1.

We study the strike price as a function of the parameters of the model. The effects of the time to maturity are obvious and we fix it at $T = 5$. The initial price of the stock is 100. Besides the optimal strike price, we also record the optimal initial effort level $\widehat{u}_0$ and the optimal initial choice of projects $\widehat{v}_0$ (expressed in %). The last two columns are helpful for intuitive purposes.

The first general observation is that for strictly positive values of the parameters, the optimal strike price is out-of-the money. Next, we analyze the type of executive.

Table 6.1. Optimal strike price, effort, and volatility.

$S_0 = 100$; $T = 5$					
δ	α	λ	$\widehat{K}$	$\widehat{u}_0$	$\widehat{v}_0(\%)$
1	0.05	0.5	140.01	9.56	2.41
1	0.1	0.5	111.77	2.38	1.26
1	0.25	0.5	103.78	0.85	1.54
1	0.5	0.5	108.70	0.98	5.38
1	0.75	0.5	143.14	1.57	33.31
1	0.25	0.25	103.40	0.79	1.48
1	0.25	0.5	103.78	0.85	1.54
1	0.25	1	104.70	0.99	1.70
1	0.25	2	107.88	1.46	2.31
1	0.25	5	120.16	3.28	5.18
1	0.5	0.1	108.60	0.97	5.33
1	0.5	0.25	108.64	0.97	5.35
1	0.5	0.5	108.70	0.98	5.35
1	0.5	1	108.85	0.99	5.43
1	0.5	2	109.16	1.02	5.56
1	0.5	5	110.35	1.13	6.06
5	0.05	0.5	146.35	1.94	2.57
5	0.1	0.5	116.24	0.85	2.79
5	0.25	0.5	116.02	0.76	7.26
5	0.5	0.5	142.98	0.95	26.66
5	0.25	0.1	115.52	0.75	7.19
5	0.25	0.25	115.70	0.76	7.22
5	0.25	0.5	116.02	0.76	7.26
5	0.25	1	116.67	0.78	7.35
5	0.25	2	118.12	0.82	7.56
5	0.25	5	123.88	0.98	8.50
5	0.5	0.1	142.87	0.96	26.62
5	0.5	0.25	142.91	0.96	26.64
5	0.5	0.5	142.98	0.96	26.66
5	0.5	1	143.11	0.97	26.72
5	0.5	2	143.39	0.97	26.83
5	0.5	5	144.24	0.99	27.17

The higher the type of the executive (higher δ), the higher the optimal strike price. A higher type can have more impact with lower effort and will also tend to choose more volatile projects, since she will be able to counteract negative shocks more easily.

The quality of the projects α is very important and has an effect on the optimal strike price that is not monotonic: the optimal strike price is very high for low quality projects, then it is decreasing as α increases, up to a certain point. Beyond that point, the optimal strike price is increasing in α. For very low α, a high strike price is needed to force the executive to choose a high effort level. As the quality of the projects increases, less effort is required. However, when volatility becomes very effective (that is, large α), it is necessary again to set a higher strike price to induce higher effort in order to guarantee that the executive keeps an adequate mix

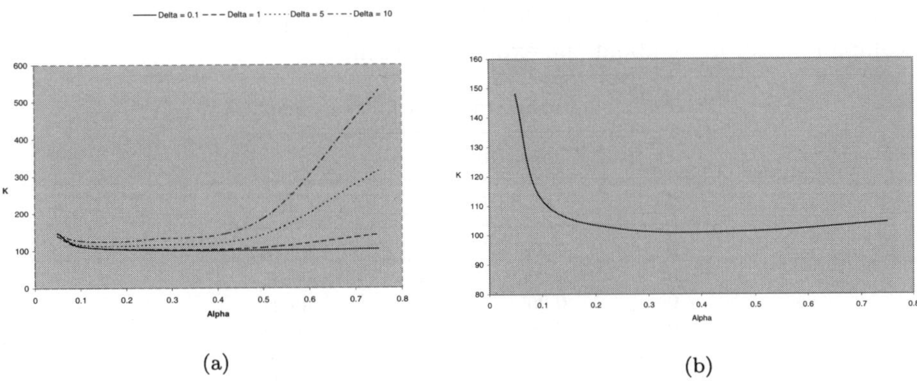

(a) (b)

Fig. 6.1. The optimal strike price as a function of the quality of projects (a) Different values of δ (b) $\delta = 0.1$

of effort u and choice of projects v, and does not rely on the choice of projects v exclusively. It is clear from our argument that the type of the executive plays a major role in this trade-off. This point is illustrated in Figures 6.1(a) and 6.1(b). In Figure 6.1(a) we observe that the pattern described above holds for any level of type of the executive, but the inflection point at which it is necessary to start raising the strike price again comes faster the higher the type of the executive. Figure 6.1(b) is the lowest of the four lines in Figure 6.1(a) measured in a different scale.

The effect of λ in this range is clear: the less the company cares about overtime volatility, the higher the strike price it should set, regardless of the values of other parameters. In order to get a higher expected price of the stock, the executive can either exercise more effort or choose riskier projects. Since the company is concerned about volatility it will prefer more effort, but the only way to induce it is by setting a higher strike price. Of course, the impact of additional volatility on expected return (α) affects the effectiveness of the strike price to attain company goals.

As we mentioned above, an important conclusion of our results is that for strictly positive values of the parameters, the optimal strike price is always out-of-the-money. Our results shed some light on the debate about whether options should be at-the-money or out-of-the-money (Hall and Murphy [6]). As we see in Table 6.2, options in-the-money would be optimal in the case of an executive whose effort does not have any effect on the stock return ($\delta = 0$) and when the company is primarily concerned about the price of the stock (high λ). We also recall from our discussion about the dynamics of the stock of equation (2.1) that the case $\delta = 0$ could correspond to the situation of perfect information about the effect of the actions of the executive on the dynamics of the stock. Options at-the-money or close to at-the-money are optimal (as we see asymptotically in Table 6.2) when the executive is of very low type (low δ) and the company is mainly interested in price volatility (low λ). We also see that for $\delta = 0$ the optimal strike price gets lower as the company cares more about the expected return.

Table 6.2. Optimal strike price for "very small" δ.

$S_0 = 100$; $T = 5$			
δ	α	λ	$\widehat{K}$
0	0.25	1	99.86
0	0.25	10	98.63
0	0.25	100	86.31
0	0.5	100	96.23
0	0.5	1000	62.26
0.005	0.25	100	510.21
0.005	0.25	10	141.02
0.005	0.25	0.1	100.41

5. Price of the Options

The executive of the problem considered in this paper faces incomplete markets as a result of the fact that he cannot sell the options and cannot trade in the underlying security. In practice, the executive will not be allowed to sell the option or take a short position in the underlying stock (as he would optimally do for diversification purposes). On the other hand, we can assume that the company faces complete markets and use, therefore, arbitrage arguments to price the option. There is no obvious way of computing the exact difference between the value of the option for the executive and for the company. The approach taken in the literature (that we follow here) is to compute the certainty equivalent of the executive: a constant amount of money instead of the option that would leave the executive at the same utility level as the option. We do not interpret the difference in the value of the no-arbitrage price and the certainty equivalent as the actual dollar amount. Rather, we consider the ratio of those two numbers as meaningful: a higher ratio of the price for the executive to the price for the company indicates that the options are more appropriate as compensation. A low ratio will question the optimality of using options as incentive.

Our setting allows us to compute the price of the option, both for the company and for the executive. We first introduce the price of the option for the company.

Proposition 5.1. *Consider the problem of the company described in section 3. Assume that the company faces complete markets, i.e., it can replicate the option with the underlying security and a risk-free asset that pays a constant interest rate r and whose price satisfies*

$$dB_t = B_t r dt \quad \text{and} \quad B_0 = 1. \tag{5.1}$$

The price for the company of the call option granted to the executive as compensation is given by

$$C_0 = \frac{1}{\bar{z}}E\left[\exp\left\{\int_0^T (\alpha - \theta_s)dW_s + \frac{1}{2}\int_0^T (\alpha^2 - 2r - \theta_s^2)ds\right\}\right]. \quad (5.2)$$

Here, θ is the market price of risk given by

$$\theta_t := (\bar{\mu}_t - r)/\bar{\nu}_t, \quad (5.3)$$

where $r \in [0, \infty)$ is the interest rate of (5.1), and $\bar{\mu}$ and $\bar{\nu}$ are defined by

$$\bar{\mu}_t := \frac{\bar{H}_t\alpha^2 + \bar{z}\delta^2 Z_t e^{\alpha^2(T-t)}}{\bar{H}_t + K - \bar{z}\delta^2 Z_t \bar{T}_t} \quad (5.4)$$

$$\bar{\nu}_t := \frac{\alpha\bar{H}_t + \alpha\bar{z}\delta^2 Z_t \bar{T}_t}{\bar{H}_t + K - \bar{z}\delta^2 Z_t \bar{T}_t}. \quad (5.5)$$

In addition, Z and $\bar{T}$ are defined in (2.4) and (2.5), $\bar{z}$ is defined in Theorem 3.1, and $\bar{H}$ is defined by

$$\bar{H}_t := \frac{1}{\bar{z}Z_t}. \quad (5.6)$$

Proof. The company faces complete markets and prices the option by arbitrage. We introduce the notation

$$\bar{H}_t = \frac{1}{\bar{z}Z_t}$$

where Z is given by equation (2.4) and $\bar{z}$ is defined in Theorem 3.1. Applying Itô's formula, we see that

$$\frac{1}{Z_t} = 1 + \int_0^t \frac{\alpha^2}{Z_s}ds + \int_0^t \frac{\alpha}{Z_s}dW_s,$$

or equivalently

$$\bar{H}_t = \frac{1}{\bar{z}} + \int_0^t \alpha^2 \bar{H}_s ds + \int_0^t \alpha\bar{H}_s dW_s.$$

Furthermore,

$$Z_t\bar{T}_t = \frac{e^{\alpha^2 T} - 1}{\alpha^2} - \int_0^t Z_s e^{\alpha^2(T-s)}ds - \int_0^t \alpha\bar{T}_s Z_s dW_s.$$

Thus, according to (2.10),

$$dS_t = d(\bar{H}_t + K - \bar{z}\delta^2 Z_t\bar{T}_t) = (\bar{H}_t\alpha^2 + \bar{z}\delta^2 Z_t e^{\alpha^2(T-t)})dt + (\alpha\bar{H}_t + \alpha\delta^2\bar{z}Z_t\bar{T}_t)dW_t,$$

where $\bar{T}$ is given by equation (2.5). According to equations (5.4)-(5.5), we have

$$dS_t/S_t = \bar{\mu}_t dt + \bar{\nu}_t dW_t.$$

The market price of risk is then given by (5.3), where $\bar{\mu}$ and $\bar{\nu}$ are defined in (5.4)-(5.5). The risk neutral density is

$$Z_T^* = \exp\left\{-\int_0^T \theta_s dW_s - \frac{1}{2}\int_0^T \theta_s^2 ds\right\}.$$

The no-arbitrage price is then

$$\begin{aligned} C_0 &= E[e^{-rT} Z_T^*(S_T - K)^+] = E[e^{-rT} Z_T^* \bar{H}_T] \\ &= \frac{1}{\bar{z}} E\left[e^{\int_0^T (\alpha-\theta_s)dW_s + \frac{1}{2}\int_0^T (\alpha^2 - 2r - \theta_s^2)ds}\right]. \end{aligned}$$

The executive, however, cannot replicate the option given his trading constraints. In order to compute the value of the option for him, we use the certainty equivalent suggested in Hall and Murphy [6]. This is the fixed amount of money, paid instead of the option, that would provide the executive the same level of utility as the option.

Proposition 5.2. *Consider the problem of the executive described in section 2. The fixed amount of money CE that would provide the executive the same utility level as the option is*

$$CE = \frac{1}{\bar{z}} \exp\left\{\frac{1}{2}\left[\alpha^2 T - \delta^2 \alpha^{-2}(e^{\alpha^2 T} - 1)\bar{z}^2\right]\right\}. \tag{5.7}$$

Proof. We compute the fixed amount x that provides the executive with the same utility level as the option. The certainty equivalent $CE = x$ is computed from

$$\log x = E\left[\log(S_T - K)^+ - \frac{1}{2}\int_0^T \widehat{u}_t^2 dt\right].$$

Substituting $S_T = \bar{H}_T + K$, $\widehat{u}_t = \delta \bar{z} Z_t$, and computing expectations, we obtain

$$\log x = -\log \bar{z} + \frac{1}{2}[\alpha^2 T - \delta^2 \alpha^{-2}(e^{\alpha^2 T} - 1)\bar{z}^2].$$

□

We now perform some comparative statics. The numerical results are presented in Table 6.3. It is similar to Table 6.1, but it also includes the prices of the option (we do not report the optimal effort level in Table 6.3). We assume that the company chooses the optimal strike price, recorded in the column $\widehat{K}$. C_0 is the price of the option for the company, that we compute using equation (5.2). The expression of equation (5.2) does not have a closed form solution, but is easy to compute numerically using Monte Carlo simulation (see Boyle, Broadie and Glasserman [17] for a review of the method). Column CE is the price of the option for the executive, the certainty equivalent, that we compute using equation (5.7). We compare the values in the columns C_0 and CE. The options are worth less for the executive than for the company. However, the ratio will vary depending on the values of the parameters. When the ratio of the price of the option for the executive and the price of the option for the company is too low, options might not be the optimal form of compensation. Finally, in the column denoted BS we record the Black and Scholes price (Black and Scholes [18]). For volatility, we use the optimal initial volatility, that is, $\widehat{v}_0$ as given by equation (2.9) and reported in the column "Vol."The only extra parameter required for Table 6.3 is the interest rate of equation (5.1). For concreteness, we assume that the interest rate is zero.

Table 6.3. Option price.

$S_0 = 100;\ T = 5$							
δ	α	λ	$\widehat{K}$	Vol	C_0	CE	BS
0.1	0.25	0.5	102.02	0.52	0	0	0.020
0.1	0.5	0.5	101.35	0.61	0.061	0	0.118
0.1	0.75	0.5	104.40	3.39	0.173	0	1.375
1	0.25	0.5	103.78	1.54	0.589	0.165	0.25
1	0.5	0.5	108.70	5.38	0.713	0.016	1.807
1	0.75	0.5	143.14	33.31	1.720	0	17.52
1	0.25	0.25	103.40	1.48	0.323	0.237	0.275
1	0.25	1	104.70	1.70	0.120	0.068	0.206
1	0.25	2	107.88	2.31	0.216	0.002	0.170
1	0.5	0.5	108.70	5.38	0.713	0.016	1.807
1	0.5	2	109.16	5.56	0.685	0.010	1.839
5	0.25	0.5	116.02	7.26	1.800	1.372	1.704
5	0.5	0.5	142.98	26.66	3.655	0.110	11.745
5	0.25	0.1	115.52	7.19	1.854	1.453	1.740
5	0.25	0.25	115.70	7.22	1.762	1.412	1.730
5	0.25	1	116.67	7.35	1.607	1.221	1.658

Table 6.3 complements nicely our conclusions from Table 6.1. It shows that when the quality of the projects is very high (large α) and the executive is of a low quality (small δ), the optimal strike price will be high in order to force the executive to exercise more effort. As a result, the value of the option for the executive will be low relative to the complete market price of the option. In that case it might not be optimal to grant options as an incentive mechanism. In summary, it seems that options are more appropriate the higher the quality of the executive and the lower the quality of the projects.

We also observe that in the cases in which the ratio of the certainty equivalent with respect to the price of the option is high, the Black and Scholes price is a relatively good approximation to the true price of the option, with our choice of volatility. However, in general, the quality of the approximation using Black and Scholes is questionable.

6. The Case of Additional Cash Compensation

In this section we consider the case in which the executive receives, in addition to the stock options, some cash compensation. The problem of the executive is to select $(\widehat{u}, \widehat{v}) \in \mathcal{A}$ that maximize the criterion $\tilde{J}$ defined by

$$\tilde{J}(u,v) := E\left[\log\{w + n(S_T - K)^+\} - \frac{1}{2}\int_0^T u_t^2 dt\right]. \tag{6.1}$$

Here $w > 0$ represents some cash compensation independent of the price of the stock of the company. The rest of the parameters are as in section 2.

Alternatively, this case can be interpreted as the situation in which the executive has some wealth whose value is independent of the price of the stock. The main difference between this case and the case considered in section 2 is that now the number of options received in the compensation package becomes relevant. The problem becomes considerably more difficult, but we can still compute numerically the optimal effort and volatility. We include in this section the main result and some numerical examples.

We start by considering the auxiliary function g defined by

$$g(t,y) := \left(K - \frac{w}{n}\right) N(d_2(t,y)) + yN(d_1(t,y)), \tag{6.2}$$

where

$$d_1(t,y) := \frac{\log\{\frac{yn}{cw}\} + \frac{1}{2}\alpha^2(T-t)}{\alpha\sqrt{T-t}}, \tag{6.3}$$

$$d_2(t,y) := \frac{\log\{\frac{yn}{cw}\} - \frac{1}{2}\alpha^2(T-t)}{\alpha\sqrt{T-t}}, \tag{6.4}$$

and

$$N(x) := \int_{-\infty}^{x} \frac{1}{\sqrt{2\pi}} \exp\left\{-\frac{z^2}{2}\right\} dz. \tag{6.5}$$

We observe that

$$\frac{\partial}{\partial y} g(t,y) = \left(K - \frac{w}{n}\right) \frac{1}{\sqrt{2\pi}} \exp\left\{-\frac{1}{2}(d_2(t,y))^2\right\} \frac{1}{y} \frac{1}{\alpha\sqrt{T-t}}$$
$$+ \frac{1}{\sqrt{2\pi}} \exp\left\{-\frac{1}{2}(d_1(t,y))^2\right\} \frac{1}{\alpha\sqrt{T-t}} + N(d_1(t,y)).$$

Let us consider the function $f : [1,\infty) \mapsto \mathbf{R}$ defined by

$$f(x) := \log\{x\} - 1 - \left(\frac{nK}{w} - 1\right)\frac{1}{x}.$$

We observe that $f(1) = -\frac{nK}{w} < 0$, $f(\infty) = \infty$, and f is strictly increasing (indeed, for every $y > 1$: $f'(y) = \frac{1}{y} + (\frac{nK}{w} - 1)\frac{1}{y^2} > 1/y - 1/y^2 > 0$). Thus, there exists a constant $c = c(n, K, w) \in (1, \infty)$ which satisfies the following two properties:

$$f(c) = 0 \quad \text{and} \quad y > c \iff f(y) > 0. \tag{6.6}$$

Let $\tilde{z}$ be the positive solution of the nonlinear equation

$$S_0 = g\left(0, \frac{1}{z}\right) - z\delta^2 \bar{T}_0. \tag{6.7}$$

Theorem 6.1. *Consider an executive who wants to solve (6.1). Consider the stochastic process Z and the function $\bar{T}$ defined in section 2. Then the optimal effort is given by*

$$\widehat{u}_t = \delta \tilde{z} Z_t \tag{6.8}$$

and the optimal choice of projects is given by

$$\widehat{v}_t S_t = \alpha \tilde{H}_t \frac{\partial}{\partial y} g(t, \tilde{H}_t) + \alpha \tilde{z} \delta^2 Z_t \bar{T}_t. \tag{6.9}$$

Here, $\tilde{H}$ is defined by

$$\tilde{H}_t = \frac{1}{\tilde{z} Z_t}. \tag{6.10}$$

Proof. We want to find $\widehat{u}$ and $\widehat{v}$ that solve the problem

$$\max_{(u,v)\in\mathcal{A}} E\left[F(S_T) - \int_0^T G(u_s)ds\right],$$

where

$$F(s) = \log\{w + n(s-K)^+\} \quad \text{and} \quad G(u) = \frac{u^2}{2}.$$

We observe that the stochastic process Z of (2.4) satisfies

$$dZ_t = -\alpha Z_t dW_t \quad \text{and} \quad Z(0) = 1.$$

Applying the formula of integration by parts, we see that

$$Z_t S_t = S_0 + \int_0^t Z_s \delta u_s ds + \int_0^t S_s Z_s (v_s - \alpha) dW_s.$$

Let us consider the stochastic process M defined by

$$M_t := Z_t S_t - \delta \int_0^t Z_s u_s ds = S_0 + \int_0^t S_s Z_s (v_s - \alpha) dW_s.$$

Obviously, M is a local martingale, but we would like to prove that M is also a martingale. For that purpose, it is sufficient to verify the condition

$$E\left[\sup_{0\le t\le T} |M_t|\right] < \infty. \tag{6.11}$$

According to the Burkholder-Davis-Gundy inequality (see, e.g., Theorem 3.3.28 of Karatzas and Shreve [19]), it is sufficient to verify

$$E\left[\left(\int_0^T (v_t - \alpha)^2 S_t^2 Z_t^2 dt\right)^{1/2}\right] < \infty. \tag{6.12}$$

We observe that, according to Theorem 1.6.16 of Yong and Zhou [20], $E\left[\sup_{0\le t\le T} Z_t^2\right] < \infty$. Since $E[\int_0^T |v_t S_t|^2 dt] < \infty$ by Definition 2.1, that theorem applied to equation (2.1) or (2.2), gives $E\left[\sup_{0\le t\le T} S_t^2\right] < \infty$. Applying Hölder's inequality (see, e.g., Theorem 4.2 of Chow and Teicher [21]) and again the condition $E[\int_0^T |v_t S_t|^2 dt] < \infty$, we note that

$$\begin{aligned} E\left[\left(\int_0^T (v_t S_t Z_t)^2\, dt\right)^{1/2}\right] &\le E\left[\left(\sup_{0\le t\le T} Z_t^2 \int_0^T (v_t S_t)^2\, dt\right)^{1/2}\right] \\ &= E\left[\left(\sup_{0\le t\le T} Z_t^2\right)^{1/2}\left(\int_0^T (v_t S_t)^2\, dt\right)^{1/2}\right] \\ &\le \left(E\left[\sup_{0\le t\le T} Z_t^2\right]\right)^{1/2}\left(E\left[\int_0^T (v_t S_t)^2\, dt\right]\right)^{1/2} \\ &< \infty. \end{aligned} \tag{6.13}$$

This implies that

$$E\left[\left(\int_0^T ((v_t-\alpha) S_t Z_t)^2\, dt\right)^{1/2}\right] < \infty, \tag{6.14}$$

and therefore M is a martingale. Thus,

$$E[M_T] \;=\; E[M_0] \;=\; S_0. \tag{6.15}$$

Next, consider the dual function $F : (0,\infty) \mapsto \mathbf{R}$ defined by

$$\tilde{F}(z) \;=\; \max_{s\ge 0}[F(s) - sz].$$

The maximum is attained at the points of the form

$$\begin{aligned} \widehat{s} &= \widehat{s}(z,a) \\ &= \left(K + \max\left(\frac{1}{z} - \frac{w}{n}, 0\right)\right) I_{\{\log\{w+\max(\frac{n}{z}-w,0)\}-zK-\max(1-\frac{wz}{n},0)>\log\{w\}\}} \\ &\quad + aI_{\{\log\{w+\max(\frac{n}{z}-w,0)\}-zK-\max(1-\frac{wz}{n},0)=\log\{w\}\}} \\ &= \left(K + \max\left(\frac{1}{z} - \frac{w}{n}, 0\right)\right) I_{\{\log\{w+n\max(\frac{1}{z}-\frac{w}{n},0)\}-zK-z\max(\frac{1}{z}-\frac{w}{n},0)>\log\{w\}\}} \\ &\quad + aI_{\{\log\{w+n\max(\frac{1}{z}-\frac{w}{n},0)\}-zK-z\max(\frac{1}{z}-\frac{w}{n},0)=\log\{w\}\}} \end{aligned}$$

where a is either 0 or $K + \max\left(\frac{1}{z} - \frac{w}{n}, 0\right)$. Consider also the dual function $G : (0,\infty) \mapsto \mathbf{R}$ defined by

$$\tilde{G}(z) \;=\; \max_{u}[-G(u) + \delta uz],$$

where the maximum is attained at

$$\widehat{u} = \widehat{u}(z) = \delta z.$$

By definitions, we get

$$E\left[F(S_T) - \int_0^T G(u_s)ds\right] \le E\left[\tilde{F}(zZ_T) + \int_0^T \tilde{G}(zZ_s)ds\right] + zE[M_T].$$

Since M is a martingale, $E[M(T)] = S_0$. Therefore the above inequality gives an upper bound for our maximization problem. The upper bound will be attained if the maximums are attained, and if $E[M_T] = S(0)$. In other words, the optimal solution is given by

$$\widehat{S}_T = \widehat{s}(\tilde{z}Z_T, A) \quad \text{and} \quad \widehat{u}_t = \delta\tilde{z}Z_t,$$

where A and $\tilde{z}$ are chosen so that A is any $\mathcal{F}_T$ measurable random variable taking only two possible values, 0 and $K + \max\left(\frac{1}{z} - \frac{w}{n}, 0\right)$, and so that $E[M_T] = S(0)$. The optimal $\widehat{v}$ is obtained from the martingale representation of

$$M_t = E[M_T|\mathcal{F}_t] = E\left[Z_T\widehat{S}_T - \delta\int_0^T Z_s\widehat{u}_s ds|\mathcal{F}_t\right] = S_0 + \int_0^t (\widehat{v}_s - \alpha)S_s Z_s dW_s.$$

For the case $\alpha > 0$ that we are considering, we will set $A \equiv 0$. Introduce the stochastic process $\tilde{W}$ defined by

$$\tilde{W}_t = W_t + \alpha t.$$

We observe that $\tilde{W}$ is a Brownian motion under the measure $\tilde{P}$ defined by $d\tilde{P}/dP = Z_T$. Introduce also the notation

$$\tilde{H}(t) := \frac{1}{\tilde{z}Z_t} \quad \text{and} \quad \tilde{M}_t := S_t - \tilde{z}\delta^2\int_0^t Z_s ds.$$

We observe that

$$\tilde{H}(t) = \frac{1}{\tilde{z}}\exp\left\{\frac{\alpha^2}{2}t + \alpha W(t)\right\} = \frac{1}{\tilde{z}}\exp\left\{-\frac{\alpha^2}{2}t + \alpha\tilde{W}(t)\right\}.$$

Thus,

$$d\tilde{H}_t = \tilde{H}_t\alpha d\tilde{W}.$$

Hence, for every $0 \le t \le s$:

$$\tilde{H}_s = \tilde{H}_t\exp\left\{-\frac{1}{2}\alpha^2(s-t) + \alpha(\tilde{W}_s - \tilde{W}_t)\right\}.$$

We also note that

$$S_t - \tilde{z}\delta^2\int_0^t Z_s ds = S_0 + \int_0^t \delta u_s ds + \int_0^t v_s S_s d\tilde{W}_s - \tilde{z}\delta^2\int_0^t Z_s ds = S_0 + \int_0^t v_s S_s d\tilde{W}_s.$$

Thus, the stochastic process $\tilde{M}$ is a $\tilde{P}-$martingale. Hence,

$$\tilde{M}_t = \tilde{E}[\tilde{M}_T|\ \mathcal{F}_t],$$

or equivalently

$$S_t = \tilde{E}\left[\widehat{s}(\tilde{z}Z_T,0) - \tilde{z}\delta^2 \int_t^T Z_s ds \mid \mathcal{F}_t\right].$$

We note that for every $0 \leq t \leq s$:

$$Z_s = \exp\left\{-\frac{\alpha^2}{2}s - \alpha W(s)\right\} = \exp\left\{\frac{\alpha^2}{2}s - \alpha \tilde{W}(s)\right\},$$

so for every $0 \leq t \leq s$:

$$\begin{aligned}\tilde{E}[Z_s \mid \mathcal{F}_t] &= \tilde{E}\left[Z_t \exp\left\{\frac{\alpha^2}{2}(s-t) - \alpha(\tilde{W}(s) - \tilde{W}(t))\right\} \mid \mathcal{F}_t\right] \\ &= Z_t \exp\left\{\alpha^2(s-t)\right\}.\end{aligned}$$

Denoting

$$\tilde{T}_t = \frac{e^{\alpha^2(T-t)} - 1}{\alpha^2},$$

we obtain

$$S_t \;=\; \tilde{E}[\widehat{s}(\tilde{z}Z_T,0) \mid \mathcal{F}_t] - \tilde{z}\delta^2 Z_t \tilde{T}_t.$$

Here,

$$\begin{aligned}\widehat{s}(\tilde{z}Z_T,0) = &\left(K + \max\left(\frac{1}{\tilde{z}Z_T} - \frac{w}{n}, 0\right)\right) \\ &\times I_{\{\log\{w + n\max(\frac{1}{\tilde{z}Z_T} - \frac{w}{n},0)\} - \tilde{z}Z_T K - \tilde{z}Z_T \max(\frac{1}{\tilde{z}Z_T} - \frac{w}{n},0) > \log\{w\}\}} \\ = &\left(K + \max\left(\tilde{H}(T) - \frac{w}{n}, 0\right)\right) \\ &\times I_{\{\log\{w + n\max(\tilde{H}(T) - \frac{w}{n},0)\} - \frac{1}{\tilde{H}(T)}K - \frac{1}{\tilde{H}(T)}\max(\tilde{H}(T) - \frac{w}{n},0) > \log\{w\}\}}.\end{aligned}$$

Thus,

$$\begin{aligned}&\tilde{E}[\widehat{s}(\tilde{z}Z_T,0) \mid \mathcal{F}_t] \\ &= K\tilde{P}\left\{\log\{w + n\max(\tilde{H}(T) - \frac{w}{n},0)\} - \frac{1}{\tilde{H}(T)}K \right.\\ &\quad \left. - \frac{1}{\tilde{H}(T)}\max(\tilde{H}(T) - \frac{w}{n},0) > \log\{w\} \mid \mathcal{F}_t\right\} \\ &\quad + \tilde{E}\Big[\max\left(\tilde{H}(T) - \frac{w}{n}, 0\right) \\ &\quad \times I_{\{\log\{w + n\max(\tilde{H}(T) - \frac{w}{n},0)\} - \frac{1}{\tilde{H}(T)}K - \frac{1}{\tilde{H}(T)}\max(\tilde{H}(T) - \frac{w}{n},0) > \log\{w\}\}} \Big| \mathcal{F}_t\Big].\end{aligned}$$

To continue with the computations, we are going to use $c = c(n, K, w) \in (1, \infty)$ defined in (6.6). We note that

$$
\begin{aligned}
&\left\{\log\left\{w + n\max\left(\tilde{H}(T) - \frac{w}{n}, 0\right)\right\} - \frac{1}{\tilde{H}(T)}K\right.\\
&\left.-\frac{1}{\tilde{H}(T)}\max\left(\tilde{H}(T) - \frac{w}{n}, 0\right) > \log\{w\}\right\}\\
&= \left\{\left\{\log\left\{w + n\max\left(\tilde{H}(T) - \frac{w}{n}, 0\right)\right\} - \frac{1}{\tilde{H}(T)}K\right.\right.\\
&\left.-\frac{1}{\tilde{H}(T)}\max\left(\tilde{H}(T) - \frac{w}{n}, 0\right) > \log\{w\}\right\}\\
&\qquad\qquad \left.\cap\left\{\tilde{H}(T) - \frac{w}{n} > 0\right\}\right\}\\
&\quad\bigcup\left\{\left\{\log\left\{w + n\max\left(\tilde{H}(T) - \frac{w}{n}, 0\right)\right\} - \frac{1}{\tilde{H}(T)}K\right.\right.\\
&\left.-\frac{1}{\tilde{H}(T)}\max\left(\tilde{H}(T) - \frac{w}{n}, 0\right) > \log\{w\}\right\}\\
&\qquad\qquad \left.\cap\left\{\tilde{H}(T) - \frac{w}{n} \le 0\right\}\right\}\\
&= \left\{\left\{\log\left\{w + n\left(\tilde{H}(T) - \frac{w}{n}\right)\right\} - \frac{1}{\tilde{H}(T)}K - \frac{1}{\tilde{H}(T)}\left(\tilde{H}(T) - \frac{w}{n}\right) > \log\{w\}\right\}\right.\\
&\qquad\qquad \left.\cap\left\{\tilde{H}(T) - \frac{w}{n} > 0\right\}\right\}\\
&\quad\bigcup\left\{\left\{\log\{w + n(0)\} - \frac{1}{\tilde{H}(T)}K - \frac{1}{\tilde{H}(T)}(0) > \log\{w\}\right\}\right.\\
&\qquad\qquad \left.\cap\left\{\tilde{H}(T) - \frac{w}{n} \le 0\right\}\right\}\\
&= \left\{\left\{\log\{n\tilde{H}(T)\} - \frac{1}{\tilde{H}(T)}K - 1 + \frac{w}{n\tilde{H}(T)} > \log\{w\}\right\}\cap\left\{\tilde{H}(T) - \frac{w}{n} > 0\right\}\right\}\\
&\quad\bigcup\left\{\left\{\log\{w\} - \frac{1}{\tilde{H}(T)}K > \log\{w\}\right\}\cap\left\{\tilde{H}(T) - \frac{w}{n} \le 0\right\}\right\}\\
&= \left\{\left\{\log\left\{\frac{n\tilde{H}(T)}{w}\right\} > \frac{nK}{w\left(\frac{n\tilde{H}(T)}{w}\right)} + 1 - \frac{w}{n\tilde{H}(T)}\right\}\cap\left\{\tilde{H}(T) - \frac{w}{n} > 0\right\}\right\}\\
&= \left\{\left\{f\left(\frac{n\tilde{H}(T)}{w}\right) > 0\right\}\cap\left\{\tilde{H}(T) - \frac{w}{n} > 0\right\}\right\}\\
&= \left\{\left\{\frac{n\tilde{H}(T)}{w} > c\right\}\cap\left\{\tilde{H}(T) - \frac{w}{n} > 0\right\}\right\}\\
&= \left\{\tilde{H}(T) > c\frac{w}{n}\right\}.
\end{aligned}
$$

Thus,

$$\begin{aligned}\tilde{E}[\widehat{s}(\tilde{z}Z_T,0)|\ \mathcal{F}_t] &= K\tilde{P}\left\{\tilde{H}_T > c\frac{w}{n}|\ \mathcal{F}_t\right\} + \tilde{E}\left[\left(\tilde{H}_T - \frac{w}{n}\right) I_{\{\tilde{H}_T > c\frac{w}{n}\}}|\ \mathcal{F}_t\right] \\ &= K\int_{\frac{1}{\alpha}\log\{\frac{cw}{n}\frac{1}{\tilde{H}_t}\}+\frac{1}{2}\alpha(T-t)}^{\infty} \frac{1}{\sqrt{2\pi(T-t)}}\exp\left\{-\frac{1}{2}\frac{x^2}{(T-t)}\right\}dx \\ &\quad + \int_{\frac{1}{\alpha}\log\{\frac{cw}{n}\frac{1}{\tilde{H}_t}\}+\frac{1}{2}\alpha(T-t)}^{\infty} \left(\tilde{H}_t\exp\left\{-\frac{1}{2}\alpha^2(T-t)+\alpha x\right\} - \frac{w}{n}\right) \\ &\qquad\qquad \frac{1}{\sqrt{2\pi(T-t)}}\exp\left\{-\frac{1}{2}\frac{x^2}{(T-t)}\right\}dx.\end{aligned}$$

To continue with the computations, we are going to return to the original probability measure P. We can write

$$\frac{dP}{d\tilde{P}} = \exp\left\{\alpha\tilde{W}_T - \frac{1}{2}\alpha^2 T\right\} = \frac{\tilde{H}(T)}{\tilde{H}(0)}$$

and

$$W_t = \tilde{W}_t - \alpha t.$$

We note that according to Bayes' formula,

$$\begin{aligned}\tilde{E}\left[\tilde{H}_T I_{\{\tilde{H}_T > c\frac{w}{n}\}}|\ \mathcal{F}_t\right] &= \exp\left\{\alpha\tilde{W}_t - \frac{1}{2}\alpha^2 t\right\} \\ &\quad \times E\left[\tilde{H}_T \frac{1}{\exp\left\{\alpha\tilde{W}_T - \frac{1}{2}\alpha^2 T\right\}} I_{\{\tilde{H}_T > c\frac{w}{n}\}}|\ \mathcal{F}_t\right] \\ &= \tilde{H}_t E\left[I_{\{\tilde{H}_T > c\frac{w}{n}\}}|\ \mathcal{F}_t\right] \\ &= \tilde{H}_t P\left\{\tilde{H}_T > c\frac{w}{n}|\ \mathcal{F}_t\right\} \\ &= \tilde{H}_t P\left\{\tilde{H}_t\exp\left\{\alpha(W_T - W_t) + \frac{1}{2}\alpha^2(T-t)\right\} > c\frac{w}{n}|\ \mathcal{F}_t\right\} \\ &= \tilde{H}_t P\left\{\alpha(W_T - W_t) + \frac{1}{2}\alpha^2(T-t) > \log\left\{\frac{1}{\tilde{H}_t}\frac{cw}{n}\right\}|\ \mathcal{F}_t\right\}.\end{aligned}$$

We know that under P, the random variable $W_T - W_t$ has, given $\mathcal{F}_t$, a normal distribution with mean zero and variance $T - t$. Hence,

$$\begin{aligned}\tilde{E}[\widehat{s}(\tilde{z}Z_T,0)|\ \mathcal{F}_t] &= \left(K - \frac{w}{n}\right)\tilde{P}\left\{\tilde{H}_T > c\frac{w}{n}|\ \mathcal{F}_t\right\} + \tilde{E}\left[\tilde{H}_T I_{\{\tilde{H}_T > c\frac{w}{n}\}}|\ \mathcal{F}_t\right] \\ &= \left(K - \frac{w}{n}\right)N(d_2(t,\tilde{H}_t)) + \tilde{H}_t N(d_1(t,\tilde{H}_t)),\end{aligned}$$

where

$$d_1(t,\tilde{H}_t) := \frac{\log\{\frac{\tilde{H}_t n}{cw}\} + \frac{1}{2}\alpha^2(T-t)}{\alpha\sqrt{T-t}}$$

$$d_2(t,\tilde{H}_t) := \frac{\log\{\frac{\tilde{H}_t n}{cw}\} - \frac{1}{2}\alpha^2(T-t)}{\alpha\sqrt{T-t}},$$

and

$$N(x) := \int_{-\infty}^{x} \frac{1}{\sqrt{2\pi}} \exp\left\{-\frac{z^2}{2}\right\} dz.$$

This means that we can write

$$\tilde{E}[\widehat{s}(\tilde{z}Z_T, 0)|\, \mathcal{F}_t] \;=\; g(t, \tilde{H}_t),$$

where the function g is defined by

$$\begin{aligned} g(t,y) := K &\int_{\frac{1}{\alpha}\log\{\frac{cw}{n}\frac{1}{y}\}+\frac{1}{2}\alpha(T-t)}^{\infty} \frac{1}{\sqrt{2\pi(T-t)}} \exp\left\{-\frac{1}{2}\frac{x^2}{(T-t)}\right\} dx \\ &+ \int_{\frac{1}{\alpha}\log\{\frac{cw}{n}\frac{1}{y}\}+\frac{1}{2}\alpha(T-t)}^{\infty} \left(y\exp\{-\frac{1}{2}\alpha^2(T-t)+\alpha x\} - \frac{w}{n}\right) \\ &\qquad\qquad \frac{1}{\sqrt{2\pi(T-t)}} \exp\left\{-\frac{1}{2}\frac{x^2}{(T-t)}\right\} dx \\ = &\left(K - \frac{w}{n}\right) N(t, d_2(y)) + yN(t, d_1(y)). \end{aligned}$$

Thus,

$$S_t \;=\; g(t, \tilde{H}_t) - \tilde{z}\delta^2 Z_t \bar{T}_t. \tag{6.16}$$

If we take $t = 0$ in the above equation, we get

$$S_0 \;=\; g(0, \tilde{H}_0) - \tilde{z}\delta^2 Z_0 \bar{T}_0 \;=\; g\left(0, \frac{1}{\tilde{z}}\right) - \tilde{z}\delta^2 \bar{T}_0. \tag{6.17}$$

Although this does not give an explicit solution for $\tilde{z}$, it is possible to obtain it numerically. Applying Itô's formula in (6.16), and comparing equations (2.2) and (6.16), we obtain

$$\widehat{v}_t S_t \;=\; \alpha\tilde{H}_t \frac{\partial}{\partial y} g(t, \tilde{H}_t) + \alpha\tilde{z}\delta^2 Z_t \bar{T}_t. \tag{6.18}$$

Here,

$$\begin{aligned} \frac{\partial}{\partial y} g(t,y) = &\left(K - \frac{w}{n}\right) \frac{1}{\sqrt{2\pi}} \exp\left\{-\frac{1}{2}(d_2(t,y))^2\right\} \frac{1}{y}\frac{1}{\alpha\sqrt{T-t}} \\ &+ \frac{1}{\sqrt{2\pi}} \exp\left\{-\frac{1}{2}(d_1(t,y))^2\right\} \frac{1}{\alpha\sqrt{T-t}} + N(d_1(t,y)). \end{aligned}$$

We observe that the $\widehat{u}$ and $\widehat{v}$ defined above are adapted stochastic processes with $E[\int_0^T |\widehat{u}_t|^2 dt] < \infty$ and $E[\int_0^T |\widehat{v}_t S_t|^2 dt] < \infty$. Therefore, if $w > 0$ and $\alpha > 0$, the optimal effort $\widehat{u}$ and optimal choice of projects $\widehat{v}$ are given by equations (6.8)-(6.9). □

From the discussion above, it is clear that both the optimal drift $\widehat{u}$ and the optimal choice of projects $\widehat{v}$ are completely determined by $\tilde{z}$. This has to be found numerically from equation (6.7). We present several examples in Table 6.4. We see that the optimal effort increases with the number of options granted; it does not change much when the strike price changes; and it may be decreasing or increasing with respect to the time to maturity, depending on the number of options granted.

Table 6.4. Optimal effort.

$S_0 = 100$, $w = 50$, $\delta = 0.5$, $\alpha = 0.5$			
n	K	T	$\tilde{z}$
1	90	1	0.0062743
5	90	1	0.0114899
1	90	5	0.0071296
5	90	5	0.0106822
1	100	1	0.0061745
5	100	1	0.0113005
1	100	5	0.0071088
5	100	5	0.0106929
1	110	1	0.0060784
5	110	1	0.0111076
1	110	5	0.0070867
5	110	5	0.0106937

7. Conclusions

We present a model where an executive is granted stock options as compensation. His decisions can affect the dynamics of the stock of the company in two ways: first, he can increase the expected return of the stock through his effort; second, he can choose the level of risk of the stock price by selecting different projects. The effort produces disutility, and higher level of risk will result in higher expected return. The executive is risk averse. We obtain closed form solutions for both the optimal effort and choice of projects. The company chooses the strike price of the call options. The company cares both about the mean and the volatility of the stock price. We find that there is a large range of optimal strike prices, depending on the values of the parameters of the model, but for most interesting cases it will be optimal to issue options out-of-the-money. We find that the optimal strike price is increasing with the type of the executive and the emphasis on the mean stock price, rather than volatility. The relationship between the optimal strike price and the quality of the projects is not monotonic.

In our setting, we can price the options for the company (assuming complete markets) and for the executive (through the certainty equivalent). Although options are always worth less for the executive than for the company, we show that the ratio is more favorable when the projects are of low quality and the executive is of high quality.

In order to derive closed form solutions for the optimal policies of the executive we have to assume that the utility is logarithmic and the only source of wealth of the executive is the package of options granted as compensation. As a result, the number of options is irrelevant. It is possible to extend our method to more general utility functions, as done in the proof of Theorem 6.1. In that case, the optimal solution would typically depend on the number of options. That would allow us to address some important issues such as policies for resetting the option contracts. However, it seems that in the general case we cannot get results as explicit as for the logarithmic utility. We can also extend some of the results to the case in which the executive receives cash, besides options. In this case, we can verify that the number of options becomes relevant. However, a procedure to derive the optimal strike price does not seem to be explicit. Another possible extension of the present work is to include tax effects in the model.

Acknowledgements

The research of A. Cadenillas was supported by the Natural Sciences and Engineering Research Council of Canada grant 194137. The research of J. Cvitanić was supported in part by the National Science Foundation, under NSF grant DMS 04-03575.

References

[1] B. Hall and J. Leibman, Are CEOs really paid like bureaucrats?, *Quarterly Journal of Economics.* **113**(3), 653–691, (1998).

[2] K. J. Murphy. Executive compensation. In *Handbook of Labor Economics, Vol III*, pp. 2485–2563. North Holland, (1999).

[3] J. Carpenter, The exercise and valuation of executive stock options, *Journal of Financial Economics.* **48**(2), 127–158, (1998).

[4] B. Hall and K. J. Murphy, Stock options for undiversified executives, *Journal of Accounting and Economics.* **33**, 3–42, (2002).

[5] J. Detemple and S. Sundaresan, Nontraded asset valuation with portfolio constraints: a binomial approach, *Review of Financial Studies.* **12**(4), 835–872, (1999).

[6] B. Hall and K. J. Murphy, Optimal exercise prices for executive stock options, *American Economic Review.* **2**, 209–214, (2000).

[7] N. Stoughton and K. Wong. Option compensation, accounting choice and industrial competition, (2003). Working Paper, University of California.

[8] J. G. Aseff and M. S. Santos, Stock options and managerial optimal contracts, *Econom. Theory.* **26**(4), 813–837, (2005). ISSN 0938-2259.

[9] O. Kadan and J. Swinkels. Moral hazard with bounded payments, (2005). Working paper.

[10] O. Kadan and J. Swinkels. Stocks or options? moral hazard, firm viability and the design of compensation contracts, (2005). Working paper.

[11] S. Johnson and Y. Tian, The value and incentive effects of nontraditional executive stock option plans, *Journal of Financial Economics.* **57**, 3–34, (2000).

[12] S. Johnson and Y. Tian, Indexed executive stock options, *Journal of Financial Economics.* pp. 35–64, (2000).

[13] A. Cadenillas, J. Cvitanić, and F. Zapatero, Leverage decision and manager compensation with choice of effort and volatility, *Journal of Financial Economics.* **73**(1), 71–92, (2004).

[14] J. Carpenter, Does option compensation increase managerial risk appetite?, *Journal of Finance.* **55**, 2311–2331, (2000).

[15] P. Bolton and C. Harris. The continuous-time principal-agent problem: Frequent-monitoring contracts, (2001). Working Paper, Princeton University, Princeton.
[16] A. Cadenillas, J. Cvitanić, and F. Zapatero. Executive stock options as a screening mechanism. Working paper.
[17] P. Boyle, M. Broadie, and P. Glasserman, Monte Carlo methods for security pricing, *Journal of Economic Dynamics and Control.* (1997).
[18] F. Black and M. Scholes, The pricing of options and corporate liabilities, *Journal of Political Economy.* **3**, 637–654, (1973).
[19] I. Karatzas and S. E. Shreve, *Brownian Motion and Stochastic Calculus.* (Springer, New York, 1991).
[20] J. Yong and X. Y. Zhou, *Stochastic Controls: Hamiltonian Systems and HJB Equations.* (Springer, New York, 1999).
[21] Y. S. Chow and H. Teicher, *Probability Theory: Independence, Interchangeability, Martingales. Second Edition.* (Springer, New York, 1988).

Chapter 7

The Freidlin-Wentzell LDP with Rapidly Growing Coefficients

Pavel Chigansky and Robert Liptser*

Department of Mathematics, The Weizmann Institute of Science
Rehovot 76100, Israel
pavel.chigansky@weizmann.ac.il

The Large Deviations Principle (LDP) is verified for a homogeneous diffusion process whose coefficients are locally Lipschitz functions with super linear growth. It is assumed that the drift is directed towards the origin and the growth rates of the drift and diffusion terms are properly balanced. Nonsingularity of the diffusion matrix is not required.

Contents

1. Introduction

In this paper we extend the set of conditions, under which Freidlin-Wentzell's Large Deviation Principle (LDP) for a homogeneous diffusion process remains valid. We consider a family $\{(X_t^\varepsilon)_{t\ge 0}\}_{\varepsilon\to 0}$ of diffusions, where $X_t^\varepsilon \in \mathbb{R}^d$, $d \ge 1$ is defined by

*Department of Electrical Engineering Systems, Tel Aviv University, 69978 Tel Aviv, Israel, liptser@eng.tau.ac.il

the Itô equation

$$X_t^\varepsilon = x_0 + \int_0^t b(X_s^\varepsilon)ds + \varepsilon \int_0^t \sigma(X_s^\varepsilon)dB_s, \tag{1.1}$$

with respect to a standard Brownian motion B_t, where $b(x)$ and $\sigma(x)$ are vector and matrix valued continuous functions of dimensions d and $d \times d$ respectively, guaranteeing existence of the unique weak solution.

The classical Freidlin-Wentzell setting [1] (see also Dembo and Zeitouni [2]) is applicable to the model (1.1) with bounded $b(x)$ and $\sigma(x)$ and uniformly positive definite diffusion matrix $a(x) = \sigma\sigma^*(x)$. Various LDP versions can be found in Dupuis and Ellis [3], Feng [4], Feng and Kurtz [5], Friedman [6], Liptser and Pukhalskii [7], Mikami [8], Narita [9], Stroock [10], Ren and Zhang [11]. In the recent paper [12], Puhalskii extends LDP to (1.1) with continuous and unbounded coefficients and singular $a(x)$, assuming $b(x)$ and $a(x)$ are Lipschitz continuous functions (concerning singular $\sigma(x)$ see also Liptser et al, [13]). Being Lipschitz continuous, the entries of b, σ grow not faster than linearly and, thereby, *automatically* guarantee one of the necessary conditions for LDP ($\|\cdot\|$ denotes the Euclidean norm in $\mathbb{R}^d$)

$$\lim_{C\to\infty} \overline{\lim_{\varepsilon\to 0}}\, \varepsilon^2 \log \mathsf{P}\Big(\sup_{t\le T} \|X_t^\varepsilon\| > C\Big) = -\infty, \ \forall\, T > 0. \tag{1.2}$$

Relinquishing the linear growth condition for b, σ would require additional assumptions providing (1.2).

This paper is inspired by Puhalskii's remark in Ref. [12]:

> *If the drift is directed towards the origin, then no restrictions are needed on the growth rate of the drift coefficient.*

In particular, in this case the LDP holds, regardless of the growth rate of $b(x)$, for a constant diffusion matrix (not necessarily nonsingular).

In this paper, we show that in fact LDP remains valid for (1.1) with non-constant diffusion term, if its growth rate is properly balanced relatively to the drift (see (2.1) of Theorem 2.1 below). Our result is formulated in terms of Khasminskii-Veretennikov's condition (2.1) (see Refs. [14–16])

The rest of the paper is organized as follows. In Sections 2 and 3, the main result, notations and preliminary facts on the LDP are given. Sections 4 - 6 contain the proof of the main result. Auxiliary technical details are gathered in Appendices A.1 - A.3.

2. Notations and the main result

The following notations and conventions are used throughout the paper.

- * denotes the transposition symbol
- all vectors are columns (unless explicitly stated otherwise)
- $|x|$ and $\|x\|$ denote the ℓ_1 and ℓ_2 (Euclidean) norms of $x \in \mathbb{R}^d$

- (x,y) denotes the scalar product of $x, y \in \mathbb{R}^d$
- $\|x\|_\Gamma^2 = (x, \Gamma x)$ with an nonnegative definite matrix Γ
- $a(x) = \sigma(x)\sigma^*(x)$
- $a^\oplus(x)$ denotes the Moore-Penrose pseudoinverse matrix of $a(x)$ (see Ref. [17])
- $\nabla V(x)$ is the gradient (row) vector of $V(x)$:

$$\nabla V(x) := \Big(\frac{\partial V(x)}{\partial x_1}, \dots, \frac{\partial V(x)}{\partial x_d}\Big)$$

- $\langle M, N\rangle_t$ is the joint quadratic variation process of continuous martingales M_t and N_t; for brevity $\langle M, M\rangle_t = \langle M\rangle_t$
- a.s. abbreviates "almost surely"; when the corresponding measure is not specified the Lebesgue measure on $\mathbb{R}_+$ is understood
- ϱ is the locally uniform metric on $\mathbb{C}_{[0,\infty)}(\mathbb{R}^d)$
- $\mathbf{I}$ denotes $d \times d$ identity matrix
- the convention $0/0 = 0$ is kept throughout
- $X^\varepsilon = (X^\varepsilon_t)_{t\ge 0}$
- $\inf\{\varnothing\} = \infty$.

We study the LDP for the family $\{X^\varepsilon\}_{\varepsilon\to 0}$ in the metric space $(\mathbb{C}_{[0,\infty)}(\mathbb{R}^d), \varrho)$ with $\varrho(x,y) = \sum_{k=1}^\infty 2^{-k}\big(1 \vee \sup_{t\le k}\|x_t - y_t\|\big)$, $x, y \in \mathbb{C}_{[0,\infty)}(\mathbb{R}^d)$. Recall that $\{X^\varepsilon\}_{\varepsilon\to 0}$ satisfies the LDP with the good rate function $J(u) : \mathbb{C}_{[0,\infty)}(\mathbb{R}^d) \mapsto [0,\infty]$ and the rate ε^2, if the level sets of $J(u)$ are compact and for any closed set F and open set G in $\mathbb{C}_{[0,\infty)}(\mathbb{R}^d)$,

$$\overline{\lim_{\varepsilon\to 0}}\, \varepsilon^2 \log \mathsf{P}\big(X^\varepsilon \in F\big) \le -\inf_{u\in F} J(u),$$

$$\underline{\lim_{\varepsilon\to 0}}\, \varepsilon^2 \log \mathsf{P}\big(X^\varepsilon \in G\big) \ge -\inf_{u\in G} J(u).$$

Our main result is

Theorem 2.1. *Assume:*

(H-1) the entries of $b(x)$ and $\sigma(x)$ are locally Lipschitz continuous functions,

(H-2) $\lim_{\|x\|\to\infty} \dfrac{(x, b(x))}{\|x\|} = -\infty$,

(H-3) for some positive constants K and L, $\dfrac{(x, a(x)x)}{\|x\|\,|(x, b(x))|} \le K$, $\forall\ \|x\| > L$.

Then $\{X^\varepsilon_t\}_{\varepsilon\to 0}$ obeys the LDP in the metric space $(\mathbb{C}_{[0,\infty)}(\mathbb{R}^d), \varrho)$ with the rate ε^2 and the rate function

$$J(u) = \begin{cases} \frac{1}{2}\int_0^\infty \|\dot u_t - b(u_t)\|^2_{a^\oplus(u_t)} dt, & u \in \Gamma \\ \infty, & u \notin \Gamma, \end{cases}$$

where

$$\Gamma = \Big\{u \in \mathbb{C}_{[0,\infty)} : \begin{matrix} u_0 = x_0,\ du_t \ll dt,\ \int_0^\infty \|\dot u_t\|^2 dt < \infty \\ a(u_t)a^\oplus(u_t)[\dot u_t - b(u_t)] = [\dot u_t - b(u_t)] \ a.s. \end{matrix}\Big\}.$$

Remark 2.2. In the scalar case (recall 0/0=0)

$$J(u) = \begin{cases} \dfrac{1}{2}\int_0^\infty \dfrac{(\dot{u}_t - b(u_t))^2}{\sigma^2(u_t)}dt, & du_t = \dot{u}_t dt,\ u_0 = x_0,\ \int_0^\infty \dot{u}_t^2 dt < \infty \\ \infty, & \text{otherwise.} \end{cases}$$

Example 2.3. A typical example within the scope of Theorem 2.1 is

$$X_t^\varepsilon = x_0 - \int_0^t (X_s^\varepsilon)^3 ds + \varepsilon \int_0^t |X_s^\varepsilon|^{3/2} dB_s.$$

3. Preliminaries

We follow the framework, set up by A. Puhalskii (see Refs. [18, 19]):

$$\left.\begin{array}{l}\text{Exponential tightness}\\ \text{Local LDP}\end{array}\right\} \Longleftrightarrow \text{LDP}$$

The exponential tightness in the metric space $(\mathbb{C}_{[0,\infty)}, \varrho)$ is convenient to verify in terms of, so called, $\mathbb{C}$-*exponential tightness* conditions introduced by A. Puhalskii (see e.g. Ref. [7]), based on the stopping times technique introduced by D. Aldous in Refs. [20, 21]). To this end, let us assume that the diffusion processes are defined on a stochastic basis $(\Omega, \mathcal{F}, \mathbf{F}^\varepsilon = (\mathscr{F}_t^\varepsilon)_{t\ge 0}, \mathsf{P})$, satisfying the usual conditions, where the filtration $\mathbf{F}^\varepsilon$ may depend on ε.

Recall (see Ref. [7]) that the family of diffusion processes is $\mathbb{C}$-exponentially tight if for any $T > 0$, $\eta > 0$ and any $\mathbf{F}^\varepsilon$-stopping time θ,

$$\lim_{C\to\infty} \varlimsup_{\varepsilon\to 0} \varepsilon^2 \log \mathsf{P}\Big(\sup_{t\le T} \|X_t^\varepsilon\| > C\Big) = -\infty, \tag{3.1}$$

$$\lim_{\triangle\to 0} \varlimsup_{\varepsilon\to 0} \varepsilon^2 \log \sup_{\theta\le T} \mathsf{P}\Big(\sup_{t\le\triangle} \|X_{\theta+t}^\varepsilon - X_\theta^\varepsilon\| > \eta\Big) = -\infty. \tag{3.2}$$

The family of diffusion processes obeys *the local LDP* in $(\mathbb{C}_{[0,\infty)}(\mathbb{R}^d), \varrho)$ if for any $T > 0$ there exists a local rate function $J_T(u)$ such that

$$\varlimsup_{\delta\to 0} \varlimsup_{\varepsilon\to 0} \varepsilon^2 \log \mathsf{P}\Big(\sup_{t\le T} \|X_t^\varepsilon - u_t\| \le \delta\Big) \le -J_T(u) \tag{3.3}$$

$$\varliminf_{\delta\to 0} \varliminf_{\varepsilon\to 0} \varepsilon^2 \log \mathsf{P}\Big(\sup_{t\le T} \|X_t^\varepsilon - u_t\| \le \delta\Big) \ge -J_T(u). \tag{3.4}$$

Under the conditions (3.1)-(3.4), the family of diffusion processes obeys the LDP with the rate ε^2 and the good rate function

$$J(u) = \sup_T J_T(u),\ u \in \mathbb{C}_{[0,\infty)}(\mathbb{R}^d),$$

where

$$J_T(u) = \begin{cases} \frac{1}{2}\int_0^T \|\dot{u}_t - b(u_t)\|^2_{a^\oplus(u_t)} dt, & u \in \Gamma_T \\ \infty, & u \notin \Gamma_T, \end{cases}$$

with

$$\Gamma_T = \left\{u \in \mathbb{C}_{[0,T]} : \begin{array}{c} u_0 = x_0,\ du_t \ll dt,\ \int_0^T \|\dot{u}_t\|^2 dt < \infty \\ a(u_t)a^{\oplus}(u_t)[\dot{u}_t - b(u_t)] = [\dot{u}_t - b(u_t)] \text{ a.s.} \end{array}\right\}.$$

Thus the proof of Theorem 2.1 reduces to establishing (3.1)–(3.4).

4. The proof of $\mathbb{C}$-exponential tightness

4.1. *Auxiliary lemma*

Let $\mathfrak{D}$ be a nonlinear operator acting on continuously differentiable functions $V(x) : \mathbb{R}^d \to \mathbb{R}$ as follows:

$$\mathfrak{D}V(x) = (\nabla V(x), b(x)) + \frac{1}{2}(\nabla V(x), a(x)\nabla V(x)).$$

Lemma 4.1. *Assume there exists twice continuously differentiable nonnegative function $V(x)$ such that*

(a-1) $\lim_{C\to\infty} \inf_{\|x\|\geq C} V(x) = \infty$
(a-2) for some $L > 0$, $\mathfrak{D}V(x) \leq 0$, $\forall\ \|x\| > L$.

Then (3.1) *holds.*

Proof. Notice that (3.1) is equivalent to

$$\lim_{C\to\infty} \overline{\lim_{\varepsilon\to 0}}\, \varepsilon^2 \log \mathsf{P}(\Theta_C \leq T) = -\infty, \tag{4.1}$$

where

$$\Theta_C = \inf\{t : \|X_t^\varepsilon\| \geq C\}, \quad C > 0 \tag{4.2}$$

are stopping times relative to $\mathbf{F}^\varepsilon$.

We use (A.1.1) of Proposition A.1.1 to estimate $\log \mathsf{P}(\Theta_C \leq T)$. An appropriate martingale M_t^ε is constructed with the help of function $V(x)$. Let $\Psi(x)$ be the Hessian of V, namely a matrix with the entries $V_{ij}(x) = \frac{\partial^2 V(x)}{\partial x_i \partial x_j}$. By the Itô formula

$$\begin{aligned}\varepsilon^{-2}V(X^\varepsilon_{\Theta_C\wedge t}) &= \varepsilon^{-2}V(x_0) + \int_0^{\Theta_C\wedge t} \varepsilon^{-2}(\nabla V(X_s^\varepsilon), b(X_s^\varepsilon))ds \\ &\quad + \int_0^{\Theta_C\wedge t} \varepsilon^{-1}(\nabla V(X_s^\varepsilon), \sigma(X_s^\varepsilon)dB_s) + \int_0^{\Theta_C\wedge t} \frac{1}{2}\operatorname{trace}\Big(\Psi(X_s^\varepsilon)a(X_s^\varepsilon)\Big)ds.\end{aligned}$$

We choose $M_t^\varepsilon = \int_0^t \varepsilon^{-1}(\nabla V(X_s^\varepsilon), \sigma(X_s^\varepsilon)dB_s)$, which has the variation process $\langle M^\varepsilon\rangle_t = \int_0^t \varepsilon^{-2}(\nabla V(X_s^\varepsilon), a(X_s^\varepsilon)\nabla V(X_s^\varepsilon))ds$. Clearly

$$\begin{aligned}M^\varepsilon_{\Theta_C^\beta\wedge t} &= \varepsilon^{-2}V(X^\varepsilon_{\Theta_C\wedge t}) - \varepsilon^{-2}V(x_0) \\ &\quad - \int_0^{\Theta_C\wedge t} \varepsilon^{-2}(\nabla V(X_s^\varepsilon), b(X_s^\varepsilon))ds - \int_0^{\Theta_C\wedge t} \frac{1}{2}\operatorname{trace}\Big(\Psi(X_s^\varepsilon)a(X_s^\varepsilon)\Big)ds.\end{aligned}$$

Hence, by the definition of $\mathfrak{D}$, one gets

$$M^\varepsilon_{\Theta_C\wedge T} - \frac{1}{2}\langle M^\varepsilon\rangle_{\Theta_C\wedge T} = \varepsilon^{-2}V(X^\varepsilon_{\Theta_C\wedge T}) - \varepsilon^{-2}V(x_0) - \int_0^{\Theta_C\wedge T}\frac{1}{2}\operatorname{trace}\Big(\Psi(X^\varepsilon_s)a(X^\varepsilon_s)\Big)ds - \int_0^{\Theta_C\wedge T}\varepsilon^{-2}\mathfrak{D}V(X^\varepsilon_s)ds. \tag{4.3}$$

On the set $\{\Theta_C \le T\}$, we have

$$\varepsilon^{-2}V(X^\varepsilon_{\Theta_C\wedge T}) - \varepsilon^{-2}V(x_0) \ge \varepsilon^{-2}\inf_{\|x\|\ge C}V(x) - \varepsilon^{-2}V(x_0),$$

and

$$\Big|\int_0^{\Theta_C\wedge T}\frac{1}{2}\operatorname{trace}\big(\Psi(X^\varepsilon_s)a(X^\varepsilon_s)\big)ds\Big| \le \frac{T}{2}\sup_{\|x\|\le C}\big|\operatorname{trace}\big(\Psi(x)a(x)\big)\big|,$$

and, by (a-2),

$$-\int_0^{\Theta\wedge T}\varepsilon^{-2}\mathfrak{D}V(X^\varepsilon_s)ds \ge -\Big|\int_0^{\Theta_C\wedge T}\varepsilon^{-2}I_{\{\|X^\varepsilon_s\|\le L\}}\mathfrak{D}V(X^\varepsilon_s)ds\Big| \ge -\varepsilon^2 T\sup_{\|x\|\le L}|\mathfrak{D}V(x)|.$$

These inequalities and (4.3) imply

$$M^\varepsilon_{\Theta_C} - \frac{1}{2}\langle M^\varepsilon\rangle_{\Theta_C} \ge \varepsilon^{-2}\inf_{\|x\|\ge C}V(x) - \varepsilon^{-2}V(x_0) - \frac{T}{2}\sup_{\|x\|\le C}\big|\operatorname{trace}\big(\Psi(x)a(x)\big)\big| - \varepsilon^{-2}T\sup_{\|x\|\le L}\big|\mathfrak{D}V(x)\big|$$

on the set $\{\Theta_C \le T\}$. Hence, due to (A.1.1) of Proposition A.1.1

$$\varepsilon^2\log\mathsf{P}\big(\Theta_C\le T\big) \le - \inf_{\|x\|\ge C}V(x) + V(x_0) + \frac{T\varepsilon^2}{2}\sup_{\|x\|\le C}\big|\operatorname{trace}\big(\Psi(x)a(x)\big)\big| + T\sup_{\|x\|\le L}\big|\mathfrak{D}V(x)\big| \xrightarrow[\varepsilon\to 0]{} -\inf_{\|x\|\ge C}V(x) + V(x_0) + T\sup_{\|x\|\le L}\big|\mathfrak{D}V(x)\big|$$

and it remains to recall that by (a-1) $\lim_{C\to\infty}\inf_{\|x\|\ge C}V(x) = \infty$. □

4.2. *The proof of (3.1)*

We apply Lemma 4.1 to

$$V(x) = \frac{c\|x\|^2}{1+\|x\|},$$

with a positive parameter $c \le \frac{1}{K}$ for K from (2.1) of Theorem 2.1. The function $V(x)$ is twice continuously differentiable and satisfies (a-1). It is left to show that $V(x)$ satisfies (a-2) as well.

Direct computations give $\nabla V(x) = c\frac{(2+\|x\|)\|x\|}{(1+\|x\|)^2}\frac{x}{\|x\|}$. Denote

$$r(x) := \frac{(2+\|x\|)\|x\|}{(1+\|x\|)^2}$$

and notice that $r(x) \le 1$. By assumption (2.1) of Theorem 2.1, one can choose $L > 0$ sufficiently large so that $(x, b(x)) < 0$ for any $\|x\| \ge L$. On the other hand, by assumption (2.1) of Theorem 2.1, $-1 + \frac{c}{2}\frac{(x,a(x)x)}{\|x\|\,|(x,b(x))|} \le -\frac{1}{2}$ for $\|x\| \ge L$ and

$$\begin{aligned}
\mathfrak{D}V(x) &= \left(c\frac{r(x)}{\|x\|}(x,b(x)) + \frac{c^2r^2(x)}{2}\frac{(x,a(x)x)}{\|x\|^2}\right) \\
&= \left(-c\frac{r(x)}{\|x\|}|(x,b(x))| + \frac{c^2r^2(x)}{2}\frac{(x,a(x)x)}{\|x\|^2}\right) \\
&= cr(x)\frac{|(x,b(x))|}{\|x\|}\left(-1 + \frac{c}{2}r(x)\frac{(x,a(x)x)}{\|x\|\,|(x,b(x))|}\right) \\
&\le cr(x)\frac{|(x,b(x))|}{\|x\|}\left(-1 + \frac{c}{2}\frac{(x,a(x)x)}{\|x\|\,|(x,b(x))|}\right) \\
&\le -\frac{1}{2}cr(x)\frac{|(x,b(x))|}{\|x\|}
\end{aligned}$$

and (a-2) follows. □

4.3. *The proof of (3.2)*

The obvious inclusion

$$\Big\{\sup_{t\le\triangle}\|X^\varepsilon_{\theta+t} - X^\varepsilon_\theta\| > \eta\Big\} \subseteq \Big\{\sup_{t\le\triangle}\|X^\varepsilon_{\theta+t} - X^\varepsilon_\theta\| > \eta,\ \Theta_C = \infty\Big\}\bigcup\Big\{\Theta_C \le T\Big\}$$

reduces the proof to verifying

$$\varlimsup_{\triangle\to0}\varlimsup_{\varepsilon\to0}\varepsilon^2\log\sup_{\theta\le T}\mathsf{P}\Big(\sup_{t\le\triangle}\|X^\varepsilon_{\theta+t} - X^\varepsilon_\theta\| > \eta,\ \Theta_C = \infty\Big) = -\infty \tag{4.4}$$

for any fixed C. Indeed if (4.4) holds, then

$$\begin{aligned}
&\lim_{\triangle\to0}\varlimsup_{\varepsilon\to0}\varepsilon^2\log\sup_{\theta\le T}\mathsf{P}\Big(\sup_{t\le\triangle}\|X^\varepsilon_{\theta+t} - X^\varepsilon_\theta\| > \eta\Big) \\
&\le \lim_{\triangle\to0}\varlimsup_{\varepsilon\to0}\varepsilon^2\log\sup_{\theta\le T}\mathsf{P}\Big(\sup_{t\le\triangle}\|X^\varepsilon_{\theta+t} - X^\varepsilon_\theta\| > \eta,\ \Theta_C = \infty\Big) \\
&\qquad\qquad\qquad\bigvee\varlimsup_{C\to\infty}\varlimsup_{\varepsilon\to0}\varepsilon^2\log\mathsf{P}(\Theta_C \le T)
\end{aligned}$$

and, thus, (3.2) is implied by (4.4) and (4.1). So, it is left to check (4.4) for any entry x^ε_t of X^ε_t:

$$\lim_{\triangle\to0}\varlimsup_{\varepsilon\to0}\varepsilon^2\log\sup_{\theta\le T}\mathsf{P}\Big(\sup_{t\le\triangle}|x^\varepsilon_{\theta+t} - x^\varepsilon_\theta| > \eta,\ \Theta_C = \infty\Big) = -\infty.$$

An entry of X^ε_t satisfies

$$x^\varepsilon_t = x^\varepsilon_0 + \int_0^t \gamma^\varepsilon_s ds + \varepsilon m^\varepsilon_t,$$

where γ^ε_t is $\mathbf{F}^\varepsilon$-adapted continuous random process and m_t is $\mathbf{F}^\varepsilon$-continuous martingale with $\langle m^\varepsilon\rangle_t = \int_0^t \mu^\varepsilon_s ds$. Since b and σ are locally Lipschitz continuous functions, there is a constant l_C, such that $|\gamma^\varepsilon_{\Theta_C\wedge t}| \le l_C$ and $\mu^\varepsilon_{\Theta_C\wedge t} \le l_C$. Taking into account that

$$\Big\{\sup_{t\le\triangle}\Big|\int_\theta^{\theta+t}\gamma^\varepsilon_s ds\Big| \ge \eta,\ \Theta_C = \infty\Big\} \subseteq \{l_C\triangle \ge \eta\} = \varnothing, \text{ for } \triangle < \eta/l_C,$$

it is left to verify

$$\lim_{\triangle\to 0}\overline{\lim_{\varepsilon\to 0}}\,\varepsilon^2\log\sup_{\theta\le T}\mathsf{P}\Big(\sup_{t\le\triangle}|\varepsilon m^\varepsilon_{\theta+t} - \varepsilon m^\varepsilon_\theta| > \eta,\ \Theta_C = \infty\Big) = -\infty.$$

Due to the obvious inclusion

$$\begin{aligned}\Big\{\sup_{t\le\triangle}|\varepsilon m^\varepsilon_{\theta+t} - \varepsilon m^\varepsilon_\theta| > \eta,\ \Theta_C = \infty\Big\} = &\\ \Big\{\sup_{t\le\triangle}|\varepsilon m^\varepsilon_{\Theta_C\wedge(\theta+t)} - \varepsilon m^\varepsilon_{\Theta_C\wedge\theta}| > \eta,\ \Theta_C = \infty\Big\}&\\ \subseteq \Big\{\sup_{t\le\triangle}|\varepsilon m^\varepsilon_{\Theta_C\wedge(\theta+t)} - \varepsilon m^\varepsilon_{\Theta_C\wedge\theta}| > \eta\Big\},&\end{aligned}$$

we shall verify

$$\overline{\lim_{\triangle\to 0}}\,\overline{\lim_{\varepsilon\to 0}}\,\varepsilon^2\log\sup_{\theta\le T}\mathsf{P}\Big(\sup_{t\le\triangle}|\varepsilon m^\varepsilon_{\Theta_C\wedge(\theta+t)} - \varepsilon m^\varepsilon_{\Theta_C\wedge\theta}| > \eta\Big) = -\infty.$$

Notice that $n^\varepsilon_t := \varepsilon m^\varepsilon_{\Theta_C\wedge(\theta+t)} - \varepsilon m^\varepsilon_{\Theta_C\wedge\theta}$ is a continuous martingale relative to $(\mathscr{F}^\varepsilon_{\Theta_C\wedge\theta+t})_{t\ge 0}$ (see e.g. Ch. 4, §7 in Ref. [22]) with $\langle n^\varepsilon\rangle_t = \varepsilon^2\int_{\Theta_C\wedge\theta}^{\Theta_C\wedge(\theta+t)}\mu^\varepsilon_s ds \le \varepsilon^2 l_C t$. By the statement (A.1.1) of Proposition A.1.1, $\mathsf{P}\big(\sup_{t\le\triangle}|n^\varepsilon_t| \ge \eta\big) \le 2e^{-\eta^2/(2l_C\varepsilon^2\triangle)}$, so that $\overline{\lim}_{\varepsilon\to 0}\varepsilon^2\log\mathsf{P}\big(\sup_{t\le\triangle}|n^\varepsilon_t|\ge\eta\big) \le -\frac{\eta^2}{2l_C\triangle}\xrightarrow[\triangle\to 0]{} -\infty.$ □

5. Local LDP upper bound

We start with the observation that (3.3) holds if for any $T > 0$

$$\overline{\lim_{\delta\to 0}}\,\overline{\lim_{\varepsilon\to 0}}\,\varepsilon^2\log\mathsf{P}\Big(\sup_{t\le T}\|X^\varepsilon_t - u_t\| \le \delta,\ \Theta_C = \infty\Big) \le -J_T(u), \tag{5.1}$$

since by the inclusion

$$\Big\{\sup_{t\le T}\|X^\varepsilon_t - u_t\|\le\delta\Big\} \subseteq \Big\{\sup_{t\le T}\|X^\varepsilon_t - u_t\|\le\delta,\ \Theta_C = \infty\Big\}\bigcup\Big\{\Theta_C\le T\Big\}$$

we have

$$\varlimsup_{\delta\to 0}\varlimsup_{\varepsilon\to 0}\varepsilon^2\log\mathsf{P}\Big(\sup_{t\le T}\|X_t^\varepsilon-u_t\|\le\delta\Big)$$
$$\le\varlimsup_{\delta\to 0}\varlimsup_{\varepsilon\to 0}\varepsilon^2\log\mathsf{P}\Big(\sup_{t\le T}\|X_t^\varepsilon-u_t\|\le\delta,\ \Theta_C=\infty\Big)\bigvee\varlimsup_{\varepsilon\to 0}\varepsilon^2\log\mathsf{P}(\Theta_C\le T),$$

and, by (4.1), the last term goes to $-\infty$ as $C\to\infty$.

The proof for $u_0\ne x_0$ or $du_t\not\ll dt$ is standard (see e.g. Ref. [2]) and is omitted. The rest of the proof is split into two steps.

Step 1: $\mathbf{u_0=x_0}$, $\mathbf{du_t\ll dt}$, $\int_0^{\mathbf{T}}\|\dot{\mathbf{u}}_{\mathbf{s}}\|^{\mathbf{2}}\mathbf{ds}<\infty$. Define the set

$$\mathfrak{A}=\Big\{\sup_{t\le T}\|X_t^\varepsilon-u_t\|\le\delta,\ \Theta_C=\infty\Big\}.$$

With a continuously differentiable vector-valued function $\lambda(s)$ of dimension d, let us introduce a continuous local martingale $U_t=\int_0^t(\lambda(s),\varepsilon\sigma(X_s^\varepsilon)dB_s)$ and its martingale exponential $\mathfrak{z}_t=e^{U_t-0.5\langle U\rangle_t}$, where

$$\langle U\rangle_t=\int_0^t\varepsilon^2(\lambda(s),a(X_s^\varepsilon)\lambda(s))ds.$$

It is well known that $\mathfrak{z}_t$ is a continuous positive local martingale, as well as a supermartingale. Consequently, $\mathsf{E}\mathfrak{z}_T\le 1$ and, therefore,

$$1\ge\mathsf{E}I_{\{\mathfrak{A}\}}\mathfrak{z}_T. \tag{5.2}$$

The required upper bound for $\mathsf{P}(\mathfrak{A})$ is obtained by estimating $\mathfrak{z}_T$ from below on $\mathfrak{A}$. Since $U_t=\int_0^t(\lambda(s),dX_s^\varepsilon-b(X_s^\varepsilon)ds)$,

$$\begin{aligned}
&U_T-0.5\langle U\rangle_T=\\
&\int_0^T\Big[(\lambda(s),dX_s^\varepsilon-b(X_s^\varepsilon)ds)-\frac{\varepsilon^2}{2}(\lambda(s),a(X_s^\varepsilon)\lambda(s))ds\Big]=\\
&\int_0^T\Big[(\lambda(s),\dot u_s-b(u_s))-\frac{\varepsilon^2}{2}(\lambda(s),a(u_s)\lambda(s))\Big]ds\\
&+\int_0^T(\lambda(s),dX_s^\varepsilon-\dot u_sds)+\int_0^T(\lambda(s),b(u_s)-b(X_s^\varepsilon)ds\\
&+\int_0^T\frac{\varepsilon^2}{2}(\lambda(s),[a(u_s)-a(X_s^\varepsilon)]\lambda(s))ds.
\end{aligned}\tag{5.3}$$

We derive lower bounds on the set $\mathfrak{A}$ for each term in the right hand side of (5.3). Applying the Itô formula to $(\lambda(t),X_t^\varepsilon-u_t)$, and taking into account that $X_0^\varepsilon=u_0$, we find that

$$(\lambda(T),X_T^\varepsilon-u_T)=\int_0^T(\lambda(s),dX_s^\varepsilon-\dot u_sds)+\int_0^T(\dot\lambda(s),X_s^\varepsilon-u_s)ds.$$

Therefore,

$$\int_0^T (\lambda(s), dX_s^\varepsilon - \dot u_s ds)$$

$$\geq -\Big|(\lambda(T), X_T^\varepsilon - u_T)\Big| - \Big|\int_0^T (\dot\lambda(s), X_s^\varepsilon - u_s)ds\Big| \geq -r_1\delta,$$

with $r_1 := r_1(\lambda, T, C) \geq 0$, independent of ε.

Further, with $r_i := r_i(\lambda, T, C) \geq 0$, $i = 2, 3$, due to the local Lipschitz continuity of σ and a, we find that

$$\int_0^T (\lambda(s), b(u_s) - b(X_s^\varepsilon))ds \geq -r_2(\lambda, C, T)\delta$$

$$\int_0^T \frac{\varepsilon^2}{2}(\lambda(s), [a(u_s) - a(X_s^\varepsilon)]\lambda(s))ds \geq -\varepsilon^2 r_3(\lambda, C, T)\delta.$$

Hence with $r := r_1 + r_2 + \varepsilon^2 r_3$,

$$\log \mathfrak{z}_T \geq \int_0^T \Big[(\lambda(s), \dot u_s - b(u_s)) - \frac{\varepsilon^2}{2}(\lambda(s), a(u_s)\lambda(s))\Big]ds - r(\lambda, T, C)\delta.$$

Set $\nu(s) = \varepsilon^2\lambda(s)$ and rewrite the above inequality as:

$$\log \mathfrak{z}_T \geq \frac{1}{\varepsilon^2}\int_0^T \Big[(\nu(s), \dot u_s - b(u_s)) - \frac{1}{2}(\nu(s), a(u_s)\nu(s))\Big]ds - r\Big(\frac{\nu}{\varepsilon^2}, T, C\Big)\delta.$$

This lower bound, along with (5.2), provides the following upper bound

$$\varepsilon^2 \log \mathsf{P}(\mathfrak{A}) \leq -\int_0^T \Big[(\nu(s), \dot u_s - b(u_s)) - \frac{1}{2}(\nu(s), a(u_s)\nu(s))\Big]ds + \varepsilon^2 r\Big(\frac{\nu}{\varepsilon^2}, T, C\Big)\delta.$$

Clearly $\overline{\lim}_{\varepsilon\to 0}\varepsilon^2 r\big(\frac{\nu}{\varepsilon^2}, T, C\big) < \infty$ and, hence,

$$\overline{\lim_{\delta\to 0}}\, \overline{\lim_{\varepsilon\to 0}}\, \varepsilon^2 \log \mathsf{P}(\mathfrak{A}) \leq -\int_0^T \Big[(\nu(s), \dot u_s - b(u_s)) - \frac{1}{2}(\nu(s), a(u_s)\nu(s))\Big]ds. \tag{5.4}$$

Since the left hand side of (5.4) is independent of $\nu(s)$, (5.1) is derived by minimizing the right hand side of (5.4) with respect to $\nu(s)$. Two difficulties arise on the way to direct minimization:

- the matrix $a(u_s)$ may be singular
- the entries of $\nu(s)$ should be continuously differentiable functions.

Assume first $a(u_s)$ is a positive definite matrix, uniformly in s, and write

$$(\nu(s), \dot u_s - b(u_s)) - \frac{1}{2}(\nu(s), a(u_s)\nu(s)) = \frac{1}{2}\|\dot u_s - b(u_s)\|^2_{a^{-1}(u_s)} - \frac{1}{2}\Big\|a^{1/2}(u_s)\big(\nu(s) - a^{-1}(u_s)[\dot u_s - b(u_s)]\big)\Big\|^2.$$

If the entries of $a^{-1}(u_s)[\dot{u}_s - b(u_s)]$ are continuously differentiable functions, then, by taking $\nu(s) \equiv -a^{-1}(u_s)[\dot{u}_s - b(u_s)]$ we find that

$$\overline{\lim_{\delta\to 0}}\,\overline{\lim_{\varepsilon\to 0}}\,\varepsilon^2 \log \mathsf{P}(\mathfrak{A}) \le -\frac{1}{2}\int_0^T \|\dot{u}_s - b(u_s)\|^2_{a^{-1}(u_s)} ds. \tag{5.5}$$

In the general case, due to $\int_0^T \|\dot{u}_s\|^2 ds < \infty$, the entries of $a^{-1}(u_s)[\dot{u}_s - b(u_s)]$ are square integrable with respect to the Lebesgue measure on $[0,T]$. Choose a maximizing sequence $\nu_n(s)$, $n \ge 1$, of continuously differentiable functions such that $\lim_{n\to\infty}\int_0^T \big\|\nu_n(s) - a^{-1}(u_s)[\dot{u}_s - b(u_s)]\big\|^2 ds = 0$. Since all the entries of $a(u_s)$ are uniformly bounded on $[0,T]$

$$\lim_{n\to\infty}\int_0^T \big\|a^{1/2}(u_s)\big(\nu_n(s) - a^{-1}(u_s)[\dot{u}_s - b(u_s)]\big)\big\|^2 ds = 0$$

and (5.5) holds too.

Now we drop the uniform nonsingularity assumption of $a(u_s)$. The upper bound in (5.5) remains valid with $a(u_s)$ replaced by $a_\beta(u_s) \equiv a(u_s) + \beta\mathbf{I}$, where β is a positive number and $\mathbf{I}$ is $(d\times d)$-unit matrix:

$$\overline{\lim_{\delta\to 0}}\,\overline{\lim_{\varepsilon\to 0}}\,\varepsilon^2 \log \mathsf{P}(\mathfrak{A}) \le -\frac{1}{2}\int_0^T \|\dot{u}_s - b(u_s)\|^2_{[a(u_s)+\beta\mathbf{I}]^{-1}} ds.$$

For any fixed s, the function $\|\dot{u}_s - b(u_s)\|^2_{[a(u_s)+\beta\mathbf{I}]^{-1}}$ increases with $\beta \downarrow 0$ and by Lemma A.2.1 possesses the limit

$$\lim_{\beta\to 0}\|\dot{u}_s - b(u_s)\|^2_{[a(u_s)+\beta\mathbf{I}]^{-1}} = \begin{cases} \|\dot{u}_s - b(u_s)\|^2_{a^{\oplus}(u_s)}, & \begin{aligned} a(u_s)a^{\oplus}(u_s)[\dot{u}_s - b(u_s)] \\ = [\dot{u}_s - b(u_s)] \end{aligned} \\ \infty, & \text{otherwise.} \end{cases}$$

Thus the required upper bound

$$\overline{\lim_{\delta\to 0}}\,\overline{\lim_{\varepsilon\to 0}}\,\varepsilon^2 \log \mathsf{P}(\mathfrak{A}) \le \begin{cases} -\int_0^T \frac{1}{2}\|\dot{u}_s - b(u_s)\|^2_{a^{\oplus}(u_s)} ds, & \begin{aligned} a(u_s)a^{\oplus}(u_s)[\dot{u}_s - b(u_s)] \\ = [\dot{u}_s - b(u_s)], \text{ a.s.} \end{aligned} \\ \infty, & \text{otherwise} \end{cases}$$

follows by the monotone convergence theorem.

Step 2. $\boldsymbol{u_0 = x_0,\ du_t \ll dt,\ \int_0^T \|\dot{u}_s\|^2 ds = \infty}$**.**

We emphasize that $du_t \ll dt$ on $[0,T]$ implies $\int_0^T \|\dot{u}_s\| ds < \infty$ and return to the upper bound from (5.4). Since b and σ are locally Lipschitz, one can choose a constant L (depending on $u(s)$), so that, $|(\nu(s), b(u_s))| \le \|b(u_s)\|\|\nu(s)\| \le L\|\nu(s)\|$ and $(\nu(s), a(u_s)\nu(s)) \le L\|\nu(s)\|^2$. Then, (5.4) implies

$$\overline{\lim_{\delta\to 0}}\,\overline{\lim_{\varepsilon\to 0}}\,\varepsilon^2 \log \mathsf{P}(\mathfrak{A}) \le -\int_0^T \Big[(\nu(s), \dot{u}_s) - L\|\nu(s)\| - \frac{L}{2}\|\nu(s)\|^2\Big] ds.$$

Let $\nu_n(s)$ be a sequence of continuously differentiable functions, approximating the bounded (for each fixed $p > 0$) function $L^{-1}\dot u_s I_{\{\|\dot u_s\| \le p\}}$ in the sense that $\lim_{n\to\infty} \int_0^T \|\frac{1}{L}\dot u_s I_{\{\|\dot u_s\|\le p\}} - \nu_n(s)\|^2 ds = 0$. Thus,

$$\overline{\lim_{\delta\to 0}}\,\overline{\lim_{\varepsilon\to 0}}\,\varepsilon^2 \log \mathsf{P}(\mathfrak{A}) \le -\frac{1}{2L}\underbrace{\int_0^T \|\dot u_s\|^2 I_{\{\|\dot u_s\|\le p\}} ds}_{\uparrow\infty \text{ as } p\uparrow\infty} + \underbrace{\int_0^T \|\dot u_s\| ds}_{<\infty} \xrightarrow[p\to\infty]{} -\infty$$

□

6. Local LDP lower bound

If $\overline{\lim}_{\delta\to 0}\overline{\lim}_{\varepsilon\to 0}\,\varepsilon^2 \log \mathsf{P}\big(\sup_{t\le\Theta_C\wedge T}|X^\varepsilon_t - u_t| \le \delta\big) \le -J_T(u) = -\infty$, then the corresponding local LDP lower bound is $-\infty$ as well and hence only the case $J_T(u) < \infty$ is to be considered, i.e. we may restrict ourselves to analyzing test functions with the properties:

$$\begin{aligned}
&\textbf{(i)} \quad u_0 = x_0\\
&\textbf{(ii)} \quad du_t \ll dt\\
&\textbf{(iii)} \quad a(u_t)a^{\oplus}(u_t)[\dot u_t - b(u_t)] = [\dot u_t - b(u_t)] \quad \text{a.s.}\\
&\textbf{(iv)} \quad \int_0^T \|\dot u_t - b(u_t)\|^2_{a^{\oplus}(u_t)} dt < \infty,\ \forall\, T > 0\\
&\textbf{(v)} \quad \int_0^T \|\dot u_t\|^2 dt < \infty.
\end{aligned} \tag{6.1}$$

Another helpful observation is that (3.4) holds if for any $C > 0$

$$\underline{\lim}_{\delta\to 0}\,\underline{\lim}_{\varepsilon\to 0}\,\varepsilon^2 \log \mathsf{P}\Big(\sup_{t\le\Theta_C\wedge T}\|X^\varepsilon_t - u_t\| \le \delta\Big) \ge -J_T(u) \tag{6.2}$$

due to

$$\Big\{\sup_{t\le\Theta_C\wedge T}\|X^\varepsilon_t - u_t\| \le \delta\Big\} \subseteq \Big\{\sup_{t\le T}\|X^\varepsilon_t - u_t\| \le \delta\Big\}\bigcup\Big\{\Theta_C \le T\Big\}$$

and (4.1).

6.1. *Nonsingular $a(x)$*

In this section, the matrix $a(x)$ is assumed to be uniformly nonsingular in $x \in \mathbb{R}$, in the sense that $a(x) \ge \beta\mathbf{I}$ for a positive number β. Let $\lambda(s) := \sigma^{-1}(X^\varepsilon_s)\big[\dot u_s - b(X^\varepsilon_s)\big]$ and introduce a martingale $U_t = \int_0^{\Theta_C\wedge t}\frac{1}{\varepsilon}(\lambda(s), dB_s)$ and its martingale exponential $\mathfrak{z}_t = e^{U_t - 0.5\langle U\rangle_t}$, $t \le T$, where $\langle U\rangle_t = \int_0^{\Theta_C\wedge t}\frac{1}{\varepsilon^2}\|\lambda(s)\|^2 ds$.

By **(iv)** and **(v)** of (6.1), $\langle U\rangle_T \le \text{const.}$ and hence $\mathsf{E}\mathfrak{z}_T = 1$. We use this fact in order to define a new probability measure Q^ε by $d\mathsf{Q}^\varepsilon = \mathfrak{z}_T d\mathsf{P}$. Since $\mathfrak{z}_T$ is positive P-a.s., $\mathsf{P} \ll \mathsf{Q}^\varepsilon$ as well and $d\mathsf{P} = \mathfrak{z}_T^{-1} d\mathsf{Q}^\varepsilon$.

We proceed with the proof of (6.2) by applying

$$\mathsf{P}(\widetilde{\mathfrak{A}}) = \int_{\widetilde{\mathfrak{A}}} \mathfrak{z}_T^{-1} d\mathsf{Q}^\varepsilon \tag{6.3}$$

to the set $\widetilde{\mathfrak{A}} = \big\{\sup_{t\le\Theta_C\wedge T}\|X_t^\varepsilon - u_t\| \le \delta\big\}$, and estimating from below the right hand side in (6.3). In order to realize this program, it is convenient to have a semimartingale description of the process $X^\varepsilon_{\Theta_C\wedge t}$ under Q^ε. Recall that the random process $B_{\Theta_C\wedge t}$ is a martingale under P with the variation process $\langle B\rangle_{\Theta_C\wedge t} \equiv (\Theta_C\wedge t)\mathbf{I}$. It is well known (see e.g. Theorem 2, Ch. 4, §5 in Ref. [22]) that $B_{\Theta_C\wedge t}$ is a continuous semimartingale under Q^ε with the decomposition $B_{\Theta_C\wedge t} = \widetilde{B}_t + A_t^B$, where $\widetilde{B}_t$ is a martingale (under Q^ε) with $\langle\widetilde{B}\rangle_t \equiv \langle B\rangle_{\Theta_C\wedge t}$ and, by the Girsanov theorem,

$$A_t^B = \int_0^{\Theta_C\wedge t} \frac{1}{\varepsilon}\sigma^{-1}(X_s^\varepsilon)[\dot{u}_s - b(X_s^\varepsilon)]ds.$$

In particular,

$$X^\varepsilon_{\Theta_C\wedge t} = u_{\Theta_C\wedge t} + \varepsilon\int_0^{\Theta_C\wedge t}\sigma(X_s^\varepsilon)d\widetilde{B}_s,\ t\le T, \quad \mathsf{Q}^\varepsilon\text{-a.s.}$$

As the next preparatory step we derive the semimartingale decomposition of U_t under Q^ε. As before, the continuous martingale U_t under P is transformed to a semimartingale under Q^ε:

$$U_t = \widetilde{U}_t + A_t^U$$

with continuous Q^ε-martingale $\widetilde{U}_t$, having the variation process $\langle\widetilde{U}\rangle_t \equiv \langle U\rangle_t$, P- and Q^ε-a.s., and a continuous drift $A_t^U \equiv \langle U\rangle_t$.

Thus, $U_t = \widetilde{U}_t + \langle U\rangle_t$, $t\le T$, Q^ε-a.s. and, thereby, $\mathfrak{z}_T^{-1} = e^{-\widetilde{U}_T - \frac{1}{2}\langle U\rangle_T}$. Consequently, (6.3) is transformed to

$$\begin{aligned}\mathsf{P}(\widetilde{\mathfrak{A}}) &= \int_{\widetilde{\mathfrak{A}}}\exp\Big(-\widetilde{U}_T - \frac{1}{2}\langle U\rangle_T\Big)d\mathsf{Q}^\varepsilon \\ &= \int_{\widetilde{\mathfrak{A}}}\exp\Big(-\widetilde{U}_T - \frac{1}{2\varepsilon^2}\int_0^{\Theta_C\wedge T}\|\dot{u}_s - b(X_s^\varepsilon)\|^2_{a^{-1}(X_s^\varepsilon)}ds\Big)d\mathsf{Q}^\varepsilon.\end{aligned}$$

We are now in the position to derive a lower bound for the right hand side. Replacing $\widetilde{\mathfrak{A}}$ with a smaller set $\widetilde{\mathfrak{A}}\cap\mathfrak{B}$, where $\mathfrak{B} = \big\{|\varepsilon^2\widetilde{U}_T| \le \eta\big\}$, write

$$\mathsf{P}(\widetilde{\mathfrak{A}}) \ge \int_{\widetilde{\mathfrak{A}}\cap\mathfrak{B}}\exp\Big(-\frac{\eta}{\varepsilon^2} - \frac{1}{2\varepsilon^2}\int_0^{\Theta_C\wedge T}\|\dot{u}_s - b(X_s^\varepsilon)\|^2_{a^{-1}(X_s^\varepsilon)}ds\Big)d\mathsf{Q}^\varepsilon.$$

By the local Lipschitz continuity of b,σ and the uniform nonsingularity of $a(x)$,

$$\Big|\|\dot{u}_s - b(X_s^\varepsilon)\|^2_{a^{-1}(X_s^\varepsilon)} - \|\dot{u}_s - b(u_s)\|^2_{a^{-1}(u_s)}\Big| \le l_C(\|\dot{u}_s\|+1)^2\delta,\ \delta\le 1,$$

on the set $\widetilde{\mathfrak{A}} \cap \mathfrak{B}$ for any $s \le \Theta_C \wedge T$. Then,

$$\begin{aligned}\mathsf{P}(\widetilde{\mathfrak{A}}) &\ge \int_{\widetilde{\mathfrak{A}}\cap\mathfrak{B}} \exp\Big(-\frac{\eta}{\varepsilon^2} - \frac{\delta l_C}{\varepsilon^2}\int_0^T (\|\dot u_s\|+1)^2 ds \\ &\qquad\qquad - \frac{1}{2\varepsilon^2}\int_0^{\Theta_C\wedge T} \|\dot u_s - b(u_s)\|^2_{a^{-1}(u_s)} ds\Big) d\mathsf{Q}^\varepsilon \\ &\ge \int_{\widetilde{\mathfrak{A}}\cap\mathfrak{B}} \exp\Big(-\frac{\eta}{\varepsilon^2} - \frac{\delta l_C}{\varepsilon^2}\int_0^T (\|\dot u_s\|+1)^2 ds \\ &\qquad\qquad - \frac{1}{2\varepsilon^2}\int_0^{T} \|\dot u_s - b(u_s)\|^2_{a^{-1}(u_s)} ds\Big) d\mathsf{Q}^\varepsilon.\end{aligned}$$

Consequently,

$$\varliminf_{\varepsilon\to 0} \varepsilon^2 \log \mathsf{P}(\widetilde{\mathfrak{A}}) \ge -\eta - \delta l_C \int_0^T (\|\dot u_s\|+1)^2 ds - J_T(u) + \varliminf_{\varepsilon\to 0} \varepsilon^2 \log \mathsf{Q}^\varepsilon\big(\widetilde{\mathfrak{A}}\cap\mathfrak{B}\big).$$

We prove now that $\varliminf_{\varepsilon\to 0} \varepsilon^2 \log \mathsf{Q}^\varepsilon\big(\widetilde{\mathfrak{A}}\cap\mathfrak{B}\big) = 0$ by showing

$$\lim_{\varepsilon\to 0} \mathsf{Q}^\varepsilon\big(\Omega\setminus\widetilde{\mathfrak{A}}\big) = 0 \quad\text{and}\quad \lim_{\varepsilon\to 0} \mathsf{Q}^\varepsilon\big(\Omega\setminus\mathfrak{B}\big) = 0.$$

To this end, recall that

$$\begin{aligned}\Omega\setminus\widetilde{\mathfrak{A}} &= \Big\{\varepsilon \sup_{t\le T}\Big\| \int_0^{\Theta_C\wedge t} \sigma(X^\varepsilon_s) d\widetilde{B}_s\Big\| > \delta\Big\} \\ \Omega\setminus\mathfrak{B} &= \Big\{\varepsilon\Big\| \int_0^{\Theta_C\wedge T} \sigma^{-1}(X^\varepsilon_s)[\dot u_s - b(X^\varepsilon_s)] d\widetilde{B}_s\Big\| > \eta\Big\}.\end{aligned} \tag{6.4}$$

We verify (6.4) componentwise. Let L^ε_t denote any entry of $\int_0^{\Theta_C\wedge t}\sigma(X^\varepsilon_s)d\widetilde{B}_s$ or $\int_0^{\Theta_C\wedge t}\sigma^{-1}(X^\varepsilon_s)[\dot u_s - b(X^\varepsilon_s)]d\widetilde{B}_s$. We show that

$$\lim_{\varepsilon\to 0} \mathsf{Q}^\varepsilon\big(\varepsilon \sup_{t\le T}|L^\varepsilon_t| > \delta\big) = 0 \text{ and } \lim_{\varepsilon\to 0} \mathsf{Q}^\varepsilon\big(\varepsilon|L^\varepsilon_T| > \delta\big) = 0. \tag{6.5}$$

In both cases, L^ε_t is a continuous Q^ε-martingale with $\langle L^\varepsilon\rangle_t = \int_0^t g(s)ds$ and $\int_\Omega \int_0^T g(s)ds d\mathsf{Q}^\varepsilon < \infty$. Then (6.5) holds by Doob's inequality:

$$\lim_{\varepsilon\to 0} \mathsf{Q}^\varepsilon\big(\varepsilon \sup_{t\le T}|L^\varepsilon_t| > \delta\big) \le \frac{4\varepsilon^2}{\delta^2}\int_\Omega\int_0^T g(s)dsd\mathsf{Q}^\varepsilon \xrightarrow[\varepsilon\to 0]{} 0.$$

Now, for any fixed δ and η,

$$\varliminf_{\varepsilon\to 0} \varepsilon^2 \log \mathsf{P}(\widetilde{\mathfrak{A}}) \ge -\eta - \delta l_C \int_0^T (\|\dot u_s\|+1)^2 ds - J_T(u).$$

The required lower bound

$$\varliminf_{\delta\to 0}\varliminf_{\varepsilon\to 0} \varepsilon^2 \log \mathsf{P}(\widetilde{\mathfrak{A}}) \ge -J_T(u)$$

follows by taking $\lim_{\eta\to 0}\lim_{\delta\to 0}$. □

6.2. *General $a(x)$*

This part of the proof requires perturbation arguments. The idea is to use the already obtained local LDP lower bound for the uniformly nonsingular $a(x)$. Let W_t be a standard d dimensional Brownian motion, independent of B_t, defined on the same stochastic basis. Since b and σ are assumed to be locally Lipschitz continuous, one can introduce the perturbed diffusion process controlled by a free parameter $\beta \in (0,1]$:

$$X_t^{\varepsilon,\beta} = x_0 + \int_0^t b(X_s^{\varepsilon,\beta})ds + \varepsilon \int_0^t [\sigma(X_s^{\varepsilon,\beta})dB_s + \sqrt{\beta}dW_s]. \tag{6.6}$$

The process $X_t^{\varepsilon,\beta}$, defined in (6.6), solves the Itô equation $X_t^{\varepsilon,\beta} = x_0 + \int_0^t b(X_s^{\varepsilon,\beta})ds + \varepsilon \int_0^t [a(X_s^{\varepsilon,\beta}) + \beta \mathbf{I}]^{1/2} dB_s^\beta$ with respect to a standard Brownian motion $B_t^\beta = \int_0^t [a(X_s^{\varepsilon,\beta}) + \beta\mathbf{I}]^{-1/2}[\sigma(X_s^{\varepsilon,\beta})dB_s + \sqrt{\beta}dW_s]$. Then the family $\{(X_t^{\varepsilon,\beta})_{t\le T}\}_{\varepsilon\to 0}$ satisfies the local LDP lower bound. Indeed, the matrix $a_\beta(x)$ is uniformly nonsingular, its entries are locally bounded and satisfy the assumption (2.1) of Theorem 2.1 since

$$\frac{(x, a_\beta(x)x)}{\|x\|\,|(x,b(x))|} = \frac{(x,a(x)x)}{\|x\|\,|(x,b(x))|} + \beta\frac{\|x\|}{|(x,b(x))|}$$

and $\frac{\|x\|}{|(x,b(x))|}$ converges to zero as $\|x\|\to\infty$ by (2.1). In particular, with $\Theta_C^\beta = \inf\{t : \|X_t^{\varepsilon,\beta}\| \ge C\}$ and $u_0 = x_0$, $du_t \ll dt$, $\int_0^T \|\dot u_t\|^2 dt < \infty$, we have

$$\varliminf_{\delta\to 0}\varliminf_{\varepsilon\to 0} \varepsilon^2 \log \mathsf{P}\Big(\sup_{t\le \Theta_C^\beta\wedge T} \|X_t^{\varepsilon,\beta} - u_t\| \le \delta\Big) \ge -\frac{1}{2}\int_0^T \|\dot u_s - b(u_s)\|^2_{(a(u_s)+\beta\mathbf{I})^{-1}} ds. \tag{6.7}$$

Further, we will use (6.7) to establish

$$\varliminf_{\delta\to 0}\varliminf_{\varepsilon\to 0} \varepsilon^2 \log \mathsf{P}\Big(\sup_{t\le T} \|X_t^\varepsilon - u_t\| \le \delta\Big) \ge -\frac{1}{2}\int_0^T \|\dot u_s - b(u_s)\|^2_{a^\oplus(u_s)} ds. \tag{6.8}$$

To this end, we introduce the filtration $\mathbf{G}^\varepsilon = (\mathscr{G}_t^\varepsilon)_{t\ge 0}$, with the general conditions, generated by $(X_t^\varepsilon, X_t^{\varepsilon,\beta})_{t\ge 0}$ and notice that both Θ_C (see (4.2)) and Θ_C^β are stopping times relative to $\mathbf{G}^\varepsilon$. Hence,

$$\tau_C^\beta = \Theta_C \wedge \Theta_C^\beta \tag{6.9}$$

is a stopping time as well relative to $\mathbf{G}^\varepsilon$. Obviously,

$$\lim_{C\to\infty}\varlimsup_{\varepsilon\to 0} \varepsilon^2 \log \mathsf{P}\big(\tau_C^\beta \le T\big) = -\infty.$$

However, the proof of (6.8) requires a stronger property:

$$\lim_{C\to\infty}\varlimsup_{\varepsilon\to 0} \varepsilon^2 \log \sup_{\beta\in(0,1]} \mathsf{P}\Big(\tau_C^\beta \le T\Big) = -\infty. \tag{6.10}$$

It is clear, that (6.10) is valid if it is valid with τ_C^β replaced by Θ_C^β. The latter is verified along the lines of Lemma 4.1 proof:

$$\begin{aligned}
&\varepsilon^2 \log \sup_{\beta\in(0,1]} \mathsf{P}\big(\Theta_C^\beta \le T\big) \le -\inf_{\|x\|\ge C} V(x) + V(x_0) \\
&\quad + \frac{T\varepsilon^2}{2} \sup_{\beta\in(0,1]} \sup_{\|x\|\le C} \big|\mathrm{trace}\,\big(\Psi(x)[a(x)+\beta\mathbf{I}]\big)\big| + T \sup_{\beta\in(0,1]} \sup_{\|x\|\le L} \big|\mathfrak{D}_\beta V(x)\big| \\
&\quad \xrightarrow[\varepsilon\to 0]{} -\inf_{\|x\|\ge C} V(x) + V(x_0) + T \sup_{\beta\in(0,1]} \sup_{\|x\|\le L} \big|\mathfrak{D}_\beta V(x)\big| \xrightarrow[C\to\infty]{} -\infty,
\end{aligned}$$

where $\mathfrak{D}_\beta V(x) = (\nabla V(x), b(x)) + \frac{1}{2}(\nabla V(x), a_\beta(x)\nabla V(x))$.

We are now in the position to prove (6.8). With $\delta \le \beta^{1/4}$, write

$$\begin{aligned}
&\Big\{\sup_{t\le\tau_C^\beta\wedge T} \|X_t^{\varepsilon,\beta} - u_t\| \le \delta\Big\} \\
&= \Big\{\sup_{t\le\tau_C^\beta\wedge T} \|X_t^{\varepsilon,\beta} - u_t\| \le \delta\Big\} \bigcap \Big\{\sup_{t\le\tau_C^\beta\wedge T} \|X_t^{\varepsilon} - X_t^{\varepsilon,\beta}\| \le \beta^{1/4}\Big\} \\
&\quad \bigcup \Big\{\sup_{t\le\tau_C^\beta\wedge T} \|X_t^{\varepsilon,\beta} - u_t\| \le \delta\Big\} \bigcap \Big\{\sup_{t\le\tau_C^\beta\wedge T} \|X_t^{\varepsilon} - X_t^{\varepsilon,\beta}\| > \beta^{1/4}\Big\} \\
&\subseteq \Big\{\sup_{t\le\tau_C^\beta\wedge T} \|X_t^{\varepsilon,\beta} - u_t\| \le \beta^{1/4}\Big\} \bigcap \Big\{\sup_{t\le\tau_C^\beta\wedge T} \|X_t^{\varepsilon} - X_t^{\varepsilon,\beta}\| \le \beta^{1/4}\Big\} \\
&\quad \bigcup \Big\{\sup_{t\le\tau_C^\beta\wedge T} \|X_t^{\varepsilon} - X_t^{\varepsilon,\beta}\| > \beta^{1/4}\Big\} \\
&\subseteq \Big\{\sup_{t\le\tau_C^\beta\wedge T} \|X_t^{\varepsilon} - u_t\| \le 2\beta^{1/4}\Big\} \bigcup \Big\{\sup_{t\le\tau_C^\beta\wedge T} \|X_t^{\varepsilon} - X_t^{\varepsilon,\beta}\| > \beta^{1/4}\Big\} \\
&\subseteq \Big\{\sup_{t\le T} \|X_t^{\varepsilon} - u_t\| \le 2\beta^{1/4}\Big\} \bigcup \Big\{\sup_{t\le\tau_C^\beta\wedge T} \|X_t^{\varepsilon} - X_t^{\varepsilon,\beta}\| > \beta^{1/4}\Big\} \\
&\qquad \bigcup \Big\{\tau_C^\beta \le T\Big\}
\end{aligned}$$

Hence,

$$\begin{aligned}
\mathsf{P}\Big(\sup_{t\le\tau_C^\beta\wedge T} \|X_t^{\varepsilon,\beta} - u_t\| \le \delta\Big) &\le 3\Big\{\mathsf{P}\Big(\sup_{t\le T} \|X_t^{\varepsilon} - u_t\| \le 2\beta^{1/4}\Big) \\
&\quad \bigvee \mathsf{P}\Big(\sup_{t\le\tau_C^\beta\wedge T} \|X_t^{\varepsilon} - X_t^{\varepsilon,\beta}\| > \beta^{1/4}\Big) \bigvee \mathsf{P}\big(\tau_C^\beta \le T\big)\Big\}.
\end{aligned}$$

Clearly, Θ_C^β can be replaced by τ_C^β, and so

$$-\frac{1}{2}\int_0^T \|\dot{u}_s - b(u_s)\|^2_{(a(u_s)+\beta\mathbf{I})^{-1}} ds \leq \varliminf_{\varepsilon\to 0} \varepsilon^2 \log \mathsf{P}\Big(\sup_{t\leq T}\|X_t^\varepsilon - u_t\| \leq 2\beta^{1/4}\Big)$$
$$\bigvee \varlimsup_{\varepsilon\to 0} \varepsilon^2 \log \mathsf{P}\Big(\sup_{t\leq \tau_C^\beta\wedge T}\|X_t^\varepsilon - X_t^{\varepsilon,\beta}\| > \beta^{1/4}\Big)$$
$$\bigvee \varlimsup_{\varepsilon\to 0} \varepsilon^2 \log \sup_{\beta\in(0,1]} \mathsf{P}\Big(\tau_C^\beta \leq T\Big). \quad (6.11)$$

Next we use the following facts:

(1) by Lemma A.2.1 and (6.1),

$$\lim_{\beta\to 0}\int_0^T \|\dot{u}_s - b(u_s)\|^2_{(a(u_s)+\beta\mathbf{I})^{-1}} ds = \int_0^T \|\dot{u}_s - b(u_s)\|^2_{a^\oplus(u_s)} ds;$$

(2) by Lemma A.3.2,

$$\lim_{\beta\to 0}\varlimsup_{\varepsilon\to 0} \varepsilon^2 \log \mathsf{P}\Big(\sup_{t\leq \tau_C^\beta\wedge T}\|X_t^\varepsilon - X_t^{\varepsilon,\beta}\| > \beta^{1/4}\Big) = -\infty;$$

(3) by (6.10), $\lim_{C\to\infty}\varlimsup_{\varepsilon\to 0}\varepsilon^2 \log \sup_{\beta\in(0,1]} \mathsf{P}\Big(\tau_C^\beta \leq T\Big) = -\infty$.

Hence, passing to the limit $\beta\to 0$ and then $C\to\infty$ in (6.11) and taking into account (1)-(3), one gets the required lower bound

$$\varliminf_{\beta\to 0}\varliminf_{\varepsilon\to 0} \varepsilon^2 \log \mathsf{P}\Big(\sup_{t\leq T}\|X_t^\varepsilon - u_t\| \leq 2\beta^{1/4}\Big) \geq -\frac{1}{2}\int_0^T \|\dot{u}_s - b(u_s)\|^2_{a^\oplus(u_s)} ds.$$

□

A.1. Exponential estimates for martingales

Proposition A.1.1. (Lemma A.1 in Ref. [23]) *Let $M = (M_t)_{t\geq 0}$, $M_t \in \mathbb{R}$, be a continuous local martingale with $M_0 = 0$ and the predictable variation process $\langle M\rangle_t$ defined on some stochastic basis with general conditions. Let τ be a stopping time, α and B positive constants and $\mathfrak{A}$ some measurable set.*

(PA-1) if $M_\tau - \frac{1}{2}\langle M\rangle_\tau \geq \alpha$ on $\mathfrak{A}$, then $\mathsf{P}(\mathfrak{A}) \leq e^{-\alpha}$;
(PA-2) if $M_\tau \geq \alpha$ and $\langle M\rangle_\tau \leq B$ on $\mathfrak{A}$, then $\mathsf{P}(\mathfrak{A}) \leq e^{-\frac{\alpha^2}{2B}}$;
(PA-3) $\mathsf{P}(\sup_{t\leq T}|M_t| \geq \alpha, \langle M\rangle_T \leq B) \leq 2e^{-\frac{\alpha^2}{2B}}$;
(PA-4) $\mathsf{P}(\sup_{t\leq T}|M_t| \geq \alpha) \leq 2e^{-\frac{\alpha^2}{2B}} \bigvee \mathsf{P}(\langle M\rangle_T > B)$.

A.2. Pseudoinverse of nonnegative definite matrices

Let $A^{\oplus}$ be the Moore-Penrose pseudoinverse matrix of A (see Ref. [17]).

Lemma A.2.1. *For $d \times d$ nonnegative definite matrix A and $x \in \mathbb{R}^d$,*

$$\lim_{\beta\to 0}(x, (A+\beta\mathbf{I})^{-1}x) = \begin{cases} \|x\|^2_{A^{\oplus}}, & AA^{\oplus}x = x \\ \infty, & \textit{otherwise}. \end{cases}$$

Proof. Let S be an orthogonal matrix, $S^*S = \mathbf{I}$, such that $D := S^*AS$ is a diagonal matrix. Then, due to $S^*(A+\beta\mathbf{I})S = D + \beta\mathbf{I}$, we have $S^*(A+\beta\mathbf{I})^{-1}S = (D+\beta\mathbf{I})^{-1}$ and $S(D+\beta\mathbf{I})^{-1}S^* = (A+\beta\mathbf{I})^{-1}$. Write $(y := S^*x)$

$$\begin{aligned}(x, (A+\beta\mathbf{I})^{-1}x) &= (x, S(D+\beta\mathbf{I})^{-1}S^*x) = (S^*x, (D+\beta\mathbf{I})^{-1}S^*x) \\ &= (y, (D+\beta\mathbf{I})^{-1}y) = (y, (D+\beta\mathbf{I})^{-1}DD^{\oplus}y) \\ &\quad + (y, (D+\beta\mathbf{I})^{-1}(\mathbf{I}-DD^{\oplus})y).\end{aligned}$$

Since $\lim_{\beta\to 0}(D+\beta\mathbf{I})^{-1}DD^{\oplus} = D^{\oplus}$, one gets

$$\lim_{\beta\to 0}(y, (D+\beta\mathbf{I})^{-1}DD^{\oplus}y) = \|y\|^2_{D^{\oplus}} = \|x\|^2_{A^{\oplus}}$$

while $\lim_{\beta\to 0}(y, (D+\beta\mathbf{I})^{-1}(\mathbf{I}-DD^{\oplus})y) \neq \infty$ only if $(\mathbf{I}-DD^{\oplus})y = 0$. Since the latter condition is nothing but $(\mathbf{I}-AA^{\oplus})x = 0$, the desired statement holds. □

A.3. Exponential negligibility of $X^{\varepsilon,\beta}_t - X^{\varepsilon}_t$

We start with an auxiliary result.

Proposition A.3.1. *Let Y_t be a nonnegative continuous semimartingale defined on a stochastic basis (with general conditions):*

$$Y_t = \int_0^t h_1(s)Y_s ds + \varepsilon\int_0^t h_2(s)Y_s dM'_s + \varepsilon\sqrt{\beta}\int_0^t h_3(s)\sqrt{Y_s}dM''_s + \varepsilon^2\beta\int_0^t h_4(s)ds, \quad \text{(A.1)}$$

where $h_i(s), i = 1,\dots,4$, are bounded predictable processes and M'_t, M''_t are continuous martingales, $d\langle M'\rangle_t = m'(t)dt$, $d\langle M''\rangle_t = m''(t)dt$, $\langle M', M''\rangle_t \equiv 0$ with bounded $m'(t)$ and $m''(t)$. Assume that for any $T > 0$ and $\beta > 0$,

$$\lim_{L\to\infty}\overline{\lim_{\varepsilon\to 0}}\,\varepsilon^2\log\mathsf{P}\Big(\sup_{t\le T}\sqrt{Y_t} > L\Big) = -\infty. \quad \text{(A.2)}$$

Then, for any $T > 0$,

$$\lim_{\beta\to 0}\overline{\lim_{\varepsilon\to 0}}\,\varepsilon^2\log\mathsf{P}\Big(\sup_{t\le T}|Y_t| > \beta^{1/4}\Big) = -\infty.$$

Proof. Obviously Y_t solves the integral equation

$$Y_t = \mathcal{E}_t \int_0^t \mathcal{E}_s^{-1} \Big[\varepsilon \sqrt{\beta} h_3(s) \sqrt{Y_s} dM''_s + \varepsilon^2 \beta h_4(s) ds \Big],$$

where $\mathcal{E}_t = \exp\big(\int_0^t [h_1(s) - \varepsilon^2 0.5 h_2^2(s)] ds + \int_0^t \varepsilon h_2(s) dM'_s \big)$. Let for definiteness $|h_i| \le r$, where r is a constant. Then, with $\varepsilon \le 1$,

$$\sup_{t\le T} |\log \mathcal{E}_t| \le T(r + 0.5r^2) + \sup_{t\le T} \Big| \varepsilon \int_0^t h_2(s) dM'_s \Big|.$$

Hence the random variable $\sup_{t\le T} |\log \mathcal{E}_t|$ is bounded on the set

$$\Big\{ \sup_{t\le T} \Big| \varepsilon \int_0^t h_2(s) dM'_s \Big| \le C \Big\}.$$

Moreover, it is exponentially tight in the sense that

$$\lim_{C\to\infty} \overline{\lim_{\varepsilon\to 0}} \varepsilon^2 \log \mathsf{P}\Big(\sup_{t\le T} |\log \mathcal{E}_t| > C \Big) = -\infty. \tag{A.3}$$

The latter is implied by

$$\lim_{C\to\infty} \overline{\lim_{\varepsilon\to 0}} \varepsilon^2 \log \mathsf{P}\Big(\sup_{t\le T} \Big| \varepsilon \int_0^t h_2(s) dM'_s \Big| > C \Big) = -\infty \tag{A.4}$$

since the martingale $N_t = \varepsilon \int_0^t h_2(s) dM'_s$ has the quadratic variation process $\langle N \rangle_t = \varepsilon^2 \int_0^t h^2(s) m'(s) ds$ and, with some positive number r_1, we have $\varepsilon^2 h^2(s) m'(s) \le \varepsilon^2 r_1$. Then, by taking into account that $\mathsf{P}\big(\langle N \rangle_T > \varepsilon^2 r_1 T\big) = 0$ and applying the statement (PA-4) of Proposition A.1.1, we obtain $\mathsf{P}\big(\sup_{t\le T} |N_t| > C\big) \le 2e^{-C^2/(2\varepsilon^2 r_1 T)}$ providing (A.4).

Now we estimate $\sup_{t\le T} |Y_t|$ on the set $\big\{ \sup_{t\le T} |\log \mathcal{E}_t| \le C \big\}$. Write

$$\sup_{t\le T} |Y_t| \le e^C T r \varepsilon^2 \beta + e^C \sup_{t\le T} \Big| \int_0^t \mathcal{E}_s^{-1} \varepsilon \sqrt{\beta} h_3(s) \sqrt{Y_s} dM''_s \Big|.$$

This upper bound and (A.2), (A.3) reduce the proof of Proposition A.3.1 to:

$$\lim_{\beta\to 0} \overline{\lim_{\varepsilon\to 0}} \varepsilon^2 \log \mathsf{P}\Big(\sup_{t\le T} \Big| \int_0^t \mathcal{E}_s^{-1} \varepsilon \sqrt{\beta} h_3(s) \sqrt{Y_s} dM''_s \Big| > \beta^{1/4},$$

$$\sup_{t\le T} \sqrt{Y_t} \le L, \ \sup_{t\le T} |\log \mathcal{E}_t| \le C \Big) = \infty$$

for any $C > 0$ and $L > 0$. Introduce the martingale

$$N''_t = \int_0^t \mathcal{E}_s^{-1} \varepsilon \sqrt{\beta} h_3(s) \sqrt{Y_s} dM''_s \text{ with } \langle N'' \rangle_t = \int_0^t \mathcal{E}_s^{-2} \varepsilon^2 \beta h_3^2(s) Y_s m''(s) ds$$

and denote $\mathfrak{C} = \big\{ \sup_{t\le T} \sqrt{Y_t} \le L, \ \sup_{t\le T} |\log \mathcal{E}_t| \le C \big\}$. With $r_2 \ge h_3^2(s) L m''(s)$, we find that

$$\langle N'' \rangle_T \le e^{2C} r_2 T \varepsilon^2 \beta.$$

Hence,

$$\mathsf{P}\Big(\sup_{t\le T}|N''_t| > \beta^{1/4}, \mathfrak{C}\Big) = \mathsf{P}\Big(\sup_{t\le T}|N''_t| > \beta^{1/4},\ \langle N''\rangle_T \le e^{2C} r_2 T\varepsilon^2\beta, \mathfrak{C}\Big)$$
$$\le \mathsf{P}\Big(\sup_{t\le T}|N''_t| > \beta^{1/4}, \langle N''\rangle_T \le e^{2C} r_2 T\varepsilon^2\beta\Big).$$

By (PA-3) from Proposition A.1.1 the latter term is upper bounded by

$$2\exp\Big(\frac{\beta^{1/2}}{2e^{2C}r_2T\varepsilon^2\beta}\Big).$$

Then we obtain

$$\varlimsup_{\varepsilon\to\infty}\varepsilon^2\log\mathsf{P}\Big(\sup_{t\le T}|N''_t| > \beta^{1/4}, \mathfrak{C}\Big) \le -\frac{1}{2e^{2C}r_2T\beta^{1/2}} \xrightarrow[\beta\to 0]{} -\infty.$$

□

We apply Proposition A.3.1 in order to prove

Lemma A.3.2. *For any $T>0$ and $C>0$,*

$$\lim_{\beta\to 0}\varlimsup_{\varepsilon\to 0}\varepsilon^2\log\mathsf{P}\Big(\sup_{t\le\tau_C^\beta\wedge T}\|X_t^{\varepsilon,\beta} - X_t^\varepsilon\| > \beta^{1/4}\Big) = -\infty.$$

Proof.
Recall that X_t^ε and $X_t^{\varepsilon,\beta}$ solve (1.1) and (6.6) respectively and τ_C^β is given in (6.9). Set $\triangle_t^{\varepsilon,\beta} = X^{\varepsilon,\beta}_{\tau_C^\beta\wedge t} - X^\varepsilon_{\tau_C^\beta\wedge t}$. By (1.1) and (6.6),

$$\triangle_t^{\varepsilon,\beta} = \int_0^{\tau_C^\beta\wedge t}\big(b(X^{\varepsilon,\beta}_{\tau_C^\beta\wedge s}) - b(X^\varepsilon_{\tau_C^\beta\wedge s})\big)ds +$$
$$+\varepsilon\int_0^{\tau_C^\beta\wedge t}\big(\sigma(X^{\varepsilon,\beta}_{\tau_C^\beta\wedge s}) - \sigma(X^\varepsilon_{\tau_C^\beta\wedge s})\big)dB_s + \varepsilon\sqrt{\beta}W_{\tau_C^\beta\wedge t}.$$

Due to the local Lipschitz continuity of b and σ and with $0/0=0$, the vector-valued and matrix-valued functions:

$$f(s) = \frac{b(X^{\varepsilon,\beta}_{\tau_C^\beta\wedge s}) - b(X^\varepsilon_{\tau_C^\beta\wedge t})}{\|\triangle_s^{\varepsilon,\beta}\|} \quad\text{and}\quad g(s) = \frac{\sigma(X^{\varepsilon,\beta}_{\tau_C^\beta\wedge s}) - \sigma(X^\varepsilon_{\tau_C^\beta\wedge s})}{\|\triangle_s^{\varepsilon,\beta}\|}$$

are well defined and their entries are bounded by a constant depending on C. Hence

$$\triangle_t^{\varepsilon,\beta} = \int_0^{\tau_C^\beta\wedge t}\|\triangle_s^{\varepsilon,\beta}\|f(s)ds + \varepsilon\int_0^{\tau_C^\beta\wedge t}\|\triangle_s^{\varepsilon,\beta}\|g(s)dB_s + \varepsilon\sqrt{\beta}W_{\tau_C^\beta\wedge t}.$$

Since $\|\triangle_t^{\varepsilon,\beta}\|^2 = (\triangle_t^{\varepsilon,\beta}, \triangle_t^{\varepsilon,\beta})$, by the Itô formula, we find that

$$\begin{aligned}\|\triangle_t^{\varepsilon,\beta}\|^2 = &\int_0^t 2\|\triangle_s^{\varepsilon,\beta}\|(\triangle_s^{\varepsilon,\beta}, f(s))ds \\ &+ \varepsilon\int_0^{\tau_C^\beta\wedge t} 2\|\triangle_s^{\varepsilon,\beta}\|(\triangle_s^{\varepsilon,\beta}, g(s)dB_s) \\ &+ \varepsilon\sqrt{\beta}\int_0^{\tau_C^\beta\wedge t} 2(\triangle_s^{\varepsilon,\beta}, dW_s) \\ &+ \varepsilon^2\int_0^{\tau_C^\beta\wedge t} \|\triangle_s^{\varepsilon,\beta}\|^2 \operatorname{trace}\big[g(s)g^*(s)\big]ds \\ &+ \varepsilon^2\beta(\tau_C^\beta\wedge t)d.\end{aligned} \tag{A.5}$$

Now, by letting $\varphi(s) = \frac{2(\triangle_s^{\varepsilon,\beta}, f(s))}{\|\triangle_s^{\varepsilon,\beta}\|}$ and $d\widehat{B}_s = \frac{2(\triangle_s^{\varepsilon,\beta}, g(s)dB_s)}{\|\triangle_s^{\varepsilon,\beta}\|}$, we rewrite (A.5) as:

$$\begin{aligned}\|\triangle_t^{\varepsilon,\beta}\|^2 = &\int_0^{\tau_C^\beta\wedge t} \|\triangle_s^{\varepsilon,\beta}\|^2\Big(\varphi(s) + \varepsilon^2 \operatorname{trace}[g(s)g^*(s)]\Big)ds \\ &+ \varepsilon\int_0^{\tau_C^\beta\wedge t} \|\triangle_s^{\varepsilon,\beta}\|^2 d\widehat{B}_s + \varepsilon\sqrt{\beta}\int_0^{\tau_C^\beta\wedge t} \|\triangle_s^{\varepsilon,\beta}\|\frac{2(\triangle_s^{\varepsilon,\beta}, dW_s)}{\|\triangle_s^{\varepsilon,\beta}\|} \\ &+ \varepsilon^2\beta(\tau_C^\beta\wedge t)d.\end{aligned} \tag{A.6}$$

With the notations

- $Y_t = \|\triangle_t^{\varepsilon,\beta}\|^2$
- $h_1(s) = I_{\{\tau_C^\beta\le s\}}\{\varphi(s) + \varepsilon^2 \operatorname{trace}[g(s)g^*(s)]\}$
- $h_2(s) \equiv 1$
- $h_4(s) = I_{\{\tau_C^\beta\le s\}}d$
- $M_t' = \widehat{B}_t$, $m'(s) = \dfrac{4(\triangle_s^{\varepsilon,\beta}, g(s)g^*(s)\triangle_s^{\varepsilon,\beta})}{\|\triangle_s^{\varepsilon,\beta}\|^2}$
- $M_t'' = \int_0^{\tau_C^\beta\wedge t} 2\dfrac{(\triangle_s^{\varepsilon,\beta}, dW_s)}{\|\triangle_s^{\varepsilon,\beta}\|}$, $m''(s) \equiv 4$,

the equation (A.6) is in the form of (A.1). Since $h_i(s)$, $i = 1, \dots, 4$ are bounded and $\sqrt{Y}_t \equiv \|X_{\tau_C^\beta\wedge t}^{\varepsilon,\beta} - X_{\tau_C^\beta\wedge t}^{\varepsilon}\| \le \|X_{\tau_C^\beta\wedge t}^{\varepsilon,\beta}\| + \|X_{\tau_C^\beta\wedge t}^{\varepsilon}\| \le 2C$, i.e., (A.2) holds too, the statement of the lemma follows from Proposition A.3.1. □

Acknowledgement

Research of P. Chigansky is supported by a grant from the Israel Science Foundation.

References

[1] M. I. Freidlin and A. D. Wentzell, *Random perturbations of dynamical systems.* vol. 260, *Grundlehren der Mathematischen Wissenschaften [Fundamental Principles of*

Mathematical Sciences], (Springer-Verlag, New York, 1984). ISBN 0-387-90858-7. Translated from the Russian by Joseph Szücs.

[2] A. Dembo and O. Zeitouni, *Large deviations techniques and applications.* vol. 38, *Applications of Mathematics (New York)*, (Springer-Verlag, New York, 1998), second edition. ISBN 0-387-98406-2.

[3] P. Dupuis and R. S. Ellis, *A weak convergence approach to the theory of large deviations.* Wiley Series in Probability and Statistics: Probability and Statistics, (John Wiley & Sons Inc., New York, 1997). ISBN 0-471-07672-4. , A Wiley-Interscience Publication.

[4] J. Feng, Martingale problems for large deviations of Markov processes, *Stochastic Process. Appl.* **81**(2), 165–216, (1999). ISSN 0304-4149.

[5] J. Feng and T. G. Kurtz, *Large deviations for stochastic processes.* vol. 131, *Mathematical Surveys and Monographs*, (American Mathematical Society, Providence, RI, 2006). ISBN 978-0-8218-4145-7; 0-8218-4145-9.

[6] A. Friedman, *Stochastic differential equations and applications. Vol. 2.* (Academic Press [Harcourt Brace Jovanovich Publishers], New York, 1976). Probability and Mathematical Statistics, Vol. 28.

[7] R. S. Liptser and A. A. Pukhalskii, Limit theorems on large deviations for semimartingales, *Stochastics Stochastics Rep.* **38**(4), 201–249, (1992). ISSN 1045-1129.

[8] T. Mikami, Some generalizations of Wentzell's lower estimates on large deviations, *Stochastics.* **24**(4), 269–284, (1988). ISSN 0090-9491.

[9] K. Narita, Large deviation principle for diffusion processes, *Tsukuba J. Math.* **12**(1), 211–229, (1988). ISSN 0387-4982.

[10] D. W. Stroock, *An introduction to the theory of large deviations.* Universitext, (Springer-Verlag, New York, 1984). ISBN 0-387-96021-X.

[11] J. Ren and X. Zhang, Freidlin-Wentzell's large deviations for homeomorphism flows of non-Lipschitz SDEs, *Bull. Sci. Math.* **129**(8), 643–655, (2005). ISSN 0007-4497.

[12] A. A. Puhalskii, On some degenerate large deviation problems, *Electron. J. Probab.* **9**, no. 28, 862–886 (electronic), (2004). ISSN 1083-6489.

[13] R. Liptser, V. Spokoiny, and A. Y. Veretennikov, Freidlin-Wentzell type large deviations for smooth processes, *Markov Process. Related Fields.* **8**(4), 611–636, (2002). ISSN 1024-2953.

[14] R. Z. Has′minskiĭ, *Stochastic stability of differential equations.* vol. 7, *Monographs and Textbooks on Mechanics of Solids and Fluids: Mechanics and Analysis*, (Sijthoff & Noordhoff, Alphen aan den Rijn, 1980). ISBN 90-286-0100-7. Translated from the Russian by D. Louvish.

[15] E. Pardoux and A. Y. Veretennikov, On the Poisson equation and diffusion approximation. I, *Ann. Probab.* **29**(3), 1061–1085, (2001). ISSN 0091-1798.

[16] È. Pardoux and A. Y. Veretennikov, On Poisson equation and diffusion approximation. II, *Ann. Probab.* **31**(3), 1166–1192, (2003). ISSN 0091-1798.

[17] A. Albert, *Regression and the Moore-Penrose pseudoinverse.* (Academic Press, New York, 1972). Mathematics in Science and Engineering, Vol 94.

[18] A. Puhalskii. On functional principle of large deviations. In *New trends in probability and statistics, Vol. 1 (Bakuriani, 1990)*, pp. 198–218. VSP, Utrecht, (1991).

[19] A. Puhalskii, *Large deviations and idempotent probability.* vol. 119, *Chapman & Hall/CRC Monographs and Surveys in Pure and Applied Mathematics*, (Chapman & Hall/CRC, Boca Raton, FL, 2001). ISBN 1-58488-198-4.

[20] D. Aldous, Stopping times and tightness, *Ann. Probability.* **6**(2), 335–340, (1978).

[21] D. Aldous, Stopping times and tightness. II, *Ann. Probab.* **17**(2), 586–595, (1989). ISSN 0091-1798.

[22] R. S. Liptser and A. N. Shiryayev, *Theory of martingales.* vol. 49, *Mathematics and its Applications (Soviet Series)*, (Kluwer Academic Publishers Group, Dordrecht, 1989). ISBN 0-7923-0395-4. Translated from the Russian by K. Dzjaparidze [Kacha Dzhaparidze].

[23] A. Guillin and R. Liptser, MDP for integral functionals of fast and slow processes with averaging, *Stochastic Process. Appl.* **115**(7), 1187–1207, (2005). ISSN 0304-4149.

Chapter 8

On the Convergence Rates of a General Class of Weak Approximations of SDEs

Dan Crisan and Saadia Ghazali*

Department of Mathematics, Imperial College London
180 Queen's Gate, London SW7 2BZ
d.crisan@imperial.ac.uk

In this paper, the convergence analysis of a class of weak approximations of solutions of stochastic differential equations is presented. This class includes recent approximations such as Kusuoka's moment similar families method and the Lyons-Victoir cubature of Wiener Space approach. We show that the rate of convergence depends intrinsically on the smoothness of the chosen test function. For smooth functions (the required degree of smoothness depends on the order of the approximation), an equidistant partition of the time interval on which the approximation is sought is optimal. For functions that are less smooth (for example Lipschitz functions), the rate of convergence decays and the optimal partition is no longer equidistant. Our analysis rests upon Kusuoka-Stroock's results on the smoothness of the distribution of the solution of a stochastic differential equation. Finally the results are applied to the numerical solution of the filtering problem.

Contents

1. Introduction

Stochastic differential equations (SDEs) constitute an ideal mathematical model for a multitude of phenomena and processes encountered in areas such as filtering, optimal stopping, stochastic control, signal processes and mathematical finance, most notably in option pricing (see for example Oksendal [1] and Kloeden & Platen [2]). Unlike their deterministic counterparts, SDEs do not have explicit solutions, apart from in a few exceptional cases, hence the necessity for a sound theory of their numerical approximation.

*saadia.ghazali@imperial.ac.uk

In this paper we will be concerned with SDEs written in Stratonovich form, in other words, we will look at equations of the form,

$$X_t = X_0 + \int_0^t V_0(X_s)ds + \sum_{j=1}^{k} \int_0^t V_j(X_s) \circ dW_s^j, \tag{1.1}$$

where the last term is a stochastic integral of Stratonovich type. There are two classes of numerical methods for approximating SDEs. The objective of the first is to produce a pathwise approximation of the solution (strong approximation). The second method involves approximating the distribution of the solution at a particular instance in time (weak approximation). For example when one is only interested in the expectation $\mathbb{E}[\varphi(X_t)]$ for some function φ, it is sufficient to have a good approximation of the distribution of the random variable X_t rather than of its sample paths. This observation was first made by Milstein [3] who showed that pathwise schemes and L^2 estimates of the corresponding errors are irrelevant in this context since the objective is to approximate the law of X_t. This paper contains approximations that belong to this second class of algorithms.

Classical results in this area concentrate on solving numerically SDEs for which the so-called 'ellipticity condition', or more generally the 'Uniform Hörmander condition' (UH), is satisfied. For a survey of such schemes see, for example, Kloeden & Platen [2] or Burrage, Burrage & Tian [4]. Under this condition, for any bounded measurable function φ, the semigroup of operators $\{P_t\}_{t\in[0,\infty)}$defined,

$$(P_t\varphi)(x) = \mathbb{E}[\varphi(X_t(x))], \tag{1.2}$$

where $X(x) = \{X_t(x)\}_{t\in[0,\infty)}$ solves (1.1) with initial condition $X_0 = x$, is smooth for any $t > 0$. It is this property upon which the majority of these schemes rely.

For example, the classical Euler-Maruyama scheme requires $P_t\varphi$ to be four times differentiable in order to obtain the optimal rate of convergence. Talay([5], [6]) and, independently, Milstein [7] introduced the appropriate methodology to analyse this scheme. They express the error as a difference including a sum of terms involving the solution of a parabolic PDE. Their analysis also shows the relationship between the smoothness of φ and the corresponding error. Talay & Tubaro [8] prove an even more precise result showing that, under the same conditions, the errors corresponding to the Euler-Maruyama and many other schemes can be expanded in terms of powers of the discretisation step. Furthermore, Bally & Talay [9] show the existence of such an expansion under a much weaker hypothesis on φ: that φ need only be measurable and bounded (even the boundedness condition can be relaxed). Higher order schemes require additional smoothness properties of $P_t\varphi$ (see for example, Platen & Wagner [10]).

In the eighties, Kusuoka & Stroock([11], [12], [13]) studied the properties of $P_t\varphi$ under a weaker condition, the so-called UFG condition (see (2.3) in Section 2). Essentially, this condition states that the Lie algebra generated by the vector fields $\{V_i\}_{i=1}^k \in C_b^\infty(\mathbb{R}^d;\mathbb{R}^d)$ is finite dimensional as a $C_b^\infty(\mathbb{R}^d)$-module. Kusuoka &

Stroock conclude Malliavin's undertaking to recover Hörmander's hypo-ellipticity theory for degenerate second order elliptic operators. They show that under the UFG condition, $P_t\varphi$ retains certain regularity properties, in particular, that $V_{i_1}V_{i_2}...V_{i_n}P_t\varphi$ is well defined for any vector fields V_{i_r}, where $i_r \in \{1,...,k\}$. They also give upper bounds for the supremum norm of $V_{i_1}V_{i_2}...V_{i_n}P_t\varphi$. Besides the UFG condition, our analysis rests upon a second assumption which we call the **V0** condition. It states that V_0 can be expressed in terms of $\{V_1,...V_k\}\cup\{[V_i,V_j],1\leq i<j\leq k\}$. This premise is weaker than the ellipticity assumption and has been used, for example, by Jerison and Sánchez-Calle ([14], [15]) to obtain estimates for the heat kernel. This second condition enables one to control the supremum norm of $V_{i_1}V_{i_2}...V_{i_n}P_t\varphi$ for $i_r \in \{0,1,...,k\}$ (see Corollary 5.4 in Section 5).

A number of schemes have recently been developed to work under these weaker conditions rather than the ellipticity condition, their convergence depending intrinsically on the above estimates of $V_{i_1}V_{i_2}...V_{i_n}P_t\varphi$. A further advantage of this new generation of schemes is a consequence of the classical result stating that the support of $X(x)$ is the closure of the set $S=\{x^\varphi:[0,\infty)\to\mathbb{R}^d\}$ where x^φ solves the ODE,

$$x_t^\varphi = x+\int_0^t V_0(x_s^\varphi)ds+\sum_{j=1}^k\int_0^t V_j(x_s^\varphi)\varphi(s)\circ ds,$$

and $\varphi:[0,\infty)\to\mathbb{R}^d$ is an arbitrary smooth function (see Stroock & Varadhan [16], [17], [18], Millet & Sanz-Sole [19]). These schemes attempt to keep the support of the approximating process on the set S. In this way, stability problems that are known to affect classical schemes can be avoided. For example, Ninomiya & Victoir [20] give an explicit example where the Euler-Maruyama approximation fails whilst their algorithm succeeds (see Example 3.4 below for the algorithm). Their example involves an SDE related to the Heston stochastic volatility model in finance.

In this paper we give a general criterion for the convergence of a class of weak approximations incorporating this new category of schemes. This criterion is based upon the stochastic Stratonovich-Taylor expansion of $P_t\varphi$ and demonstrates how the rate of convergence depends on the smoothness of the test function selected.

Our plan is as follows. In section 2 we set out some essential terminology, adopted partly from Kusuoka [21], which is required to index the stochastic Stratonovich-Taylor expansions that follow and to state the UFG and V0 conditions imposed on the given vector fields (see (2.3) & (2.4) below). Section 3 contains our main results on the convergence analysis of a class of weak approximations of solutions of SDEs, characterised by introducing the concept of an *m-perfect family* (Definition 3.1). The Lyons-Victoir and Ninomiya-Victoir approximations are both members of this class. Although the Kusuoka approximation is not within this family, it can effectively be categorised in the same way (see Example 3.6 and the subsequent comment). In our main Theorem 3.7, we show that the rate of convergence depends intrinsically on the smoothness of the chosen test function; the higher the order of the approximation, the smoother the test function required.

For a smooth function, an equidistant partition of the time interval on which the approximation is sought is optimal. For less smooth functions, this is no longer true. We emphasise that the UFG+V0 conditions are not required for a smooth test function. Furthermore, the Kusuoka approximation does not require the V0 condition. Finally, in Section 4 we present an application of this theory to the numerical solution of the filtering problem.

2. Preliminaries

Let $(\Omega, \mathcal{F}, \mathbb{P})$ be a probability space satisfying the usual conditions and $W = \{W_t\}_{t\in[0,\infty)}$ be a k-dimensional Brownian motion defined on it. We also set $W_t^0 = t$ for $t \in [0, \infty)$. Let $C_b^\infty(\mathbb{R}^d, \mathbb{R}^d)$ denote the space of smooth functions $\varphi : \mathbb{R}^d \to \mathbb{R}^d$ with bounded derivatives, that is, bounded partial derivatives (of all orders) of the d component functions $\{\varphi^i : \mathbb{R}^d \to \mathbb{R}\}_{i=1}^d$ exist. We have a dual interpretation of this space, in that we also regard its elements as vector fields. In other words, a smooth function $V : C_b^\infty(\mathbb{R}^d) \to C_b^\infty(\mathbb{R}^d)$ is equivalent to an operator on $C_b^\infty(\mathbb{R}^d) \equiv C_b^\infty(\mathbb{R}^d, \mathbb{R})$ defined by,

$$V\varphi = \sum_{i=1}^{d} V^i \frac{\partial \varphi}{\partial x_i}$$

where $V^i \in C_b^\infty(\mathbb{R}^d)$ is the i-th component of L for $i = 1, \ldots, d$.

We next consider the Stratonovich SDE with drift vector $\{V_0^i(x)\}_{i=1}^d$ and dispersion matrix $\{V_j^i(x)\}_{i,j=1}^{d,k}$ for $x \in \mathbb{R}^d$, for some $V_0, \ldots, V_k \in C_b^\infty(\mathbb{R}^d, \mathbb{R}^d)$. This is written componentwise as,

$$X_t^i = X_0^i + \int_0^t V_0^i(X_s)ds + \sum_{j=1}^{k} \int_0^t V_j^i(X_s) \circ dW_s^j \tag{2.1}$$

for $i = 1, \ldots, d$.

The following essential terminology has been adopted from Kusuoka [21]. Let $\mathcal{A}$ be the set of multi-indices,

$$\mathcal{A} = \{\emptyset\} \cup \bigcup_{m=1}^{\infty} \{0, 1, \ldots, k\}^m$$

where k is the dimension of the Brownian Motion introduced above. This set will be used to index the Stratonovich-Taylor expansions that follow. Let $|\cdot|$ and $\|\cdot\|$ be the following two norms defined on $\mathcal{A}$ by

$$|\emptyset| = 0, \ |\alpha| = r \ if \ \alpha = (i_1, \ldots, i_r) \in \{0, 1, \ldots, k\}^r \ \text{ for } r \in \mathbb{N}$$

and $\|\alpha\| = |\alpha| + card\{1 \le j \le |\alpha| : \ i_j = 0\}$. Furthermore, let $\mathcal{A}_0 = \mathcal{A} \setminus \{\emptyset\}$, $\mathcal{A}_1 = \mathcal{A} \setminus \{\emptyset, (0)\}$ and correspondingly, $\mathcal{A}(m) = \{\alpha \in \mathcal{A} : \|\alpha\| \le m\}$, $\mathcal{A}_0(m) = \{\alpha \in \mathcal{A}_0 :$

$\|\alpha\| \leq m\}$ and $\mathcal{A}_1(m) = \{\alpha \in \mathcal{A}_1 : \|\alpha\| \leq m\}$ where the integer $m \in \mathbb{N}$ above corresponds to the level of the truncation in the Stratonovich-Taylor expansions that follow (see (3.4)).

We also require the following concatenation on $\mathcal{A}$,

$$(i_1, \ldots, i_r) * (j_1, \ldots, j_s) = (i_1, \ldots, i_r, j_1, \ldots, j_s)$$

and need to define a further operation on the vector fields $\{V_j \in C_b^\infty(\mathbb{R}^d, \mathbb{R}^d)\}_{j=0}^k$ introduced above.

Definition 2.1. For $\alpha \in \mathcal{A}$, the vector field $V_{[\alpha]}$ is defined inductively by,

$$V_{[\varphi]} = 0;\ V_{[(j)]} = V_j;\ V_{[\alpha*(j)]} = [V_\alpha, V_j]$$

for $j = 0, 1, \ldots, k$ where $[V_i, V_j] := V_iV_j - V_jV_i$ for $V_i, V_j \in C_b^\infty(\mathbb{R}^d, \mathbb{R}^d)$.

In the following we will make use of the semi-norm,

$$\|\varphi\|_{V,i} = \sum_{u=1}^{i} \sum_{\substack{\alpha_1,\ldots,\alpha_u \in \mathcal{A}_0 \\ \|\alpha_1 * \ldots * \alpha_u\| = i}} \left\|V_{[\alpha_1]} \cdots V_{[\alpha_u]}\varphi\right\|_\infty .$$

for $i \in \mathbb{N}$. Furthermore, we introduce the semi-norm,

$$\|\varphi\|_p = \sum_{i=1}^{p} \left\|\nabla^i \varphi\right\|_\infty \tag{2.2}$$

for $p \in \mathbb{N}, \varphi \in C_b^p(\mathbb{R}^d)$ where,

$$\left\|\nabla^i \varphi\right\|_\infty = \max_{j_1,\ldots,j_i \in \{1,\ldots,d\}} \left\| \frac{\partial^i \varphi}{\partial x_{j_1} \ldots \partial x_{j_i}} \right\|_\infty$$

and note that it can easily be deduced from the chain rule, for $\alpha \in \mathcal{A}_0$, $\varphi \in C_b^{|\alpha|}(\mathbb{R}^d)$, that $\|V_\alpha \varphi\|_\infty \leq C \|\varphi\|_{|\alpha|}$ and hence $\|\varphi\|_{V,i} \leq C \|\varphi\|_i$ for $i \in \mathbb{N}$. Also let $\|\cdot\|_{p,\infty}$ be the norm $\|\varphi\|_{p,\infty} := \|\varphi\|_\infty + \|\varphi\|_p$ for $\varphi \in C_b^p(\mathbb{R}^d)$. We define the space,

$$C_b^{V,i}(\mathbb{R}^d) = \left\{\varphi : \|\varphi\|_{V,i} < \infty\right\}$$

which appears in the definition of an *m-perfect* family below (see Definition 3.1).

We now introduce the two conditions on the vector fields $\{V_j \in C_b^\infty(\mathbb{R}^d, \mathbb{R}^d)\}_{j=0}^k$ required to treat the case when the test function φ is not smooth. We emphasise that the smooth case requires neither the UFG nor the V0 condition. Furthermore, the Kusuoka approximation (Example 3.6) does not require the V0 condition as the corresponding error is controlled in terms of $\{V_{[\beta]} : \beta \in \mathcal{A}_1\}$ alone.

The UFG condition [Kusuoka & Stroock [13, 21]]. There exists some $l \in \mathbb{N}$ such that for any $\alpha \in \mathcal{A}_1$ we have,

$$V_{[\alpha]} = \sum_{\beta \in \mathcal{A}_1(l)} \varphi_{\alpha,\beta} V_{[\beta]}. \tag{2.3}$$

where $\varphi_{\alpha,\beta} \in C_b^\infty(\mathbb{R}^d)$ for all $\beta \in \mathcal{A}_1(l)$.

The V0 condition. There exist $\varphi_\beta \in C_b^\infty(\mathbb{R}^d)$, $\beta \in \mathcal{A}_1(2)$ such that,

$$V_0 = \sum_{\beta \in \mathcal{A}_1(2)} \varphi_\beta V_{[\beta]}. \tag{2.4}$$

As mentioned in the introduction, the UFG condition states that the Lie algebra generated by the vector fields $\{V_i\}_{i=1}^k \in C_b^\infty(\mathbb{R}^d;\mathbb{R}^d)$ is finite dimensional as a $C_b^\infty(\mathbb{R}^d)$-module. It is implied by the Uniform Hörmander Condition which states that,

$$\exists l \in \mathbb{N} \text{ and } c > 0 \text{ s.t. } \sum_{\alpha \in \mathcal{A}_1(l)} \left\langle V_{[\alpha]}(x), \xi \right\rangle^2 \geq c\,|\xi|^2 \tag{2.5}$$

for all $x, \xi \in \mathbb{R}^d$, where $\langle V, \xi \rangle = \sum_{i=1}^d V^i \xi^i$ for $V \in C_b^\infty(\mathbb{R}^d, \mathbb{R}^d)$.

Under conditions (2.3)+(2.4), one can show that for any $r \in \mathbb{Z}^+$, $\{\alpha_i \in \mathcal{A}_0\}_{i=1}^r$ and $p = 1, ..., \|\alpha_1 * \ldots * \alpha_r\|$, there exists a constant $C_p^T > 0$ such that,

$$\left\| V_{[\alpha_1]} \cdots V_{[\alpha_r]} P_t \varphi \right\|_\infty \leq C_p^T t^{(p - \|\alpha_1 * \ldots * \alpha_r\|)/2} \left\|\varphi\right\|_p, \qquad t \in [0,T]. \tag{2.6}$$

where $\{P_t\}_{t \in [0,\infty)}$ is the semigroup associated with X defined in (1.2). We remark that condition (2.3) will only give us (2.6) for $\alpha_i \in \mathcal{A}_1$, $i = 1, ...r$. Under both (2.3)+(2.4) we have,

$$\left\|V_0 P_t \varphi\right\|_\infty \leq C t^{-1} \left\|\varphi\right\|_\infty.$$

However, under condition (2.3) alone, $\left\|V_0 P_t \varphi\right\|_\infty$ may be of higher order. Kusuoka has given an explicit example in which,

$$c t^{-\frac{l}{2}} \left\|\varphi\right\|_\infty \leq \left\|V_0 P_t \varphi\right\|_\infty \leq C t^{-\frac{l}{2}} \left\|\varphi\right\|_\infty$$

for some constants $c, C > 0$ and where l is the constant appearing in (2.3) (see Proposition 14 and Proposition 16 in Ref. [22]). This coarser bound on $\left\|V_0 P_t \varphi\right\|_\infty$ results in lower rates of convergence. The authors believe that (2.4) is the most general condition required to preserve the same rates of convergence as those obtained when $\{V_i\}_{i=1}^k$ satisfy the ellipticity condition (for non-smooth test functions φ).

Result (2.6) is a corollary of a certain representation theorem proved in Kusuoka-Stroock [13]. For completeness, we state the representation theorem (Theorem 5.2) in Section 5 where we also sketch a proof of inequality (2.6) in Corollary 5.4. We note that the case $p = 1$ has been proved in Ref. [13] for $\{\alpha_i \in \mathcal{A}_1\}_{i=1}^r$. Inequality (2.6) proves to be crucial in obtaining upper bounds on the error of of the class algorithms that we study below (see Theorem 3.7).

3. The Main Theorem

In this section we introduce the concept of an *m-perfect* family. Such families correspond to various weak approximations of SDEs, including the Lyons-Victoir and Ninomiya-Victoir schemes. The main result appears in Theorem 3.7 and Corollary 3.8.

For $\alpha = (i_1, \ldots, i_r) \in \mathcal{A}_0$ and $\varphi \in C_b^r(\mathbb{R}^d)$ let $f_{\alpha,\varphi}$ be defined as $f_{(i_1,\ldots,i_r),\varphi} := V_{i_1} \ldots V_{i_r}\varphi$. We also need to define the iterated Stratonovich integral

$$I_{f_{\alpha,\varphi}}(t) := \int_0^t \int_0^{s_0} \cdots \left(\int_0^{s_{r-2}} f_{\alpha,\varphi}(X_{s_{r-1}}) \circ dW_{s_{r-1}}^{i_1} \right) \circ \cdots \circ dW_{s_1}^{i_{r-1}} \circ dW_{s_0}^{i_r},$$

for $t \geq 0$. If $i_1 = 0$ then $I_{f_{\alpha,\varphi}}(t)$ is well defined for $\varphi \in C_b^r(\mathbb{R}^d)$. However, if $i_1 \neq 0$ then $I_{f_{\alpha,\varphi}}(t)$ is well defined provided $\varphi \in C_b^{r+2}(\mathbb{R}^d)$, since the semimartingale property of $f_{\alpha,\varphi}(X)$ is required in the definition of the first Stratonovich integral $\int_0^{s_{r-2}} f_{\alpha,\varphi}(X_{s_{r-1}}) \circ dW_{s_{r-1}}^{i_1}$. Note that the Stratonovich integrals are evaluated innermost first. Finally let

$$I_\alpha(t) := \int_0^t \int_0^{s_0} \cdots \left(\int_0^{s_{r-2}} 1 \circ dW_{s_{r-1}}^{i_1} \right) \circ \cdots \circ dW_{s_1}^{i_{r-1}} \circ dW_{s_0}^{i_r}.$$

Let $\alpha = (i_1, \ldots, i_r) \in \mathcal{A}_0$ be an arbitrary multi-index such that $\|\alpha\| = m \in \mathbb{N}$ (and $|\alpha| = r \in \mathbb{N}$). If m is odd, then $\mathbb{E}[I_\alpha(t)] = 0$ and if m is even then

$$\mathbb{E}[I_\alpha(t)] = \begin{cases} \frac{t^{\frac{m}{2}}}{2^{r-\frac{m}{2}}(\frac{m}{2})!} & \text{if } \alpha \in \mathcal{A}_0^{m,r} \\ 0 & \text{otherwise} \end{cases}, \tag{3.1}$$

where $\mathcal{A}_0^{m,r}$ is the set of multi-indices $\alpha = \alpha_1 * \cdots * \alpha_{\frac{m}{2}} \in \mathcal{A}_0(m)$ such that each $\alpha_i = (0)$ or (j,j) for some $j \in \{1, \ldots, k\}$. Note that $r - \frac{m}{2}$ is equal to the number of pairs of indices (j,j) occurring in α. A proof of this result can be found in Ref. [23].

The set of iterated Stratonovich integrals plays a central role in the theory of approximation of solutions of SDEs and there are numerous papers that study its structure. Here, we adopt the hierarchical set approach introduced by Kloeden & Platen [24]. An alternative method can be found in Gaines [25] where it is shown how Lyndon words provide a basis for iterated Stratonovich integrals and also how shuffle products may be used to obtain moments of stochastic integrals. Pettersson [26] gives a notationally and computationally convenient Stratonovich-Taylor expansion. Furthermore, Burrage & Burrage [27] use rooted-tree theory to describe the aforementioned set and Burrage & Burrage [28] presents an approach based on B-series.

We state three further results in (3.2), (3.3) and (3.5). The proofs are all elementary and can be found in Ref. [23]. The first two give an upper bound on the L^2 norm of $I_{f_{\alpha,\varphi}}(t)$ for smooth φ. The third provides an explicit form for the remainder of $\varphi(X_t)$ when expanded in terms of iterated integrals.

For $\varphi \in C_b^{||\alpha||+2}(\mathbb{R}^d)$ and any multi-index $\alpha = (i_1, \dots, i_r) \in \mathcal{A}_0$ such that $i_1 \neq 0$, we have

$$\left\|I_{f_{\alpha,\varphi}}(t)\right\|_2 \leq c_1 \left\|f_{\alpha,\varphi}\right\|_\infty t^{\frac{\|\alpha\|}{2}} + c_2 \sum_{i=1}^{k} \left\|V_i f_{\alpha,\varphi}\right\|_\infty t^{\frac{\|\alpha\|+1}{2}} \tag{3.2}$$

for some constants $c_1 \equiv c_1(\alpha), c_2 \equiv c_2(\alpha) \geq 0$. For $\varphi \in C_b^{||\alpha||}(\mathbb{R}^d)$ and any multi-index $\alpha = (i_1, \dots, i_r) \in \mathcal{A}_0$ such that $i_1 = 0$ we have

$$\left\|I_{f_{\alpha,\varphi}}(t)\right\|_2 \leq c_1 \left\|f_{\alpha,\varphi}\right\|_\infty t^{\frac{\|\alpha\|}{2}}. \tag{3.3}$$

For $m \in \mathbb{N}$, $\varphi \in C_b^{m+3}(\mathbb{R}^d)$ and $x \in \mathbb{R}^d$, we define the truncation,

$$\varphi_t^m(x) := \varphi(x) + \sum_{\alpha \in \mathcal{A}_0(m)} f_{\alpha,\varphi}(x) I_\alpha(t). \tag{3.4}$$

Then for $t \geq 0$ the remainder is

$$R_{m,t,\varphi}(x) := \varphi(X_t) - \varphi_t^m(x) = \Big(\sum_{\|\alpha\|=m+1} + \sum_{\|\alpha\|=m+2, \alpha=0*\beta, \|\beta\|=m}\Big) I_{f_{\alpha,\varphi}}(t). \tag{3.5}$$

In the following, we define a class of approximations of X expressed in terms of certain families of stochastic processes, $\bar{X}(x) = \{\bar{X}_t(x)\}_{t\in[0,\infty)}$ for $x \in \mathbb{R}^d$, which are explicitly solvable. In particular, we can explicitly compute the operator,

$$(Q_t\varphi)(x) = \mathbb{E}[\varphi(\bar{X}_t(x))]. \tag{3.6}$$

The semigroup P_T will then be approximated by $Q_{h_n}^m Q_{h_{n-1}}^m \cdots Q_{h_1}^m$ where $\{h_j := t_j - t_{j-1}\}_{j=1}^n$ and $\pi_n = \{t_j := (\frac{j}{n})^\gamma T\}_{j=0}^n$ for $n \in \mathbb{N}$, is a sufficiently fine partition of the interval $[0, T]$. In particular $h_j \in [0, 1)$ for $j = 1, ..., n$. The underlying idea is that $Q_t\varphi$ will have the same truncation as $P_t\varphi$.

So let $\bar{X}(x) = \{\bar{X}_t(x)\}_{t\in[0,\infty)}$, where $x \in \mathbb{R}^d$, be a family of progressively measurable stochastic processes such that, $\lim_{y\to x_0} \bar{X}_t(y) = \bar{X}_t(x_0)$ $\mathbb{P}-$almost surely, for any $t \geq 0$ and $x_0 \in \mathbb{R}^d$. As a result, the operator Q_t defined in (3.6) has the property that $Q_t\varphi \in C_b(\mathbb{R}^d)$ for any $\varphi \in C_b(\mathbb{R}^d)$. In particular, $Q_t : C_b(\mathbb{R}^d) \to C_b(\mathbb{R}^d)$ is a Markov operator.

Definition 3.1. For $m \in \mathbb{N}$, the family $\bar{X}(x) = \{\bar{X}_t(x)\}_{t\in[0,\infty)}$ where $x \in \mathbb{R}^d$, is said to be **m-perfect** for the process X if there exist constants $C > 0$ and $M \geq m+1$ such that for $\varphi \in C_b^{V,M}(\mathbb{R}^d)$,

$$\sup_{x\in\mathbb{R}^d} |Q_t\varphi(x) - \mathbb{E}[\varphi_t^m(x)]| \leq C \sum_{i=m+1}^{M} t^{i/2} \|\varphi\|_{V,i}. \tag{3.7}$$

As we can see from (3.7), the quantity $\mathbb{E}[\varphi_t^m(x)]$ plays the same role as the classical truncation in the standard Taylor expansion of a function. Using (3.1) we deduce that,

$$\begin{aligned}
\mathbb{E}[\varphi_t^0(x)] &= \varphi(x) \\
\mathbb{E}[\varphi_t^2(x)] &= \varphi(x) + V_0\varphi(x)t + \sum_{i=1}^k V_i^2\varphi(x)\frac{t}{2} = \varphi(x) + L\varphi(x)t \\
\mathbb{E}[\varphi_t^4(x)] &= \mathbb{E}[\varphi_t^2(x)] + \sum_{i=1}^k V_0^2\varphi(x)\frac{t^2}{2} + \sum_{i=1}^k V_0V_i^2\varphi(x)\frac{t^2}{4} \\
&\quad + \sum_{i=1}^k V_i^2V_0\varphi(x)\frac{t^2}{4} + \sum_{i,j=1}^k V_j^2V_i^2\varphi(x)\frac{t^2}{8} \\
&= \varphi(x) + L\varphi(x)t + L^2\varphi(x)\frac{t^2}{2},
\end{aligned}$$

where $L = V_0 + \frac{1}{2}\sum_{i=1}^k V_i^2$. Furthermore, since $\mathbb{E}[I_\alpha(t)] = 0$ for odd $\|\alpha\|$, it follows that $\mathbb{E}[\varphi_t^1(x)] = \mathbb{E}[\varphi_t^0(x)], \mathbb{E}[\varphi_t^3(x)] = \mathbb{E}[\varphi_t^2(x)]$ and $\mathbb{E}[\varphi_t^5(x)] = \mathbb{E}[\varphi_t^4(x)]$.

There now follow some examples of $m-$perfect families corresponding to $\{P_t\}_{t\in[0,\infty)}$ as described in (1.2), the Lyons-Victoir method and the Ninomiya-Victoir algorithm.

Example 3.2. The family of stochastic processes $\{X_t(x)\}_{t\in[0,\infty)}$, where $x \in \mathbb{R}^d$, is $m-$perfect. More precisely there exists a constant $c_3 > 0$ such that for $\varphi \in C_b^{V,m+2}(\mathbb{R}^d)$,

$$\sup_x |P_t\varphi(x) - \mathbb{E}[\varphi_t^m(x)]| \le c_3 \sum_{i=m+1}^{m+2} t^{i/2} \|\varphi\|_{V,i}, \tag{3.8}$$

Proof. For $\varphi \in C_b^{V,m+3}(\mathbb{R}^d)$,

$$|P_t\varphi(x) - \mathbb{E}[\varphi_t^m(x)]| = |\mathbb{E}[R_{m,t,\varphi}(x)]| = \left|\mathbb{E}[(\sum_{\|\alpha\|=m+1} + \sum_{\|\alpha\|=m+2,\alpha=0*\beta,\|\beta\|=m})I_{f_{\alpha,\varphi}}(t)]\right|$$

Applying inequality (3.2) to the first sum,

$$\begin{aligned}
\sum_{\|\alpha\|=m+1} \left\|I_{f_{\alpha,\varphi}}(t)\right\|_2 &\le \sum_{\|\alpha\|=m+1} \{c_1(\alpha)\|f_{\alpha,\varphi}\|_\infty t^{\frac{m+1}{2}} + c_2(\alpha)\sum_{i=1}^k \|V_i f_{\alpha,\varphi}\|_\infty t^{\frac{m+2}{2}}\} \\
&\le c_4 \sum_{i=m+1}^{m+2} t^{i/2} \|\varphi\|_{V,i}
\end{aligned} \tag{3.9}$$

for some constant $c_4 > 0$. Applying result (3.3) to the second sum,

$$\begin{aligned}
\sum_{\|\alpha\|=m+2,\alpha=0*\beta,\|\beta\|=m} \left\|I_{f_{\alpha,\varphi}}(t)\right\|_2 &\le \sum_{\|\alpha\|=m+2,\alpha=0*\beta,\|\beta\|=m} c_1(\alpha)\|f_{\alpha,\psi}\|_\infty t \\
&\le c_5 \|\varphi\|_{V,m+2} t^{\frac{m+2}{2}}
\end{aligned} \tag{3.10}$$

for some $c_5 > 0$. The result for $\varphi \in C_b^{V,m+3}(\mathbb{R}^d)$ follows from combining (3.9) and (3.10). Since none of the terms in (3.8) depend on partial derivatives of order $m+3$, the inequality is also valid for any $\varphi \in C_b^{V,m+2}(\mathbb{R}^d)$ (a standard approximation method can be used). □

In the following example, the family of processes $\bar{X}(x) = \{\bar{X}_t(x)\}_{t\in[0,1]}$, where $x \in \mathbb{R}^d$, corresponds to the Lyons-Victoir approximation (see Ref. [29]). The example involves a set of l finite variation paths, $\omega_1, \ldots, \omega_l \in C_0^0([0,1], \mathbb{R}^k)$, for some $l \in \mathbb{N}$, together with some weights $\lambda_1, \ldots, \lambda_l \in \mathbb{R}^+$ such that $\sum_{j=1}^{l} \lambda_j = 1$. These paths are said to define a **cubature formula on Wiener Space of degree m** if, for any $\alpha \in \mathcal{A}_0(m)$,

$$\mathbb{E}[I_\alpha(1)] = \sum_{j=1}^{l} \lambda_j I_\alpha^{\omega_j}(1)$$

where,

$$I_{(i_1,\ldots,i_r)}^{\omega_j}(1) := \int_0^1 \int_0^{s_0} \cdots \left(\int_0^{s_{r-2}} d\omega_j^{i_1}(s_{r-1}) \right) \cdots d\omega_j^{i_{r-1}}(s_1) d\omega_j^{i_r}(s_0).$$

From the scaling properties of the Brownian motion we can deduce, for $t \geq 0$,

$$\mathbb{E}[I_\alpha(t)] = \sum_{j=1}^{l} \lambda_j I_\alpha^{\omega_{t,j}}(t)$$

where $\omega_{t,1}, \ldots, \omega_{t,l} \in C_0^0([0,t], \mathbb{R}^k)$ is defined by $\omega_{t,j}(s) = \sqrt{t}\omega_j\left(\frac{s}{t}\right)$, $s \in [0,t]$. In other words, the expectation of the iterated Stratonovich integrals $I_\alpha(t)$ is the same under the Wiener measure as it is under the measure,

$$\mathbb{Q}_t := \sum_{j=1}^{l} \lambda_j \delta_{\omega_{t,j}}.$$

Example 3.3. If we choose $\bar{X}$ to satisfy the evolution equation (2.1) but with the driving Brownian motion replaced by the paths $\omega_{t,1}, \ldots, \omega_{t,l}$ defined above then the family of processes, $\{\overline{X}_t(x)\}_{t\in[0,1]}$, with corresponding operator $(\overline{Q}_t\varphi)(x) := \mathbb{E}_{\mathbb{Q}_t}[\varphi(\overline{X}_t(x))]$, is m-perfect. More precisely, there exists a constant $c_6 > 0$ such that for $\varphi \in C_b^{V,m+2}(\mathbb{R}^d)$,

$$\sup_x \left| \overline{Q}_t\varphi(x) - \mathbb{E}[\varphi_t^m(x)] \right| \leq c_6 \sum_{i=m+1}^{m+2} t^{i/2} \|\varphi\|_{V,i}$$

For example, if $(\lambda_j, \omega_{t,j})$ are chosen such that for $l = 2^k$ the paths are $\omega_{t,j} : t \mapsto t(1, z_j^1, .., z_j^k)$ for $j = 1, \ldots, 2^k$ with points $z_j \in \{-1,1\}^k$ and weights $\lambda_j = 2^{-k}$, we obtain a cubature formula of degree 3 and a corresponding 3-perfect family.

Proof. Let us first observe that $I_\alpha^{\omega_{t,j}}(t) = t^{\frac{|\alpha|}{2}} I_\alpha^{\omega_j}(1)$ Hence, for $\varphi \in C_b^{V,m+2}(\mathbb{R}^d)$,

$$\begin{aligned} \left|\overline{Q}_t\varphi(x) - \mathbb{E}[\varphi_t^m(x)]\right| &= |\mathbb{E}_{\mathbb{Q}_t}[R_{m,t,\varphi}(x)]| \\ &\leq \Big(\sum_{\|\alpha\|=m+1} + \sum_{\|\alpha\|=m+2, \alpha=0*\beta, \|\beta\|=m} \Big) \|f_{\alpha,\varphi}\|_\infty \|\mathbb{E}_{\mathbb{Q}_t}[I_\alpha(t)]\|_2 \\ &\leq \Big(\sum_{\|\alpha\|=m+1} + \sum_{\|\alpha\|=m+2, \alpha=0*\beta, \|\beta\|=m} \Big) \|f_{\alpha,\varphi}\|_\infty \sum_{j=1}^{l} \lambda_j \|I_\alpha^{\omega_{t,j}}(t)\|_2 \\ &\leq \Big(\sum_{\|\alpha\|=m+1} + \sum_{\|\alpha\|=m+2, \alpha=0*\beta, \|\beta\|=m} \Big) k_\alpha \|f_{\alpha,\varphi}\|_\infty t^{\frac{|\alpha|}{2}}. \end{aligned}$$

where $k_\alpha = \sum_{j=1}^{l} \lambda_j \left\|I_\alpha^{\omega_j}(1)\right\|_2$.

□

Remarks

(i) There has been no change to the underlying measure in the example above. Merely a representation in terms of the measure $\mathbb{Q}_t$ has been introduced to ease the computation of $\overline{Q}_t$. More precisely, the family of processes $\{\overline{X}_t(x)\}_{t\in[0,1]}$ where $x \in \mathbb{R}^d$ is constructed as follows. We take,

$$\overline{X}_0(x) = x$$

and then randomly choose a path $\omega_{t,r}$ from the set $\{\omega_{t,1}, \ldots, \omega_{t,l}\}$ with corresponding probabilities $(\lambda_1, \ldots, \lambda_l)$. Each process then follows a deterministic trajectory driven by the solution of the ordinary differential equation,

$$d\overline{X}_t = V_0(\overline{X}_t)dt + \sum_{j=1}^{k} V_j(\overline{X}_t)d\omega_{t,k}^j$$

for some $V_0, \ldots, V_k \in C_b^\infty(\mathbb{R}^d, \mathbb{R}^d)$ as in (2.1). We can therefore compute the expected values of a functional of $X_t(x)$ as integrals on the path space with respect to the Radon measure $\mathbb{Q}_t$. Hence the identities,

$$\overline{Q}_t\varphi(x) = \mathbb{E}\left[\varphi(\overline{X}_t(x))\right] = \mathbb{E}_{\mathbb{Q}_t}\left[\varphi(\overline{X}_t(x))\right]$$

(ii) The approach adopted by Lyons and Victoir to construct the above approximation resembles the ideas developed by Clark and Newton in a series of papers ([30], [31], [32], [33]). Heuristically, Clark and Newton constructed strong approximations of SDEs using flows driven by vector fields which were measurable with respect to the filtration generated by the driving Wiener process. In a similar vein, Castell & Gaines [34] provide a method of strongly approximating the solution of an SDE by means of exponential Lie series.

For the following example, we will denote by $\exp(Vt)f$ the value at time t of the solution of the ODE $y' = V(y)$, $y(0) = f$ where $V \in C_b^\infty(\mathbb{R}^d, \mathbb{R}^d)$. In particular,

$\exp(Vt)(x)$ is $\exp(Vt)f$ for f being the identity function. The family of processes $Y(x) = \{Y_t(x)\}_{t\in[0,1]}$ below corresponds to the Ninomiya-Victoir approximation (see Ref. [20]).

Example 3.4. Let Λ and Z be two independent random variables such that Λ is Bernoulli distributed $\mathbb{P}(\Lambda = 1) = \mathbb{P}(\Lambda = -1) = \frac{1}{2}$ and $Z = (Z^i)_{i=1}^k$ is a standard normal $k-$ dimensional random variable. Consider the family of processes $Y(x) = \{Y_t(x)\}_{t\in[0,1]}$ defined by

$$Y_t(x) = \begin{cases} \exp(\frac{V_0}{2}t)\prod_{i=1}^{k}\exp(Z^iV_it^{1/2})\exp(\frac{V_0}{2}t)(x) & \text{if } \Lambda = 1 \\ \exp(\frac{V_0}{2}t)\prod_{i=1}^{k}\exp(Z^{k+1-i}V_{k+1-i}t^{1/2})\exp(\frac{V_0}{2}t)(x) & \text{if } \Lambda = -1 \end{cases}$$

with the corresponding operator $(Q_t\varphi)(x) := \mathbb{E}[\varphi(Y_t(x))]$. Then there exists a constant $c_7 > 0$ such that for $\varphi \in C_b^{V,8}(\mathbb{R}^d)$

$$\sup_x \left|\overline{Q}_t\varphi(x) - \mathbb{E}[\varphi_t^5(x)]\right| \leq c_7 t^3 \left\|\varphi\right\|_{V,6}$$

Hence $\{Y_t(x)\}_{t\in[0,1]}$ is 5-perfect.

Proof. We first consider the case $\Lambda = 1$. Let $\{Y^i\}_{i=0}^{k+1}$ be defined,

$$\begin{aligned} \frac{d\varphi}{ds}(Y_s^0) &= V_0\varphi(Y_s^0) \text{ for } s \in [0, \frac{t}{2}], Y_0^0 = x \\ \frac{d\varphi}{ds}(Y_s^1) &= Z^1V_1\varphi(Y_s^1) \text{ for } s \in [0, \sqrt{t}], Y_0^1 = Y_{\frac{t}{2}}^0 \\ \frac{d\varphi}{ds}(Y_s^i) &= Z^iV_i\varphi(Y_s^i) \text{ for } s \in [0, \sqrt{t}], Y_0^i = Y_{\sqrt{t}}^{i-1},\ i = 2, \ldots, k \\ \frac{d\varphi}{ds}(Y_s^{k+1}) &= V_0\varphi(Y_s^{k+1}) \text{ for } s \in [0, \frac{t}{2}], Y_0^{k+1} = Y_{\sqrt{t}}^k \end{aligned}$$

It follows from the definition of the algorithm and by Itô's Formula that,

$$\varphi(Y_{\frac{t}{2}}^{k+1}) = \varphi(Y_{\sqrt{t}}^k) + V_0\varphi(Y_{\sqrt{t}}^k)\frac{t}{2} + V_0^2\varphi(Y_{\sqrt{t}}^k)\frac{t^2}{8} + \int_0^{t/2}\int_0^{s_1}\int_0^{s_2} V_0^3\varphi(Y_{s_3}^{k+1})ds_3ds_2ds_1. \tag{3.11}$$

We need to expand the right hand side of (3.11) and divide the resulting expansion into two parts: the required truncation and a remainder whose expected value should be bounded by $Ct^3\left\|\varphi\right\|_{V,6}$. The final term in (3.11) belongs to the remainder and indeed,

$$\mathbb{E}\left[\left|\int_0^{t/2}\int_0^{s_1}\int_0^{s_2} V_0^3\varphi(Y_{s_3}^{k+1})ds_3ds_2ds_1\right|\right] \leq \left\|V_0^3\varphi\right\|_\infty \frac{(t/2)^3}{6} \leq \frac{t^3}{48}\left\|\varphi\right\|_{V,6}. \tag{3.12}$$

Expanding the third term in (3.11),

$$\begin{aligned}
V_0^2\varphi(Y^k_{\sqrt{t}}) &= V_0^2\varphi(Y^{k-1}_{\sqrt{t}}) + \int_0^{\sqrt{t}} Z^k V_k V_0^2\varphi(Y^k_s)ds \\
&= V_0^2\varphi(Y^{k-1}_{\sqrt{t}}) + Z^k V_k V_0^2\varphi(Y^{k-1}_{\sqrt{t}})\sqrt{t} + \int_0^{\sqrt{t}}\int_0^{s_1} \left(Z^k\right)^2 V_k^2 V_0^2\varphi(Y^k_{s_2})ds_2 ds_1 \\
&= \ldots \\
&= V_0^2\varphi(Y^0_{\frac{t}{2}}) + Z^1 V_1 V_0^2\varphi(Y^0_{\frac{t}{2}})\sqrt{t} + \sum_{i=2}^{k} Z^i V_i V_0^2\varphi(Y^{i-1}_{\sqrt{t}})\sqrt{t} \\
&\quad + \sum_{i=1}^{k}\int_0^{\sqrt{t}}\int_0^{s_1} \left(Z^i\right)^2 V_i^2 V_0^2\varphi(Y^i_{s_2})ds_2 ds_1
\end{aligned}$$

So

$$\begin{aligned}
V_0^2\varphi(Y^k_{\sqrt{t}}) &= V_0^2\varphi(x) + \int_0^{\frac{t}{2}} V_0^3\varphi(Y^0_s)ds + Z^1 V_1 V_0^2\varphi(Y^0_{\frac{t}{2}})\sqrt{t} + \sum_{i=2}^{k} Z^i V_i V_0^2\varphi(Y^i_{\sqrt{t}})\sqrt{t} \\
&\quad + \sum_{i=1}^{k}\int_0^{\sqrt{t}}\int_0^{s_1} \left(Z^i\right)^2 V_i^2 V_0^2\varphi(Y^i_{s_2})ds_2 ds_1.
\end{aligned}$$

Now the last four terms are all $O(t)$ since $\mathbb{E}\left[Z^1 V_1 V_0^2\varphi(Y^0_{\frac{t}{2}})\sqrt{t}\right] = \mathbb{E}\left[Z^i V_i V_0^2\varphi(Y^{i-1}_{\sqrt{t}})\sqrt{t}\right] = 0$ because Z is normal, $\mathbb{E}\left[\left|\int_0^{\frac{t}{2}} V_0^3\varphi(Y^0_s)ds\right|\right] \le \left\|V_0^3\varphi\right\|_\infty \frac{t}{2}$ and finally $\mathbb{E}\left[\left|\int_0^{\sqrt{t}}\int_0^{s_1} \left(Z^i\right)^2 V_i^2 V_0^2\varphi(Y^i_{s_2})ds_2 ds_1\right|\right] \le \left\|V_i^2 V_0^2\varphi\right\|_\infty \frac{t}{2}$.

So for the third term in (3.11) we have established,

$$\begin{aligned}
\left|E\left[V_0^2\varphi(Y^k_{\sqrt{t}})\frac{t^2}{8}\right] - V_0^2\varphi(x)\frac{t^2}{8}\right| &\le \frac{t^3}{16}\left(\left\|V_0^3\varphi\right\|_\infty + \sum_{i=1}^{k}\left\|V_i^2 V_0^2\varphi\right\|_\infty\right) \\
&\le \frac{t^3}{16}\left\|\varphi\right\|_{V,6}. \qquad (3.13)
\end{aligned}$$

Similarly, for the second term on the RHS of (3.11),

$$\left|E\left[V_0\varphi(Y^k_{\sqrt{t}})\frac{t}{2}\right] - \left(V_0\varphi(x) + V_0^2\varphi(x)\frac{t}{2} + \sum_{i=1}^{k} V_i^2 V_0\varphi(x)\frac{t}{2}\right)\frac{t}{2}\right| \le \frac{t^3}{8}\left\|\varphi\right\|_{V,6} \tag{3.14}$$

Finally, for the first term on the RHS of (3.11),

$$\left|E\left[\varphi(Y^k_{\sqrt{t}})\right] - \left(\begin{array}{c}\varphi(x) + V_0\varphi(x)\frac{t}{2} + V_0^2\varphi(x)\frac{t^2}{8} + \sum_{i=1}^{k} V_i^4\varphi(x)\frac{t^2}{4!} \\ + \sum_{i=1}^{k}\left(V_i^2\varphi(x)\frac{t}{2} + V_0 V_i^2\varphi(x)\frac{t^2}{4}\right) + \sum_{i=2}^{k}\sum_{j=1}^{i-1} V_j^2 V_i^2\varphi(x)\frac{t^2}{4}\end{array}\right)\right|$$

$$\le \frac{t^3}{16}\left\|\varphi\right\|_{V,6} \tag{3.15}$$

Substituting (3.12), (3.13), (3.14) and (3.15) in (3.11) gives the bound for the case $\Lambda = 1$. An analogous bound is then established for the case $\Lambda = -1$. The final result is obtained by taking the average of the two cases. See [23] for details. □

The following Lemma is required to prove the main theorem below.

Lemma 3.5. *For* $0 < s \le t \le 1$ *and any* $m-$*perfect family* $\{\overline{X}_t(x)\}_{t\in(0,1]}$ *with corresponding operator* $Q = \{Q_t\}_{t\in(0,1]}$ *we have,*

$$\|P_t(P_s\varphi) - Q_t(P_s\varphi)\|_\infty \le c_8 \|\varphi\|_p \sum_{j=m+1}^{M} \frac{t^{j/2}}{s^{\frac{j-p}{2}}}, \tag{3.16}$$

where $\varphi \in C_b^p(\mathbb{R}^d)$ *for* $0 \le p < \infty$ *and some constant* $c_8 > 0$. *In particular, for* $\varphi \in C_b^M(\mathbb{R}^d)$,

$$\|P_t(P_s\varphi) - Q_t(P_s\varphi)\|_\infty \le c_8 \|\varphi\|_p t^{\frac{m+1}{2}}. \tag{3.17}$$

Proof. Since $C_b^\infty(\mathbb{R}^d)$ is dense in $C_b^p(\mathbb{R}^d)$ in the topology generated by the norm $||\cdot||_{p,\infty}$ it suffices to prove (3.16) and (3.17) only for a function $\varphi \in C_b^\infty(\mathbb{R}^d)$. By Corollary 5.4 in Section 5,

$$\begin{aligned}\|P_t\varphi\|_{V,j} &= \sum_{i=1}^{j} \sum_{\substack{\alpha_1,\dots,\alpha_i\in\mathcal{A}_0 \\ \|\alpha_1*\dots*\alpha_i\|=j}} \|V_{[\alpha_1]}\cdots V_{[\alpha_i]}P_t\varphi\|_\infty \\ &\le \sum_{i=1}^{j} \sum_{\substack{\alpha_1,\dots,\alpha_i\in\mathcal{A}_0 \\ \|\alpha_1*\dots*\alpha_i\|=j}} \frac{C_p^T \|\varphi\|_p}{t^{(\|\alpha_1*\dots*\alpha_i\|-p)/2}} \le \frac{c_9 \|\varphi\|_p}{t^{\frac{j-p}{2}}}\end{aligned}$$

for some $c_9 \equiv c_9(j,p) \ge 0$. Then (3.16) and (3.17) follow from the definition of an $m-$perfect family. □

The family of processes $\bar{X}(x) = \{\bar{X}_t(x)\}_{t\in[0,\infty)}$ below corresponds to the Kusuoka approximation. We recall that Kusuoka's result requires only the UFG condition.

Example 3.6. A family of random variables $\{Z_\alpha : \alpha \in \mathcal{A}_0\}$ is said to be **m-moment** similar if $\mathbb{E}[\,|Z_\alpha|^r] < \infty$ for any $r \in \mathbb{N}$, $\alpha \in \mathcal{A}_0$ and $Z_{(0)} = 1$ with,

$$\mathbb{E}[Z_{\alpha_1}\dots Z_{\alpha_j}] = \mathbb{E}[I_{\alpha_1}\dots I_{\alpha_j}]$$

for any $j = 1,\dots,m$ and $\alpha_1,\dots,\alpha_j \in \mathcal{A}_0$ such that $\|\alpha_1\| + \cdots + \|\alpha_j\| \le m$ where I_α is defined as above.

Let $\{Z_\alpha : \alpha \in \mathcal{A}_0\}$ be a family of $m-$moment similar random variables and let $\bar{X}(x) = \{\bar{X}_t(x)\}_{t\in[0,\infty)}$ be the family of processes,

$$\bar{X}_t(x) = \sum_{j=0}^{m} \frac{1}{j!} \sum_{\substack{\alpha_1,\dots,\alpha_j\in\mathcal{A}_0, \\ \|\alpha_1\|+\dots+\|\alpha_j\|\le m}} t^{\frac{\|\alpha_1\|+\dots+\|\alpha_j\|}{2}} (P^0_{\alpha_1}\dots P^0_{\alpha_j})(V_{[\alpha_1]}\dots V_{[\alpha_j]}H)(x) \tag{3.18}$$

where $H : \mathbb{R}^d \to \mathbb{R}^d$ is defined $H(x) = x$ and

$$P_\alpha^0 := |\alpha|^{-1} \sum_{j=0}^{|\alpha|} \frac{(-1)^{j+1}}{j} \sum_{\beta_1 * \ldots * \beta_j = \alpha} Z_{\beta_1} \ldots Z_{\beta_j}$$

with the corresponding operator $Q = \{Q_t\}_{t \in (0,1]}$ in $C_b(\mathbb{R}^d)$ defined by,

$$Q_t \varphi(x) = \mathbb{E}[\varphi(\bar{X}_t(x))]$$

for $\varphi \in C_b^\infty(\mathbb{R}^d)$ Then,

$$\|P_{t+s}\varphi - Q_t P_s \varphi\|_\infty \leq c_{10} \|\nabla \varphi\|_\infty \sum_{j=m+1}^{m^{m+1}} \frac{t^{j/2}}{s^{\frac{j-1}{2}}} \tag{3.19}$$

for some constant $c_{10} > 0$.

Proof. See Definition 1, Theorem 3 and Lemma 18 in Kusuoka [35] for (3.19). □

The family $\bar{X}(x)$, $x \in \mathbb{R}^d$ as defined in (3.18) is not m-perfect. However, inequality (3.19) is a particular case of (3.16) where $p = 1$ and $M = m^{m+1}$. Since (3.16) is the only result required to obtain (3.20), we deduce from the proof of Theorem 3.7 that (3.20), with $p = 1$, holds for Kusuoka's method as well. Similarly part (ii) of Corollary 3.8 holds for Kusuoka's method.

The set of vector fields appearing in (3.18) belong to the Lie algebra generated by the original vector fields $\{V_0, V_1, ..., V_k\}$. Ben Arous [36] and Burrage & Burrage [27] employ the same set of vector fields to produce *strong* approximations of solutions of SDEs. Notably, the same ideas appear much earlier in Magnus [37], in the context of approximations of the solution of linear (deterministic) differential equations. Castell [38] also gives an explicit formula for the solution of an SDE in terms of Lie brackets and iterated Stratonovich integrals.

We now prove our main result on m-perfect families, the gist of which can be conveyed by the concept of local and global order of an approximation. Local order measures how close an approximation is to the exact solution on a sub-interval of the integration, given an exact initial condition at the start of that subinterval. The global order of an approximation looks at the build up of errors over the entire integration range. The theorem below states that, in the best possible case, the global order of an approximation obtained using an m-perfect family is one less than the local order. More precisely, for a suitable partition, the global error is of order $\frac{m-1}{2}$ whilst the local error is of order $\frac{m+1}{2}$.

Let us define the function,

$$\Upsilon^p(n) = \begin{cases} n^{-\frac{1}{2}\min(\gamma p, (m-1))} & \text{if } \gamma p \neq m-1 \\ n^{-(m-1)/2} \ln n & \text{for } \gamma p = m-1 \end{cases}$$

In the following,

$$\mathcal{E}^{\gamma,n}(\varphi) := \left\| P_T \varphi - Q_{h_n}^m Q_{h_{n-1}}^m \cdots Q_{h_1}^m \varphi \right\|_\infty$$

for $\gamma \in \mathbb{R}$, $n \in \mathbb{N}$.

Theorem 3.7. *Let $T, \gamma > 0$ and $\pi_n = \{t_j = (\frac{j}{n})^\gamma T\}_{j=0}^n$ be a partition of the interval $[0,T]$ where $n \in \mathbb{N}$ is such that $\{h_j = t_j - t_{j-1}\}_{j=1}^n \subseteq (0,1]$. Then for any m-perfect family $\{\overline{X}_t(x)\}_{t\in[0,T]}$ with corresponding operator $Q = \{Q_t\}_{t\in(0,1]}$ we have, for $\varphi \in C_b^p(\mathbb{R}^d)$ where $p = 1, ..., m$,*

$$\mathcal{E}^{\gamma,n}(\varphi) \le c_{11}\Upsilon^p(n)\|\varphi\|_p + \left\|P_{h_1}\varphi - Q_{h_1}^m\varphi\right\|_\infty \tag{3.20}$$

for some constant $c_{11} \equiv c_{11}(\gamma, M, T) > 0$ where $M \ge m+1$, as in Definition 3.1. In particular, if $\gamma \ge \frac{m-1}{p}$ then,

$$\mathcal{E}^{\gamma,n}(\varphi) \le \frac{c_{11}}{n^{\frac{m-1}{2}}}\|\varphi\|_p + \left\|P_{h_1}\varphi - Q_{h_1}^m\varphi\right\|_\infty$$

Proof. We have,

$$\begin{aligned}\mathcal{E}^{\gamma,n}(\varphi) &= P_{h_n}(P_{T-h_n}\varphi) - Q_{h_n}^m(P_{T-h_n}\varphi)\\ &\quad + \sum_{j=1}^{n-1} Q_{h_n}^m \cdots Q_{h_{j+1}}^m(P_{T-h_{j+1}-\cdots-h_n}\varphi - Q_{h_j}^m P_{T-h_j-\cdots-h_n}\varphi)\\ &= P_{h_n}(P_{t_{n-1}}\varphi) - Q_{h_n}^m(P_{t_{n-1}}\varphi)\\ &\quad + \sum_{j=1}^{n-1} Q_{h_n}^m \cdots Q_{h_{j+1}}^m(P_{h_j}(P_{t_{j-1}}\varphi) - Q_{h_j}^m(P_{t_{j-1}}\varphi)).\end{aligned}$$

By Lemma 3.5, there exists a constant $c_8 > 0$ such that,

$$\left\|P_{h_n}(P_{t_{n-1}}\varphi) - Q_{h_n}^m(P_{t_{n-1}}\varphi)\right\|_\infty \le c_8\|\varphi\|_p \sum_{l=m+1}^{M} \frac{h_n^{l/2}}{t_{n-1}^{\frac{l-p}{2}}}.$$

Since P is a semigroup and $Q_{h_j}^m$ is a Markov operator for $j = 2, \ldots, n-1$,

$$\begin{aligned}\left\|Q_{h_n}^m \cdots Q_{h_{j+1}}^m(P_{h_j}(P_{t_{j-1}}\varphi) - Q_{h_j}^m(P_{t_{j-1}}\varphi))\right\|_\infty &\le \left\|P_{h_j}(P_{t_{j-1}}\varphi) - Q_{h_j}^m(P_{t_{j-1}}\varphi)\right\|_\infty\\ &\le c_{12}\|\varphi\|_p \sum_{l=m+1}^{M} \frac{h_j^{l/2}}{t_{j-1}^{\frac{l-p}{2}}}\end{aligned}$$

for some $c_{12} > 0$. Finally, since $Q_{h_j}^m$ is a Markov operator, it follows from (3.24) that,

$$\left\|Q_{h_n}^m \cdots Q_{h_2}^m(P_{h_1}\varphi - Q_{h_1}^m\varphi)\right\|_\infty \le \left\|P_{h_1}\varphi - Q_{h_1}^m\varphi\right\|_\infty.$$

Combining these last four results gives,

$$\mathcal{E}^{\gamma,n}(\varphi) = \left\|P_T\varphi - Q_{h_n}^m \cdots Q_{h_1}^m\varphi\right\|_\infty \le \left\|P_{h_1}\varphi - Q_{h_1}^m\varphi\right\|_\infty + c_{12}\|\varphi\|_p \sum_{j=2}^{n}\sum_{l=m+1}^{M} \frac{h_j^{l/2}}{t_{j-1}^{\frac{l-p}{2}}}.$$

It follows, almost immediately from the definition of h_j that,

$$h_j = \frac{\gamma T(j-1)^{\gamma-1}}{n^\gamma} \int_{j-1}^{j} \left(\frac{u}{j-1}\right)^{\gamma-1} du$$

but for $j \in \{2,\ldots,n\}, (\frac{u}{j-1})^{\gamma-1} \leq \max[(\frac{j}{j-1})^{\gamma-1}, 1] \leq \max[2^{\gamma-1}, 1]$. Hence for $l = m+1,\ldots,M$,

$$\begin{aligned} \frac{h_j^{l/2}}{t_{j-1}^{(l-p)/2}} &\leq \frac{(\frac{\gamma T(j-1)^{\gamma-1}}{n^\gamma} \max[2^{\gamma-1},1])^{l/2}}{\left(\left(\frac{j-1}{n}\right)^\gamma T\right)^{(l-p)/2}} \\ &\leq c_{13}(\frac{T}{n^\gamma})^{\frac{l}{2}-\frac{(l-p)}{2}}(j-1)^{\frac{(\gamma-1)l}{2}-\frac{\gamma(l-p)}{2}} = c_{13}(\frac{T}{n^\gamma})^{\frac{p}{2}}(j-1)^{\frac{\gamma p - l}{2}} \end{aligned}$$

where $c_{13} \equiv c_{13}(\gamma, M) = \max[1, (\gamma \max[2^{\gamma-1},1])^{M/2}]$. It follows that,

$$\sum_{l=m+1}^{M} \frac{h_j^{l/2}}{t_{j-1}^{(l-p)/2}} \leq c_{14}(\frac{1}{n})^{\frac{\gamma p}{2}} \sum_{l=m+1}^{M} (j-1)^{\frac{\gamma p - l}{2}}$$

where $c_{14} \equiv c_{14}(\gamma, M, T) = T^{p/2} c_{13}$ and since $\sum_{l=m+1}^{M}(j-1)^{\frac{\gamma p-l}{2}} = (j-1)^{\frac{\gamma p-(m+1)}{2}}$ $\times \sum_{l=0}^{M-(m+1)}(j-1)^{-\frac{l}{2}} \leq (j-1)^{\frac{\gamma p-(m+1)}{2}} M$ we have,

$$\sum_{l=m+1}^{M} \frac{h_j^{l/2}}{t_{j-1}^{(l-p)/2}} \leq c_{14} M(\frac{1}{n})^{\frac{\gamma p}{2}}(j-1)^{\frac{\gamma p-(m+1)}{2}}. \tag{3.21}$$

We now consider (3.21) for three different ranges of γ.

For $\gamma \in \left(0, \frac{m-1}{p}\right)$, $\sum_{j=2}^{n}(j-1)^{\frac{\gamma p-(m+1)}{2}} \leq \sum_{j=2}^{\infty}(j-1)^{\frac{\gamma p-(m+1)}{2}}$ and since the series on the RHS is convergent, we have,

$$n^{-\frac{\gamma p}{2}} \sum_{j=2}^{n}(j-1)^{\frac{\gamma p-(m+1)}{2}} \leq c_{15} n^{-\frac{\gamma p}{2}}$$

for some constant $c_{15} \equiv c_{15}(\gamma, M) > 0$.

For $\gamma = \frac{m-1}{p}$, $\sum_{j=2}^{n}(j-1)^{-1} \leq c_{16} \ln n$ for some constant $c_{16} \equiv c_{16}(\gamma, M) > 0$ so we have,

$$n^{-\frac{\gamma p}{2}} \sum_{j=2}^{n}(j-1)^{\frac{\gamma p-(m+1)}{2}} \leq c_{16} n^{-\frac{(m-1)}{2}} \ln n$$

For $\gamma > \frac{m-1}{p}, \sum_{j=2}^{n}(\frac{j-1}{n})^{\frac{\gamma p-(m+1)}{2}} \frac{1}{n} \leq c_{17} \int_0^1 x^{\frac{\gamma p-(m+1)}{2}} dx = c_{17} \int_0^1 x^{-1+\frac{\gamma p-(m-1)}{2}} dx$ $\leq c_{18}$ (like a Riemann integral) for some constants $c_{17} \equiv c_{17}(\gamma, M)$, $c_{18} \equiv c_{18}(\gamma, M)$ so,

$$n^{-\frac{\gamma p}{2}} \sum_{j=2}^{n}(j-1)^{\frac{\gamma p-(m+1)}{2}} = n^{-\frac{m-1}{2}} \sum_{j=2}^{n} \left(\frac{j-1}{n}\right)^{\frac{\gamma p-(m+1)}{2}} \frac{1}{n} \leq c_{18} n^{-\frac{m-1}{2}}$$

□

We observe that the rate of convergence is the controlled by the maximum between $\Upsilon(n)$ and the rate at which $\left\|P_{h_1}\varphi - Q_{h_1}^m\varphi\right\|_\infty$ converges to 0. We define $\bar{\Upsilon}^{k_1,k_2}(n) = \Upsilon^{k_1}(n) + n^{-\frac{\gamma k_2}{2}}$. Hence we have the following corollary:

Corollary 3.8.
(i) For any $\varphi \in C_b^M(\mathbb{R}^d)$,

$$\mathcal{E}^{\gamma,n}(\varphi) \leq c_{19}\bar{\Upsilon}^{m+1,m+1}(n)\left\|\varphi\right\|_M.$$

for some constant $c_{19} > 0$. *In particular, if* $\gamma \geq 1$, *then* $\mathcal{E}^{\gamma,n}(\varphi) \leq \frac{c_{19}}{n^{\frac{m-1}{2}}}\left\|\varphi\right\|_M$.
(ii) For any $\varphi \in C_b^1(\mathbb{R}^d)$,

$$\mathcal{E}^{\gamma,n}(\varphi) \leq c_{21}\bar{\Upsilon}^{1,1}(n)\left\|\varphi\right\|_1$$

for some constant $c_{21} > 0$, *if there exists a constant* $c_{20} > 0$ *independent of* t *such that,*

$$\sup_{x\in\mathbb{R}^d}\left|\bar{X}_t(x) - x\right| \leq c_{20}\sqrt{t}. \tag{3.22}$$

In particular, if $\gamma \geq m-1$, *then* $\mathcal{E}^{\gamma,n}(\varphi) \leq \frac{c_{21}}{n^{\frac{m-1}{2}}}\left\|\varphi\right\|_1$.
(iii) For any $\varphi \in C_b^l(\mathbb{R}^d)$ *where* $1 < l < M$,

$$\mathcal{E}^{\gamma,n}(\varphi) \leq c_{24}\bar{\Upsilon}^{l,c_{23}}(n)\left\|\varphi\right\|_l$$

for some constant $c_{24} > 0$, *if there exist two constants* $c_{22}, c_{23} > 0$ *independent of* t *such that,*

$$\left\|P_t\varphi - Q_t^m\varphi\right\|_\infty \leq c_{22}t^{\frac{c_{23}}{2}}\left\|\varphi\right\|_l. \tag{3.23}$$

In particular, if $\gamma \geq m-1$, *then* $\mathcal{E}^{\gamma,n}(\varphi) \leq \frac{c_{24}}{n^{\frac{m-1}{2}}}\left\|\varphi\right\|_l$.

Proof.
(i) The result follows from the theorem and the definition of an $m-$perfect family.
(ii) If $\varphi \in C_b(\mathbb{R}^d)$ is Lipschitz then,

$$\left|Q_t\varphi(x) - \varphi(x)\right| \leq c_{20}\left\|\nabla\varphi\right\|_\infty\sqrt{t} \tag{3.24}$$

hence,

$$\left\|P_{h_1}\varphi - Q_{h_1}^m\varphi\right\|_\infty \leq c_{20}\left\|\varphi\right\|_1\sqrt{t}.$$

(iii) The result follows from the theorem and (3.23). □

Finally we define μ_t to be the law of X_t :

$$\mu_t(\varphi) = E\left[\varphi(X_t)\right] \text{ for } \varphi \in C_b(\mathbb{R}^d).$$

We also define μ_t^N to be the probability measure defined by,

$$\mu_t^N(\varphi) = \mathbb{E}\left[Q_{h_n}^m Q_{h_{n-1}}^m \cdots Q_{h_1}^m\varphi(X_0)\right] = \int_{\mathbb{R}^d} Q_{h_n}^m Q_{h_{n-1}}^m \cdots Q_{h_1}^m\varphi(x)\,\mu_0(dx)$$

for $\varphi \in C_b(\mathbb{R}^d)$ and need to introduce the family of norms on the set of signed measures:

$$|\mu|_l = \sup\left\{|\mu(\varphi)|, \varphi \in C_b^l(\mathbb{R}^d), \|\varphi\|_{l,\infty} \leq 1\right\}, \quad l \geq 1.$$

Obviously, $|\mu|_l \leq |\mu|_{l'}$ if $l \leq l'$. In other words, the higher the value of l, the coarser the norm. We have the following.

Corollary 3.9.
(i) For $l \geq M$, we have $\left|\mu_t - \mu_t^N\right|_l \leq c_{19}\bar{\Upsilon}^{m+1,m+1}(n)$. In particular, if $\gamma \geq 1$, then $\left|\mu_t - \mu_t^N\right|_l \leq \frac{c_{19}}{n^{\frac{m-1}{2}}}$.
(ii) If (3.22) is satisfied then $\left|\mu_t - \mu_t^N\right|_l \leq c_{21}\bar{\Upsilon}^{1,1}(n)$. In particular, if $\gamma \geq m-1$, then $\left|\mu_t - \mu_t^N\right|_l \leq \frac{c_{21}}{n^{\frac{m-1}{2}}}$.
(iii) If (3.23) is satisfied then $\left|\mu_t - \mu_t^N\right|_l \leq c_{24}\bar{\Upsilon}^{l,c_{23}}(n)$. In particular, if $\gamma \geq m-1$, then $\left|\mu_t - \mu_t^N\right|_l \leq \frac{c_{24}}{n^{\frac{m-1}{2}}}$.

Throughout, the constants $c_{19}, c_{21}, c_{23}, c_{24} > 0$ correspond to those found in Corollary 3.8.

Remark We deduce that there is a payoff between the rate of convergence and the coarseness of the norm employed: the finer the norm the slower the rate of convergence. Hence intermediate results such as part *(iii)* of Corollaries 3.8 and 3.9 may prove useful in subsequent applications. The additional constraint (3.23) holds, for example, for the Lyons-Victoir method, as a cubature formula of degree m is also a cubature formula of degree m' for $m' \leq m$. Similarly, it holds for Kusuoka's approximation since an $m-$similar family is also $m'-$similar for any $m' \leq m$.

4. An Application to Filtering

We begin with a short description of the filtering problem. Let (X, Y) be a system of partially observed random processes. The process X satisfies the stochastic differential equation (1.1) and is the unobserved component. The process Y is the observable component and satisfies the evolution equation,

$$Y_t = \int_0^t h(X_s)ds + B_t,$$

where $\{B_t\}_{t\in[0,\infty)}$ is an l-dimensional Brownian motion independent of X and $h = \left(h^i\right)_{i=1}^l \in C_b^\infty\left(\mathbb{R}^d, \mathbb{R}^l\right)$. Let $(\mathcal{Y}_t)_{t\geq 0}$ be the filtration generated by Y, $\mathcal{Y}_t = \sigma\left(Y_s,\ 0 \leq s \leq t\right)$. The problem of stochastic filtering for the partially observed system (X, Y) involves the construction of $\pi_t(\varphi)$, where $\pi = \{\pi_t, t \geq 0\}$ is the conditional distribution of X_t given $\mathcal{Y}_t$ and φ belongs to a suitably large class of functions. If φ is square integrable with respect to the law of X_t then,

$$\pi_t(\varphi) = \mathbb{E}\left[\varphi(X_t)\,|\mathcal{Y}_t\right], \ \mathbb{P} - \text{almost surely}.$$

Using Girsanov's theorem, one can find a new probability measure $\tilde{\mathbb{P}}$ absolutely continuous with respect to $\mathbb{P}$ (and vice versa), so that Y is a Brownian motion under $\tilde{\mathbb{P}}$, independent of X and, almost surely,

$$\pi_t(\varphi) = \frac{\rho_t(\varphi)}{\rho_t(1)}, \tag{4.1}$$

where,

$$\rho_t(\varphi) = \tilde{\mathbb{E}}\left[\varphi(X_t)\exp\left(\sum_{i=1}^{l}\int_0^t h^i(X_s)dY_s^i - \frac{1}{2}\sum_{i=1}^{l}\int_0^t h^i(X_s)^2 ds\right)\bigg|\mathcal{Y}_t\right] \tag{4.2}$$

and $\tilde{\mathbb{E}}$ is the expectation with respect to $\tilde{\mathbb{P}}$. The measure ρ_t is called the *unnormalised conditional distribution* of the signal. The identity (4.1) is called the Kallianpur-Striebel Formula. In the following, we will denote by $||\cdot||_p$, the L^p-norm with respect to the probability measure $\tilde{\mathbb{P}}$, $|\xi|_p = \tilde{\mathbb{E}}\left[|\xi|^p\right]^{\frac{1}{p}}$, for any random variable ξ. The laws of the families $X(x) = \{X_t(x)\}_{t\in[0,\infty)}$, $x \in \mathbb{R}^d$ and $\bar{X}(x) = \{\bar{X}_t(x)\}_{t\in[0,\infty)}$, $x \in \mathbb{R}^d$ are not affected by the change of measure, hence, to avoid working with both $\mathbb{P}$ and $\tilde{\mathbb{P}}$ we can write,

$$(P_t\varphi)(x) = \tilde{\mathbb{E}}[\varphi(X_t(x))], \quad (Q_t\varphi)(x) = \tilde{\mathbb{E}}[\varphi(\bar{X}_t(x))].$$

In the following, we will only consider equidistant partitions and smooth functions. The method of approximation and the results closely follow the application of the classical Euler method as described in Picard [39] and Talay [5].

Let y_r, $r = 1, ..., n$ be the observation process increments $y_r = Y_{\frac{(r+1)t}{n}} - Y_{\frac{rt}{n}}$ and $h_r \in C_b^\infty\left(\mathbb{R}^d\right)$, $r = 0, ..., n-1$, be the (observation dependent) functions defined by $h_r = \sum_{i=1}^{l}(h^i y_r^i - \frac{t}{2n}\left(h^i\right)^2)$. Let $R_s^r,\ \bar{R}_s^r : C_b^\infty\left(\mathbb{R}^d\right) \to C_b^\infty\left(\mathbb{R}^d\right)$, $r = 0, 1, ..., n$ be the following operators,

$$R_s^n\varphi(x) = P_s\varphi(x), \quad \bar{R}_s^n\varphi(x) = Q_s\varphi(x) \text{ for } \varphi \in C_b^\infty\left(\mathbb{R}^d\right), \quad x \in \mathbb{R}^d$$

and, for $r = 0, 1, ..., n-1$, and for $\varphi \in C_b^\infty\left(\mathbb{R}^d\right)$, $x \in \mathbb{R}^d$,

$$\begin{aligned} R_s^r\varphi(x) &= \tilde{\mathbb{E}}\left[\varphi(X_s(x))\exp(h_r(X_s(x)))|\,\mathcal{Y}_s\right] = P_s\varphi^r(x) \\ \bar{R}_s^r\varphi(x) &= \tilde{\mathbb{E}}\left[\varphi\left(\bar{X}_s(x)\right)\exp\left(h_r\left(\bar{X}_s(x)\right)\right)\big|\,\mathcal{Y}_s\right] = Q_s\varphi^r(x), \end{aligned}$$

where $\varphi^r = \varphi\exp(h_r)$ and $s \in [0,1]$.

Firstly, one approximates ρ by replacing the (continuous) observation path with a discrete version. We choose the equidistant partition $\left\{\frac{it}{n},\, i = 0, 1, ...n\right\}$ of the interval $[0,t]$ and consider only the observation data $\{y_r,\, r = 0, 1, ..., n\}$. We define the measure,

$$\begin{aligned} \rho_t^n(\varphi) &= \tilde{\mathbb{E}}\left[\varphi(X_t)\exp\left(\sum_{i=0}^{n-1} h_i\left(X_{\frac{it}{n}}\right)\right)\bigg|\mathcal{Y}_t\right] \\ &= \tilde{\mathbb{E}}\left[R^0_{\frac{t}{n}} R^1_{\frac{t}{n}} ... R^n_{\frac{t}{n}}\varphi(X_0)\Big|\mathcal{Y}_t\right] \text{ for } \varphi \in C_b\left(\mathbb{R}^d\right). \end{aligned} \tag{4.3}$$

Following Theorem 1 from Picard [39], for any $\varphi \in C_b^\infty(\mathbb{R}^d)$ there is a constant $c \equiv c(t,\varphi)$ such that,

$$||\rho_t(\varphi) - \rho_t^n(\varphi)||_2 \leq \frac{c}{n}$$

The next step is to approximate $\prod_{i=0}^n R_{\frac{t}{n}}^i$ with $\prod_{i=0}^n \bar{R}_{\frac{t}{n}}^i$. For this we need to adapt the definition of an $m-$perfect family so that we may use functions which are parametrized by the observation path Y. Let $C_b^{Y,\infty}(\mathbb{R}^d)$ be the set of measurable functions, $f : \mathbb{R}^d \times C([0,T],\mathbb{R}^l) \to \mathbb{R}$ with the following properties:

i. for any $y \in C([0,T],\mathbb{R}^l)$ the function $x \to f(x,y)$ belongs to $C_b^\infty(\mathbb{R}^d)$.
ii. for any multi-index $\alpha \in \mathcal{A}$, any $x \in \mathbb{R}^d$ and $p \geq 1$, $|D_\alpha f(x,Y)|_p < \infty$.
iii. for any multi-index $\alpha \in \mathcal{A}$ and $p \geq 1$, $|||D_\alpha f|||_{p,\infty} = \sup_{x\in\mathbb{R}^d} |D_\alpha f(x,Y)|_p < \infty$.

For $f \in C_b^{Y,\infty}(\mathbb{R}^d)$ we define the norm $|||f|||_p^m = \sum_{\alpha\in\mathcal{A}(m)} |||D_\alpha f|||_{p,\infty}$. We note that if $f : \mathbb{R}^d \times C([0,T],\mathbb{R}^l) \to \mathbb{R}$ is constant in the y variable, then $||D_\alpha f(x,Y)||_p = |D_\alpha f(x,Y)|$ and $|||f|||_p^m = ||f||_\infty + \left\|\nabla^1 f\right\|_\infty + ... \|\nabla^m f\|_\infty$.

We now consider an m-perfect family $\bar{X}(x)$ that satisfies the equivalent of (3.7) extended to functions in $C_b^{Y,\infty}(\mathbb{R}^d)$; more precisely we will assume that for any $f \in C_b^{Y,\infty}(\mathbb{R}^d)$,

$$\left|\left|\left|Q_t f - \tilde{\mathbb{E}}[f_t^m|Y]\right|\right|\right|_{p,\infty} \leq C \sum_{i=m+1}^{M} t^{i/2} |||f|||_p^i, \tag{4.4}$$

for some constants $C > 0$ and $M \geq m+1$, where f_t^m is the truncation defined in (3.4). Note that the original definition (3.7) implies (4.4) due to the inequality $\|\varphi\|_{V,i} \leq C\|\varphi\|_i$ for $i \in \mathbb{N}$ first established in Section 2. Indeed, if $f \in C_b^\infty(\mathbb{R}^d)$, in other words it is constant in the y variable, then (3.7) and (4.4) actually coincide. The original Markov family $X(x)$, the family generated by the Lyons-Victoir method and the one generated by the Ninomiya-Victoir algorithm satisfy (4.4).

The following two lemmas are required to prove the main theorem below.

Lemma 4.1. *Let $\bar{X}(x)$ be an m-perfect family $\bar{X}(x)$ that satisfies (4.4). Then there is a constant $c_{25} = c_{25}(t,m,p) > 0$ such that for any $\varphi \in C_b^\infty(\mathbb{R}^d)$ and $r = 1,...,n$, we have,*

$$\left|\left|\left|\bar{R}_{\frac{t}{n}}^{r-1} R_{\frac{t}{n}}^r R_{\frac{t}{n}}^{r+1} ... R_{\frac{t}{n}}^n \varphi - R_{\frac{t}{n}}^{r-1} R_{\frac{t}{n}}^r R_{\frac{t}{n}}^{r+1} ... R_{\frac{t}{n}}^n \varphi\right|\right|\right|_{p,\infty} \leq c_{25} n^{-(m+1)/2} ||\varphi||_M.$$

Proof. Again, using the variational argument in Friedman [40] p.122 (5.17) one can check that for any $\varphi \in C_b^\infty(\mathbb{R}^d)$ and $r = 1,...,n$ the functions,

$$\exp(h_{r-1}) R_{\frac{t}{n}}^r R_{\frac{t}{n}}^{r+1} ... R_{\frac{t}{n}}^n \varphi \in C_b^{Y,\infty}(\mathbb{R}^d).$$

Moreover there is a constant $c_{26} \equiv c_{26}(M,p)$ such that for any $\varphi \in C_b^\infty(\mathbb{R}^d)$ and $r = 1,...,n$,

$$\left|\left|\left|\exp(h_{r-1}) R^r_{\frac{t}{n}} R^{r+1}_{\frac{t}{n}} ... R^n_{\frac{t}{n}}\varphi\right|\right|\right|_p^M \leq c_{26} \|\varphi\|_M \tag{4.5}$$

Let $\bar{X}(x)$ be an m-perfect family that satisfies (4.4). Then since $X(x)$ also satisfies (4.4), it follows by the triangle inequality that for $s \in [0,1]$,

$$|||Q_s f - P_s f|||_{p,\infty} \leq C s^{(m+1)/2} |||f|||_p^M$$

Hence,

$$\left|\left|\left|Q_{\frac{t}{n}} \exp(h_{r-1}) R^r_{\frac{t}{n}} R^{r+1}_{\frac{t}{n}} ... R^n_{\frac{t}{n}}\varphi - P_{\frac{t}{n}} \exp(h_{r-1}) R^r_{\frac{t}{n}} R^{r+1}_{\frac{t}{n}} ... R^n_{\frac{t}{n}}\varphi\right|\right|\right|_{p,\infty}$$
$$\leq C\left(\frac{t}{n}\right)^{(m+1)/2} \left|\left|\left|\exp(h_{r-1}) R^r_{\frac{t}{n}} R^{r+1}_{\frac{t}{n}} ... R^n_{\frac{t}{n}}\varphi\right|\right|\right|_p^M \tag{4.6}$$

The result now follows from (4.5) and (4.6) and the fact that $\bar{R}^{r-1}_{\frac{t}{n}} R^r_{\frac{t}{n}} R^{r+1}_{\frac{t}{n}} ... R^n_{\frac{t}{n}}\varphi = Q_{\frac{t}{n}} \exp(h_{r-1}) R^r_{\frac{t}{n}} R^{r+1}_{\frac{t}{n}} ... R^n_{\frac{t}{n}}\varphi$ and $R^{r-1}_{\frac{t}{n}} R^r_{\frac{t}{n}} R^{r+1}_{\frac{t}{n}} ... R^n_{\frac{t}{n}}\varphi = P_{\frac{t}{n}} \exp(h_{r-1}) R^r_{\frac{t}{n}} R^{r+1}_{\frac{t}{n}} ... R^n_{\frac{t}{n}}\varphi$. □

Lemma 4.2. *Let $\bar{X}(x)$ be an m-perfect family that satisfies (4.4). Then there is a constant $c_{27} \equiv c_{27}(t,m,p) > 0$ such that for any $\varphi \in C_b^\infty(\mathbb{R}^d)$ we have,*

$$\left|\left|\left|R^0_{\frac{t}{n}} R^1_{\frac{t}{n}} ... R^n_{\frac{t}{n}}\varphi - \bar{R}^0_{\frac{t}{n}} \bar{R}^1_{\frac{t}{n}} ... \bar{R}^n_{\frac{t}{n}}\varphi\right|\right|\right|_{p,\infty} \leq c_{27} n^{-(m-1)/2} \|\varphi\|_M.$$

Proof. Let us observe that,

$$\begin{aligned} R^0_{\frac{t}{n}} R^1_{\frac{t}{n}} ... R^n_{\frac{t}{n}}\varphi - \bar{R}^0_{\frac{t}{n}} \bar{R}^1_{\frac{t}{n}} ... \bar{R}^n_{\frac{t}{n}}\varphi &= R^0_{\frac{t}{n}} R^1_{\frac{t}{n}} ... R^n_{\frac{t}{n}}\varphi - \bar{R}^0_{\frac{t}{n}} R^1_{\frac{t}{n}} ... R^n_{\frac{t}{n}}\varphi \\ &+ \sum_{j=1}^{n-1} \bar{R}^0_{\frac{t}{n}} ... \bar{R}^{j-1}_{\frac{t}{n}} \left(\bar{R}^j_{\frac{t}{n}} R^{j+1}_{\frac{t}{n}} ... R^n_{\frac{t}{n}}\varphi - R^j_{\frac{t}{n}} R^{j+1}_{\frac{t}{n}} ... R^n_{\frac{t}{n}}\varphi\right) \\ &+ \bar{R}^0_{\frac{t}{n}} ... \bar{R}^{n-1}_{\frac{t}{n}} \left(\bar{R}^n_{\frac{t}{n}}\varphi - R^n_{\frac{t}{n}}\varphi\right) \end{aligned}$$

Also note that for $p \geq 1$ and $r = 1,...,n$,

$$\begin{aligned} &\left|\left|\left|\bar{R}^0_{\frac{t}{n}} ... \bar{R}^{j-1}_{\frac{t}{n}} \left(\bar{R}^j_{\frac{t}{n}} R^{j+1}_{\frac{t}{n}} ... R^n_{\frac{t}{n}}\varphi - R^j_{\frac{t}{n}} R^{j+1}_{\frac{t}{n}} ... R^n_{\frac{t}{n}}\varphi\right)\right|\right|\right|_{p,\infty}^p \\ &\leq c_{28} \tilde{\mathbb{E}}\left[\left|\left|\left|\bar{R}^{r-1}_{\frac{t}{n}} R^r_{\frac{t}{n}} R^{r+1}_{\frac{t}{n}} ... R^n_{\frac{t}{n}}\varphi - R^{r-1}_{\frac{t}{n}} R^r_{\frac{t}{n}} R^{r+1}_{\frac{t}{n}} ... R^n_{\frac{t}{n}}\varphi\right|\right|\right|_{p,\infty}^p\right] \\ &\leq c_{28} n^{-(m+1)/2} \|\varphi\|_M. \end{aligned}$$

where $c_{28} \equiv c_{28}(t,m,p) > 0$ is a constant independent of φ and r. The claim follows. □

Finally let us define now the measures,

$$\bar{\rho}^n_t(\varphi) = \tilde{\mathbb{E}}\left[\bar{R}^0_{\frac{t}{n}} \bar{R}^1_{\frac{t}{n}} ... \bar{R}^n_{\frac{t}{n}}\varphi(X_0)\middle| \mathcal{Y}_t\right] \quad \text{for } \varphi \in C_b(\mathbb{R}^d)$$

and let $\bar{\pi}^n_t = \bar{\rho}^n_t / \bar{\rho}^n_t(1)$ be its normalized version.

Theorem 4.3. *Let $\bar{X}(x)$ be an m-perfect family that satisfies (4.4). Then there is a constant $c_{29} \equiv c_{29}(t,m,p) > 0$ such that for any $\varphi \in C_b^\infty(\mathbb{R}^d)$ we have,*

$$||\rho_t^n(\varphi) - \bar{\rho}_t^n(\varphi)||_p \leq c_{29} n^{-(m-1)/2} ||\varphi||_M \cdot$$

Proof. We have for $p \geq 1$,

$$\begin{aligned} ||\rho_t^n(\varphi) - \bar{\rho}_t^n(\varphi)||_p &\leq \tilde{\mathbb{E}}\left[\tilde{\mathbb{E}}\left[\left(R^0_{\frac{t}{n}} R^1_{\frac{t}{n}} \ldots R^n_{\frac{t}{n}} \varphi - \bar{R}^0_{\frac{t}{n}} \bar{R}^1_{\frac{t}{n}} \ldots \bar{R}^n_{\frac{t}{n}} \varphi\right)(X_0)\middle| \mathcal{Y}_t\right]^p\right]^{\frac{1}{p}} \\ &\leq \left|\left|\left|R^0_{\frac{t}{n}} R^1_{\frac{t}{n}} \ldots R^n_{\frac{t}{n}} \varphi - \bar{R}^0_{\frac{t}{n}} \bar{R}^1_{\frac{t}{n}} \ldots \bar{R}^n_{\frac{t}{n}} \varphi\right|\right|\right|_{p,\infty}, \end{aligned}$$

hence the result. □

Corollary 4.4. *Let $\bar{X}(x)$ be an m-perfect family that satisfies (4.4) with $m \geq 3$ and assume that X_0 has all moments finite. Then for any $\varphi \in C_b^\infty(\mathbb{R}^d)$ there is a constant $c_{30} \equiv c_{30}(t,m,p,\varphi) > 0$ such that,*

$$||\pi_t^n(\varphi) - \bar{\pi}_t^n(\varphi)||_p \leq \frac{c_{30}(\varphi)}{n}.$$

Proof. Since,

$$|\pi_t(\varphi) - \bar{\pi}_t^n(\varphi)| \leq \frac{1}{\rho_t(1)} |\rho_t(\varphi) - \bar{\rho}_t^n(\varphi)| + \frac{||\varphi||}{\rho_t(1)} |\rho_t(1) - \bar{\rho}_t^n(1)|$$

the Corollary follows by applying Theorem 4.3 and the fact that $\left|\left|\rho_t(1)^{-1}\right|\right|_p$ is finite for any $p \geq 0$. □

5. Some Auxiliary Results

The main aim of this section is to deduce inequality (5.3). In the following we will adopt the framework of Kusuoka-Stroock [13]. Let $(\Theta, \mathcal{B}, \mathcal{W})$ be the standard Wiener space with continuous paths $\theta : [0,\infty) \to \mathbb{R}^d$ satisfying $\theta(0) = 0$. Then Θ with the topology of uniform convergence on compact intervals is a Polish space. Also let $H \subset \Theta$ be the Hilbert space of absolutely continuous functions $h \in \Theta$ such that $\|h\|_H = (\int_0^\infty |h'(t)|^2\, dt)^{1/2} < \infty$.

Let $W^1(\mathbb{R})$ denote the space of measurable $\Phi : \Theta \to \mathbb{R}$ with the following two properties:

(i) For all $h \in H$, there exists a measurable function $\overline{\Phi}_h : \Theta \to \mathbb{R}$ such that $\overline{\Phi}_h = \Phi$, $\mathcal{W} - a.s.$ and $t \in \mathbb{R} \to \overline{\Phi}_h(\theta + th) \in \mathbb{R}$ is strictly absolutely continuous for all θ.

(ii) There exist a measurable map, $D\Phi : \Theta \to \mathbb{L}(H;\mathbb{R})$ such that, for all $h \in H$ and $\varepsilon > 0$,

$$\lim_{|t|\downarrow 0} \mathcal{W}\left(\theta : \left|\frac{\Phi(\theta + th) - \Phi(t)}{t} - D\Phi(\theta)(h)\right| \geq \varepsilon\right) = 0.$$

On $W^1(\mathbb{R})$ the norm $\|\cdot\|_{q;\mathbb{R}}^{(n)}$ is defined as follows,

$$\|\Phi\|_{q;\mathbb{R}}^{(n)} := \left\| \|\Phi\|_{\mathbb{R}}^{(n)} \right\|_{L^q(\mathcal{W})},$$

where $\|\Phi\|_{\mathbb{R}}^{(n)} = \sum_{m=0}^{n} \|D^m\Phi\|_{\mathcal{H}^m(\mathbb{R})}$ for $q \in [2,\infty)$. Note that $D^0\Phi = \Phi$ with $\|D^0\Phi\|_{\mathcal{H}^0(\mathbb{R})} = |\Phi|$ so $\mathbb{E}[|\Phi|] \leq \|\Phi\|_{q;\mathbb{R}}^{(n)}$. Finally let $\mathcal{G}(\mathcal{L})$ be the set of all $\Phi \in W^1(\mathbb{R})$ to which D and the Ornstein-Uhlenbeck operator $\mathcal{L}$ as defined in Ref. [11] can be applied infinitely often.

Definition 5.1. We say that $f \in \eta_r(\mathbb{R}^d;\mathbb{R})$ for $r \in \mathbb{Z}$ if f is a measurable map from $(0,\infty) \times \mathbb{R}^d \times \Omega$ into $\mathbb{R}$ such that,

(i) $f(t,\cdot,\omega) : \mathbb{R}^d \to \mathbb{R}$ is smooth for each $t \in (0,\infty)$ and $\mathcal{W} - a.e\ \omega \in \Omega$.
(ii) $f(\cdot,x,\cdot) : (0,\infty) \times \Omega \to \mathbb{R}$ is progressively measurable for each $x \in \mathbb{R}^d$.
(iii) $\frac{\partial^\alpha}{\partial x^\alpha} f(t,x,\cdot) \in \mathcal{G}(\mathcal{L})$ and is continuous in $t \in (0,\infty)$ for any $\alpha \in \mathcal{A}_1, x \in \mathbb{R}^d$
(iv) $\sup_{0<t\leq T} \sup_{x\in\mathbb{R}^d} \frac{1}{t^{r/2}} \left\| \frac{\partial^\alpha}{\partial x^\alpha} f(t,x,\omega) \right\|_{q;\mathbb{R}}^{(k)} < \infty$ for any $\alpha \in \mathcal{A}_1$, $k \in \mathbb{N}$, $T > 0$ and $2 \leq q < \infty$.

For a proof of the following Theorem see Kusuoka-Stroock [13], Theorem 2.15 (p. 405).

Theorem 5.2. *(Kusuoka-Stroock) Under the UFG condition, for any $\Phi \in \eta_r(\mathbb{R}^d;\mathbb{R})$, $r \in \mathbb{Z}$ and $\alpha \in \mathcal{A}_1$ there exists $\bar{\Phi}_\alpha \in \eta_{r-\|\alpha\|}(\mathbb{R}^d;\mathbb{R})$ such that,*

$$V_{[\alpha]}\mathbb{E}[\Phi(t,x)f(X_t(x))] = \mathbb{E}[\bar{\Phi}_\alpha(t,x)f(X_t(x))] \tag{5.1}$$

for any $f \in C_b^\infty(\mathbb{R}^d)$, $t > 0$, $x \in \mathbb{R}^d$.

The following immediate extension of Theorem 5.2 holds under the additional condition V0.

Lemma 5.3. *Under the UFG+V0 condition, for any $\Phi \in \eta_r(\mathbb{R}^d;\mathbb{R})$ there exists $\bar{\Phi}_\alpha \in \eta_{r-2}(\mathbb{R}^d;\mathbb{R})$ such that,*

$$V_0\mathbb{E}[\Phi(t,x)f(X_t(x))] = \mathbb{E}[\bar{\Phi}_0(t,x)f(X_t(x))] \tag{5.2}$$

for any $f \in C_b^\infty(\mathbb{R}^d)$, $t > 0$, $x \in \mathbb{R}^d$.

Proof. From Theorem 5.2 we get that for any $\Phi \in \eta_r(\mathbb{R}^d;\mathbb{R})$, $r \in \mathbb{Z}$ and $\beta \in \mathcal{A}_1(2)$ there exists $\bar{\Phi}_\beta \in \eta_{r-\|\beta\|}(\mathbb{R}^d;\mathbb{R})$ such that,

$$V_{[\beta]}\mathbb{E}[\Phi(t,x)f(X_t(x))] = \mathbb{E}[\bar{\Phi}_\beta(t,x)f(X_t(x))]$$

for any $f \in C_b^\infty(\mathbb{R}^d)$, $t > 0$, $x \in \mathbb{R}^d$. Hence,

$$\begin{aligned} V_0\mathbb{E}[\Phi(t,x)f(X_t(x))] &= \sum_{\beta\in\mathcal{A}_1(2)} \varphi_\beta V_{[\beta]}\mathbb{E}[\Phi(t,x)f(X_t(x))] \\ &= \mathbb{E}[\bar{\Phi}_0(t,x)f(X_t(x))] \end{aligned}$$

where $\bar{\Phi}_0(t,x) = \sum_{\beta\in\mathcal{A}_1(2)} \varphi_\beta \bar{\Phi}_\beta(t,x) \in \eta_{r-2}(\mathbb{R}^d;\mathbb{R})$. □

The following Corollary is a generalization of Corollary 2.19 (p.407) in Kusuoka-Stroock [13].

Corollary 5.4. *For any $r \in \mathbb{Z}^+$, $\{\alpha_i \in \mathcal{A}_0\}_{i=1}^r$ and $\|\alpha_1 * \ldots * \alpha_r\|$, there exists a constant $C_p^T > 0$ such that,*

$$\left\|V_{[\alpha_1]}\cdots V_{[\alpha_r]}P_t\varphi\right\|_\infty \le C_p^T t^{(p-\|\alpha_1*\ldots*\alpha_r\|)/2}\left\|\varphi\right\|_p, \qquad t\in[0,T]. \tag{5.3}$$

for $\varphi \in C_b^\infty(\mathbb{R}^d)$ where $\|\cdot\|_p$ for $p \in \mathbb{N}$ is defined in (2.2).

Proof.
For the case $p = 1$, let $Y(x)$ be the matrix valued process $Y_t^{i,j}(x) := \frac{\partial}{\partial x_j}(X^i(x))$ where $i,j = 1,...,d$. (see Ikeda & Watanabe [41], Chapter V, for details). Then for any $i,j = 1,...,d$ we have $Y^{i,j}(t,x) \in \eta_0(\mathbb{R}^d;\mathbb{R})$, in particular,

$$\sup_{t\in[0,T]}\sup_{x\in\mathbb{R}^d}\mathbb{E}\left[\sum_{i=1}^d Y^{i,j}(x)\right] \le c(T). \tag{5.4}$$

Differentiating under the integral sign (see Friedman [40], p.122 (5.17)) gives,

$$\frac{\partial}{\partial x_j}\mathbb{E}[\varphi(X_t(x))] = \sum_{i=1}^d \mathbb{E}\left[\frac{\partial\varphi}{\partial x_i}(X_t(x))Y_t^{i,j}(x)\right]. \tag{5.5}$$

Hence,

$$V_{[\alpha_r]}P_t\varphi(x) = \sum_{i=1}^d \mathbb{E}\left[\frac{\partial\varphi}{\partial x_i}(X_t(x))\Phi^i(t,x)\right]$$

with $\Phi^i(t,x) = \sum_{j=1}^d V_{[\alpha_r]}^j(x)\,Y_t^{i,j}(x) \in \eta_0(\mathbb{R}^d;\mathbb{R})$ for $i = 1,...,d$. Then by Theorem 5.2 & Lemma 5.3 there exists $\bar{\Phi}_\alpha^i \in \eta_{-(\|\alpha_1*\cdots*\alpha_{r-1}\|)}(\mathbb{R}^d;\mathbb{R})$, $i = 1,...,d$ such that, for any $\varphi \in C_b^1(\mathbb{R}^d)$,

$$\begin{aligned} V_{[\alpha_1]}\cdots V_{[\alpha_{r-1}]}V_{[\alpha_r]}P_t\varphi(x) &= \sum_{i=1}^d V_{[\alpha_1]}\cdots V_{[\alpha_{r-1}]}\mathbb{E}\left[\frac{\partial\varphi}{\partial x_i}(X_t(x))\Phi^i(t,x)\right] \\ &= \sum_{i=1}^d \mathbb{E}\left[\bar{\Phi}_\alpha^i(t,x)\frac{\partial\varphi}{\partial x_i}(X_t(x))\right] \end{aligned}$$

for any $\varphi \in C_b^\infty(\mathbb{R}^d)$. Hence,

$$\left|V_{[\alpha_1]}\cdots V_{[\alpha_r]}P_t\varphi(x)\right| \le \|\varphi\|_1 \sum_{i=1}^d \mathbb{E}[|\bar{\Phi}_\alpha^i(t,x)|]$$

and (5.3) follows from the uniform bound (in (t,x)) on the L^1-norm of $\bar{\Phi}_\alpha^i(t,x)$, $i = 1,...,d$.

For $p > 1$ one can obtain an analogue of (5.5) with higher derivatives of $P_t\varphi$ in terms of derivatives of φ and a set of processes analogous to Y. The proof follows in a similar manner.

□

Acknowledgments

The authors wish to thank the anonymous referee for his comments. We would also like to thank Alan Bain for carefully reading the paper and his many pertinent remarks.

References

[1] B. Øksendal, *Stochastic differential equations.* Universitext, (Springer-Verlag, Berlin, 1998), fifth edition. ISBN 3-540-63720-6. An introduction with applications.

[2] P. E. Kloeden and E. Platen, *Numerical solution of stochastic differential equations.* vol. 23, *Applications of Mathematics (New York)*, (Springer-Verlag, Berlin, 1992). ISBN 3-540-54062-8.

[3] G. N. Milstein, A method with second order accuracy for the integration of stochastic differential equations, *Theory Probab. Appl.* **23**(2), 396–401, (1978). ISSN 0040-361x.

[4] K. Burrage, P. M. Burrage, and T. Tian, Numerical methods for strong solutions of stochastic differential equations: an overview, *Proc. R. Soc. Lond. Ser. A Math. Phys. Eng. Sci.* **460**(2041), 373–402, (2004). ISSN 1364-5021. Stochastic analysis with applications to mathematical finance.

[5] D. Talay. Efficient numerical schemes for the approximation of expectations of functionals of the solution of a SDE and applications. In *Filtering and control of random processes (Paris, 1983)*, vol. 61, *Lecture Notes in Control and Inform. Sci.*, pp. 294–313. Springer, Berlin, (1984).

[6] D. Talay, Discrétisation d'une équation différentielle stochastique et calcul approché d'espérances de fonctionnelles de la solution, *RAIRO Modél. Math. Anal. Numér.* **20** (1), 141–179, (1986). ISSN 0764-583X.

[7] G. N. Milstein, Weak approximation of solutions of systems of stochastic differential equations, *Theory Probab. Appl.* **30**(4), 750–766, (1985). ISSN 0040-361X.

[8] D. Talay and L. Tubaro, Expansion of the global error for numerical schemes solving stochastic differential equations, *Stochastic Anal. Appl.* **8**(4), 483–509 (1991), (1990). ISSN 0736-2994.

[9] V. Bally and D. Talay, The law of the Euler scheme for stochastic differential equations. I. Convergence rate of the distribution function, *Probab. Theory Related Fields.* **104**(1), 43–60, (1996). ISSN 0178-8051.

[10] E. Platen and W. Wagner, On a Taylor formula for a class of Itô processes, *Probab. Math. Statist.* **3**(1), 37–51 (1983), (1982). ISSN 0208-4147.

[11] S. Kusuoka and D. Stroock. Applications of the Malliavin calculus. I. In *Stochastic analysis (Katata/Kyoto, 1982)*, vol. 32, *North-Holland Math. Library*, pp. 271–306. North-Holland, Amsterdam, (1984).

[12] S. Kusuoka and D. Stroock, Applications of the Malliavin calculus. II, *J. Fac. Sci. Univ. Tokyo Sect. IA Math.* **32**(1), 1–76, (1985). ISSN 0040-8980.

[13] S. Kusuoka and D. Stroock, Applications of the Malliavin calculus. III, *J. Fac. Sci. Univ. Tokyo Sect. IA Math.* **34**(2), 391–442, (1987). ISSN 0040-8980.

[14] D. S. Jerison and A. Sánchez-Calle, Estimates for the heat kernel for a sum of squares of vector fields, *Indiana Univ. Math. J.* **35**(4), 835–854, (1986). ISSN 0022-2518.

[15] A. Sánchez-Calle, Fundamental solutions and geometry of the sum of squares of vector fields, *Invent. Math.* **78**(1), 143–160, (1984). ISSN 0020-9910.

[16] D. W. Stroock and S. R. S. Varadhan, Diffusion processes with continuous coefficients. I, *Comm. Pure Appl. Math.* **22**, 345–400, (1969). ISSN 0010-3640.

[17] D. W. Stroock and S. R. S. Varadhan. On the support of diffusion processes with applications to the strong maximum principle. In *Proceedings of the Sixth Berkeley Symposium on Mathematical Statistics and Probability (Univ. California, Berkeley, Calif., 1970/1971), Vol. III: Probability theory*, pp. 333–359, Berkeley, Calif., (1972). Univ. California Press.

[18] D. Stroock and S. R. S. Varadhan, On degenerate elliptic-parabolic operators of second order and their associated diffusions, *Comm. Pure Appl. Math.* **25**, 651–713, (1972). ISSN 0010-3640.

[19] A. Millet and M. Sanz-Solé. A simple proof of the support theorem for diffusion processes. In *Séminaire de Probabilités, XXVIII*, vol. 1583, *Lecture Notes in Math.*, pp. 36–48. Springer, Berlin, (1994).

[20] S. Ninomiya and N. Victoir. Weak approximations of stochastic differential equations and application to derivative pricing. `www.crest.fr/pageperso/elie/bachelier\_fichiers/Programme\_Fichiers/oxf%ord2004.pdf`.

[21] S. Kusuoka. Approximation of expectation of diffusion process and mathematical finance. In *Taniguchi Conference on Mathematics Nara '98*, vol. 31, *Adv. Stud. Pure Math.*, pp. 147–165. Math. Soc. Japan, Tokyo, (2001).

[22] S. Kusuoka, Malliavin calculus revisited, *J. Math. Sci. Univ. Tokyo.* **10**(2), 261–277, (2003). ISSN 1340-5705.

[23] S. Ghazali. *The Global Error in Weak Approximations of Stochastic Differential Equations.* PhD thesis, Imperial College London, (2006).

[24] P. E. Kloeden and E. Platen, Stratonovich and Itô stochastic Taylor expansions, *Math. Nachr.* **151**, 33–50, (1991). ISSN 0025-584X.

[25] J. G. Gaines, The algebra of iterated stochastic integrals, *Stochastics Stochastics Rep.* **49**(3-4), 169–179, (1994). ISSN 1045-1129.

[26] R. Pettersson, Stratonovich-Taylor expansion and numerical methods, *Stochastic Anal. Appl.* **10**(5), 603–612, (1992). ISSN 0736-2994.

[27] K. Burrage and P. M. Burrage, High strong order methods for non-commutative stochastic ordinary differential equation systems and the Magnus formula, *Phys. D.* **133**(1-4), 34–48, (1999). ISSN 0167-2789. Predictability: quantifying uncertainty in models of complex phenomena (Los Alamos, NM, 1998).

[28] K. Burrage and P. M. Burrage, Order conditions of stochastic Runge-Kutta methods by *B*-series, *SIAM J. Numer. Anal.* **38**(5), 1626–1646 (electronic), (2000). ISSN 0036-1429.

[29] T. Lyons and N. Victoir, Cubature on Wiener space, *Proc. R. Soc. Lond. Ser. A Math. Phys. Eng. Sci.* **460**(2041), 169–198, (2004). ISSN 1364-5021. Stochastic analysis with applications to mathematical finance.

[30] J. M. C. Clark. An efficient approximation scheme for a class of stochastic differential equations. In *Advances in filtering and optimal stochastic control (Cocoyoc, 1982)*, vol. 42, *Lecture Notes in Control and Inform. Sci.*, pp. 69–78. Springer, Berlin, (1982).

[31] J. M. C. Clark. Asymptotically optimal quadrature formulae for stochastic integrals. In *Proceedings of the 23rd Conference on Decision and Control, Las Vegas, Nevada*, pp. 712–715. IEEE, (1984).

[32] N. J. Newton, An asymptotically efficient difference formula for solving stochastic

differential equations, *Stochastics.* **19**(3), 175–206, (1986). ISSN 0090-9491.

[33] N. J. Newton, An efficient approximation for stochastic differential equations on the partition of symmetrical first passage times, *Stochastics Stochastics Rep.* **29**(2), 227–258, (1990). ISSN 1045-1129.

[34] F. Castell and J. Gaines, An efficient approximation method for stochastic differential equations by means of the exponential Lie series, *Math. Comput. Simulation.* **38**(1-3), 13–19, (1995). ISSN 0378-4754. Probabilités numériques (Paris, 1992).

[35] S. Kusuoka. Approximation of expectation of diffusion processes based on Lie algebra and Malliavin calculus. In *Advances in mathematical economics. Vol. 6*, vol. 6, *Adv. Math. Econ.*, pp. 69–83. Springer, Tokyo, (2004).

[36] G. Ben Arous, Flots et séries de Taylor stochastiques, *Probab. Theory Related Fields.* **81**(1), 29–77, (1989). ISSN 0178-8051.

[37] W. Magnus, On the exponential solution of differential equations for a linear operator, *Comm. Pure Appl. Math.* **7**, 649–673, (1954). ISSN 0010-3640.

[38] F. Castell, Asymptotic expansion of stochastic flows, *Probab. Theory Related Fields.* **96**(2), 225–239, (1993). ISSN 0178-8051.

[39] J. Picard. Approximation of nonlinear filtering problems and order of convergence. In *Filtering and control of random processes (Paris, 1983)*, vol. 61, *Lecture Notes in Control and Inform. Sci.*, pp. 219–236. Springer, Berlin, (1984).

[40] A. Friedman, *Stochastic differential equations and applications. Vol. 1.* (Academic Press [Harcourt Brace Jovanovich Publishers], New York, 1975). Probability and Mathematical Statistics, Vol. 28.

[41] N. Ikeda and S. Watanabe, *Stochastic differential equations and diffusion processes.* vol. 24, *North-Holland Mathematical Library*, (North-Holland Publishing Co., Amsterdam, 1989), second edition. ISBN 0-444-87378-3.

Chapter 9

Flow Properties of Differential Equations Driven by Fractional Brownian Motion

Laurent Decreusefond and David Nualart*

GET, Ecole Nationale Supérieure des Télécommunications
LTCI-UMR 5141, CNRS, 46, rue Barrault, 75634 Paris, France
Laurent.Decreusefond@enst.fr

We prove that solutions of stochastic differential equations driven by fractional Brownian motion for $H > 1/2$ define flows of homeomorphisms on $\mathbb{R}^d$.

Contents

1. Introduction

Suppose that $B^H = \{B_t^H, t \geq 0\}$ is an m-dimensional *fractional Brownian motion* with Hurst parameter $H \in (0, 1)$, defined in a complete probability space $(\Omega, \mathcal{F}, P)$. That is, the components $B^{H,j}$, $j = 1, \ldots, m$, are independent zero mean Gaussian processes with the covariance function

$$R_H(t, s) = \frac{1}{2}\left(t^{2H} + s^{2H} - |t - s|^{2H}\right). \tag{1.1}$$

For $H = \frac{1}{2}$, the process B^H is an m-dimensional ordinary Brownian motion. On the other hand, from (1.1) it follows that

$$E(\left|B_t^{H,j} - B_s^{H,j}\right|^2) = |t - s|^{2H}.$$

As a consequence, the processes $B^{H,j}$ have stationary increments, and for any $\alpha < H$ we can select versions with Hölder continuous trajectories of order α on a compact interval $[0, T]$.

*Department of Mathematics, University of Kansas, 1460 Jayhawk Blvd, Lawrence, Kansas 66045-7523, nualart@math.ku.edu

This process was first studied by Kolmogorov [1] and later by Mandelbrot and Van Ness [2], where a stochastic integral representation in terms of an ordinary Brownian motion was established.

The fractional Brownian motion has the following *self-similar* property: For any constant $a > 0$, the processes $\{a^{-H}B^H_{at}, t \geq 0\}$ and $\{B^H_t, t \geq 0\}$ have the same distribution. For $H = \frac{1}{2}$ the process B^H has independent increments, but for $H \neq \frac{1}{2}$, this property is no longer true. In particular, if $H > \frac{1}{2}$, the fractional Brownian motion has the long range dependence property, which means that for each $j = 1, \ldots, m$

$$\sum_{n=1}^{\infty} \mathrm{Corr}(B^{H,j}_{n+1} - B^{H,j}_n, B^{H,j}_1) = \infty.$$

The self-similar and long range dependence properties make the persistent fractional Brownian motion a convenient model for some input noises in a variety of topics from finance to telecommunication networks, where the Markov property is not required. This fact has motivated the recent development of the stochastic calculus with respect to the fractional Brownian motion. See Refs. [3–5], and the references therein.

In this paper we are interested in stochastic differential equations on $\mathbb{R}^d$ driven by a multi-dimensional fractional Brownian motion with Hurst parameter $H > \frac{1}{2}$, that is, equations of the form

$$X^i_t = X^i_0 + \sum_{j=1}^{m} \int_0^t \sigma^{i,j}(s, X_s) dB^{H,j}_s + \int_0^t b^i(s, X_s) ds, \tag{1.2}$$

$i = 1, \ldots, d$. The stochastic integral appearing in (1.2) is a path-wise Riemann-Stieltjes integral. In fact, under suitable conditions on σ, the processes $\sigma(s, X_s)$ and B^H_s have trajectories which are Hölder continuous of order strictly larger than $\frac{1}{2}$ and we can use the approach introduced by Young [6]. A first result on the existence and uniqueness of a solution for this kind of equations was obtained by Lyons [7], using the notion of p-variation. On the other hand, the theory of rough path analysis introduced by Lyons [7] (see also the monograph by Lyons and Qian [8]), has allowed Coutin and Qian [9] to establish the existence of strong solutions and a Wong-Zakai type approximation limit for the stochastic differential equations of the form (1.2) driven by a fractional Brownian motion with parameter $H > \frac{1}{4}$. In Ref. [9] sufficient conditions on the vector fields b and σ are given to ensure existence and uniqueness of the solution of (1.2) even when vector fields do not commute.

Zähle [10] has introduced a generalized Stieltjes integral using the techniques of fractional calculus. This integral is expressed in terms of fractional derivative operators and it coincides with the Riemann-Stieltjes integral $\int_a^b f dg$, when the functions f and g are Hölder continuous of orders λ and μ, respectively and $\lambda + \mu > 1$. Using this formula for the Riemann-Stieltjes integral, Nualart and Răşcanu [11] have obtained the existence of a unique solution for the stochastic differential equations (1.2) under general conditions on the coefficients.

Later on, Nualart and Saussereau [12] have studied the regularity in the sense of Malliavin Calculus of the solution of Equation (1.2), and they have established the absolute continuity of the law of the random variable X_t under some non-degeneracy conditions on the coefficient σ.

The main result of this paper is the flow and homeomorphic properties of X as a function of the initial condition x. Since the solution of (1.2) is defined path-wise, ordinary (i.e., deterministic) methods are in use here. Namely, we use the estimates found in Ref. [11] and approximate fractional Brownian motion by a sequence of regular processes to prove that the solution $\{X_{rt}(x),\, 0 \le r \le t \le T,\, x \in \mathbb{R}^d\}$ of

$$X_{rt}(x) = x + \int_r^t \sigma\left(s, X_{rs}(x)\right) dB^H(s) + \int_r^t b(s, X_{rs}(x))ds,$$

defines a flow of $\mathbb{R}^d$-homeomorphisms.

The paper is organized as follows. Section 2 contains some preliminaries on fractional calculus. In Section 3 we review some results on the properties of the solution of Equation (1.2) and we establish some continuity estimates as a function of the initial condition and the driven input, which are needed later. Finally in Section 4 we show that Equation (1.2) defines a flow of homeomorphisms.

2. Preliminaries

Let $a, b \in \mathbb{R}$, $a < b$. Let $f \in L^1(a,b)$ and $\alpha > 0$. The left-sided and right-sided fractional Riemann-Liouville integrals of f of order α are defined for almost all $x \in (a,b)$ by

$$I_{a+}^{\alpha} f(x) = \frac{1}{\Gamma(\alpha)} \int_a^x (x-y)^{\alpha-1} f(y)\, dy$$

and

$$I_{b-}^{\alpha} f(x) = \frac{(-1)^{-\alpha}}{\Gamma(\alpha)} \int_x^b (y-x)^{\alpha-1} f(y)\, dy,$$

respectively, where $(-1)^{-\alpha} = e^{-i\pi\alpha}$ and $\Gamma(\alpha) = \int_0^\infty r^{\alpha-1} e^{-r} dr$ is the Euler function. Let $I_{a+}^{\alpha}(L^p)$ (resp. $I_{b-}^{\alpha}(L^p)$) be the image of $L^p(a,b)$ under the action of operator I_{a+}^{α} (resp. I_{b-}^{α}). If $f \in I_{a+}^{\alpha}(L^p)$ (resp. $f \in I_{b-}^{\alpha}(L^p)$) and $0 < \alpha < 1$ then the Weyl derivative

$$D_{a+}^{\alpha} f(x) = \frac{1}{\Gamma(1-\alpha)} \left(\frac{f(x)}{(x-a)^{\alpha}} + \alpha \int_a^x \frac{f(x) - f(y)}{(x-y)^{\alpha+1}} dy \right) 1_{(a,b)}(x) \tag{2.1}$$

$$\left(\text{resp.} \quad D_{b-}^{\alpha} f(x) = \frac{(-1)^{\alpha}}{\Gamma(1-\alpha)} \left(\frac{f(x)}{(b-x)^{\alpha}} + \alpha \int_x^b \frac{f(x) - f(y)}{(y-x)^{\alpha+1}} dy \right) 1_{(a,b)}(x) \right) \tag{2.2}$$

is defined for almost all $x \in (a,b)$.

For any $0 < \lambda \leq 1$, denote by $C^\lambda(0,T)$ the space of λ-Hölder continuous functions $f : [0,T] \to \mathbb{R}$, equipped with the norm $\|f\|_\infty + \|f\|_\lambda$, where

$$\|f\|_\lambda = \sup_{0\leq s<t\leq T} \frac{|f(t) - f(s)|}{(t-s)^\lambda}.$$

Recall from Ref. [13] that we have:

- If $\alpha < \frac{1}{p}$ and $q = \frac{p}{1-\alpha p}$ then $I^\alpha_{a+}(L^p) = I^\alpha_{b-}(L^p) \subset L^q(a,b)$.
- If $\alpha > \frac{1}{p}$ then $I^\alpha_{a+}(L^p) \cup I^\alpha_{b-}(L^p) \subset C^{\alpha-\frac{1}{p}}(a,b)$.

The linear spaces $I^\alpha_{a+}(L^p)$ are Banach spaces with respect to the norms

$$\|f\|_{I^\alpha_{a+}(L^p)} = \|f\|_{L^p} + \left\|D^\alpha_{a+}f\right\|_{L^p} \sim \left\|D^\alpha_{a+}f\right\|_{L^p},$$

and the same is true for $I^\alpha_{b-}(L^p)$.

Suppose that $f \in C^\lambda(a,b)$ and $g \in C^\mu(a,b)$ with $\lambda + \mu > 1$. Then, from the classical paper by Young [6], the Riemann-Stieltjes integral $\int_a^b f dg$ exists. The following proposition can be regarded as a fractional integration by parts formula, and provides an explicit expression for the integral $\int_a^b f dg$ in terms of fractional derivatives (see Ref. [10]).

Proposition 2.1. *Suppose that $f \in C^\lambda(a,b)$ and $g \in C^\mu(a,b)$ with $\lambda + \mu > 1$. Let $\lambda > \alpha$ and $\mu > 1-\alpha$. Then the Riemann-Stieltjes integral $\int_a^b f dg$ exists and it can be expressed as*

$$\int_a^b f dg = (-1)^\alpha \int_a^b D^\alpha_{a+} f(t)\, D^{1-\alpha}_{b-} g_{b-}(t)\, dt, \tag{2.3}$$

where $g_{b-}(t) = g(t) - g(b)$.

Zähle [10] introduced a generalized Stieltjes integral of f with respect to g defined by the right-hand side of (2.3), assuming that f and g are functions such that $g(b-)$ exists, $f \in I^\alpha_{a+}(L^p)$ and $g_{b-} \in I^{1-\alpha}_{b-}(L^q)$ for some $p, q \geq 1$, $1/p + 1/q \leq 1$, $0 < \alpha < 1$.

Let $\alpha < \frac{1}{2}$ and $d \in \mathbb{N}^*$. Denote by $W_0^{\alpha,\infty}(0,T;\mathbb{R}^d)$ the space of measurable functions $f : [0,T] \to \mathbb{R}^d$ such that

$$\|f\|_{\alpha,\infty} := \sup_{t\in[0,T]} \left(|f(t)| + \int_0^t \frac{|f(t) - f(s)|}{(t-s)^{\alpha+1}} ds \right) < \infty.$$

We have, for all $0 < \varepsilon < \alpha$

$$C^{\alpha+\varepsilon}(0,T;\mathbb{R}^d) \subset W_0^{\alpha,\infty}(0,T;\mathbb{R}^d) \subset C^{\alpha-\varepsilon}(0,T;\mathbb{R}^d).$$

Denote by $W_T^{1-\alpha,\infty}(0,T;\mathbb{R}^m)$ the space of measurable functions $g:[0,T]\to\mathbb{R}^m$ such that

$$\|g\|_{1-\alpha,\infty,T} := \sup_{0<s<t<T}\left(\frac{|g(t)-g(s)|}{(t-s)^{1-\alpha}}+\int_s^t \frac{|g(y)-g(s)|}{(y-s)^{2-\alpha}}dy\right)<\infty.$$

Clearly, for all $\varepsilon>0$ we have

$$C^{1-\alpha+\varepsilon}(0,T;\mathbb{R}^m)\subset W_T^{1-\alpha,\infty}(0,T;\mathbb{R}^m)\subset C^{1-\alpha}(0,T;\mathbb{R}^m). \tag{2.4}$$

Moreover, if g belongs to $W_T^{1-\alpha,\infty}(0,T;\mathbb{R}^m)$, its restriction to $(0,t)$ belongs to $I_{t-}^{1-\alpha}(L^\infty(0,t;\mathbb{R}^m))$ for all t and

$$\begin{aligned}\Lambda_\alpha(g) &:= \frac{1}{\Gamma(1-\alpha)}\sup_{0<s<t<T}\left|\left(D_{t-}^{1-\alpha}g_{t-}\right)(s)\right|\\ &\le \frac{1}{\Gamma(1-\alpha)\Gamma(\alpha)}\|g\|_{1-\alpha,\infty,T}<\infty.\end{aligned} \tag{2.5}$$

The integral $\int_0^t f dg$ can be defined for all $t\in[0,T]$ if g belongs to $W_T^{1-\alpha,\infty}(0,T)$ and f satisfies

$$\|f\|_{\alpha,1} := \int_0^T \frac{|f(s)|}{s^\alpha}ds+\int_0^T\int_0^s \frac{|f(s)-f(y)|}{(s-y)^{\alpha+1}}dy\,ds<\infty.$$

Furthermore the following estimate holds

$$\left|\int_0^T f dg\right|\le \Lambda_\alpha(g)\,\|f\|_{\alpha,1}\,.$$

3. Stochastic differential equations driven by an fBm

We are going to consider first the case of a deterministic equation. Let $0<\alpha<\frac{1}{2}$ be fixed. Let $g\in W_T^{1-\alpha,\infty}(0,T;\mathbb{R}^m)$. Consider the deterministic differential equation on $\mathbb{R}^d$

$$\xi_t^i = x_0^i+\int_0^t b^i(s,\xi_s)ds+\sum_{j=1}^m\int_0^t \sigma^{i,j}(s,\xi_s)\,dg_s^j,\ \ t\in[0,T], \tag{3.1}$$

$i=1,...,d$, where $x_0\in\mathbb{R}^d$, and the coefficients $\sigma^{i,j},b^i:[0,T]\times\mathbb{R}^d\to\mathbb{R}$ are measurable functions. Set $\sigma=\left(\sigma^{i,j}\right)_{d\times m}$, $b=\left(b^i\right)_{d\times 1}$ and for a matrix $A=\left(a^{i,j}\right)_{d\times m}$ and a vector $y=\left(y^i\right)_{d\times 1}$ denote $|A|^2=\sum_{i,j}\left|a^{i,j}\right|^2$ and $|y|^2=\sum_i\left|y^i\right|^2$.

Let us consider the following assumptions on the coefficients.

(H1) $\sigma(t,x)$ is differentiable in x, and there exist some constants $0<\beta,\delta\le 1$,

$M_1, M_2, M_3 > 0$ such that the following properties hold:

$$\begin{cases} i) \;\; \text{Lipschitz continuity} \\ |\sigma(t,x)-\sigma(t,y)| \le M_1|x-y|, \quad \forall\, x \in \mathbb{R}^d, \forall\, t \in [0,T] \\ ii) \;\; \text{Hölder continuity} \\ |\partial_{x_i}\sigma(t,x)-\partial_{x_i}\sigma(t,y)| \le M_2|x-y|^\delta, \\ \qquad\qquad \forall\, |x|, |y| \in \mathbb{R}^d, \forall t \in [0,T], i=1,\dots,d, \\ iii) \;\; \text{Hölder continuity in time} \\ |\sigma(t,x)-\sigma(s,x)| + |\partial_{x_i}\sigma(t,x)-\partial_{x_i}\sigma(s,x)| \le M_3|t-s|^\beta \\ \qquad\qquad \forall\, x \in \mathbb{R}^d, \forall\, t,s \in [0,T]. \end{cases}$$

(H2) There exist constants $L_1, L_2 > 0$ such that the following properties hold:

$$\begin{cases} i) \;\; \text{Local Lipschitz continuity} \\ \quad |b(t,x)-b(t,y)| \le L_1|x-y|, \ \forall\, |x|, |y| \in \mathbb{R}^d, \forall t \in [0,T], \\ ii) \;\; \text{Linear growth} \\ \quad |b(t,x)| \le L_2(1+|x|), \ \forall x \in \mathbb{R}^d, \forall t \in [0,T]. \end{cases}$$

Set

$$\alpha_0 = \min\left\{\frac{1}{2}, \beta, \frac{\delta}{1+\delta}\right\}.$$

The following existence and uniqueness result has been proved in Ref. [11].

Theorem 3.1. *Suppose that the coefficients $\sigma(t,x)$ and $b(t,x)$ satisfy assumptions (H1) and (H2). Then, if $\alpha < \alpha_0$ there exists a unique solution of Equation (3.1) in the space $C^{1-\alpha}(0,T;\ \mathbb{R}^d)$.*

Actually, these conditions can be slightly relaxed. For instance, the Hölder continuity of the partial derivatives of σ and the Hölder continuity of the coefficient b may hold only locally (see Ref. [11] for the details).

We now state two theorems which are consequences of the estimates found in Ref. [11].

For any $\lambda \ge 0$ we introduce the equivalent norm in the space $W_0^{\alpha,\infty}(0,T;\mathbb{R}^d)$ defined by

$$\|f\|_{\alpha,\lambda} = \sup_{t\in[0,T]} e^{-\lambda t}\left(|f(t)| + \int_0^t \frac{|f(t)-f(s)|}{(t-s)^{\alpha+1}}\, ds\right).$$

Theorem 3.2. *Let $\xi_t(x_0)$ denote the solution of (3.1) at time t with initial condition x_0. Fix $R > 1$. Then there exists a constant C such that for any x_0 and x_1 in the ball $B(0,\ R) = \{x \in \mathbb{R}^d,\ |x| \le R\}$ and for any $\lambda > \left[R\exp\left(C\,(1+\Lambda_\alpha(g))^{\frac{1}{1-2\alpha}}\right)\right]^{\frac{1}{1-2\alpha}}$ we have*

$$\|\xi(x_0)-\xi(x_1)\|_{\alpha,\lambda} \le \left(1 - R\exp\left(C\,(1+\Lambda_\alpha(g))^{\frac{1}{1-2\alpha}}\right)\lambda^{2\alpha-1}\right)^{-1}|x_0-x_1|.$$

Proof. It is proved in Ref. [11] that there exists a constant C_1 such that if $\lambda_0 = C_1 (1 + \Lambda_\alpha(g))^{\frac{1}{1-2\alpha}}$ then for any initial condition x_0 in the ball of radius R we have

$$\|\xi(x_0)\|_{\alpha,\lambda_0} \leq 2(1 + |x_0|) \leq 4R. \tag{3.2}$$

Given a function $f \in W_0^{\alpha,\infty}(0,T;\mathbb{R}^d)$ we define as in Ref. [11]

$$\begin{aligned} F_t^{(b)}(f) &= \int_0^t b(s,\, f(s))ds, \\ G_t^{(\sigma)}(g,f) &= \int_0^t \sigma(s,\, f(s))dg(s), \\ \Delta(f) &= \sup_{r\in[0,T]} \int_0^r \frac{|f(r)-f(s)|^\delta}{(r-s)^{\alpha+1}} ds. \end{aligned}$$

If $f, h \in W_0^{\alpha,\infty}(0,T;\mathbb{R}^d)$ (see Ref. [11]) there exist constants C_2 and C_3 such that

$$\left\| F^{(b)}(f) - F^{(b)}(h) \right\|_{\alpha,\lambda} \leq C_2 \lambda^{\alpha-1} \|f-h\|_{\alpha,\lambda}, \tag{3.3}$$

$$\begin{aligned} \left\| G^{(\sigma)}(g,f) - G^{(\sigma)}(g,h) \right\|_{\alpha,\lambda} &\leq C_3 \Lambda_\alpha(g) \lambda^{2\alpha-1} \\ &\quad \times (1 + \Delta(f) + \Delta(h)) \|f-h\|_{\alpha,\lambda}, \end{aligned} \tag{3.4}$$

$$\left\| G^{(\sigma)}(g,f) \right\|_{\alpha,\lambda} \leq C_4 \Lambda_\alpha(g) \lambda^{2\alpha-1} \left(1 + \|f\|_{\alpha,\lambda}\right) \tag{3.5}$$

Also, if $f \in W_0^{\alpha,\infty}(0,T;\mathbb{R}^d)$ and h is a bounded measurable function, then

$$\left\| F^{(b)}(h) \right\|_{1-\alpha} \leq C_5 (1 + \|h\|_\infty), \tag{3.6}$$

$$\left\| G^{(\sigma)}(g,f) \right\|_{1-\alpha} \leq C_6 \Lambda_\alpha(g) \left(1 + \|f\|_{\alpha,\infty}\right). \tag{3.7}$$

We have the following estimate

$$\begin{aligned} \Delta(\xi(x_0)) &= \sup_{r\in[0,T]} \int_0^r \frac{|\xi_r(x_0) - \xi_s(x_0)|^\delta}{(r-s)^{\alpha+1}} ds \\ &\leq \frac{T^{\delta-\alpha(1+\delta)}}{\delta - \alpha(1+\delta)} \|\xi(x_0)\|_{1-\alpha}, \end{aligned} \tag{3.8}$$

and using (3.6), (3.7), and (3.2) we obtain

$$\begin{aligned} \|\xi(x_0)\|_{1-\alpha} &\leq |x_0| + \left\| F^{(b)}(\xi(x_0)) \right\|_{1-\alpha} + \left\| G^{(\sigma)}(\xi(x_0)) \right\|_{1-\alpha} \\ &\leq |x_0| + C_5 (1 + \|\xi(x_0)\|_\infty) + \Lambda_\alpha(g) C_6 \left(1 + \|\xi(x_0)\|_{\alpha,\infty}\right) \\ &\leq C_7 e^{\lambda_0 T} (1 + |x_0|)(1 + \Lambda_\alpha(g)) \\ &\leq 2C_7 e^{\lambda_0 T} R (1 + \Lambda_\alpha(g)) \\ &= 2C_7 \exp\left(TC_1 (1 + \Lambda_\alpha(g))^{\frac{1}{1-2\alpha}}\right) R(1 + \Lambda_\alpha(g)). \end{aligned}$$

Hence, from (3.8) we get

$$\Delta(\xi(x_0)) \le C_8 \exp\left(TC_1\left(1+\Lambda_\alpha(g)\right)^{\frac{1}{1-2\alpha}}\right) R(1+\Lambda_\alpha(g)). \tag{3.9}$$

Using (3.3), (3.4), and (3.9) we obtain for x_0 and x_1 in the ball of radius R

$$\begin{aligned}
\|\xi(x_0)-\xi(x_1)\|_{\alpha,\lambda} &\le |x_0-x_1| + \|F^{(b)}(\xi(x_0)) - F^{(b)}(\xi(x_1))\|_{\alpha,\lambda} \\
&\quad + \left\|G^{(\sigma)}(g,\xi(x_0)) - G^{(\sigma)}(g,\xi(x_1))\right\|_{\alpha,\lambda} \\
&\le |x_0-x_1| + C_1\lambda^{\alpha-1}\|\xi(x_0)-\xi(x_1)\|_{\alpha,\lambda} \\
&\quad + C_2\Lambda_\alpha(g)\lambda^{2\alpha-1}\left(1+\Delta\left(\xi(x_0)\right)+\Delta\left(\xi(x_1)\right)\right)\|\xi(x_0)-\xi(x_1)\|_{\alpha,\lambda} \\
&\le |x_0-x_1| + C_1\lambda^{\alpha-1}\|\xi(x_0)-\xi(x_1)\|_{\alpha,\lambda} \\
&\quad + C_2\Lambda_\alpha(g)\lambda^{2\alpha-1}\Big(1+2C_8\exp\left(TC_1\left(1+\Lambda_\alpha(g)\right)^{\frac{1}{1-2\alpha}}\right) \\
&\quad \times R(1+\Lambda_\alpha(g))\Big)\|\xi(x_0)-\xi(x_1)\|_{\alpha,\lambda}.
\end{aligned}$$

As a consequence,

$$\begin{aligned}
&\|\xi(x_0)-\xi(x_1)\|_{\alpha,\lambda} \\
&\le |x_0-x_1| + \lambda^{2\alpha-1}\left(\exp\left(C\left(1+\Lambda_\alpha(g)\right)^{\frac{1}{1-2\alpha}}\right)R\right)\|\xi(x_0)-\xi(x_1)\|_{\alpha,\lambda}
\end{aligned}$$

for some constant C, which implies the result. □

Theorem 3.3. *The map*

$$\begin{aligned}
\xi\ :\ W_T^{1-\alpha,\infty}(0,T;\mathbb{R}^m) &\longrightarrow W_0^{\alpha,\infty}(0,T;\mathbb{R}^d) \\
g &\longmapsto \xi,
\end{aligned}$$

where ξ is the solution of (3.1) with $x_0 \in B(0,R)$, is continuous. Namely, for $\lambda > 0$ large enough we have

$$\|\xi(g)-\xi(h)\|_{\alpha,\lambda} \le \frac{C_1\lambda^{2\alpha-1}\|\xi(g)\|_{\alpha,\lambda}}{1-C_2\lambda^{2\alpha-1}\left(1+\Lambda_\alpha(h)\right)}\Lambda_\alpha(g-h).$$

Proof. With the previous notations, we have

$$\xi(g) = x_0 + F^{(b)}(\xi(g)) + G^{(\sigma)}(g,\xi(g))$$

and

$$\xi(h) = x_0 + F^{(b)}(\xi(h)) + G^{(\sigma)}(h,\xi(h)).$$

Thus,

$$\begin{aligned}
\xi(g)-\xi(h) &= F^{(b)}(\xi(g)) - F^{(b)}(\xi(h)) \\
&\quad + G^{(\sigma)}(g-h,\xi(g)) + \left(G^{(\sigma)}(h,\xi(g)) - G^{(\sigma)}(h,\xi(h))\right).
\end{aligned}$$

According to (3.5), for λ sufficiently large,

$$\begin{aligned}\|\xi(g)-\xi(h)\|_{\alpha,\lambda} \leq C_2\lambda^{\alpha-1}\,\|\xi(g)-\xi(h)\|_{\alpha,\lambda} \\ + C_4\lambda^{2\alpha-1}(1+\Lambda_\alpha(g-h))\|\xi(g)\|_{\alpha,\lambda} \\ + C_3\lambda^{2\alpha-1}\Lambda_\alpha(h)\|\xi(g)-\xi(h)\|_{\alpha,\lambda}.\end{aligned}$$

Hence the result. □

Consider Equation (1.2) on $\mathbb{R}^d$, for $t \in [0,T]$, where X_0 is a d-dimensional random variable, and the coefficients $\sigma^{i,j}, b^i : \Omega \times [0,T] \times \mathbb{R}^d \rightarrow \mathbb{R}$ are measurable functions.

Theorem 3.4. *Suppose that X_0 is an $\mathbb{R}^d$-valued random variable, the coefficients $\sigma(t,x)$ and $b(t,x)$ satisfy assumptions (H1) and (H2), where the constants might depend on ω, with $\beta > 1-H$, $\delta > 1/H-1$. Then if $\alpha \in (1-H, \alpha_0)$, then there exists a unique stochastic process, whose trajectories belong to the space $W_0^{\alpha,\infty}(0,T;\mathbb{R}^d)$, solution of the stochastic equation (1.2) and, moreover, for $\mathbb{P}$-almost all $\omega \in \Omega$*

$$X(\omega,\cdot) \in C^{1-\alpha}\left(0,T;\mathbb{R}^d\right).$$

Consider the particular case where $b = 0$ and σ is time independent, that is,

$$X_t = X_0 + \int_0^t \sigma(X_s)dB_s^H. \tag{3.10}$$

By the above theorem this equation has a unique solution provided σ is continuously differentiable, and σ' is bounded and Hölder continuous of order $\delta > \frac{1}{H} - 1$.

Hu and Nualart [14] have established the following estimates. Choose $\theta \in \left(\frac{1}{2}, H\right)$. Then, the solution to Equation (3.10) satisfies

$$\sup_{0\leq t\leq T} |X_t| \leq 2^{1+kT\left(\|\sigma'\|_\infty \vee |\sigma(0)|\right)^{1/\theta}\|B^H\|_\theta^{1/\theta}}\left(|X_0|+1\right), \tag{3.11}$$

where k is a constant depending only on θ. Moreover, if σ is bounded and $\|\sigma'\|_\infty \neq 0$ this estimate can be improved in the following way

$$\sup_{0\leq t\leq T} |X_t| \leq |X_0| + k\|\sigma\|_\infty \left(T^\theta\|B^H\|_\theta \vee T\|\sigma'\|_\infty^{\frac{1-\theta}{\theta}}\|B^H\|_\theta^{\frac{1}{\theta}}\right), \tag{3.12}$$

where again k is a constant depending only on θ.

These estimates improve those obtained by Nualart and Răşcanu [11] based on a suitable version of Gronwall's lemma. The estimates (3.11) and (3.12) lead to the following integrability properties for the solution of Equation (3.10).

Theorem 3.5. *Consider the stochastic differential equation (3.10), and assume that $E(|X_0|^p) < \infty$ for all $p \geq 2$. If σ' is bounded and Hölder continuous of order $\delta > \frac{1}{H} - 1$, then*

$$E\left(\sup_{0\leq t\leq T} |X_t|^p\right) < \infty \tag{3.13}$$

for all $p \geq 2$. If furthermore σ is bounded and $E(\exp(\lambda|X_0|^\gamma)) < \infty$ for any $\lambda > 0$ and $\gamma < 2H$, then

$$E\left(\exp \lambda \left(\sup_{0\leq t\leq T} |X_t|^\gamma\right)\right) < \infty \tag{3.14}$$

for any $\lambda > 0$ and $\gamma < 2H$.

Nualart and Saussereau [12] have proved that the random variable X_t belongs locally to the space $\mathbb{D}^\infty$ if the function σ is infinitely differentiable and bounded together with all its partial derivatives. As a consequence, they have derived the absolute continuity of the law of X_t for any $t > 0$ assuming that the initial condition is constant and the vector space spanned by $\{(\sigma^{i,j}(x_0))_{1\leq i\leq d}, 1 \leq j \leq m\}$ is $\mathbb{R}^d$.

Applying Theorem 3.5 Hu and Nualart have proved [14] that if the function σ is infinitely differentiable and bounded together with all its partial derivatives, then for any $t \in [0, T]$ the random variable X_t belongs to the space $\mathbb{D}^\infty$. As a consequence, if the matrix $a(x) = \sigma\sigma^T(x)$ is uniformly elliptic, then, for any $t > 0$ the probability law of X_t has a C^∞ density. In a recent paper, Baudoin and Coutin [15] have extended this result and derived the regularity of the density under Hörmander hypoellipticity conditions.

4. Flow of homeomorphisms

Let $\pi = \{0 = t_0 < t_1 < \cdots < t_n = T\}$ be the uniform partition of the interval $[0, T]$. That is $t_k = \frac{kT}{n}$, $k = 0, \ldots, n$. We denote by $B^{n,H}$ the polygonal approximation of the fractional Brownian motion defined by

$$B_t^{n,H} = \sum_{k=0}^{n-1} \left(B_{t_k}^H + \frac{n}{T}(t - t_k)\left(B_{t_{k+1}}^H - B_{t_k}^H\right)\right) \mathbf{1}_{(t_k,t_{k+1}]}(t).$$

In order to get a precise rate for these approximations we will make use of the following exact modulus of continuity of the fractional Brownian motion (see Ref. [16]). There exists a random variable G such that almost surely for any $s, t \in [0, T]$ we have

$$\left|B_t^H - B_s^H\right| \leq G|t - s|^H \sqrt{\log\left(|t - s|^{-1}\right)}. \tag{4.1}$$

Fix $\theta < H$. We have the following result, which provides the rate of convergence of these approximations in Hölder norm.

Lemma 4.1. *There exist a random variable $C_{T,\beta}$ such that*

$$\|B^H - B^{n,H}\|_{C^\theta(0,T;\mathbb{R}^m)} \leq C_{T,\beta} n^{\theta - H}\sqrt{\log n}. \tag{4.2}$$

Proof. To simplify the notation we will assume that $m = 1$. Fix $0 < s < t < T$ and assume that $s \in [t_l, t_{l+1}]$ and $t \in [t_k, t_{k+1}]$. Let us first estimate

$$h_1(s, t) = \frac{1}{(t - s)^\theta}\left|B_t^{n,H} - B_t^H - (B_s^{n,H} - B_s^H)\right|.$$

If $t - s \geq \frac{T}{n}$, then using (4.1) we obtain

$$\begin{aligned}|h_1(s,t)| &\leq T^{-\beta} n^{\beta} \left[\left| B_{t_k}^H - B_t^H + \frac{n}{T}(t - t_k) \left(B_{t_{k+1}}^H - B_{t_k}^H \right) \right| \right. \\ &\quad \left. + \left| B_{t_l}^H - B_s^H + \frac{n}{T}(s - t_l) \left(B_{t_{l+1}}^H - B_{t_l}^H \right) \right| \right] \\ &\leq 4GT^{-\theta+H} n^{-H+\theta} \sqrt{\log (n/T)}.\end{aligned}$$

If $t - s < \frac{T}{n}$, then there are two cases. Suppose first that $s, t \in [t_k, t_{k+1}]$. In this case, if n is large enough we obtain using (4.1)

$$\begin{aligned}|h_1(s,t)| &\leq \frac{|B_t^H - B_s^H|}{(t-s)^{\theta}} + \frac{n}{T} \frac{|B_{t_{k+1}}^H - B_{t_k}^H|}{(t-s)^{\theta}} (t-s) \\ &\leq G|t-s|^{H-\theta} \sqrt{\log |t-s|^{-1}} + GT^{-1+H} \sqrt{\log(n/T)}\, n^{1-H} (t-s)^{1-\theta} \\ &\leq 2GT^{-\theta+H} n^{-H+\theta} \sqrt{\log (n/T)}.\end{aligned}$$

On the other hand, if $s \in [t_{k-1}, t_k]$ and $t \in [t_k, t_{k+1}]$ we have, again if n is large enough

$$\begin{aligned}|h_1(s,t)| &\leq \frac{1}{(t-s)^{\theta}} \left| B_{t_k}^H - B_t^H + \frac{n}{T}(t - t_k) \left(B_{t_{k+1}}^H - B_{t_k}^H \right) \right. \\ &\quad \left. - \left\{ B_{t_k}^H - B_s^H - \frac{n}{T}(t_k - s) \left(B_{t_k}^H - B_{t_{k-1}}^H \right) \right\} \right| \\ &\leq \frac{1}{(t-s)^{\theta}} \left[|B_t^H - B_s^H| + \frac{n}{T}(t-s) \left(|B_{t_k}^H - B_{t_{k-1}}^H| + |B_{t_{k+1}}^H - B_{t_k}^H| \right) \right] \\ &\leq \frac{G}{(t-s)^{\theta}} \left[|t-s|^H \sqrt{\log |t-s|^{-1}} + 2(t-s) \left(\frac{n}{T} \right)^H \sqrt{\log (n/T)} \right] \\ &\leq 3GT^{-\theta+H} n^{-H+\theta} \sqrt{\log (n/T)}.\end{aligned}$$

This proves (4.2). □

Corollary 4.2. *For any $\alpha \in (1 - H, 1/2)$, we have:*

$$\sup_n \Lambda_{\alpha}(B^{n,H}) < +\infty \text{ and } \lim_{n \to +\infty} \Lambda_{\alpha}(B^{n,H} - B^H) = 0.$$

Proof. Choose $\eta > 0$ in such a way that $1 - \alpha + \eta < H$. According to (2.4) and (2.5), we have

$$\Lambda_{\alpha}(B^{n,H}) \leq c_{\eta} \|B^{n,H}\|_{C^{1-\alpha+\eta}(0,T;\mathbb{R}^m)}$$

and

$$\Lambda_{\alpha}(B^{n,H} - B^H) \leq c_{\eta} \|B^{n,H} - B^H\|_{C^{1-\alpha+\eta}(0,T;\mathbb{R}^m)}.$$

Then, Lemma 4.1 implies that the sequence $B^{n,H}$ converges to B^H in the norm of $C^{1-\alpha+\eta}(0,T;\mathbb{R}^m)$ which yields the results. □

Consider for any $0 \leq r \leq t \leq T$ and any natural number $n \geq 1$ the following equations

$$X^n_{rt}(x) = x + \int_r^t \sigma\left(s, X^n_{rs}(x)\right) dB^{n,H}(s) + \int_r^t b(s, X^n_{rs}(x))ds, \tag{4.3}$$

and

$$Y^n_{rt}(x) = x + \int_r^t \sigma\left(s, Y^n_{st}(x)\right) dB^{n,H}(s) + \int_r^t b(s, Y^n_{st}(x))ds. \tag{4.4}$$

We know from standard results on ordinary differential equations that for any $n \geq 1$,

(1) Equations (4.3) and (4.4) have a unique solution.
(2) For any $x \in \mathbb{R}^d$, for any $0 \leq r \leq \tau \leq t \leq T$, $X^n_{\tau t}(X^n_{r\tau}(x)) = X^n_{rt}(x)$.
(3) For any $x \in \mathbb{R}^d$, for any $0 \leq r \leq \tau \leq t \leq T$, $Y^n_{r\tau}(Y^n_{\tau t}(x)) = Y^n_{rt}(x)$.
(4) The maps $(x \mapsto X^n_{rt}(x))$ and $(x \mapsto Y^n_{rt}(x))$ are $\mathbb{R}^d$-homeomorphisms inverse of each other:

$$X^n_{rt}(Y^n_{rt}(x)) = x \text{ and } Y^n_{rt}(X^n_{rt}(x)) = x.$$

We are then in position to prove our main theorem:

Theorem 4.3. *Assume that Hypotheses (H1) and (H2) hold. Then, claims 1, 2, 3 and 4 also hold for the equations*

$$X_{rt}(x) = x + \int_r^t \sigma\left(s, X_{rs}(x)\right) dB^H(s) + \int_r^t b(s, X_{rs}(x))ds, \tag{4.5}$$

and

$$Y_{rt}(x) = x + \int_r^t \sigma\left(s, Y_{st}(x)\right) dB^H(s) + \int_r^t b(s, Y_{st}(x))ds. \tag{4.6}$$

Proof. Point 1 is proved in Ref. [11]. As to the second claim, proceed as follows:

$$\begin{aligned} X^n_{\tau t}(X^n_{r\tau}(x)) - X_{\tau t}(X_{r\tau}(x)) &= X^n_{\tau t}(X^n_{r\tau}(x)) - X^n_{\tau t}(X_{r\tau}(x)) \\ &\quad + (X^n_{\tau t} - X_{\tau t})(X_{r\tau}(x)). \end{aligned}$$

Fix $\varepsilon > 0$ and α such that $1-H < \alpha < \frac{1}{2}$. Fix a trajectory $\omega \in \Omega$. Choose n_0 so that $\Lambda_\alpha(B^{n,H}-B^H) \leq \varepsilon$ for all $n \geq n_0$ and choose λ such that $\lambda^{2\alpha-1}C_2 \sup_n \Lambda_\alpha(B^{n,H}) \leq \frac{1}{2}$. Then, according to Theorem 3.3, for any $n \geq n_0$,

$$\begin{aligned} \|X^n_{r\cdot} - X_{r\cdot}\|_{\alpha,\lambda} &\leq \frac{C_1\lambda^{2\alpha-1}\|X_{r\cdot}\|_{\alpha,\lambda}}{1 - C_2\lambda^{2\alpha-1}\left(1 + \Lambda_\alpha(B^{n,H})\right)} \Lambda_\alpha(B^{n,H} - B^H) \\ &\leq 2C_1\lambda^{2\alpha-1}\|X_{r\cdot}\|_{\alpha,\lambda}\varepsilon. \end{aligned}$$

Hence, for $n \geq n_0$,

$$|(X^n_{\tau t} - X_{\tau t})(X_{r\tau}(x))| \leq c\varepsilon.$$

The convergence of $X^n_{r\cdot}$ implies that there exists R such that for any $\tau \in [r, t]$ and for any $n \geq n_0$, $X^n_{r\tau}(x) \in B(0, R)$. Then, Theorem 3.2 implies that for λ large enough

$$\begin{aligned}
&|X^n_{\tau t}(X^n_{r\tau}(x)) - X^n_{\tau t}(X_{r\tau}(x))| \\
&\leq \left(1 - R\exp\left(C\left(1 + \sup_n \Lambda_\alpha(B^{n,H})\right)^{\frac{1}{1-2\alpha}}\right)\lambda^{2\alpha-1}\right)^{-1} \\
&\quad \times |X^n_{r\tau}(x) - X_{r\tau}(x)| \\
&\leq c|X^n_{r\tau}(x) - X_{r\tau}(x)|.
\end{aligned}$$

We have thus proved that

$$\begin{aligned}
0 &= \lim_{n\to+\infty} X^n_{\tau t}(X^n_{r\tau}(x)) - X_{\tau t}(X_{r\tau}(x)) \\
&= \lim_{n\to+\infty} X^n_{rt}(x) - X_{\tau t}(X_{r\tau}(x)) \\
&= X_{rt}(x) - X_{\tau t}(X_{r\tau}(x)).
\end{aligned}$$

Other points are handled similarly. □

Acknowledgement

This work was carried out during a stay of Laurent Decreusefond at Kansas University, Lawrence KS. He would like to thank KU for warm hospitality and generous support.

References

[1] A. N. Kolmogorov, Wienershe Spiralen und einige andere interessante Kurven im Hilbertschen Raum, *C. R. (Doklady) Acd. Sci. URSS (N. S.).* **26**, 115–118, (1940).

[2] B. B. Mandelbrot and J. W. Van Ness, Fractional Brownian motions, fractional noises and applications, *SIAM Rev.* **10**, 422–437, (1968). ISSN 1095-7200.

[3] L. Decreusefond and A. S. Üstünel, Stochastic analysis of the fractional Brownian motion, *Potential Anal.* **10**(2), 177–214, (1999). ISSN 0926-2601.

[4] D. Feyel and A. de La Pradelle, On fractional Brownian processes, *Potential Anal.* **10**(3), 273–288, (1999). ISSN 0926-2601.

[5] D. Nualart. Stochastic integration with respect to fractional Brownian motion and applications. In *Stochastic models (Mexico City, 2002)*, vol. 336, *Contemp. Math.*, pp. 3–39. Amer. Math. Soc., Providence, RI, (2003).

[6] L. C. Young, An inequality of the Hölder type connected with Stieltjes integration, *Acta Math.* **67**, 251–282, (1936).

[7] T. Lyons, Differential equations driven by rough signals. I. An extension of an inequality of L. C. Young, *Math. Res. Lett.* **1**(4), 451–464, (1994). ISSN 1073-2780.

[8] T. Lyons and Z. Qian, *System control and rough paths.* Oxford Mathematical Monographs, (Oxford University Press, Oxford, 2002). ISBN 0-19-850648-1. Oxford Science Publications.

[9] L. Coutin and Z. Qian, Stochastic analysis, rough path analysis and fractional Brownian motions, *Probab. Theory Related Fields.* **122**(1), 108–140, (2002). ISSN 0178-8051.

[10] M. Zähle, Integration with respect to fractal functions and stochastic calculus. I, *Probab. Theory Related Fields.* **111**(3), 333–374, (1998). ISSN 0178-8051.

[11] D. Nualart and A. Răşcanu, Differential equations driven by fractional Brownian motion, *Collect. Math.* **53**(1), 55–81, (2002). ISSN 0010-0757.

[12] D. Nualart and B. Saussereau. Malliavin calculus for stochastic differential equations driven by a fractional Brownian motion, (2005). Preprint.

[13] S. G. Samko, A. A. Kilbas, and O. I. Marichev, *Fractional integrals and derivatives.* (Gordon and Breach Science Publishers, Yverdon, 1993). ISBN 2-88124-864-0. Theory and applications, Edited and with a foreword by S. M. Nikol′skiĭ, Translated from the 1987 Russian original, Revised by the authors.

[14] Y. Hu and D. Nualart. Differential equations driven by Hölder continuous functions of order greater than 1/2, (2006). Preprint.

[15] F. Baudoin and M. Hairer. A version of hörmander's theorem for the fractional brownian motion, (2006). Preprint.

[16] W. Wang, On a functional limit result for increments of a fractional Brownian motion, *Acta Math. Hungar.* **93**(1-2), 153–170, (2001). ISSN 0236-5294.

Chapter 10

Regularity of Transition Semigroups Associated to a 3D Stochastic Navier-Stokes Equation

Franco Flandoli and Marco Romito*

Dipartimento di Matematica Applicata, Universit di Pisa
via Bonanno Pisano, 25/b, 56126 Pisa, Italy
flandoli@dma.unipi.it

A 3D stochastic Navier-Stokes equation with a suitable non degenerate additive noise is considered. The regularity in the initial conditions of every Markov transition kernel associated to the equation is studied by a simple direct approach. A by-product of the technique is the equivalence of all transition probabilities associated to every Markov transition kernel.

Contents

1. Introduction

An old dream in stochastic fluid dynamics is to prove the well posedness of a stochastic version of the 3D Navier-Stokes equations, taking advantage of the noise, as one can do for finite dimensional stochastic equations with non regular drift (see for instance Stroock & Varadhan [1]). The problem is still open, although some intriguing results have been recently proved, see for instance Da Prato & Debussche [2], Mikulevicius & Rozovski [3], Flandoli & Romito [4, 5]. We recall here the framework constructed in Ref. [4] and prove some additional results.

*Dipartimento di Matematica, Universit di Firenze, viale Morgagni 67/a, 50134 Firenze, Italy, romito@math.unifi.it

We consider a viscous, incompressible, homogeneous, Newtonian fluid described by the stochastic Navier-Stokes equations on the torus $\mathcal{T} = [0, L]^3$, $L > 0$,

$$\frac{\partial u}{\partial t} + (u \cdot \nabla) u + \nabla p = \nu \Delta u + \sum_{i=1}^{\infty} \sigma_i h_i(x) \dot{\beta}_i(t) \tag{1.1}$$

with $\operatorname{div} u = 0$ and periodic boundary conditions, with suitable fields $h_i(x)$ and independent Brownian motions $\beta_i(t)$. The 3D random vector field $u = u(t, x)$ is the velocity of the fluid and the random scalar field $p = p(t, x)$ is the pressure. To simplify the exposition, we avoid generality and focus on one of the simplest set of assumptions:

$$\sigma_i^2 = \lambda_i^{-3}$$

where λ_i are the eigenvalues of the Stokes operator (see the next section). This assumption also allows us to compare more closely the results in Da Prato & Debussche [2] and Flandoli [6]. However, following Flandoli & Romito [4], we could treat any power law for σ_i. Under this assumption, one can associate a transition probability kernel $P(t, x, \cdot)$ to equation (2.2), which is the abstract version of (1.1), in $D(A)$ (see the definitions in Section 2.1 below), satisfying the Chapman-Kolmogorov equation. In other words, there exists a Markov selection in $D(A)$ for equation (2.2). To avoid misunderstandings, this does not mean that equation (2.2) has been solved in $D(A)$ with continuous trajectories: this would imply well posedness. What has been proved is that the law of weak martingale solutions is supported on $D(A)$ for all times, with a number of related additional properties, but a priori the typical trajectory may sometimes blow-up in the topology of $D(A)$.

The transition probabilities $P(t, x, \cdot)$ are irreducible and strong Feller, hence equivalent, in $D(A)$. These results and the existence of $P(t, x, \cdot)$ have been proved first in Da Prato & Debussche [2] and Debussche & Odasso [7] by a careful selection from the Galerkin scheme. Then another proof by an abstract selection principle and the local-in-time regularity of equation (2.2) has been given in Flandoli & Romito [4]. More precisely, first one proves the existence of a Markov kernel $P(t, x, \cdot)$ by means of a general and abstract method, then one proves that *any* such kernel is irreducible and strong Feller, hence equivalent, in $D(A)$.

We complement here the approach of Ref. [4] with two results. First, the simple idea used in Ref. [4] to prove the strong Feller property is here developed further, to show a weak form of Lipschitz continuity of $P(t, x, \cdot)$ in $x \in D(A)$. More precisely, we prove the estimate

$$|P(t, x_0 + h, \Gamma) - P(t, x_0, \Gamma)| \leq \frac{C_T}{t \wedge 1}(1 + |Ax_0|^6)|Ah| \log(|Ah|^{-1}) \tag{1.2}$$

for $t \in (0, T]$, x_0, $h \in D(A)$, with $|Ah| \leq 1$. This result has been proved in a stronger version in Da Prato & Debussche [2] for the transition kernel constructed from the Galerkin scheme, and also in Flandoli [6] for any Markov kernel associated to equation (2.2). In both cases the proof is based on the very powerful approach introduced in Ref. [2] which however requires a considerable amount of technical work. Here we give a rather elementary proof along the lines of Flandoli & Romito [4], based on the following simple idea: given x_0, $h \in D(A)$, for a short random time the solution is regular, unique and differentiable in the initial conditions; then the propagation of regularity in x from small time to arbitrary time is due to the Markov property. Unfortunately we cannot prove in this way the stronger estimate obtained in Ref. [2] (where the right-hand-side of (1.2) has the form $t^{-1+\varepsilon}(1+|Ax_0|^2)|Ah|$), so our first result here has mostly a pedagogical character, since the proof is conceptually very easy.

The second result, which follows from the same main estimates used to prove (1.2), is the equivalence

$$P^{(1)}(t,x,\cdot) \sim P^{(2)}(t',x',\cdot)$$

for any t, $t' > 0$ and x, $x' \in D(A)$, when $P^{(i)}(t,x,\cdot)$, $i=1,2$, are any two Markov transition kernels associated to equation (2.2) in $D(A)$. We have not proved yet the existence of invariant measures associated to such kernels[a], but if we assume to have such invariant measures, it also follows that they are equivalent. This result and the gradient estimates discussed above could be steps to understand better the open question of well posedness for equation (2.2). In particular, it seems to be not so easy to produce examples of stochastic differential equations without uniqueness but where all Markov solutions are equivalent.

Among the open problems related to this research we mention the relation between the regularity results for $P(t,x,\cdot)$ in the initial condition discussed above and the properties of Malliavin derivatives, investigated for stochastic 3D Navier-Stokes equations by Mikulevicius and Rozovsky [3, 8].

2. Preliminaries

2.1. *Notations*

Denote by $\mathcal{T} = [0,1]^3$ the three-dimensional torus, and let $\mathbb{L}^2(\mathcal{T})$ be the space of vector fields $u : \mathcal{T} \to \mathbb{R}^3$ with $L^2(\mathcal{T})$-components. For every $\alpha > 0$, let $\mathbb{H}^\alpha(\mathcal{T})$ be the space of fields $u \in \mathbb{L}^2(\mathcal{T})$ with components in the Sobolev space $H^\alpha(\mathcal{T}) = W^{\alpha,2}(\mathcal{T})$.

Let $\mathcal{D}^\infty$ be the space of infinitely differentiable divergence free periodic fields u on $\mathcal{T}$, with zero mean. Let H be the closure of $\mathcal{D}^\infty$ in the topology of $\mathbb{L}^2(\mathcal{T})$: it is the space of all zero mean fields $u \in \mathbb{L}^2(\mathcal{T})$ such that $\operatorname{div} u = 0$ and $u \cdot n$ on the boundary is periodic. We denote by $\langle .,. \rangle_H$ and $|.|_H$ (or simply by $\langle .,. \rangle$ and $|.|$) the usual L^2-inner product and norm in H. Let V (resp. $D(A)$) be the closure of $\mathcal{D}^\infty$ in the topology of $\mathbb{H}^1(\mathcal{T})$ (in the topology of $\mathbb{H}^2(\mathcal{T})$, respectively): it is the space of divergence free, zero mean, periodic elements of $\mathbb{H}^1(\mathcal{T})$ (respectively of $\mathbb{H}^2(\mathcal{T})$).

[a]This is apparently due to technical reasons and it is the subject of a work in progress.

The spaces V and $D(A)$ are dense and compactly embedded in H. From Poincaré inequality we may endow V with the norm $\|u\|_V^2 := \int_T |Du(x)|^2\, dx$.

Let $A : D(A) \subset H \to H$ be the operator $Au = -\Delta u$ (component wise). Since A is a selfadjoint positive operator in H, there is a complete orthonormal system $(h_i)_{i\in\mathbb{N}} \subset H$ of eigenfunctions of A, with eigenvalues $0 < \lambda_1 \le \lambda_2 \le \ldots$ (that is, $Ah_i = \lambda_i h_i$). The fields h_i in equation (2.2) will be these eigenfunctions. We have

$$\langle Au, u\rangle_H = \|u\|_V^2$$

for every $u \in D(A)$.

Let V' be the dual of V; with proper identifications we have $V \subset H \subset V'$ with continuous injections, and the scalar product $\langle\cdot,\cdot\rangle_H$ extends to the dual pairing $\langle\cdot,\cdot\rangle_{V,V'}$ between V and V'. We may enlarge this scheme to $D(A) \subset V \subset H \subset V' \subset D(A)'$. Let $B(\cdot,\cdot) : V \times V \to V'$ be the bi linear operator defined as

$$\langle w, B(u,v)\rangle_{V,V'} = \sum_{i,j=1}^{3} \int_T u_i \frac{\partial v_j}{\partial x_i} w_j\, dx$$

for every $u, v, w \in V$. We shall repeatedly use the following inequality:

$$\left|A^{1/2} B(u,v)\right|_H \le C_0 |Au|\, |Av| \tag{2.1}$$

for $u, v \in D(A)$. The proof is elementary (see Flandoli [9]).

2.2. *Definitions, assumptions and known results*

We (formally) rewrite equations (1.1) as an abstract stochastic evolution equation in H,

$$du(t) + [\nu Au(t) + B(u(t), u(t))]\, dt = \sum_{i=1}^{\infty} \sigma_i h_i\, d\beta_i(t)\,. \tag{2.2}$$

Let us set

$$\Omega = C([0,\infty); D(A)')$$

and denote by $(\xi_t)_{t\ge 0}$ the canonical process on Ω, defined as $\xi_t(\omega) = \omega(t)$, by $\mathcal{F}$ the Borel σ-algebra in Ω and by $\mathcal{F}_t$ the σ-algebra generated by the events $\{\xi_s \in \Gamma\}$ with $s \in [0,t]$ and Γ a Borel set of $D(A)'$. Finally, denote by $\mathcal{B}(D(A))$ the Borel σ-algebra of $D(A)$ and by $B_b(D(A))$ the set of all real valued bounded measurable functions on $D(A)$).

Definition 2.1. Given a probability measure μ_0 on H, we say that a probability measure P on $(\Omega, \mathcal{F})$ is a solution to the martingale problem associated to equation (2.2) with initial law μ_0 if

(MP1) $P[\xi \in L^\infty_{\text{loc}}([0,\infty); H) \cap L^2_{\text{loc}}([0,\infty); V)] = 1$,

(MP2) for each $\varphi \in \mathcal{D}^\infty$ the process $(M_t^\varphi, \mathcal{F}_t, P)_{t\geq 0}$, defined P-a. s. on $(\Omega, \mathcal{F})$ as

$$M_t^\varphi := \langle \xi_t - \xi_0, \varphi \rangle_H + \int_0^t \nu \langle \xi_s, A\varphi \rangle_H \, ds - \int_0^t \langle B(\xi_s, \varphi), \xi_s \rangle_H \, ds$$

is a continuous square integrable martingale with quadratic variation

$$[M^\varphi]_t = t \sum_{i\in\mathbb{N}} \sigma_i^2 |\langle \varphi, h_i \rangle|^2,$$

(MP3) the marginal of P at time 0 is μ_0.

Remark 2.2. Among all test functions in property $[MP2]$, we can choose $\varphi = h_i$. Set for all i, $\beta_i(t) = \frac{1}{\sigma_i} M_t^{h_i}$ (and 0 if $\sigma_i = 0$). The $(\beta_i)_{i\in\mathbb{N}}$ are a sequence of independent standard Brownian motions. Under the assumption $\sum_i \sigma_i^2 < \infty$, the series $\sum_{i=1}^\infty \sigma_i h_i \beta_i(t)$ defines an H-valued Brownian motion on $(\Omega, \mathcal{F}, \mathcal{F}_t, P)$, that we shall denote by $W(t)$. The canonical process (ξ_t) is a weak martingale solution of (2.2), in the sense that it satisfies (2.2) in the following weak form: there exists a Borel set $\Omega_0 \subset \Omega$ with $P(\Omega_0) = 1$ such that on Ω_0 for every $\varphi \in \mathcal{D}^\infty$ and $t \geq 0$ we have

$$\langle \xi_t - \xi_0, \varphi \rangle_H + \int_0^t \nu \langle \xi_s, A\varphi \rangle_H \, ds - \int_0^t \langle B(\xi_s, \varphi), \xi_s \rangle_H \, ds = \langle W(t), \varphi \rangle_H . \quad (2.3)$$

The following theorem is well known, see for instance the survey paper of Flandoli [9] and the reference therein.

Theorem 2.3. *Assume $\sum_i \sigma_i^2 < \infty$. Let μ be a probability measure on H such that $\int_H |x|_H^2 \mu(dx) < \infty$. Then there exists at least one solution to the martingale problem with initial condition μ.*

Definition 2.4. We say that $P(\cdot,\cdot,\cdot) : [0,\infty) \times D(A) \times \mathcal{B}(D(A)) \to [0,1]$ is a Markov kernel in $D(A)$ of transition probabilities associated to equation (1.1) if $P(\cdot,\cdot,\Gamma)$ is Borel measurable for every $\Gamma \in \mathcal{B}(D(A))$, $P(t,x,\cdot)$ is a probability measure on $\mathcal{B}(D(A))$ for every $(t,x) \in [0,\infty) \times D(A)$, the Chapman-Kolmogorov equation

$$P(t+s,x,\Gamma) = \int_{D(A)} P(t,x,dy)\, P(s,y,\Gamma)$$

holds for every $t, s \geq 0$, $x \in D(A)$, $\Gamma \in \mathcal{B}(D(A))$, and for every $x \in D(A)$ there is a solution P_x on (Ω, F) of the martingale problem associated to equation (2.2) with initial condition x such that

$$P(t,x,\Gamma) = P_x[\xi_t \in \Gamma] \text{ for all } t \geq 0.$$

We recall the following result from Da Prato & Debussche [2], Debussche & Odasso [7] or Flandoli & Romito [4]:

Theorem 2.5. *There exists at least one Markov kernel $P(t,x,\Gamma)$ in $D(A)$ of transition probabilities associated to equation (1.1).*

We recall that a $P(t,x,\Gamma)$ is called *irreducible* in $D(A)$ if for every $t>0$, x_0, $x_1 \in D(A)$, $\varepsilon > 0$, we have

$$P(t, x_0, B_A(x_1, \varepsilon)) > 0,$$

where $B_A(x_1,\varepsilon)$ is the ball in $D(A)$ of centre x_1 and radius ε.

We say that $P(t,x,\Gamma)$ is *strong Feller* in $D(A)$ if

$$x \mapsto \int_{D(A)} \varphi(y)\, P(t,x,dy)$$

is continuous on $D(A)$ for every bounded measurable function $\varphi : D(A) \to \mathbb{R}$ and for every $t > 0$. It is well known (see for example Da Prato & Zabczyk [10, Proposition 4.1.1]) that irreducibility and strong Feller in $D(A)$ imply that the laws $P(t,x,\cdot)$ are all mutually equivalent, as (t,x) varies in $(0,\infty)\times D(A)$. Because of this equivalence property, we say that $P(t,x,\Gamma)$ is *regular.*

We recall also that $P(t,x,\Gamma)$ is called *stochastically continuous* in $D(A)$ if $\lim_{t\to 0} P(t,x,B_A(x,\varepsilon)) = 1$ for every $x \in D(A)$ and $\varepsilon > 0$.

In Da Prato & Debussche [2], the transition probability kernel constructed by Galerkin approximations is proved to be stochastically continuous, irreducible and strong Feller in $D(A)$, hence regular. More generally (see Flandoli & Romito [4]):

Theorem 2.6. *Every Markov kernel $P(t,x,\Gamma)$ in $D(A)$ of transition probabilities associated to equation (1.1) is stochastically continuous, irreducible and strong Feller in $D(A)$, hence regular.*

3. The Log-Lipschitz estimate

Theorem 3.1. *Let $P(t,x,\Gamma)$ be a Markov kernel in $D(A)$ of transition probabilities associated to equation (1.1). Then, given $T > 0$, there is a constant C_T such that the inequality*

$$|P(t,x_0+h,\Gamma) - P(t,x_0,\Gamma)| \le \frac{C_T}{t\wedge 1}(1+|Ax_0|^6)|Ah|\log(|Ah|^{-1})$$

holds for every $t \in (0,T]$, x_0, $h \in D(A)$, with $|Ah| \le 1$, and $\Gamma \in \mathcal{B}(D(A))$.

We explain here only the logical skeleton of the proof, which is very simple. The two main technical ingredients will be treated in the next two separate subsections. The first idea is to decompose:

$$\begin{aligned} &P(t,x_0+h,\Gamma) - P(t,x_0,\Gamma) \\ &\quad = \int_{D(A)} [P(\varepsilon, x_0+h, dy) - P(\varepsilon, x_0, dy)]\, P(t-\varepsilon, y, \Gamma). \end{aligned}$$

To shorten some notation, let us write

$$(P_t\varphi)(x) = \int_{D(A)} \varphi(y)\, P(t,x,dy)$$

so, with the function $\varphi(x) = \mathbf{1}_{\{x \in \Gamma\}}$ the previous identity reads

$$(P_t\varphi)(x_0 + h) - (P_t\varphi)(x_0) = (P_\varepsilon(P_{t-\varepsilon}\varphi))(x_0 + h) - (P_\varepsilon(P_{t-\varepsilon}\varphi))(x_0). \quad (3.1)$$

It is now sufficient to estimate

$$(P_\varepsilon\psi)(x_0 + h) - (P_\varepsilon\psi)(x_0)$$

uniformly in $\psi \in B_b(D(A))$. The value of ε has to be chosen depending on the size of x_0 and h, as we shall see.

The second idea is to use an *initial coupling*: we introduce the equation with cut-off $\chi_R(|Au|^2)$, where $\chi_R(r) : [0,\infty) \to [0,1]$ is a non-increasing smooth function equal to 1 over $[0, R]$, to 0 over $[R+2,\infty)$, and with derivative bounded by 1. The equation is

$$\begin{aligned} &du + \left[Au + B(u,u)\chi_R\left(|Au|^2\right)\right] dt = \textstyle\sum_{i=1}^\infty \sigma_i h_i \, d\beta_i(t), \\ &u(0) = x. \end{aligned} \quad (3.2)$$

The definition of martingale problem for this equation is the same (with obvious adaptations) as the definition given above for equation (1.1). Let $\tau_R : \Omega \to [0,\infty]$ be defined as

$$\tau_R(\omega) = \inf\{t \ge 0 : |A\omega(t)| \ge R\}.$$

We recall the following result from Flandoli & Romito [4, Lemma 5.11]:

Lemma 3.2. *For every $x \in D(A)$ there is a unique solution $P_x^{(R)}$ of the martingale problem associated to equation (3.2), with the additional property*

$$P_x^{(R)}[\xi \in C([0,\infty); D(A))] = 1.$$

Let P_x be any solution on (Ω, F) of the martingale problem associated to equation (2.2) with initial condition x. Then

$$\mathbb{E}^{P_x^{(R)}}\left[\varphi(\xi_t)\mathbf{1}_{\{\tau_R \ge t\}}\right] = \mathbb{E}^{P_x}\left[\varphi(\xi_t)\mathbf{1}_{\{\tau_R \ge t\}}\right]$$

for every $t \ge 0$ and $\varphi \in B_b(D(A))$.

Introduce the notation

$$(P_t^{(R)}\varphi)(x) = \mathbb{E}^{P_x^{(R)}}[\varphi(\xi_t)].$$

The previous lemma implies that for every $\psi \in B_b(D(A))$ we have

$$|(P_\varepsilon\psi)(x) - (P_\varepsilon^{(R)}\psi)(x)| \le 2P_x[\tau_R < \varepsilon]\,\|\psi\|_\infty. \quad (3.3)$$

Summarising:

Corollary 3.3. *For every x_0, $h \in D(A)$ and $\psi \in B_b(D(A))$ we have*

$$\begin{aligned} |(P_\varepsilon\psi)(x_0 + h) - (P_\varepsilon\psi)(x_0)| &\le 2\left(P_{x_0+h}[\tau_R < \varepsilon] + P_{x_0}[\tau_R < \varepsilon]\right)\|\psi\|_\infty \\ &\quad + \left|(P_\varepsilon^{(R)}\psi)(x_0 + h) - (P_\varepsilon^{(R)}\psi)(x_0)\right|. \end{aligned}$$

Let us give now the proof of Theorem 3.1. Assume $t \in (0,T]$, x_0, $h \in D(A)$ be given, with $|Ah| \leq 1$. Let $K > 0$ be such that $|Ax_0| + 1 \leq K$. We have $|A(x_0+h)| \leq K$, so we may apply Proposition 3.5 below to both x_0 and $x_0 + h$. We thus get, for $\varepsilon \in (0, \frac{1}{5C^* K^2})$, where $C^* > 0$ is the constant defined by (A.3) in the Appendix, we have

$$P_{x_0+h}\left[\tau_{2K} < \varepsilon\right] + P_{x_0}\left[\tau_{2K} < \varepsilon\right] \leq 2C_{\#}\mathrm{e}^{-\eta_{\#}\frac{K^2}{4\varepsilon}}.$$

Given h, K and t as above, let us look for a value $\varepsilon \in (0, \frac{1}{5C^* K^2})$ such that $\varepsilon \leq t$ and the latter exponential quantity is smaller than $|Ah|$. We impose

$$\eta_{\#}\frac{K^2}{4\varepsilon} \geq \log(|Ah|^{-1})$$

hence it is sufficient to take

$$\varepsilon \leq \frac{\eta_{\#} K^2}{4\log(|Ah|^{-1})} \wedge \frac{t}{2} \wedge \frac{1}{5C^* K^2}. \tag{3.4}$$

We have proved so far the first claim of the following lemma. The second claim is a simple consequence of (3.1) and the previous corollary.

Lemma 3.4. *Given $t > 0$, x_0, $h \in D(A)$, with $|Ah| \leq 1$, and $\Gamma \in \mathcal{B}(D(A))$, if ε is chosen as in (3.4), then*

$$P_{x_0+h}\left[\tau_{2K} < \varepsilon\right] + P_{x_0}\left[\tau_{2K} < \varepsilon\right] \leq 2C_{\#}|Ah|$$

and for $\varphi(x) = \mathbf{1}_{\{x\in\Gamma\}}$ and $\psi = P_{t-\varepsilon}\varphi$,

$$|P_t\varphi(x_0+h) - P_t\varphi(x_0)| \leq 4C_{\#}|Ah|\,\|\varphi\|_\infty + |P_\varepsilon^{(2K)}\psi(x_0+h) - P_\varepsilon^{(2K)}\psi(x_0)|.$$

Finally, from Proposition 3.6 below, renaming the constant C, with $\varphi(x) = \mathbf{1}_{\{x\in\Gamma\}}$ and $\psi = P_{t-\varepsilon}\varphi$,

$$\left|(P_\varepsilon^{(2K)}\psi)(x_0+h) - (P_\varepsilon^{(2K)}\psi)(x_0)\right| \leq \frac{C}{\varepsilon}\,|Ah|\,e^{CK^6\varepsilon}.$$

Thus, for ε as in (3.4), we get

$$|P_t\varphi(x_0+h) - P_t\varphi(x_0)| \leq 4C_{\#}|Ah| + \frac{C}{\varepsilon}|Ah|\mathrm{e}^{CK^6\varepsilon}.$$

Let us further restrict ourselves to

$$\varepsilon \leq \frac{\eta_{\#} K^2}{4\log(|Ah|^{-1})} \wedge \frac{t}{2} \wedge \frac{1}{5C^* K^2} \wedge \frac{1}{K^6},$$

so that we have

$$|P_t\varphi(x_0+h) - P_t\varphi(x_0)| \leq 4C_{\#}|Ah| + \frac{C}{\varepsilon}|Ah|.$$

The choice

$$\varepsilon = C\frac{t\wedge 1}{K^6\log(|Ah|^{-1})}$$

is admissible for a suitable constant $C > 0$, and we finally get (1.2). The proof of Theorem 3.1 is complete.

3.1. *Probability of blow-up*

Proposition 3.5. *Let $K \geq 1$ and assume that $x_0 \in D(A)$ and $\varepsilon > 0$ are given such that $|Ax_0| \leq K$ and $\varepsilon \leq \frac{1}{5C^*K^2}$, where C^* is the constant defined in (A.3). Then*

$$P_{x_0}\left[\tau_{2K} < \varepsilon\right] \leq C_{\#} \mathrm{e}^{-\eta_{\#} \frac{K^2}{4\varepsilon}},$$

for suitable universal constants $\eta_{\#} > 0$ and $C_{\#} > 0$.

Proof. From Corollary A.2.2 we know that if $\varepsilon \leq \frac{1}{5C^*K^2}$ and $|Ax_0| \leq K$, then one has

$$\theta_\varepsilon^2 \leq \frac{1}{4}K^2 \quad \Rightarrow \quad |Au(s)| < 2K \text{ for } s \in [0,\varepsilon] \quad \Rightarrow \quad \tau_{2K} \geq \varepsilon,$$

where θ_ε is defined in (A.2) in the Appendix. Therefore, with the constraints $|Ax_0| \leq K$ and $\varepsilon \leq \frac{1}{5C^*K^2}$, by Proposition A.1.1 one gets

$$P_{x_0}\left[\tau_{2K} < \varepsilon\right] \leq P_{x_0}\left[\Theta_\varepsilon^2 > \frac{1}{4}K^2\right] \leq C_{\#} \mathrm{e}^{-\eta_{\#} \frac{K^2}{4\varepsilon}}.$$

□

3.2. *Derivative of the regularised problem*

Here we show the regularity of the transition semigroup associated to the regularised problem (3.2).

Proposition 3.6. *For every $R \geq 1$ and $x_0,\ h \in D(A)$,*

$$\left|\left(P_\varepsilon^{(R)}\psi\right)(x_0+h) - \left(P_\varepsilon^{(R)}\psi\right)(x_0)\right| \leq \frac{C\,\|\psi\|_\infty}{\varepsilon}\,|Ah|\,\mathrm{e}^{CR^6\varepsilon},$$

where C is a universal constant.

Proof. We write the following computations for the limit problem but the understanding is that we do it on the Galerkin approximations. For every $\psi \in B_b(H)$, $\varepsilon > 0$, from the Bismut-Elworthy-Li formula (see Da Prato & Zabczyk [10]),

$$\begin{aligned}&\left|\left(P_\varepsilon^{(R)}\psi\right)(x_0+h) - \left(P_\varepsilon^{(R)}\psi\right)(x_0)\right| \\ &\qquad\qquad \leq \frac{C\,\|\psi\|_\infty}{\varepsilon} \sup_{\eta\in[0,1]} \mathbb{E}\left[\left(\int_0^\varepsilon \left|A^{\frac{3}{2}} D_h u_{x_0+\eta h}^{(R)}(s)\right|^2 ds\right)^{\frac{1}{2}}\right],\end{aligned}$$

where, for each $R \geq 1$ and $x \in D(A)$, $u_x^{(R)}$ is the solution, starting at x, of problem (3.2). From the regularised equation we have

$$
\begin{aligned}
&\frac{1}{2}\frac{d}{dt}\left|AD_h u_x^{(R)}(t)\right|^2 + \left|A^{\frac{3}{2}}D_h u_x^{(R)}(t)\right|^2 \\
&\leq \chi_R(|Au_x^{(R)}(t)|^2)\left|\left\langle AD_h u_x^{(R)}, AB(D_h u_x^{(R)}, u_x^{(R)}) + AB(u_x^{(R)}, D_h u_x^{(R)})\right\rangle\right| \\
&\quad + 2\chi_R'(|Au_x^{(R)}(t)|^2)\langle Au_x^{(R)}, AD_h u_x^{(R)}\rangle|\langle AD_h u_x^{(R)}, AB(u_x^{(R)}, u_x^{(R)})\rangle| \\
&\leq C\chi_R\left(|Au_x^{(R)}(t)|^2\right)\left|A^{\frac{3}{2}}D_h u_x^{(R)}(t)\right|\left|AD_h u_x^{(R)}(t)\right|\left|Au_x^{(R)}(t)\right| \\
&\quad + C\chi_R'\left(\left|Au_x^{(R)}(t)\right|^2\right)\left|Au_x^{(R)}(t)\right|^3\left|AD_h u_x^{(R)}(t)\right|\left|A^{\frac{3}{2}}D_h u_x^{(R)}(t)\right| \\
&\leq \frac{1}{2}\left|A^{\frac{3}{2}}D_h u_x^{(R)}(t)\right|^2 + C\chi_R^2\left(\left|Au_x^{(R)}(t)\right|^2\right)\left|AD_h u_x^{(R)}(t)\right|^2\left|Au_x^{(R)}(t)\right|^2 \\
&\quad + C\chi_R'\left(|Au_x^{(R)}(t)|^2\right)^2\left|AD_h u_x^{(R)}(t)\right|^2\left|Au_x^{(R)}(t)\right|^6 \\
&\leq \frac{1}{2}\left|A^{\frac{3}{2}}D_h u_x^{(R)}(t)\right|^2 + CR^6\left|AD_h u_x^{(R)}(t)\right|^2.
\end{aligned}
$$

Thus

$$\frac{1}{2}\frac{d}{dt}\left|AD_h u_x^{(R)}(t)\right|^2 + \frac{1}{2}\left|A^{\frac{3}{2}}D_h u_x^{(R)}(t)\right|^2 \leq CR^6\left|AD_h u_x^{(R)}(t)\right|^2.$$

This implies

$$\left|AD_h u_x^{(R)}(t)\right|^2 \leq \mathrm{e}^{CR^6 t}|Ah|^2$$

and

$$\int_0^\varepsilon \left|A^{\frac{3}{2}}D_h u_{x_0+\eta h}^{(R)}(s)\right|^2 ds \leq |Ah|^2\left(1 + \int_0^\varepsilon CR^6\mathrm{e}^{CR^6 s}\,ds\right) = |Ah|^2\,\mathrm{e}^{CR^6\varepsilon}.$$

Thus

$$\left|\left(P_\varepsilon^{(R)}\psi\right)(x_0+h) - \left(P_\varepsilon^{(R)}\psi\right)(x_0)\right| \leq \frac{C\,\|\psi\|_\infty}{\varepsilon}|Ah|\,\mathrm{e}^{CR^6\varepsilon}.$$

The proposition is proved. □

4. Equivalence of all transition probabilities

To make the following statement independent of previous results, we shall assume stochastic continuity, irreducibility and the strong Feller property in the theorem below, but we recall that these properties have been proved for every Markov kernel in $D(A)$ associated to equation (1.1), under the assumptions of the introduction.

Theorem 4.1. *Let $P^{(i)}(t, x, \Gamma)$ be two Markov kernels in $D(A)$ of transition probabilities associated to equation (1.1). Assume they are stochastically continuous, irreducible and strong Feller in $D(A)$. Then the probability measures $P^{(1)}(t, x, \cdot)$ and $P^{(2)}(t', x', \cdot)$ are equivalent, for any t, $t' > 0$ and x, $x' \in D(A)$.*

Proof. Step 1. Let us recall the following fact (see Ref. [10]): if a Markov kernel of transition probabilities $P(t, x, \Gamma)$ in $D(A)$ (or any Polish space) is stochastically continuous, irreducible and strong Feller, then the laws $P(t_1, x_1, .)$ and $P(t_2, x_2, .)$ are equivalent, for every $t_1, t_2 > 0$, $x_1, x_2 \in D(A)$. Therefore, under our assumptions, it is sufficient to prove that, given $t_0 > 0$, $x_0 \in D(A)$, the laws $P^{(1)}(t_0, x_0, .)$ and $P^{(2)}(t_0, x_0, .)$ are equivalent. Since the argument is symmetric between the two kernels, it is sufficient to prove the following statement: given $t_0 > 0$, $x_0 \in D(A)$, if Γ is a Borel set in $D(A)$ such that $P^{(2)}(t_0, x_0, \Gamma) = 0$ then $P^{(1)}(t_0, x_0, \Gamma) = 0$. In the following steps we prove such claim.

Let us remark that, again by the equivalence property for $P^{(2)}$, the assumption that $P^{(2)}(t_0, x_0, \Gamma) = 0$ implies that $P^{(2)}(t, x, \Gamma) = 0$ for every $t > 0$, $x \in D(A)$. Therefore what we are going to prove is the following claim: given $t_0 > 0$, $x_0 \in D(A)$, let Γ be a Borel set in $D(A)$ such that $P^{(2)}(t, x, \Gamma) = 0$ for every $t > 0$, $x \in D(A)$; then $P^{(1)}(t_0, x_0, \Gamma) = 0$.

Step 2. Since both $P^{(1)}(\cdot,\cdot,\cdot)$ and $P^{(2)}(\cdot,\cdot,\cdot)$ satisfy (3.3),

$$P^{(1)}(t,x,\Gamma) = |P^{(1)}(t,x,\Gamma) - P^{(2)}(t,x,\Gamma)| \leq 2(P_x^{(1)}[\tau_R < t] + P_x^{(2)}[\tau_R < t]).$$

Now, for every pair (ε, x), with $\varepsilon > 0$ and $x \in D(A)$, such that $5C^*(1+|Ax|)^2\varepsilon \leq 1$ (the constant C^* is defined in (A.3), in the appendix), Proposition 3.5 implies that

$$P^{(1)}(\varepsilon,x,\Gamma) \leq 2C_{\#}\mathrm{e}^{-\eta_{\#}\frac{(1+|Ax|)^2}{4\varepsilon}} \leq 2C_{\#}\mathrm{e}^{-\frac{1}{4\varepsilon}\eta_{\#}}.$$

Step 3. For every $\varepsilon < \frac{1}{5C^*}$, set $A_\varepsilon = \{x \in D(A) : 5C^*(1+|Ax|)^2\varepsilon \leq 1\}$, then by the Markov property and the previous step,

$$\begin{aligned} P^{(1)}(t_0+\varepsilon,x_0,\Gamma) &= \int_{A_\varepsilon^c} P^{(1)}(\varepsilon,x,\Gamma)\,P^{(1)}(t_0,x_0,dx) \\ &\quad + \int_{A_\varepsilon} P^{(1)}(\varepsilon,x,\Gamma)\,P^{(1)}(t_0,x_0,dx) \\ &\leq 2C_{\#}\mathrm{e}^{-\frac{1}{4\varepsilon}\eta_{\#}} + P^{(1)}(t_0,x_0,A_\varepsilon^c) \end{aligned}$$

Since $P^{(1)}(s, x_0, D(A)) = 1$, we have $P^{(1)}(t_0,x_0,A_\varepsilon^c) \longrightarrow 0$, as $\varepsilon \to 0$, and thus

$$\lim_{\varepsilon\to 0} P^{(1)}(t_0+\varepsilon, x_0, \Gamma) = 0.$$

Step 4. Again by the Markov property, for every neighborhood G of x_0 in $D(A)$,

$$\begin{aligned} P^{(1)}(t_0+\varepsilon,x_0,\Gamma) &= \int P^{(1)}(t_0,y,\Gamma)\,P^{(1)}(\varepsilon,x_0,dy) \\ &\geq P^{(1)}(\varepsilon,x_0,G)\inf_{y\in G} P^{(1)}(t_0,y,\Gamma). \end{aligned}$$

Since the kernel $P^{(1)}$ is stochastically continuous, it follows that $P^{(1)}(\varepsilon,x_0,G) \to 1$, as $\varepsilon \to 0$, and so, by the previous step, $\inf_{y\in G} P^{(1)}(t_0,y,\Gamma) \longrightarrow 0$ as $\varepsilon \to 0$. By the strong Feller property, the map $y \mapsto P^{(1)}(t_0,y,\Gamma)$ is continuous, hence in conclusion $P^{(1)}(t_0,x_0,\Gamma) = 0$. The proof is complete. □

5. Conclusion and remarks

We have proved that the transition probabilities associated to any Markov selection are all equivalent to each other. However, the problem of uniqueness of Markov selections remains open. We stress that it would imply uniqueness of solutions to the martingale problem, by the argument of Stroock & Varadhan [1, Theorem 12.2.4].

The estimates proved in this work allows us at least to state a sufficient condition for uniqueness of Markov selections. The proof is inspired by a well known proof in semigroup theory as well as by the proof of uniqueness given by Bressan and co-authors (see for instance Ref. [11]).

Proposition 5.1. *Assume that a Markov selection $(P_x)_{x\in D(A)}$ has the following property: for every $t>0$ and $x \in D(A)$,*

$$\lim_{n\to\infty}\sum_{k=1}^{n} P\left(t-\frac{k}{n}t, x, B_A\left(0,\sqrt{\frac{n}{t}}\right)^{c}\right)=0$$

where $B_A(0,n)$ is the ball in $D(A)$ of radius n. Then $(P_x)_{x\in D(A)}$ coincides with any other Markov selection.

Proof. Let $(Q_x)_{x\in D(A)}$ be another Markov selection. Let us rewrite, for $\varphi \in C_b(D(A))$:

$$\begin{aligned} P_t\varphi - Q_t\varphi &= P_{t-\frac{t}{n}}P_{\frac{t}{n}}\varphi - P_{t-\frac{t}{n}}Q_{\frac{t}{n}}\varphi \\ &\quad + P_{t-\frac{t}{n}}Q_{\frac{t}{n}}\varphi - Q_{t-\frac{t}{n}}Q_{\frac{t}{n}}\varphi \end{aligned}$$

and so on iteratively until we have

$$P_t\varphi - Q_t\varphi = \sum_{k=1}^{n} P_{t-\frac{kt}{n}}\left(P_{\frac{t}{n}}\psi_{\frac{(k-1)t}{n}} - Q_{\frac{t}{n}}\psi_{\frac{(k-1)t}{n}}\right)$$

where $\psi_s = Q_s\varphi$. We have, by using (3.3) and Proposition 3.5,

$$\begin{aligned} &\left|P_{t-\frac{kt}{n}}\left(P_{\frac{t}{n}}\psi_{\frac{(k-1)t}{n}} - Q_{\frac{t}{n}}\psi_{\frac{(k-1)t}{n}}\right)(x)\right| \\ &= \left|\mathbb{E}^{P_x}\left[\left(P_{\frac{t}{n}}\psi_{\frac{(k-1)t}{n}} - Q_{\frac{t}{n}}\psi_{\frac{(k-1)t}{n}}\right)\left(\xi_{t-\frac{kt}{n}}\right)\right]\right| \\ &\le \mathbb{E}^{P_x}\left[\left|\left(P_{\frac{t}{n}}\psi_{\frac{(k-1)t}{n}} - Q_{\frac{t}{n}}\psi_{\frac{(k-1)t}{n}}\right)\left(\xi_{t-\frac{kt}{n}}\right)\right|\mathbf{1}_{\{\xi_{t-\frac{k}{n}t}\in A_{\frac{t}{n}}\}}\right] \\ &\quad + \mathbb{E}^{P_x}\left[\left|\left(P_{\frac{t}{n}}\psi_{\frac{(k-1)t}{n}} - Q_{\frac{t}{n}}\psi_{\frac{(k-1)t}{n}}\right)\left(\xi_{t-\frac{kt}{n}}\right)\right|\mathbf{1}_{\{\xi_{t-\frac{k}{n}t}\in A^{c}_{\frac{t}{n}}\}}\right] \\ &\le 4C_{\#}\mathrm{e}^{-\frac{n}{t}\eta_{\#}} + 2P_x[\xi_{t-\frac{k}{n}t}\in A^{c}_{\frac{t}{n}}] \\ &\le 4C_{\#}\mathrm{e}^{-\frac{n}{t}\eta_{\#}} + 2P\left(t-\frac{k}{n}t, x, B_A\left(0,\sqrt{\frac{n}{t}}\right)^{c}\right), \end{aligned}$$

where we have set $A_t = \{5C^*t(1+|Ax|)^2 \le 1\}$ and, roughly, $A_{\frac{t}{n}} \approx B_A\left(0, \sqrt{\frac{n}{t}}\right)$. Hence

$$|P_t\varphi(x) - Q_t\varphi(x)| \le 4nC_\# \mathrm{e}^{-\frac{n}{t}\eta_\#} + 2\sum_{k=1}^{n} P\left(t - \frac{k}{n}t, x, B_A\left(0, \sqrt{\frac{n}{t}}\right)^c\right)$$

which completes the proof of the proposition. □

The criterion of this proposition is apparently not really useful at the present stage of our understanding. Indeed, if we apply Chebichev inequality we get the sufficient condition

$$\lim_{n\to\infty} \sum_{k=1}^{n} \left(\frac{t}{n}\right)^{1+\varepsilon} \mathbb{E}^{P_x}\left[\left|A\xi_{t-\frac{kt}{n}}\right|^{2(1+\varepsilon)}\right] = 0$$

with is implied by the condition

$$\mathbb{E}^{P_x}\left[\int_0^t |A\xi_s|^{2(1+\varepsilon)}\, ds\right] < \infty$$

which however would easily imply the well posedness of the 3D Navier-Stokes equation by direct estimates of the difference of two solutions.

A.1. An exponential tail estimate for the Stokes problem

Consider the following Stokes problem

$$dZ + AZ\, dt = A^{-\frac{3}{2}}\, dW,$$
$$Z(0) = 0,$$

and set $\Theta_t = \sup_{s\in[0,t]} |AZ(s)|$. The next result is well known, but we give a proof to keep track of the dependence on the constants of interest in this paper.

Proposition A.1.1. *There exist $\eta_\# > 0$ and $C_\# > 0$ such that for every $K \ge \frac{1}{2}$ and $\varepsilon > 0$,*

$$\mathbb{P}\big[\Theta_\varepsilon \ge K\big] \le C_\# \mathrm{e}^{-\eta_\# \frac{K^2}{\varepsilon}}.$$

Proof. Step 1. Set $y(t) = \varepsilon^{-\frac{1}{2}} Z(\varepsilon t)$, then it is easy to see that y solves the equation $dy + \varepsilon Ay\, dt = Q^{\frac{1}{2}}\, dW$. Next, fix a value $\alpha \in (\frac{1}{6}, \frac{1}{4})$, then by the *factorisation method* (see Da Prato & Zabczyk [12, Chapter 5]),

$$y(t) = \int_0^t \mathrm{e}^{-\varepsilon(t-s)A}\, dW_s = C_\alpha \int_0^t \mathrm{e}^{-\varepsilon(t-s)A}(t-s)^{\alpha-1} Y(s)\, ds,$$

where $Y(s) = \int_0^s \mathrm{e}^{-\varepsilon(s-r)A}(s-r)^{-\alpha}\, dW_r$ and C_α denotes a generic constant depending only on α (it will keep changing value along the proof). For every $t \in (0,1]$, since $\alpha > \frac{1}{6}$, it follows from Hölder's inequality that

$$|Ay(t)|_H \le C_\alpha \int_0^t (t-s)^{\alpha-1} |AY(s)|_H\, ds \le C_\alpha \left(\int_0^1 |AY(s)|_H^6\, ds\right)^{\frac{1}{6}}.$$

In conclusion, since $\varepsilon^{-1}\Theta_\varepsilon^2 = \sup_{t\in[0,1]} |Ay(t)|_H^2$, it follows by the above inequality and standard arguments that

$$\mathbb{P}[\Theta_\varepsilon \geq K] \leq \mathrm{e}^{-\frac{a}{\varepsilon}K^2}\mathbb{E}\left[\exp\left(\tilde{a}\left(\int_0^1 |AY(s)|^6\,ds\right)^{\frac{1}{3}}\right)\right], \tag{A.1}$$

with a constant a that will be specified later (and $\tilde{a} = a\,C_\alpha$).
Step 2. In order to estimate the expectation in (A.1), notice that

$$\begin{aligned}\exp\left(\tilde{a}\left(\int_0^1 |AY(s)|^6\,ds\right)^{\frac{1}{3}}\right) &= \sum_{n=0}^{\infty}\frac{\tilde{a}^n}{n!}\left[\int_0^1 |AY(s)|_H^6\,ds\right]^{\frac{n}{3}} \\ &\leq \sum_{n=0}^{2}\frac{\tilde{a}^n}{n!}\left[\int_0^1 |AY(s)|_H^6\right]^{\frac{n}{3}} + \int_0^1 \sum_{n=3}^{\infty}\frac{\tilde{a}^n}{n!}|AY(s)|_H^{2n}\end{aligned} \tag{A.2}$$

$$\leq \tilde{a}\left[\int_0^1 |AY(s)|^{10}\,ds\right]^{\frac{1}{5}} + \frac{\tilde{a}^2}{2}\left[\int_0^1 |AY(s)|^8\,ds\right]^{\frac{1}{2}} + \int_0^1 \mathrm{e}^{\tilde{a}|AY(s)|_H^2}\,ds$$

Step 3. Now, $AY(s)$ is a centered Gaussian process with covariance (cfr. proof of Theorem 5.9 in Da Prato & Zabczyk [12])

$$\tilde{Q}_s = \int_0^s (s-r)^{-2\alpha}A^{-1}\mathrm{e}^{-2\varepsilon(s-r)A}\,dr,$$

so that, by Proposition 2.16 in Ref. [12],

$$\mathbb{E}[\mathrm{e}^{\tilde{a}|AY(s)|_H^2}] = \mathrm{e}^{-\frac{1}{2}\mathrm{Tr}[\log(1-2\tilde{a}\tilde{Q}_s)]},$$

provided that $\tilde{a} \leq \inf_{\lambda\in\sigma(\tilde{Q}_s)}\frac{1}{2\lambda}$, where $\sigma(\tilde{Q}_s)$ is the spectrum of $\tilde{Q}_s$. Similarly, $\mathbb{E}|AY(s)|^{2p} = C_p(\mathrm{Tr}(\tilde{Q}_s))^p$, for all integers p.

In order to choose a suitable value of a, let $\mu \in \sigma(\tilde{Q}_s)$, then there is a eigenvalue λ of A such that $\mu = \mu(\lambda)$ is given by

$$\mu = \lambda^{-1}\int_0^s r^{-2\alpha}\mathrm{e}^{-2r\lambda\varepsilon}\,dr = \lambda^{-2+2\alpha}(2\varepsilon)^{-(1-2\alpha)}\int_0^{2\lambda\varepsilon s} r^{-2\alpha}\mathrm{e}^{-r}\,dr \leq C_\alpha\lambda_0^{-1},$$

where λ_0 is the smallest eigenvalue of A. Hence a can be chosen as $C_\alpha\lambda_0$, for a suitable C_α.
Step 4. We conclude the proof: since a is small enough, we have that $-\mathrm{Tr}[\log(1-2\tilde{a}\tilde{Q}_s)] \leq C_\alpha\mathrm{Tr}[\tilde{Q}_s]$ and, as in *step 3*,

$$Tr[Q_s] = \sum_{\lambda\in\sigma(A)}\lambda^{-2+2\alpha}(2\varepsilon)^{-(1-2\alpha)}\int_0^{2\lambda\varepsilon s} r^{-2\alpha}\mathrm{e}^{-r}\,dr \leq C_\alpha\varepsilon^{-(1-2\alpha)},$$

where the sum in λ converges since $\alpha < \frac{1}{4}$ and $\lambda_n \approx n^{\frac{2}{3}}$. Hence, by (A.1) and (A.2),

$$\begin{aligned}\mathbb{P}[\Theta_\varepsilon \geq K] &\leq e^{-\frac{aK^2}{\varepsilon}} \mathbb{E}\left[\tilde{a}\left[\int_0^1 |AY(s)|^{10}\right]^{\frac{1}{5}} + \frac{\tilde{a}^2}{2}\left[\int_0^1 |AY(s)|^8\right]^{\frac{1}{2}} + \int_0^1 e^{\tilde{a}|AY(s)|_H^2}\right] \\ &\leq C_\alpha e^{-\frac{aK^2}{\varepsilon}}\left(e^{C_\alpha \varepsilon^{-(1-2\alpha)}} + \varepsilon^{-(1-2\alpha)} + \varepsilon^{-2(1-2\alpha)}\right) \\ &\leq C_\# e^{-\eta_\# \frac{K^2}{\varepsilon}},\end{aligned}$$

where $\eta_\#$ and $C_\#$ can be easily found, since $K \geq \frac{1}{2}$. □

A.2. The deterministic equation

The basic ingredient of our approach is the bunch of regular paths that every weak solution has for a positive local (random) time, when the initial condition is regular. It was called *regular jet* in Flandoli [6]. It is based on the solutions of the following deterministic equation

$$u(t) + \int_0^t \left(Au(s) + B(u,u)\right) ds = x + w(t). \tag{A.3}$$

We say that

$$u \in C\left([0,\infty; H_\sigma\right) \cap L^2_{loc}\left([0,\infty); V\right)$$

is a weak solution of (A.3) if

$$\langle u(t), \varphi\rangle + \int_0^t \left(\langle u(s), A\varphi\rangle - \langle B(u(s), \varphi), u(s)\rangle\right) ds = \langle x, \varphi\rangle + \langle w(t), \varphi\rangle$$

for every $\varphi \in \mathcal{D}^\infty$. Notice that all terms in the above definition are meaningful, included the quadratic one in u due to the estimate

$$|\langle B(u,v), z\rangle| \leq C\,|Dv|_{L^\infty} |u|_{L^2} |z|_{L^2}.$$

We take $w \in \Omega^*$ where

$$\Omega^* = \bigcap_{\substack{\beta\in(0,\frac{1}{2})\\ \alpha\in(0,\frac{3}{4})}} C^\beta\left([0,\infty); D(A^\alpha)\right).$$

Consider also the auxiliary Stokes equations

$$z(t) + \int_0^t Az(s)\, ds = w(t)$$

having the unique mild solution

$$z(t) = e^{-tA} w(t) - \int_0^t Ae^{-(t-s)A}\left(w(s) - w(t)\right) ds.$$

From elementary arguments based on the analytic estimates $\left|A^{\alpha}e^{-tA}\right| \leq \frac{C_{\alpha,T}}{t^{\alpha}}$ for $t \in (0,T)$, we have (see for instance Flandoli [13] for details)

$$z \in C\left([0,\infty); D(A)\right).$$

Let us set

$$\theta_T = \sup_{t\in[0,T]} |Az(t)|. \tag{A.4}$$

Let $C_0 > 0$ be the constant of inequality (2.1) and let

$$C^* := 4C_0^2. \tag{A.5}$$

Lemma A.2.1. *Given $x \in D(A)$ and $w \in \Omega^*$, let $K \geq |Ax|$ and $\varepsilon > 0$ be such that*

$$\left(K^2 + \theta_\varepsilon^2\right)\left(\frac{1}{2K^2} + C^*\varepsilon\right) < 1$$

Then there exists a solution $u \in C\left([0,\varepsilon]; D(A)\right)$, which is unique in the class of weak solutions, and $|Au(s)| < 2K$ for $s \in [0,\varepsilon]$.

Proof. We show only the quantitative estimate, the other statements being standard in the theory of Navier-Stokes equations. For simplicity, all computations will be made on the limit problem, although they should be made on its Galerkin approximations. The uniqueness of local solution ensures that the procedure is nevertheless correct.

Set $v = u - z$, then

$$\frac{dv}{dt} + Av + B(u,u) = 0$$

and, by using (2.1),

$$\begin{aligned}
\frac{d}{dt}|Av|^2 + 2\|Av\|_V^2 &\leq 2\left|\langle Av, AB(u,u)\rangle\right| \leq 2\|Av\|_V\left|A^{1/2}B(u,u)\right| \\
&\leq 2C_0\|Av\|_V|Au|^2 \leq \|Av\|_V^2 + C_0^2|Au|^4 \\
&\leq \|Av\|_V^2 + C^*(|Av|^2 + |Az|^2)^2.
\end{aligned}$$

Hence on $[0,\varepsilon]$ we have that

$$\frac{d}{dt}|Av|^2 \leq C^*(|Av|^2 + \theta_\varepsilon^2)^2,$$

and so, if we set $y(t) = |Av(t)|^2 + \theta_\varepsilon^2$, it follows that

$$\frac{dy}{dt} \leq C^* y^2, \qquad \text{on } [0,\varepsilon].$$

Consequently, since $y > 0$ (except for the irrelevant case $w \equiv 0$), we have

$$y(s) \leq \frac{y(0)}{1 - C^* s y(0)},$$

namely,

$$|A(u(s)-z(s))|^2+\theta_\varepsilon^2 \le \frac{|Ax|^2+\theta_\varepsilon^2}{1-C^*s\left(|Ax|^2+\theta_\varepsilon^2\right)}$$

for $s\in[0,\varepsilon]$. Therefore

$$|Au(s)|^2 \le \frac{2(|Ax|^2+\theta_\varepsilon^2)}{1-C^*s\left(|Ax|^2+\theta_\varepsilon^2\right)} \le \frac{2(K^2+\theta_\varepsilon^2)}{1-C^*s\,(K^2+\theta_\varepsilon^2)}.$$

This result is true until $1-C^*s\left(K^2+\theta_\varepsilon^2\right)>0$, namely for $s\in[0,\frac{1}{C^*(K^2+\theta_\varepsilon^2)})$. The assumption of the lemma ensures that $[0,\varepsilon]$ is included in this interval. Thus the last inequality is true at least on $[0,\varepsilon]$. Moreover, again by the assumption of the lemma,

$$\frac{2(K^2+\theta_\varepsilon^2)}{4K^2} < 1-C^*s\left(K^2+\theta_\varepsilon^2\right)$$

that implies

$$\frac{2(K^2+\theta_\varepsilon^2)}{1-C^*s\,(K^2+\theta_\varepsilon^2)} < 4K^2,$$

and thus $|Au(s)|^2<4K^2$, for $s\in[0,\varepsilon]$. □

Corollary A.2.2. *Assume there are $K>0$ and $\varepsilon>0$ such that*

$$\varepsilon \le \frac{1}{5C^*K^2} \qquad \text{and} \qquad \theta_\varepsilon^2 \le \frac{1}{4}K^2,$$

then, for every $x\in D(A)$ such that $|Ax|\le K$, we have $|Au(s)|<2K$ for $s\in[0,\varepsilon]$.

References

[1] D. W. Stroock and S. R. S. Varadhan, *Multidimensional diffusion processes.* vol. 233, *Grundlehren der Mathematischen Wissenschaften [Fundamental Principles of Mathematical Sciences]*, (Springer-Verlag, Berlin, 1979). ISBN 3-540-90353-4.

[2] G. Da Prato and A. Debussche, Ergodicity for the 3D stochastic Navier-Stokes equations, *J. Math. Pures Appl. (9)*. **82**(8), 877–947, (2003). ISSN 0021-7824.

[3] R. Mikulevicius and B. L. Rozovskii, Global L_2-solutions of stochastic Navier-Stokes equations, *Ann. Probab.* **33**(1), 137–176, (2005). ISSN 0091-1798.

[4] F. Flandoli and M. Romito. Markov selections for the 3D stochastic Navier-Stokes equations. `http://www.arxiv.org/abs/math.PR/0602612`.

[5] F. Flandoli and M. Romito, Markov selections and their regularity for the three-dimensional stochastic Navier-Stokes equations, *C. R. Math. Acad. Sci. Paris.* To appear.

[6] F. Flandoli, On the method of Da Prato and Debussche for the 3D stochastic Navier Stokes equations, *J. Evol. Eq.* To appear.

[7] A. Debussche and C. Odasso. Markov solutions for the 3d stochastic Navier-Stokes equations with state dependent noise. `http://www.arxiv.org/abs/math.AP/0512361`.

[8] R. Mikulevicius and B. L. Rozovskii, Stochastic Navier-Stokes equations for turbulent flows, *SIAM J. Math. Anal.* **35**(5), 1250–1310 (electronic), (2004). ISSN 0036-1410.

[9] F. Flandoli. An introduction to 3D stochastic fluid dynamics. In *SPDE in hydrodynamics: recent progress and prospects (CIME course)*, Lecture Notes in Mathematics. Springer. `http://www.cime.unifi.it`.

[10] G. Da Prato and J. Zabczyk, *Ergodicity for infinite-dimensional systems*. vol. 229, *London Mathematical Society Lecture Note Series*, (Cambridge University Press, Cambridge, 1996). ISBN 0-521-57900-7.

[11] A. Bressan, *Hyperbolic systems of conservation laws*. vol. 20, *Oxford Lecture Series in Mathematics and its Applications*, (Oxford University Press, Oxford, 2000). ISBN 0-19-850700-3. The one-dimensional Cauchy problem.

[12] G. Da Prato and J. Zabczyk, *Stochastic equations in infinite dimensions*. vol. 44, *Encyclopedia of Mathematics and its Applications*, (Cambridge University Press, Cambridge, 1992). ISBN 0-521-38529-6.

[13] F. Flandoli, Stochastic differential equations in fluid dynamics, *Rend. Sem. Mat. Fis. Milano*. **66**, 121–148, (1996). ISSN 0370-7377.

Chapter 11

Rate of Convergence of Implicit Approximations for Stochastic Evolution Equations

István Gyöngy and Annie Millet*

School of Mathematics and Maxwell Institute for Mathematical Sciences, University of Edinburgh
King's Buildings, Edinburgh, EH9 3JZ, United Kingdom
gyongy@maths.ed.ac.uk

Stochastic evolution equations in Banach spaces with unbounded nonlinear drift and diffusion operators are considered. Under some regularity condition assumed for the solution, the rate of convergence of implicit Euler approximations is estimated under strong monotonicity and Lipschitz conditions. The results are applied to a class of quasilinear stochastic PDEs of parabolic type.

Contents

1. Introduction

Let $V \hookrightarrow H \hookrightarrow V^*$ be a *normal triple* of spaces with dense and continuous embeddings, where V is a separable and reflexive Banach space, H is a Hilbert space, identified with its dual by means of the inner product in H, and V^* is the dual of V. Thus $\langle v, h\rangle = (v, h)$ for all $v \in V$ and $h \in H^* = H$, where $\langle v, v^*\rangle = \langle v^*, v\rangle$ denotes the duality product of $v \in V$, $v^* \in V^*$, and (h_1, h_2) denotes the inner product of $h_1, h_2 \in H$. Let $W = \{W(t) : t \geq 0\}$ be a d_1-dimensional Brownian motion carried

*Laboratoire de Probabilités et Modèles Aléatoires (CNRS UMR 7599), Universités Paris 6-Paris 7, Boîte Courrier 188, 4 place Jussieu, 75252 Paris Cedex 05, amil@ccr.jussieu.fr, and Centre d'Economie de la Sorbonne (CNRS UMR 8174), Équipe SAMOS-MATISSE, Université Paris 1 Panthéon Sorbonne, 90 Rue de Tolbiac, 75634 Paris Cedex 13, annie.millet@univ-paris1.fr

by a stochastic basis $(\Omega, \mathcal{F}, (\mathcal{F}_t)_{t\geq 0}, P)$. Consider the stochastic evolution equation

$$u(t) = u_0 + \int_0^t A(s, u(s))\, ds + \sum_{k=1}^{d_1} \int_0^t B_k(s, u(s))\, dW^k(s)\,, \tag{1.1}$$

where u_0 is a V-valued $\mathcal{F}_0$-measurable random variable, A and B are (non-linear) adapted operators defined on $[0, \infty[\times V \times \Omega$ with values in V^* and $H^{d_1} := H \times \ldots \times H$, respectively.

It is well-known, see Refs. [1–3], that this equation admits a unique solution if the following conditions are met: There exist constants $\lambda > 0$, $K \geq 0$ and an $(\mathcal{F}_t)$-adapted non-negative locally integrable stochastic process $f = \{f_t : t \geq 0\}$ such that

(i) (Monotonicity) There exists a constant K such that

$$2\langle u - v, A(t, u) - A(t, v)\rangle + \sum_{k=1}^{d_1} |B_k(t, u) - B_k(t, v)|_H^2 \leq K|u - v|_H^2, \tag{1.2}$$

(ii) (Coercivity)

$$2\langle v, A(t, v)\rangle + \sum_{k=1}^{d_1} |B_k(t, v)|_H^2 \leq -\lambda|v|_V^2 + K|v|_H^2 + f(t),$$

(iii) (Linear growth)

$$|A(t, v)|_{V^*}^2 \leq K|v|_{V^*}^2 + f(t),$$

(iv) (Hemicontinuity)

$$\lim_{\lambda \to 0} \langle w, A(t, v + \lambda u)\rangle = \langle w, A(t, v)\rangle$$

hold for all for $u, v, w \in V$, $t \in [0, T]$ and $\omega \in \Omega$.

Under these conditions equation (1.1) has a unique solution u on $[0, T]$. (See Definition 2.3 below for the definition of the solution.) Moreover, if $E|u_0|_H^2 < \infty$ and $E\int_0^T f(t)\, dt < \infty$, then

$$E \sup_{t\leq T} |u(t)|_H^2 + E\int_0^T |u(t)|_V^2\, dt < \infty.$$

In Ref. [4] it is shown that under these conditions, approximations defined by various implicit and explicit schemes converge to u.

Our aim is to prove rate of convergence estimates for these approximations. To achieve this aim we require stronger assumptions: a strong monotonicity condition on A, B and a Lipschitz condition on B in $v \in V$. In the present paper we consider implicit time discretizations. Note that without space discretizations, in general, explicit time discretizations do not converge. Consider, for example, the heat equation $du(t) = \Delta u(t)$, with initial condition $u(0) = u_0 \in L_2(\mathbb{R}^d)$. Then the explicit time discretization on the grid $\{k/n\}_{k=0}^n$ gives the approximation $u_n(k/n) := (I + \Delta/n)^k u_0$ at time $t = k/n$. Hence clearly, if $u_0 \notin \cap_{i=1}^\infty W_2^i(\mathbb{R}^d)$, then $u(k/n)$ does not belong to the Sobolev space $W_2^l(\mathbb{R}^d)$, with any fixed negative index l, when k is sufficiently large.

The study of various space-time discretization schemes will be done in the continuation of the present paper.

We require also the following time regularity from the solution u (see condition (T2)): $E|u_0|_V^2 < \infty$, almost surely $u_t \in V$ for every $t \in T$, and there exist some constants C and $\nu > 0$ such that

$$E|u(t) - u(s)|_V^2 \le C\,|t - s|^{2\nu}\,,$$

for all $s, t \in [0, T]$. Note that unlike the solutions to stochastic differential equations, the solutions to stochastic PDEs can satisfy this condition with a variety of exponents ν, different from $1/2$, due to the interplay between space and time regularities of the solutions. (See Ref. [5] for space and time regularity of the solutions to stochastic parabolic PDEs of second order.) Note also that our general setting allows us to cover a large class of stochastic parabolic PDEs of order $2m$ for any $m \ge 1$ (see Ref. [1] for the class of stochastic parabolic SPDEs of order $2m$ and Ref. [6] for the stochastic Cahn-Hilliard equation).

In the case of time independent operators A and B we obtain the rate of convergence for the implicit approximation u^τ corresponding to the mesh size $\tau = T/m$ of the partition of $[0, T]$

$$E\max_{i\le m}|u(i\tau) - u^\tau(i\tau)|_H^2 + E\sum_{i\le m}|u(i\tau) - u^\tau(i\tau)|_V^2\tau \le C\tau^\nu,$$

where C is a constant independent of τ. If in addition to the above assumptions A is also Lipschitz continuous in $v \in V$ then the order of convergence is doubled,

$$E\max_{i\le m}|u(i\tau) - u^\tau(i\tau)|_H^2 + E\sum_{i\le m}|u(i\tau) - u^\tau(i\tau)|_V^2\tau \le C\tau^{2\nu}.$$

In the case of time dependent A and B it is natural to assume that they are Hölder continuous in t in order to control the error due to their discretization in time. However, it is possible to control this discretization error when the operator A is not even continuous in t, if we discretize it by taking the average of $A(s)$ over the intervals $[t_i, t_{i+1}]$. This explains the discretization of $A(t)$ and condition (T1) below. If both operators A and B are Hölder continuous in time then we use also the obvious discretization: $A^\tau_{t_i} = A(t_{i+1}, .)$ and $B^\tau_{k,t_i} = B(t_i, .)$.

As examples we present a class of quasi-linear stochastic partial differential equations (SPDEs) of parabolic type, and show that it satisfies our assumptions. Thus we obtain rate of convergence results also for implicit approximations of linear parabolic SPDEs, in particular, for the Zakai equation of nonlinear filtering. We refer to Refs. [3, 7–9] for basic results for the stochastic PDEs of nonlinear filtering.

We will extend these results to degenerate parabolic SPDEs, and to space-time explicit and implicit schemes for stochastic evolution equations in the continuation of this paper.

In Section 2 we give a precise description of the schemes and state the assumptions on the coefficients which ensure the convergence of these schemes to the solution u of (1.1). In Section 3 estimates for the speed of convergence of time implicit schemes are stated and proved. Finally, in the last section, we give a class of examples of quasi-linear stochastic PDEs for which all the assumptions of the main theorem, Theorem 3.4, are fulfilled.

As usual, we denote by C a constant which can change from line to line.

2. Preliminaries and the approximation scheme

Let $(\Omega, \mathcal{F}, (\mathcal{F}_t)_{t\geq 0}, P)$ be a stochastic basis, satisfying the usual conditions, i.e., $(\mathcal{F}_t)_{t\geq 0}$ is an increasing right-continuous family of sub-σ-algebras of $\mathcal{F}$ such that $\mathcal{F}_0$ contains every P-null set. Let $W = \{W(t) : t \geq 0\}$ be a d_1-dimensional Wiener martingale with respect to $(\mathcal{F}_t)_{t\geq 0}$, i.e., W is an $(\mathcal{F}_t)$-adapted Wiener process with values in $\mathbb{R}^{d_1}$ such that $W(t) - W(s)$ is independent of $\mathcal{F}_s$ for all $0 \leq s \leq t$.

Let T be a given positive number. Consider the stochastic evolution equation (1.1) for $t \in [0, T]$ in a triplet of spaces

$$V \hookrightarrow H \equiv H^* \hookrightarrow V^*,$$

satisfying the following conditions: V is a separable and reflexive Banach space over the real numbers, embedded continuously and densely into a Hilbert space H, which is identified with its dual H^* by means of the inner product $(\cdot,\cdot)$ in H, such that $(v, h) = \langle v, h\rangle$ for all $v \in V$ and $h \in H$, where $\langle\cdot,\cdot\rangle$ denotes the duality product between V and V^*, the dual of V. Such triplet of spaces is called a normal triplet.

Let us state now our assumptions on the initial value u_0 and the operators A, B in the equation. Let

$$A : [0, T] \times V \times \Omega \to V^*, \quad B : [0, T] \times V \times \Omega \to H^{d_1}$$

be such that for every $v, w \in V$ and $1 \leq k \leq d_1$, $\langle A(s, v), w\rangle$ and $(B_k(s, v), w)$ are adapted processes and the following conditions hold:

(C1) The pair (A, B) satisfies the *strong monotonicity condition*, i.e., there exist constants $\lambda > 0$ and $L > 0$ such almost surely

$$2\,\langle u - v, A(t, u) - A(t, v)\rangle + \sum_{k=1}^{d_1} |B_k(t, u) - B_k(t, v)|_H^2$$
$$+\lambda\,|u - v|_V^2 \leq L\,|u - v|_H^2 \tag{2.1}$$

for all $t \in]0, T]$, u and v in V.

(C2) (*Lipschitz condition on B*) There exists a constant L_1 such that almost surely

$$\sum_{k=1}^{d_1} |B_k(t, u) - B_k(t, v)|_H^2 \leq L_1\,|u - v|_V^2 \tag{2.2}$$

for all $t \in [0,T]$, u and v in V.

(C3) (*Lipschitz condition on* A) There exists a constant L_2 such that almost surely

$$|A(t,u) - A(t,v)|_{V^*}^2 \le L_2 \, |u-v|_V^2 \tag{2.3}$$

for all $t \in [0,T]$, u and v in V.

(C4) $u_0 : \Omega \to V$ is $\mathcal{F}_0$-measurable and $E|u_0|_V^2 < \infty$. There exist non-negative random variables K_1 and K_2 such that $EK_i < \infty$, and

$$\sum_{k=1}^{d_1} |B_k(t,0)|_H^2 \le K_1 \, , \tag{2.4}$$

$$|A(t,0)|_{V^*}^2 \le K_2 \tag{2.5}$$

for all $t \in [0,T]$ and $\omega \in \Omega$.

Remark 2.1. If $\lambda = 0$ in (2.1) then one says that (A, B) satisfies the *monotonicity condition*. Notice that this condition together with the Lipschitz condition (2.3) on A implies the Lipschitz condition (2.2) on B.

Remark 2.2. (1) Clearly, (2.3)–(2.5) and (2.2)–(2.4) imply that A and B satisfy the *growth condition*

$$\sum_{j=1}^{d_1} |B_k(t,v)|_H^2 \le 2L_1 |v|_V^2 + 2K_1, \tag{2.6}$$

and

$$|A(t,v)|_{V^*}^2 \le 2L_2 \, |v|_V^2 + 2K_2 \tag{2.7}$$

respectively, for all $t \in [0,T]$, $\omega \in \Omega$ and $v \in V$.

(2) Condition (2.3) obviously implies that the operator A is *hemicontinuous*:

$$\lim_{\varepsilon \to 0} \langle A(t, u + \varepsilon v), w \rangle = \langle A(t,u), w \rangle \tag{2.8}$$

for all $t \in [0,T]$ and $u, v, w \in V$.

(3) The strong monotonicity condition (C1), (C2), (2.4) and (2.5) yield that the pair (A, B) satisfies the following *coercivity condition*: there exists a non-negative random variable K_3 such $EK_3 < \infty$ and almost surely

$$2 \, \langle v, A(t,v) \rangle + \sum_{k=1}^{d_1} |B_k(t,v)|_H^2 + \tfrac{\lambda}{2} \, |v|_V^2 \le L|v|_H^2 + K_3 \tag{2.9}$$

for all $t \in]0,T]$, $\omega \in \Omega$ and $v \in V$.

Proof. We show only (3). By the strong monotonicity condition (C1)

$$2 \, \langle v, A(t,v) \rangle + \sum_{k=1}^{d_1} |B_k(t,v)|_H^2 + \lambda \, |v|_V^2 \le L|v|_H^2 + R_1(t) + R_2(t) \tag{2.10}$$

with

$$R_1(t) = 2\,\langle v, A(t,0)\rangle,$$

$$R_2(t) = \sum_{k=1}^{d_1} |B_k(t,0)|_H^2 + 2\sum_{k=1}^{d_1} \Big(B_k(t,v) - B_k(t,0)\,,\, B_k(t,0)\Big)\,.$$

Using (C2) and (2.5), we have

$$|R_1(t)| \le \tfrac{\lambda}{4}|v|_V^2 + \tfrac{4K_2}{\lambda}\,,$$

$$|R_2(t)| \le 2\left(\sum_{j=1}^{d_1} |B_k(t,v) - B_k(t,0)|_H^2\right)^{\frac12} \left(\sum_{k=1}^{d_1} |B_k(t,0)|_H^2\right)^{\frac12} + K_1$$
$$\le \tfrac{\lambda}{4}|v|_V^2 + CK_1.$$

Thus, (2.10) concludes the proof of (2.9). □

Definition 2.3. An H-valued adapted continuous process $u = \{u(t) : t \in [0,T]\}$ is a solution to equation (1.1) on $[0,T]$ if almost surely :
$u(t) \in V$ for almost every $t \in [0,T]$,

$$\int_0^T |u(t)|_V^2\, dt < \infty\,, \tag{2.11}$$

and

$$(u(t),v) = (u_0,v) + \int_0^t \langle A(s,u(s)),v\rangle\, ds + \sum_{k=1}^{d_1} \int_0^t (B_k(s,u(s)),v)\, dW^k(s) \tag{2.12}$$

holds for all $t \in [0,T]$ and $v \in V$. We say that the solution to (1.1) on $[0,T]$ is unique if for any solutions u and v to (1.1) on $[0,T]$ we have

$$P\Big(\sup_{t\in[0,T]} |u(t) - v(t)|_H > 0\Big) = 0.$$

The following theorem is well-known (see Refs. [1–3]).

Theorem 2.4. *Let A and B satisfy the monotonicity, coercivity, linear growth and hemicontinuity conditions (i)-(iv) formulated in the Introduction. Then for every H-valued $\mathcal{F}_0$-measurable random variable u_0, equation (1.1) has a unique solution u on $[0,T]$. Moreover, if $E|u_0|_H^2 < \infty$ and $E\int_0^T f(t)\,dt < \infty$, then*

$$E\big(\sup_{t\in[0,T]} |u(t)|_H^2\big) + E\int_0^T |u(t)|_V^2\, dt < \infty \tag{2.13}$$

holds.

Hence by the previous remarks we have the following corollary.

Corollary 2.5. *Assume that conditions (C1), (C2) hold. Then for every H-valued random variable u_0 equation* (1.1) *has a unique solution u, and if $E|u_0|_H^2 < \infty$, then* (2.13) *holds.*

Approximation scheme. For a fixed integer $m \geq 1$ and $\tau := T/m$ we define the approximation u^τ for the solution u by an implicit time discretization of equation (1.1) as follows:

$$\begin{aligned} u^\tau(t_0) &= u_0 \,, \\ u^\tau(t_{i+1}) &= u^\tau(t_i) + \tau\, A^\tau_{t_i}(u^\tau(t_{i+1})) \\ &\quad + \sum_{k=1}^{d_1} B^\tau_{k,t_i}(u^\tau(t_i))\left(W^k(t_{i+1}) - W^k(t_i)\right) \quad \text{for } 0 \leq i < m, \end{aligned} \tag{2.14}$$

where $t_i := i\tau$ and

$$A^\tau_{t_i}(v) = \frac{1}{\tau}\int_{t_i}^{t_{i+1}} A(s,v)\,ds \,, \tag{2.15}$$

$$B^\tau_{k,0}(v) = 0, \quad B^\tau_{k,t_{i+1}}(v) = \frac{1}{\tau}\int_{t_i}^{t_{i+1}} B_k(s,v)\,ds \tag{2.16}$$

for $i = 0, 1, 2, ..., m$.

A random vector $u^\tau := \{u^\tau(t_i) : i = 0, 1, 2, ..., m\}$ is called a solution to scheme (2.14) if $u^\tau(t_i)$ is a V-valued $\mathcal{F}_{t_i}$-measurable random variable such that $E|u^\tau(t_i)|_V^2 < \infty$ and (2.14) holds for every $i = 0, \cdots, m-1$.

We use the notation

$$\kappa_1(t) := i\tau \quad \text{for } t \in [i\tau, (i+1)\tau[, \text{ and } \kappa_2(t) := (i+1)\tau \quad \text{for } t \in]i\tau, (i+1)\tau] \tag{2.17}$$

for integers $i \geq 0$, and set

$$A_t(v) = A_{t_i}(v), \quad B_{k,t}(v) = B_{t_i}(v)$$

for $t \in [t_i, t_{i+1}[$, $i = 0, 1, 2, ...m-1$ and $v \in V$.

Another possible choice is

$$A^\tau_{t_i}(u) = A(t_{i+1}, u) \quad \text{and} \quad B^\tau_{k,t_i}(u) = B_k(t_i, u) \quad \text{for } i = 0, 1, \cdots, m-1. \tag{2.18}$$

The following theorem establishes the existence and uniqueness of u^τ for large enough m, and provides estimates in V and in H. We remark that in practice (2.14) should also be solved numerically. This is possible for example by Galerkin's approximations and by finite elements methods. In the continuation of this paper we consider explicit and implicit time discretization schemes together with simultaneous 'space discretizations', and we estimate the error of the corresponding approximations for (1.1).

Theorem 2.6. *Assume that A and B satisfy the monotonicity, coercivity, linear growth and hemicontinuity conditions (i)–(iv). Assume also that (C4) holds. Let A^τ and B^τ be defined either by* (2.15) *and* (2.16)*, or by* (2.18)*. Then there exist an* *integer m_0 and a constant C, such that for $m \geq m_0$ equation (2.14) has a unique solution $\{u^\tau(t_i) : i = 0, 1, ..., m\}$, and*

$$E \max_{0\leq i\leq m} \left|u^\tau(i\tau)\right|_H^2 + E\sum_{i=1}^{m} \left|u^\tau(i\tau)\right|_V^2 \tau \leq C\,. \tag{2.19}$$

Proof. For the sake of simplicity, we only give the proof in the case A^τ and B^τ are defined by (2.15) and (2.16). This theorem with estimate

$$\max_{0\leq i\leq m} E\left|u^\tau(i\tau)\right|_H^2 + E\sum_{i=1}^{m} \left|u^\tau(i\tau)\right|_V^2 \tau \leq C \tag{2.20}$$

in place of (2.19) is proved in Ref. [4] for a slightly different implicit scheme. For the above implicit scheme the same proof can be repeated without essential changes. For the convenience of the reader we recall from Ref. [4] that the existence and uniqueness of the solution $\{u^\tau(t_i) : i = 0, 1, 2, ..., m\}$ to (2.14)–(2.16) is based on the following proposition (Proposition 3.4 from Ref. [4]): Let $D : V \to V^*$ be a mapping such that

(a) D is monotone, i.e., for every $x, y \in V$, $\langle D(x) - D(y), x - y\rangle \geq 0$;
(b) D is hemicontinuous, i.e., $\lim_{\varepsilon\to 0}\langle D(x + \varepsilon y), z\rangle = \langle D(x), z\rangle$ for every $x, y, z \in V$;
(c) there exist positive constants K, C_1 and C_2, such that

$$|D(x)|_{V^*} \leq K\,(1 + |x|_V), \quad \langle D(x), x\rangle \geq C_1\,|x|_V^2 - C_2\,, \quad \forall x \in V.$$

Then for every $y \in V^*$, there exists $x \in V$ such that $D(x) = y$ and

$$|x|_V^2 \leq \frac{C_1 + 2\,C_2}{C_1} + \frac{1}{C_1^2}\,|y|_{V^*}^2\,.$$

If there exists a positive constant C_3 such that

$$\langle D(x_1) - D(x_2), x_1 - x_2\rangle \geq C_3\,|x_1 - x_2|_{V^*}^2\,, \quad \forall x_1, x_2 \in V\,, \tag{2.21}$$

then for any $y \in V^*$, the equation $D(x) = y$ has a unique solution $x \in V$.
Note that for each $i = 1, 2, ...m-1$ equation (2.14) for $x := u^\tau(t_{i+1})$ can be rewritten as $Dx = y$ with

$$D := I - \tau A_{t_i}^\tau, \quad y := u^\tau(t_i) + \sum_{k=1}^{d_1} B_{k,t_i}^\tau\left(u^\tau(t_i)\right)\left(W^k(t_{i+1}) - W^k(t_i)\right)$$

where I denotes the identity on V. It is easy to verify that due to conditions (i)–(iv) and (C4) the operator D satisfies the conditions (a), (b) and (c) for sufficiently large m. Thus a solution $\{u^\tau(t_i) : i = 0, 1, ..., m\}$ can be obtained by recursion on i for all m greater than some m_0. To show the uniqueness we need only verify (2.21). By (2.15) and by the monotonicity condition (i) we have

$$\langle D(x_1) - D(x_2), x_1 - x_2\rangle = |x_1 - x_2|_H^2 - \int_{t_i}^{t_{i+1}} \langle A(s, x_1) - A(s, x_2), x_1 - x_2\rangle\, ds$$

$$\geq |x_1 - x_2|_H^2 - K\tau|x_1 - x_2|_H^2 = (1 - K\tau)|x_1 - x_2|_H^2,$$

where the constant K is from (1.2). Hence it is clear that (2.21) holds if m is sufficiently large.

Now we show (2.20). From the definition of $u^\tau(t_{i+1})$ we have

$$|u^\tau(t_j)|_H^2 = |u_0|_H^2 + \mathcal{I}(t_j) + \mathcal{J}(t_j) + \mathcal{K}(t_j) - \sum_{i=1}^{j} |A_{t_i}^\tau(u^\tau(i\tau))|_H^2 \tau \tag{2.22}$$

for $t_j = j\tau$, $j = 0, 1, 2, ...m$, where

$$\mathcal{I}(t_j) := 2\int_0^{t_j} \langle u^\tau(\kappa_2(s)), A(s, u^\tau(\kappa_2(s)))\rangle \, ds,$$
$$\mathcal{J}(t_j) := \sum_{1\leq i<j} |\sum_k B_{k,t_i}^\tau(u^\tau(i\tau))(W^k(t_{i+1}) - W^k(t_i))|_H^2,$$
$$\mathcal{K}(t_j) := 2\sum_k \int_0^{t_j} \left(u^\tau(\kappa_1(s)), B_{k,s}^\tau(u^\tau(\kappa_1(s)))\right) dW^k(s),$$

and κ_1, κ_2 are piece-wise constant functions defined by (2.17). By Itô's formula for every $k, l = 1, 2, ..., d_1$,

$$(W^k(t_{i+1}) - W^k(t_i))(W^l(t_{i+1}) - W^l(t_i)) = \delta_{kl}(t_{i+1} - t_i) + M^{kl}(t_{i+1}) - M^{kl}(t_i),$$

where $\delta_{kl} = 1$ for $k = l$ and 0 otherwise, and

$$M^{kl}(t) := \int_0^t (W^k(s) - W^k(\kappa_1(s)))\, dW^l(s) + \int_0^t (W^l(s) - W^l(\kappa_1(s)))\, dW^k(s).$$

Thus we get

$$\mathcal{J}(t_j) = \mathcal{J}_1(t_j) + \mathcal{J}_2(t_j),$$

with

$$\mathcal{J}_1(t_j) := \sum_{1\leq i<j} \sum_k |B_{k,t_i}^\tau(u^\tau(t_i))|_H^2 \tau\,,$$
$$\mathcal{J}_2(t_j) := \int_0^{t_j} \sum_{k,l} (B_{k,s}^\tau(u^\tau(\kappa_1(s))), B_{l,s}^\tau(u^\tau(\kappa_1(s))))\, dM^{kl}(s).$$

By the Davis inequality we have

$$
\begin{aligned}
E\max_{j\le m}|\mathcal{J}_2(t_j)| &\le 3\sum_{k,l} E\left\{\int_0^T |B^\tau_{k,s}(u^\tau(\kappa_1(s)))|_H^2 |B^\tau_{l,s}(u^\tau(\kappa_1(s)))|_H^2\, d\langle M^{kl}\rangle(s)\right\}^{1/2} \\
&\le C_1\sum_{k,l} E\left\{\int_0^T |B^\tau_{k,s}(u^\tau(\kappa_1(s)))|_H^4 |W^l(s)-W^l(\kappa_1(s))|^2\, ds\right\}^{1/2} \\
&\le C_1\sum_{k,l} E\Big[\max_{j\le m}|B^\tau_{k,t_j}(u^\tau(t_j))|_H\sqrt{\tau} \\
&\quad\times\left\{\frac{1}{\tau}\int_0^T |B^\tau_{k,s}(u^\tau(\kappa_1(s)))|_H^2\,|W^l(s)-W^l(\kappa_1(s))|^2 ds\right\}^{1/2}\Big] \\
&\le d_1C_1\sum_k \tau E\max_{j\le m}|B^\tau_{k,t_j}(u^\tau(t_j))|_H^2 \\
&\quad + C_1\tau^{-1}\sum_{k,l} E\int_0^T |B^\tau_{k,s}(u^\tau(\kappa_1(s)))|_H^2\,|W^l(s)-W^l(\kappa_1(s))|^2 ds \\
&\le C_2\Big(1+E\sum_{j\le m}|u^\tau(j\tau)|_V^2\tau\Big),
\end{aligned}
$$

where C_1 and C_2 are constants, independent of τ. Here we use that by Jensen's inequality for every k

$$
\sum_{1\le i<j}|B^\tau_{k,t_i}(u^\tau(i\tau))|_H^2\tau \le \int_0^{t_j}|B_k(s,u^\tau(\kappa_2(s))|_H^2\, ds,
$$

and that the coercivity condition (ii) and the growth condition on (iii) imply the growth condition (2.6) on B with some constant L_1 and random variable K_1 satisfying $EK_1<\infty$. Hence by taking into account the coercivity condition we obtain

$$
\begin{aligned}
&E\max_{j\le m}\left[\mathcal{I}(t_j)+\mathcal{J}(t_j)\right] \\
&\quad\le E\max_{j\le m}\int_0^{t_j}\Big[2\left\langle u^\tau(\kappa_2(s)),\, A(s,u^\tau(\kappa_2(s)))\right\rangle+\sum_k|B_k(s,u^\tau(\kappa_2(s))|_H^2\Big]\,ds \\
&\quad\quad + E\max_{j\le m}|\mathcal{J}_2(t_j)| \\
&\quad\le C\Big(1+\max_{j\le m}E|u^\tau(j\tau)|_H^2+E\sum_{j=1}^m|u^\tau(j\tau)|_V^2\tau\Big) \qquad (2.23)
\end{aligned}
$$

with a constant C independent of τ. By using the Davis inequality again we obtain

$$
\begin{aligned}
E \max_{j\le m} |\mathcal{K}(t_j)| &\le 6\, E \left\{ \int_0^T \sum_k \left| \left(u^\tau(\kappa_1(s)), B^\tau_{k,s}(u^\tau(\kappa_1(s))\right)\right|^2 ds \right\}^{1/2} \\
&\le 6\, E \left[\max_{j\le m} \left|u^\tau(j\tau)\right|_H \left\{ \int_0^T \sum_k |B^\tau_{k,s}(u^\tau(\kappa_1(s)))|_H^2\, ds \right\}^{1/2} \right] \\
&\le \tfrac{1}{2}\, E \max_{j\le m} |u^\tau(j\tau)|_H^2 + 18\, E \int_0^T \sum_k \left|B^\tau_{k,s}(u^\tau(\kappa_1(s)))\right|_H^2 ds \\
&\le \tfrac{1}{2}\, E \max_{j\le m} |u^\tau(j\tau)|_H^2 + C \left(1 + E \sum_{j\le m} |u^\tau(j\tau)|_V^2 \tau \right)
\end{aligned}
\tag{2.24}
$$

with a constant C independent of τ. From (2.20)–(2.24) we get

$$
\begin{aligned}
E \max_{j\le m} |u^\tau(j\tau)|_H^2 &\le E|u_0|_H^2 + E \max_{j\le m} \left(\mathcal{I}(t_j) + \mathcal{J}(t_j)\right) + E \max_{j\le m} |\mathcal{K}(t_j)| \\
&\le \tfrac{1}{2}\, E \max_{j\le m} |u^\tau(j\tau)|_H^2 + C\,(1 + \max_{j\le m} E|u^\tau(j\tau)|_H^2 + E \sum_{j\le m} |u^\tau(j\tau)|_V^2 \tau) \\
&\le \tfrac{1}{2}\, E \max_{j\le m} |u^\tau(j\tau)|_H^2 + C\,(1+L) < \infty
\end{aligned}
$$

by virtue of (2.20), which proves the estimate (2.19). □

3. Convergence results

In order to obtain a speed of convergence, we require further properties from $B(t,v)$ and from the solution u of (1.1).

We assume that there exists a constant $\nu \in]0,1/2]$ such that:

(T1) The coefficient B satisfies the following *time-regularity*: There exists a constant C and a random variable $\eta \ge 0$ with finite first moment, such that almost surely

$$
\sum_{k=1}^{d_1} |B_k(t,v) - B_k(s,v)|_H^2 \le |t-s|^{2\nu}(\eta + C|v|_V^2) \tag{3.1}
$$

for all $s \in [0,T]$ and $v \in V$.

(T2) The solution u to equation (1.1) satisfies the following *regularity* property: almost surely $u(t) \in V$ for all $t \in [0,T]$, and there exists a constant $C > 0$ such that

$$
E|u(t) - u(s)|_V^2 \le C\,|t-s|^{2\nu} \tag{3.2}
$$

for all $s,t \in [0,T]$.

Remark 3.1. Clearly, (3.2) implies

$$
\sup_{t\in[0,T]} E|u(t)|_V^2 < \infty. \tag{3.3}
$$

Finally, in order to prove a convergence result in the H norm uniformly in time, we also have to require the following uniform estimate on the V-norm of u:

(T3) There exists a random variable ξ such that $E\xi^2 < \infty$ and

$$\sup_{t \le T} |u(t)|_V \le \xi \quad \text{(a.s.)}.$$

To establish the rate of convergence of the approximations we first suppose that the coefficients A and B satisfy the Lipschitz property.

Theorem 3.2. *Suppose that the conditions (C1)-(C4), (T1) and (T2) hold. Let A^τ and B^τ be defined by* (2.15) *and* (2.16). *Then there exist a constant C and an integer $m_0 \ge 1$ such that*

$$\sup_{0 \le l \le m} E|u(l\tau) - u^\tau(l\tau)|_H^2 + E\sum_{j=0}^{m} |u(j\tau) - u^\tau(j\tau)|_V^2 \tau \le C\tau^{2\nu} \tag{3.4}$$

for all integers $m \ge m_0$.

The following proposition plays a key role in the proof.

Proposition 3.3. *Assume assumptions (i) through (iv) from the Introduction and let A^τ and B^τ be defined by* (2.15) *and* (2.16). *Suppose, moreover condition (C4). Then*

$$\begin{aligned}
|u(t_l) - u^\tau(t_l)|_H^2 &= 2\int_0^{t_l} \big\langle u(\kappa_2(s)) - u^\tau(\kappa_2(s)), A(s,u(s)) - A(s,u^\tau(\kappa_2(s)))\big\rangle\, ds \\
&+ \sum_{i=0}^{l-1} \left| \int_{t_i}^{t_{i+1}} \sum_{k=1}^{d_1} \left[B_k(s,u(s)) - B_{k,s}^\tau(u^\tau(t_i))\right] dW^k(s) \right|_H^2 \\
&+ 2\sum_{k=1}^{d_1} \int_0^{t_l} \big(B_k(s,u(s)) - B_{k,s}^\tau(u^\tau(t_i))\,,\, u(\kappa_1(s)) - u^\tau(\kappa_1(s))\big)\, dW^k(s) \\
&- \sum_{i=0}^{l-1} \left| \int_{t_i}^{t_{i+1}} \left[A(s,u(s)) - A(s,u^\tau(t_{i+1}))\right] ds \right|_H^2
\end{aligned} \tag{3.5}$$

holds for every $l = 1, 2, ..., m$.

Proof. Using (2.14) we have for any $i = 0, \cdots, m-1$

$$
\begin{aligned}
&|u(t_{i+1}) - u^\tau(t_{i+1})|_H^2 - |u(t_i) - u^\tau(t_i)|_H^2 = \\
&\quad 2\int_{t_i}^{t_{i+1}} \big\langle u(t_{i+1}) - u^\tau(t_{i+1}), A(s,u(s)) - A(s,u^\tau(t_{i+1}))\big\rangle\, ds \\
&\quad + 2\sum_{k=1}^{d_1}\Big(\int_{t_i}^{t_{i+1}} \big[B_k(s,u(s)) - B_{k,s}^\tau(u^\tau(t_i))\big]\, dW^k(s)\,,\, u(t_{i+1}) - u^\tau(t_{i+1})\Big) \\
&\quad - \Big|\int_{t_i}^{t_{i+1}} \big[A(s,u(s)) - A(s,u^\tau(t_{i+1}))\big]\, ds \\
&\qquad + \sum_{k=1}^{d_1}\int_{t_i}^{t_{i+1}} \big[B_k(s,u(s)) - B_{k,s}^\tau(u^\tau(t_i))\big]\, dW^k(s)\Big|_H^2 \\
&= 2\int_{t_i}^{t_{i+1}} \big\langle u(t_{i+1}) - u^\tau(t_{i+1}), A(s,u(s)) - A(s,u^\tau(t_{i+1}))\big\rangle\, ds \\
&\quad + \Big|\sum_{k=1}^{d_1}\int_{t_i}^{t_{i+1}} \big[B_k(s,u(s)) - B_{k,s}^\tau(u^\tau(t_i))\big]\, dW^k(s)\Big|_H^2 \\
&\quad + 2\sum_{k=1}^{d_1}\Big(\int_{t_i}^{t_{i+1}} \big[B_k(s,u(s)) - B_{k,s}^\tau(u^\tau(t_i))\big]\, dW^k(s)\,,\, u(t_i) - u^\tau(t_i)\Big) \\
&\quad - \Big|\int_{t_i}^{t_{i+1}} \big[A(s,u(s)) - A(s,u^\tau(t_{i+1}))\big]\, ds\Big|_H^2
\end{aligned}
$$

Summing up for $i = 1, \cdots, l-1$, we obtain (3.5). □

Proof of Theorem 3.2.
Taking expectations in both sided of (3.5) and using the strong monotonicity condition (C1), we deduce that for $l = 1, \cdots, m$,

$$
\begin{aligned}
&E|u(t_l) - u^\tau(t_l)|_H^2 \\
&\le E\int_0^{t_l} 2\big\langle u(\kappa_2(s)) - u^\tau(\kappa_2(s)), A(s,u(\kappa_2(s))) - A(s,u^\tau(\kappa_2(s)))\big\rangle ds \\
&\quad + \sum_{k=1}^{d_1} E\int_0^{t_{l-1}} |B_k(s,u(\kappa_2(s))) - B_k(s,u^\tau(\kappa_2(s)))|_H^2\, ds + \sum_{k=1}^{3} R_k \\
&\le -\lambda E\int_0^{t_l} |u(\kappa_2(s)) - u^\tau(\kappa_2(s))|_V^2\, ds + L\tau E|u(t_l) - u^\tau(t_l)|_H^2 \\
&\quad + L\tau \sum_{i=1}^{l-1} E|u(t_i) - u^\tau(t_i)|_H^2\, ds + \sum_{k=1}^{3} R_k\,, \qquad (3.6)
\end{aligned}
$$

where

$$R_1 = E\int_0^{t_l} 2\left\langle u(\kappa_2(s)) - u^\tau(\kappa_2(s)), A(s,u(s)) - A(s,u(\kappa_2(s)))\right\rangle ds\,,$$

$$R_2 = \sum_{k=1}^{d_1} E\int_0^{\tau} |B_k(s,u(s))|_H^2\, ds\,,$$

$$R_3 = \sum_{k=1}^{d_1}\sum_{i=1}^{l-1} E\Big[\int_{t_i}^{t_{i+1}} ds\, \Big|B_k(s,u(s)) - \frac{1}{\tau}\int_{t_{i-1}}^{t_i} B_k(t,u^\tau(t_i))\,dt\Big|_H^2 - \int_{t_{i-1}}^{t_i} |B_k(t,u(t_i)) - B_k(t,u^\tau(t_i))|_H^2\, dt\Big].$$

The Lipschitz property of A imposed in (2.3), (3.2) and Schwarz's inequality imply

$$\begin{aligned}|R_1| &\le L_2\, E\int_0^{t_l} |u(\kappa_2(s)) - u^\tau(\kappa_2(s))|_V\, |u(s) - u(\kappa_2(s))|_V\, ds\,,\\ &\le L_2\left(E\int_0^{t_l} |u(\kappa_2(s)) - u^\tau(\kappa_2(s))|_V^2 ds\right)^{\frac12}\left(E\int_0^{t_l} |u(s) - u(\kappa_2(s))|_V^2 ds\right)^{\frac12}\\ &\le \frac{\lambda}{3}E\int_0^{t_l} |u(\kappa_2(s)) - u^\tau(\kappa_2(s))|_V^2 ds + C\tau^{2\nu}.\end{aligned} \tag{3.7}$$

A similar computation based on (2.2) yields

$$\begin{aligned}|R_3| &\le \sum_{k=1}^{d_1}\sum_{i=1}^{l-1} E\int_{t_{i-1}}^{t_i} dt \frac{1}{\tau}\int_{t_i}^{t_{i+1}} ds\Big(|B_k(s,u(s)) - B_k(t,u^\tau(t_i))|_H^2\\ &\qquad - |B_k(t,u(t_i))) - B_k(t,u^\tau(t_i)))|_H^2\Big)\\ &\le \frac{\lambda}{3}E\int_0^{t_{l-1}} |u(\kappa_2(t)) - u^\tau(\kappa_2(t))|_V^2 dt + C\,R_3'\end{aligned}$$

where

$$R_3' = \sum_{k=1}^{d_1} E\frac{1}{\tau}\int_{t_1}^{t_l} ds \int_{\kappa_1(s)-\tau}^{\kappa_1(s)} dt\, |B_k(s,u(s)) - B_k(t,u(\kappa_2(t)))|_H^2.$$

Hence, using (2.2), (3.1) and (3.2) we have

$$\begin{aligned}R_3' &\le \sum_{k=1}^{d_1} E\frac{1}{\tau}\int_{t_1}^{t_l} ds\int_{\kappa_1(s)-\tau}^{\kappa_1(s)} dt\Big[|B_k(s,u(s)) - B_k(t,u(s))|_H^2\\ &\qquad + |B_k(t,u(s)) - B_k(t,u(t))|_H^2 + |B_k(t,u(t)) - B_k(t,u(\kappa_2(t)))|_H^2\Big]\\ &\le E\int_{t_1}^{t_l} \tau^{2\nu}\, |u(s)|_V^2\, ds + CE\frac{1}{\tau}\int_0^{t_{l-1}} dt\int_{\kappa_2(t)}^{\kappa_2(t)+\tau} ds\Big[|u(s) - u(t)|_V^2\\ &\qquad + |u(t) - u(\kappa_2(t)|_V^2\Big] \le C\,\tau^{2\nu}.\end{aligned}$$

Hence

$$|R_3| \le C\,\tau^{2\nu} + \frac{\lambda}{3} E \int_0^{t_l} |u(\kappa_2(s)) - u^\tau(\kappa_2(s))|_V^2 ds\,. \tag{3.8}$$

Furthermore (2.6) and (3.3) imply

$$|R_2| \le C\tau \tag{3.9}$$

with a constant C independent of τ. By inequalities (3.6)–(3.9), for sufficiently large m,

$$E|u(t_l) - u^\tau(t_l)|_H^2 + \frac{\lambda}{3} E \int_0^{t_l} |u(\kappa_2(s)) - u^\tau(\kappa_2(s))|_V^2 ds \\ \le \sum_{i=1}^{l-1} L\,\tau\, E|u(t_i) - u^\tau(t_i)|_H^2 + C\tau^{2\nu}. \tag{3.10}$$

Since $\sup_m \sum_{i=1}^m L\,\tau < +\infty$, a discrete version of Gronwall's lemma yields that there exists $C > 0$ such that for m large enough

$$\sup_{0\le l\le m} E|u(t_l) - u^\tau(t_l)|_H^2 \le C\tau^{2\nu}.$$

This in turn with (3.2) implies

$$E \int_0^T |u(s) - u^\tau(\kappa_2(s))|_V^2 ds \le C\tau^{2\nu},$$

which completes the proof of the theorem. □

Assume now that the solution u of equation (1.1) satisfies also the condition (T3) Then we can improve the estimate (3.4) in the previous theorem.

Theorem 3.4. *Let (C1)-(C4) and (T1)–(T3) hold, and let A^τ and B^τ be defined by (2.15) and (2.16). Then for all sufficiently large m*

$$E \max_{0\le j\le m} |u(j\tau) - u^\tau(j\tau)|_H^2 + E \sum_{j=0}^m |u(j\tau) - u^\tau(j\tau)|_V^2 \tau \le C\,\tau^{2\nu} \tag{3.11}$$

holds, where C is a constant independent of τ.

Proof. For $k = 1, \cdots, d_1$, set

$$F_k(t) = B_k(t, u(t)) - B_{k,t}^\tau(u^\tau(\kappa_1(t)))$$

$$m(t) = \sum_{k=1}^{d_1} \int_0^t F_k(s)\, dW^k(s) \quad \text{and} \quad G(s) = m(s) - m(\kappa_1(s)).$$

Then by Itô's formula

$$|m(t_{i+1}) - m(t_i)|_H^2 = 2\int_{t_i}^{t_{i+1}} \sum_k (G(s)\,,\, F_k(s))\, dW^k(s) + \sum_{k=1}^{d_1} \int_{t_i}^{t_{i+1}} |F_k(s)|_H^2\, ds$$

for $i = 0, ..., m-1$. Hence by using (3.5) we deduce that for $l = 1, \cdots, m$

$$|u(t_l) - u^\tau(t_l)|_H^2 \le I_1(t_l) + I_2(t_l) + 2M_1(t_l) + 2M_2(t_l) \tag{3.12}$$

with

$$I_1(t) := 2\int_0^t \langle u(\kappa_2(s)) - u^\tau(\kappa_2(s))\,,\, A(s,u(s)) - A(s,u^\tau(\kappa_2(s)))\rangle\, ds,$$

$$I_2(t) := \sum_{k=1}^{d_1} \int_0^t |B_k(s,u(s)) - B_{k,s}^\tau(u^\tau(\kappa_1(s)))|_H^2\, ds,$$

$$M_1(t) := \sum_{k=1}^{d_1} \int_0^t (G(s)\,,\, F_k(s))\, dW^k(s),$$

$$M_2(t) := \sum_{k=1}^{d_1} \int_0^t \big(F_k(s)\,,\, u(\kappa_1(s)) - u^\tau(\kappa_1(s))\big)\, dW^k(s).$$

By (C3)

$$\sup_{0\le l\le m} |I_1(t_l)| \le \int_0^T |u(\kappa_2(s)) - u^\tau(\kappa_2(s))|_V^2\, ds + L_2 \int_0^T |u(s) - u^\tau(\kappa_2(s))|_V^2\, ds$$
$$\le (1+2L_2)\sum_{i=1}^m |u(t_i) - u^\tau(t_i)|_V^2 \tau + 2L_2 \int_0^T |u(s) - u(\kappa_2(s))|_V^2\, ds.$$

Hence by Theorem 3.2 and by condition (T2)

$$E \sup_{0\le l\le m} |I_1(t_l)| \le C\tau^{2\nu}, \tag{3.13}$$

where C is a constant independent of τ.

Using Jensen's inequality, (2.6) and condition (T1) we have for $s \le \tau$

$$\sum_k |F_k(s)|_H^2 = \sum_k |B_k(s,u(s))|_H^2 \le 2\, L_1\, |u(s)|_H^2 + 2\, K_1, \tag{3.14}$$

while for $s \in [t_i, t_{i+1}]$, $1 \le i \le m$, one has for some constant C independent of τ

$$\begin{aligned}\sum_k |F_k(s)|_H^2 &\le \frac{1}{\tau}\sum_k \int_{t_{i-1}}^{t_i} |B_k(s,u(s)) - B_k(r,u^\tau(t_i))|_H^2\, dr \\ &\le 3\frac{1}{\tau}\sum_k \int_{t_{i-1}}^{t_i} \Big[|B_k(s,u(s)) - B_k(r,u(s))|_H^2 \\ &\quad + |B_k(r,u(s)) - B_k(r,u(t_i))|_H^2 + |B_k(r,u(t_i)) - B_k(u^\tau(t_i))|_H^2\Big]\, dr \\ &\le C\Big[\tau^{2\nu}\left(\eta + |u(s)|_V^2\right) + |u(s)-u(t_i)|_V^2 + |u(t_i) - u^\tau(t_i)|_V^2\Big]. \end{aligned} \tag{3.15}$$

Thus, (3.14) and (3.15) yield

$$\begin{aligned}\sup_{0\le l\le m} |I_2(t_l)| \le\ & C\int_0^\tau |u(s)|_V^2\, ds + C\tau + C\,\tau^{2\nu}\int_0^T (\eta + C\,|u(s)|_V^2)\, ds \\ & + C\int_0^T |u(s) - u(\kappa_2(s))|_V^2\, ds + C\sum_{i=1}^m |u(t_i) - u^\tau(t_i)|_V^2\, \tau. \end{aligned}$$

Hence by Theorem 3.2 and by condition (T2)

$$E \sup_{0\le l\le m} |I_2(t_l)| \le C\tau^{2\nu}, \tag{3.16}$$

where C is a constant independent of τ. Since $\sup_{0\le s\le T}\sum_k |F_k(s)|_H^2$ need not be measurable, we denote by Γ the set of random variables ζ satisfying

$$\sup_{0\le s\le T}\sum_k |F_k(s)|_H^2 \le \zeta \quad \text{(a.s.)}.$$

For $\zeta\in\Gamma$, the Davis inequality, and the simple inequality $ab \le \frac{\tau}{2}a^2 + \frac{1}{2\tau}b^2$ yield

$$\begin{aligned}E\sup_{1\le l\le m} |M_1(t_l)| &\le 3\,E\left(\int_0^T \sum_{k=1}^{d_1} |(F_k(s)\,,\, G(s))|^2\, ds\right)^{\frac12} \\ &\le 3\,E\left(\zeta^{1/2}\left[\int_0^T |G(s)|_H^2\, ds\right]^{\frac12}\right) \\ &\le \frac{3}{2}\tau \inf_{\zeta\in\Gamma} E\zeta + \frac{3}{2\tau}E\int_0^T |G(s)|_H^2\, ds. \end{aligned} \tag{3.17}$$

By (2.6) and (3.15) we deduce

$$\begin{aligned}\sup_{0\le s\le T}\sum_k |F_k(s)|_H^2 &\le C\,\tau^{2\nu}\big(\sup_{0\le s\le T} |u(s)|_V^2 + \eta + 1\big) + C\max_{1\le i\le m} |u(t_i) - u^\tau(t_i)|_V^2 \\ &\le C\Big(1 + \xi + \max_{1\le i\le m} |u(t_i) - u^\tau(t_i)|_V^2\Big), \end{aligned}$$

where ξ is the random variable from condition (T3) and C is a constant, independent of τ. Hence Theorem 3.2 yield

$$\tau \inf_{\zeta\in\Gamma} E\zeta \le \tau\, C\,(E\eta + E\xi) + C\,\tau \sum_{i=1}^{m} E|u(t_i) - u^\tau(t_i)|_V^2 \le C_1\,\tau^{2\nu}, \tag{3.18}$$

where C_1 is a constant, independent of τ. Similarly, due to conditions (T1)-(T2) and Theorem 3.2

$$E\sum_k \int_0^T |F_k(s)|_H^2\,ds \le C\,\tau^{2\nu}\Big(1 + E\int_0^T |u(s)|_V^2\,ds\Big) + C\,\tau^{2\nu} + C\,\tau\, E\sum_{i=1}^{m} |u(t_i) - u^\tau(t_i)|_V^2 \le C\,\tau^{2\nu} \tag{3.19}$$

with a constant C, independent of τ. Furthermore, the isometry of stochastic integrals and (3.22) yield

$$\frac{1}{\tau}E\int_0^T |G(t)|_H^2\,dt \le \frac{1}{\tau}E\int_0^T \Big|\int_{\kappa_1(t)}^t \sum_k F_k(s)\,dW^k(s)\Big|_H^2\,dt \le \frac{1}{\tau}E\int_0^T dt\int_{\kappa_1(t)}^t \sum_k |F_k(s)|_H^2\,ds \le C\,\tau^{2\nu}. \tag{3.20}$$

Thus from (3.17) by (3.18) and (3.20) we have

$$E\sup_{1\le l\le m} |M_1(t_l)| \le C\tau^{2\nu} \tag{3.21}$$

Finally, the Davis inequality implies

$$E\sup_{1\le l\le m} |M_2(t_l)|_H \le 3\,E\Big(\int_0^T \sum_k |(F_k(s)\,,\, u(\kappa_1(s)) - u^\tau(\kappa_1(s)))|^2\,ds\Big)^{\frac{1}{2}} \le \frac{1}{4}\,E\sup_{1\le l\le m} |u(\kappa_1(s)) - u^\tau(\kappa_1(s)))|_H^2 + 18\,E\int_0^{t_j} |F_k(s)|_H^2\,ds. \tag{3.22}$$

Thus, from (3.12) by inequalities (3.13), (3.16), (3.21) and (3.22) we obtain

$$\frac{1}{2}E\sup_{1\le l\le m} |u(t_l) - u^\tau(t_l)|_H^2 \le C\,\tau^{2\nu},$$

with a constant C, independent of τ, which with (3.4) completes the proof of the theorem. □

We now prove that if the coefficient A does not satisfy the Lipschitz property (C3) but only the coercivity and growth conditions (2.7)-(2.9), then the order of convergence is divided by two.

Theorem 3.5. *Let A and B satisfy the conditions (C1), (C2) and (C4). Suppose that conditions (T1) and (T2) hold, and let A^τ and B^τ be defined by (2.15) and*

(2.16). *Then there exists a constant* C*, independent of* τ*, such that for all sufficiently large* m

$$\sup_{0\le j\le m} E|u(j\tau)-u^\tau(j\tau)|_H^2 + E\sum_{j=1}^{m} |u(j\tau)-u^\tau(j\tau)|_V^2\,\tau \le C\,\tau^\nu\,. \tag{3.23}$$

Proof. Using (3.5), taking expectations and using (C1) with $u(s)$ and $u^\tau(\kappa_2(s))$, we obtain for every $l=1\cdots,m$

$$\begin{aligned} E|u(t_l)-u^\tau(t_l)|_H^2 \le{}& -\lambda E\int_0^{t_l} |u(s)-u^\tau(\kappa_2(s)|_V^2\,ds \\ &+E\int_0^{t_l} K_1\,|u(s)-u^\tau(\kappa_2(s)|_H^2\,ds + \sum_{k=1}^{3}\bar R_i\,, \end{aligned}\tag{3.24}$$

where

$$\bar R_1 = \sum_{j=1}^{r} 2E\int_0^{t_l} \big\langle u(\kappa_2(s))-u(s)\,,\, A(s,u(s))-A(s,u^\tau(\kappa_2(s)))\big\rangle\,ds\,,$$

$$\bar R_2 = \sum_{k=1}^{d_1} E\int_0^{T} |B_k(s,u(s))|_H^2\,ds\,,$$

$$\begin{aligned}\bar R_3 ={}& \sum_{k=1}^{d_1}\sum_{i=1}^{l-1} E\frac{1}{\tau}\int_{t_i}^{t_{i+1}} ds\int_{t_{i-1}}^{t_i} dt\Big[|B_k(s,u(s))-B_k(t,u^\tau(t_i)))|_H^2 \\ &- |B_k(t,u(t))-B_k(t,u^\tau(t_i))|_H^2\Big]\,.\end{aligned}$$

Using (2.7), (3.2), (3.3) and Schwarz's inequality, we deduce

$$\begin{aligned} |\bar R_1| &\le C\,E\int_0^{t_l} |u(\kappa_2(s))-u(s)|_V\big[|u(s)|_V+|u^\tau(\kappa_2(s))|_V+K_2\big]\,ds \\ &\le C\left(E\int_0^{t_l}|u(s)-u(\kappa_2(s))|_V^2\,ds\right)^{\frac12}\left(E\int_0^{t_l}\left(|u(s)|_V^2+|u(\kappa_2(s))|_V^2\right)ds\right)^{\frac12} \\ &\quad + C\left(E\int_0^{t_l}|u(s)-u(\kappa_2(s))|_V^2\,ds\right)^{\frac12} \\ &\le C\tau^\nu\,. \end{aligned}\tag{3.25}$$

Furthermore, Schwarz's inequality, (C2) and computations similar to that proving (3.8) yield for any $\delta>0$ small enough

$$\begin{aligned} |\bar R_3| \le{}& \delta\sum_{k=1}^{d_1} E\int_0^{t_{l-1}} |B_k(t,u(t))-B_k(t,u^\tau(\kappa_2(t)))|_H^2\,dt \\ &+C\sum_{k=1}^{d_1}\sum_{i=1}^{l-1}\frac{1}{\tau}E\int_{t_i}^{t_{i+1}} ds\int_{t_{i-1}}^{t_i} dt\,|B_k(s,u(s)))-B_k(t,u(t))|_H^2 \\ \le{}& \frac{\lambda}{2}E\int_0^{t_{l-1}} |u(s)-u^\tau(\kappa_2(s))|_V^2\,ds + C\tau^{2\nu}\,. \end{aligned}$$

This inequality and (3.25) imply that

$$E|u(t_l) - u^\tau(t_l)|_H^2 + \frac{\lambda}{2} E \int_0^{t_l} |u(s) - u^\tau(\kappa_2(s))|_V^2 ds$$
$$\leq K_1 \int_0^{t_l} E|u(s) - u^\tau(\kappa_2(s))|_H^2 \, ds + C\,\tau^\nu.$$

Hence for any $t \in [0,T]$,

$$E|u(t) - u^\tau(\kappa_2(t)|_H^2 \leq 2\,E|u(\kappa_2(t)) - u^\tau(\kappa_2(t))|_H^2 + 2\,E|u(t) - u(\kappa_2(t))|_H^2$$
$$\leq 2\,K_1 \int_0^{\kappa_2(t)} E|u(s) - u^\tau(\kappa_2(s))|_H^2 ds + C\,\tau^\nu + 2\,E|u(t) - u(\kappa_2(t))|_H^2$$
$$\leq 2\,K_1 \int_0^{t} E|u(s) - u^\tau(\kappa_2(s))|_H^2 \, ds + C\,\tau^\nu + 2\,E|u(t) - u(\kappa_2(t))|_H^2$$
$$+ C\,\tau \Big[\sup_s E\big(|u(s)|_H^2 + |u^\tau(\kappa_2(s))|_H^2\big)\Big].$$

Itô's formula and (2.9) imply that for any $t \in [0,T]$,

$$E|u(t) - u(\kappa_2(t))|_H^2 = E \int_t^{\kappa_2(t)} \Big[2\langle A(s,u(s))\,,\,u(s)\rangle + \sum_{k=1}^{d_1} |B_k(s,u(s))|_H^2\Big]\,ds$$
$$\leq K_1\,E \int_t^{\kappa_2(t)} |u(s)|_H^2 \, ds \leq K_1\,\tau \sup_{0\leq s\leq T} E|u(s)|_H^2.$$

Hence (2.13) and (2.19) imply that

$$E|u(t) - u^\tau(\kappa_2(t))|_H^2 \leq 2\,K_1 \int_0^t E|u(s) - u^\tau(\kappa_2(s))|_H^2 \, ds + C\,\tau^\nu$$

and Gronwall's lemma yields

$$\sup_{0\leq t\leq T} E|u(t) - u^\tau(\kappa_2(t))|_H^2 \leq C\tau^\nu. \tag{3.26}$$

Therefore,

$$E \int_0^T |u(t) - u^\tau(\kappa_2(t)|_V^2 \, dt < C\tau^\nu \tag{3.27}$$

follows by (3.24). Finally taking into account that by (T2) there exists a constant C such that

$$E|u(t) - u(\kappa_2(t)|_V^2 \leq C\tau^{2\nu} \quad \text{for all } t \in [0,T],$$

from (3.26) and (3.27) we obtain (3.23). □

Using the above result one can easily obtain the following theorem in the same way as Theorem 3.2 is obtained from Theorem 3.4.

Theorem 3.6. *Let A and B satisfy the conditions (C1), (C2) and (C4). Suppose that conditions (T1)–(T3) hold and let A^τ and B^τ be defined by* (2.15) *and* (2.16). *Then there exists a constant C such that for m large enough,*

$$E \max_{0\le j\le m} |u(j\tau) - u^\tau(j\tau)|_H^2 + E\sum_{j=0}^{m} |u(j\tau) - u^\tau(j\tau)|_V^2 \tau \le C\tau^\nu . \tag{3.28}$$

Remark 3.7. By analyzing their proof, it is not difficult to see that Theorems 3.2, 3.4, 3.5 and 3.6 remain true, if instead of (2.15) and (2.16), one uses (2.18) in the definition of the implicit scheme, and requires furthermore that A satisfies the following time-regularity similar to (T1): there exist a constant $C \ge 0$ and a random variable $\eta \ge 0$ with finite expectation, such that almost surely

$$|A(t,u) - A(s,u)|_{V^*}^2 \le |t-s|^{2\nu}(\eta + C\|u\|_V^2)$$

for $0 \le s \le t \le T$ and $u \in V$.

4. Examples

4.1. *Quasilinear stochastic PDEs*

Let us consider the stochastic partial differential equation

$$\begin{aligned} du(t,x) =& \big[Lu(t,x) + F(t,x,\nabla u(t,x),u(t,x))\big]\,dt \\ &+ \sum_{k=1}^{d_1} \big[M_k u(t,x) + G_k(t,x,u(t,x))\big]\,dW^k(t), \end{aligned} \tag{4.1}$$

for $t \in (0,T]$, $x \in \mathbb{R}^d$ with initial condition

$$u(0,x) = u_0(x), \quad x \in \mathbb{R}^d, \tag{4.2}$$

where W is a d_1-dimensional Wiener martingale with respect to the filtration $(\mathcal{F}_t)_{t\ge0}$, F and G_k are Borel functions of $(\omega,t,x,p,r) \in \Omega\times[0,\infty)\times\mathbb{R}^d\times\mathbb{R}^d\times\mathbb{R}$ and of $(\omega,t,x,r) \in \Omega\times[0,\infty)\times\mathbb{R}^d\times\mathbb{R}$, respectively, and L, M_k are differential operators of the form

$$L(t)v(x) = \sum_{|\alpha|\le1,|\beta|\le1} D^\alpha(a^{\alpha\beta}(t,x)D^\beta v(x)), \quad M_k(t)v(x) = \sum_{|\alpha|\le1} b_k^\alpha(t,x)D^\alpha v(x), \tag{4.3}$$

with functions $a^{\alpha\beta}$ and b_k^α of $(\omega,t,x) \in \Omega\times[0,\infty)\times\mathbb{R}^d$, for all multi-indices $\alpha = (\alpha_1,...,\alpha_d)$, $\beta = (\beta_1,...,\beta_d)$ of length $|\alpha| = \sum_i \alpha_i \le 1$, $|\beta| \le 1$.

Here, and later on D^α denotes $D_1^{\alpha_1}...D_d^{\alpha_d}$ for any multi-indices $\alpha = (\alpha_1,...,\alpha_d) \in \{0,1,2,...\}^d$, where $D_i = \frac{\partial}{\partial x_i}$ and D_i^0 is the identity operator.

We use the notation $\nabla_p := (\partial/\partial p_1,...,\partial/\partial p_d)$. For $r \ge 0$ let $W_2^r(\mathbb{R}^d)$ denote the space of Borel functions $\varphi : \mathbb{R}^d \to \mathbb{R}$ whose derivatives up to order r are square integrable functions. The norm $|\varphi|_r$ of φ in W_2^r is defined by

$$|\varphi|_r^2 = \sum_{|\gamma|\le r} \int_{\mathbb{R}^d} |D^\gamma \varphi(x)|^2\, dx.$$

In particular, $W_0^2(\mathbb{R}^d) = L_2(\mathbb{R}^d)$ and $|\varphi|_0 := |\varphi|_{L_2(\mathbb{R}^d)}$. Let us use the notation $\mathcal{P}$ for the σ-algebra of predictable subsets of $\Omega \times [0,\infty)$, and $\mathcal{B}(\mathbb{R}^d)$ for the Borel σ-algebra on $\mathbb{R}^d$.

We fix an integer $l \ge 0$ and assume that the following conditions hold.

Assumption (A1) (Stochastic parabolicity). *There exists a constant $\lambda > 0$ such that*

$$\sum_{|\alpha|=1,|\beta|=1} \left(a^{\alpha\beta}(t,x) - \tfrac{1}{2}\sum_{k=1}^{d_1} b_k^\alpha b_k^\beta(t,x) \right) z^\alpha z^\beta \ge \lambda \sum_{|\alpha|=1} |z^\alpha|^2 \tag{4.4}$$

for all $\omega \in \Omega$, $t \in [0,T]$, $x \in \mathbb{R}^d$ and $z = (z^1, ..., z^d) \in \mathbb{R}^d$, where $z^\alpha := z_1^{\alpha_1} z_2^{\alpha_2} ... z_d^{\alpha_d}$ for $z \in \mathbb{R}^d$ and multi-indices $\alpha = (\alpha_1, \alpha_2, ..., \alpha_d)$.

Assumption (A2) (Smoothness of the linear term). *The derivatives of $a^{\alpha\beta}$ and b_k^α up to order l are $\mathcal{P} \otimes \mathcal{B}(\mathbb{R}^d)$ -measurable real functions such that for a constant K*

$$|D^\gamma a^{\alpha\beta}(t,x)| \le K, \quad |D^\gamma b_k^\alpha(t,x)| \le K, \quad \text{for all } |\alpha| \le 1,\ |\beta| \le 1,\ k = 1,\cdots,d_1, \tag{4.5}$$

for all $\omega \in \Omega$, $t \in [0,T]$, $x \in \mathbb{R}^d$ and multi-indices γ with $|\gamma| \le l$.

Assumption (A3) (Smoothness of the initial condition). *Let u_0 be a W_2^l-valued $\mathcal{F}_0$-measurable random variable such that*

$$E|u_0|_l^2 < \infty. \tag{4.6}$$

Assumption (A4) (Smoothness of the nonlinear term). *The function F and their first order partial derivatives in p and r are $\mathcal{P} \otimes \mathcal{B}(\mathbb{R}^d) \otimes \mathcal{B}(\mathbb{R}^d) \otimes \mathcal{B}(\mathbb{R})$-measurable functions, and g_k and its first order derivatives in r are $\mathcal{P}\otimes\mathcal{B}(\mathbb{R}^d)\otimes\mathcal{B}(\mathbb{R})$ -measurable functions for every $k = 1, .., d_1$. There exists a constant K such that*

$$|\nabla_p F(t,x,p,r)| + |\tfrac{\partial}{\partial r} F(t,x,p,r)| + \sum_{k=1}^{d_1} |\tfrac{\partial}{\partial r} G_k(t,x,r)| \le K \tag{4.7}$$

for all $\omega \in \Omega$, $t \in [0,T]$, $x \in \mathbb{R}^d$, $p \in \mathbb{R}^d$ and $r \in \mathbb{R}$. There exists a random variable ξ with finite first moment, such that

$$|F(t,\cdot,0,0)|_0^2 + \sum_{k=1}^{d_1} |G_k(t,\cdot,0)|_0^2 \le \xi \tag{4.8}$$

for all $\omega \in \Omega$ and $t \in [0,T]$.

Definition 4.1. An $L_2(\mathbb{R}^d)$-valued continuous $(\mathcal{F}_t)$-adapted process $u = \{u(t) : t \in [0,T]\}$ is called a generalized solution to the Cauchy problem (4.1)-(4.2) on $[0,T]$ if almost surely $u(t) \in W_2^1(\mathbb{R}^d)$ for almost every t,

$$\int_0^T |u(t)|_1^2\, dt < \infty,$$

and

$$\begin{aligned} d(u(t),\varphi) =& \Big\{ \sum_{|\alpha|\le 1, |\beta|\le 1} (-1)^{|\alpha|} \big(a^{\alpha\beta}(t)\, D^\beta u(t)\,,\, D^\alpha \varphi\big) + \big(F(t,\nabla u(t), u(t))\,,\, \varphi\big) \Big\}\, dt \\ &+ \sum_{k=1}^{d_1} \Big\{ \sum_{|\alpha|\le 1} \big(b_k^\alpha(t) D^\alpha u(t)\,,\, \varphi\big) + \big(G_k(t,u(t))\,,\,\varphi\big) \Big\}\, dW^k(t) \end{aligned}$$

holds on $[0,T]$ for every $\varphi \in C_0^\infty(\mathbb{R}^d)$, where (v,φ) denotes the inner product of v and φ in $L_2(\mathbb{R}^d)$.

Set $H = L_2(\mathbb{R}^d)$, $V = W_2^1(\mathbb{R}^d)$ and consider the normal triplet $V \hookrightarrow H \hookrightarrow V^*$ based on the inner product in $L_2(\mathbb{R}^d)$, which determines the duality $\langle\, ,\, \rangle$ between V and $V^* = W_2^{-1}(\mathbb{R}^d)$. By (4.5), (4.7) and (4.8) there exist a constant C and a random variable ξ with finite first moment, such that

$$\Big| \sum_{|\alpha|\le 1, |\beta|\le 1} (-1)^{|\alpha|} \big(a^{\alpha\beta}(t)\, D^\beta v\,,\, D^\alpha\varphi\big) \Big| \le C|v|_1 |\varphi|_1,$$

$$\sum_{k=1}^{d_1} |(b_k^\alpha(t) D^\alpha v\,,\, \varphi)|^2 \le C|v|_1^2 |\varphi|_0^2,$$

$$|(F(t,\nabla v, v)\,,\,\varphi)|^2 \le C|v|_1^2|\varphi|_1^2 + \xi|\varphi|_0^2, \quad \sum_{k=1}^{d_1} |(G_k(t,v(t))\,,\,\varphi)|^2 \le C|v|_1^2|\varphi|_0^2 + \xi|\varphi|_0^2$$

for all ω, $t \in [0,T]$ and $v, \varphi \in V$. Therefore the operators $A(t)$, $B_k(t)$ defined by

$$\begin{aligned} \langle A(t,v), \varphi\rangle &= \sum_{|\alpha|\le 1, |\beta|\le 1} (-1)^{|\alpha|} \big(a^{\alpha\beta}(t)\, D^\beta v\,,\, D^\alpha \varphi\big) + \big(F(t,\nabla v, v)\,,\,\varphi\big), \\ (B_k(t,v)\,,\,\varphi) &= \big(b_k^\alpha(t) D^\alpha v\,,\,\varphi\big) + \big(G_k(t,v)\,,\,\varphi\big), \quad v,\varphi \in V \end{aligned} \tag{4.9}$$

are mappings from V into V^* and H, respectively, for each k and ω, t, such that the growth conditions (2.6) and (2.7) hold. Thus we can cast the Cauchy problem (4.1)– (4.2) into the evolution equation (1.1), and it is an easy exercise to show that Assumptions (A1), (A2) with $l = 0$ and Assumption (A4) ensure that conditions (C1) and (C2) hold. Hence Corollary 2.5 gives the following result.

Theorem 4.2. *Let Assumptions (A1)-(A4) hold with $l = 0$. Then problem* (4.1)-(4.2) *admits a unique generalized solution u on $[0,T]$. Moreover,*

$$E\Big(\sup_{t\in[0,T]} |u(t)|_0^2\Big) + E\int_0^T |u(t)|_1^2\, dt < \infty. \tag{4.10}$$

Next we formulate a result on the regularity of the generalized solution. We need the following assumptions.

Assumption (A5) *The first order derivatives of G_k in x are $\mathcal{P}\otimes\mathcal{B}(\mathbb{R}^d)\otimes\mathcal{B}(\mathbb{R})$-measurable functions, and there exist a constant L, a $\mathcal{P}\otimes\mathcal{B}(\mathbb{R})$-measurable function K of (ω,t,x) and a random variable ξ with finite first moment, such that*

$$\sum_{k=1}^{d_1}|D^\alpha G_k(t,x,r)|\le L|r|+K(t,x),\quad |K(t)|_0^2\le\xi$$

for all multi-indices α with $|\alpha|=1$, for all $\omega\in\Omega$, $t\in[0,T]$, $x\in\mathbb{R}^d$ and $r\in\mathbb{R}$.

Assumption (A6) *The first order derivatives of F in x are $\mathcal{P}\otimes\mathcal{B}(\mathbb{R}^d)\otimes\mathcal{B}(\mathbb{R}^d)\otimes\mathcal{B}(\mathbb{R})$-measurable functions, and there exist a constant L, a $\mathcal{P}\otimes\mathcal{B}(\mathbb{R})$-measurable function K of (ω,t,x) and a random variable ξ with finite first moment, such that*

$$|\nabla_x F(t,x,p,r)|\le L(|p|+|r|)+K(t,x),\quad |K(t)|_0^2\le\xi$$

for all ω,t,x,p,r.

Assumption (A7) *There exist $\mathcal{P}\otimes\mathcal{B}(\mathbb{R})$-measurable functions g_k such that*

$$G_k(t,x,r)=g_k(t,x)\quad\text{for all } k=1,2,...,d_1,\ t,x,r,$$

and the derivatives in x of g_k up to order l are $\mathcal{P}\otimes\mathcal{B}(\mathbb{R})$-measurable functions such that

$$\sum_{k=1}^{d_1}|g_k(t)|_l^2\le\xi,$$

for all (ω,t), where ξ is a random variable with finite first moment.

Theorem 4.3. *Let Assume (A1)-(A4) with $l=1$. Then for the generalized solution u of (4.1)-(4.2) the following statements hold:*

(i) Suppose (A5). Then u is a $W_2^1(\mathbb{R}^d)$-valued continuous process and

$$E\Big(\sup_{t\le T}|u(t)|_1^2\Big)+E\int_0^T|u(t)|_2^2\,dt<\infty;\tag{4.11}$$

(ii) Suppose (A6) and (A7) with $l=2$. Then u is a $W_2^2(\mathbb{R}^d)$-valued continuous process and

$$E\Big(\sup_{t\le T}|u(t)|_2^2\Big)+E\int_0^T|u(t)|_3^2\,dt<\infty.\tag{4.12}$$

Proof. Define

$$\psi(t,x)=F(t,x,\nabla u(t,x),u(t,x)),\quad \varphi_k(t,x)=G_k(t,x,u(t,x))$$

for $t\in[0,T]$, $\omega\in\Omega$ and $x\in\mathbb{R}^d$, where u is the generalized solution of (4.1)-(4.2). Then due to (4.10)

$$E\int_0^T|\psi(t)|_0^2\,dt<\infty,\quad E\sum_k\int_0^T|\varphi_k(t)|_1^2\,dt<\infty.$$

Therefore, the Cauchy problem

$$\begin{aligned} dv(t,x) &= \big[Lv(t,x) + \psi(t,x)\big]\,dt \\ &\quad + \sum_{k=1}^{d_1} \big[M_k v(t,x) + \varphi_k(t,x)\big]\,dW^k(t), \quad t \in (0,T]\,,\ x \in \mathbb{R}^d\,, \qquad (4.13) \\ v(0,x) &= u_0(x), \quad x \in \mathbb{R}^d \qquad (4.14) \end{aligned}$$

has a unique generalized solution v on $[0,T]$. Moreover, by Theorem 1.1 in Ref. [10], v is a W_2^1-valued continuous $\mathcal{F}_t$-adapted process and

$$E\Big(\sup_{t\le T} |v(t)|_1^2\Big) + E\int_0^T |v(t)|_2^2\,dt < \infty.$$

Since u is a generalized solution to (4.13)–(4.14), by virtue of the uniqueness of the generalized solution we have $u = v$, which proves (i). Assume now (A6) and (A7) with $l = 2$. Then obviously (A5) holds, and therefore due to (4.11)

$$E\int_0^T |\psi(t)|_1^2\,dt < \infty, \quad E\sum_k \int_0^T |\varphi_k(t)|_2^2\,dt < \infty.$$

Thus by Theorem 1.1 in Ref. [10] the generalized solution $v = u$ of (4.13)–(4.14) is a $W_2^2(\mathbb{R}^d)$-valued continuous process such that (4.12) holds. The proof of the theorem is complete. □

Corollary 4.4. *Let (A1)-(A4) hold with $l = 2$. Assume also (A6) and (A7). Then there exists a constant C such that for the generalized solution u of (4.1)–(4.2) we have*

$$E|u(t) - u(s)|_1^2 \le C|t-s| \quad \text{for all } s,t \in [0,T].$$

Proof. By the theorem on Itô's formula (see Refs. [1] or [11]) from almost surely

$$u(t) = u_0 + \int_0^t \big[Lu(s) + \psi(s)\big]\,ds + \sum_{k=1}^{d_1} \int_0^t \big[M_k u(s) + g_k(s)\big]\,dW^k(s)$$

holds, as an equality in $L_2(\mathbb{R}^d)$, for all $t \in [0,T]$, where

$$\psi(s,\cdot) := F(s,\cdot,\nabla u(s,\cdot), u(s,\cdot)).$$

Due to (ii) from Theorem 4.3

$$\begin{aligned} E\Big|\int_s^t \big[L(r)u(r) + \psi(r)\big]\,dr\Big|_1^2 &\le E\Big(\int_s^t |Lu(r) + \psi(r)|_1\,dr\Big)^2 \\ &\le |t-s|\,E\int_s^t |Lu(r) + \psi(r)|_1^2\,dr \\ &\le C\,|t-s|\left(E\int_0^T |u(t)|_3^2\,dt + E\int_0^T |\psi(t)|_1^2\,dt\right) \le C|t-s| \end{aligned}$$

for all $s,t \in [0,T]$, where C is a constant. Furthermore, by Doob's inequality

$$E\left|\int_s^t [M_k u(r) + g_k(r)]\, dW^k(r)\right|_1^2 \le 4\int_s^t E|M_k u(r) + g_k(r)|_1^2\, dr$$
$$\le C_1|t-s|\Big[1 + E\int_0^T |u(t)|_2^2\, dt\Big] \le C_2|t-s|$$

for all $s,t \in [0,T]$, where C_1 and C_2 are constants. Hence

$$\begin{aligned} E|u(t)-u(s)|_1^2 \le& 2E\left|\int_s^t [Lu(r) + \psi(r))\, dr\right|_1^2 \\ &+ 2E\left|\sum_{k=1}^{d_1}\int_s^t [M_k u(r) + g_k(r)]\, dW^k(r)\right|_1^2 \le C|t-s|, \end{aligned}$$

and the proof of the corollary is complete. □

The implicit scheme (2.14) applied to problem (4.1)-(4.2) reads as follows.

$$\begin{aligned} u^\tau(t_0) &= u_0\,, \\ u^\tau(t_{i+1}) &= u^\tau(t_i) + \big(L^\tau_{t_i} u^\tau(t_{i+1}) + F^\tau_{t_i}(u^\tau(t_{i+1}))\big)\,\tau \\ &\quad + \sum_{k=1}^{d_1}\big(M^\tau_{k,t_i} u^\tau(t_i) + G^\tau_{k,t_i}(u^\tau(t_i))\big)\,\big(W^k(t_{i+1}) - W^k(t_i)\big)\,, \end{aligned} \tag{4.15}$$

for $0 \le i < m$, where

$$L^\tau_{t_i} v := \sum_{|\alpha|\le 1, |\beta|\le 1} D^\alpha(a^{\alpha\beta}_{t_i}(x) D^\beta v), \quad M^\tau_{k,t_i} := \sum_{|\alpha|\le 1} b^\alpha_{k,t_i} D^\alpha v,$$

$$a^{\alpha\beta}_{t_i}(x) := \frac{1}{\tau}\int_{t_i}^{t_{i+1}} a^{\alpha\beta}(s,x)\, ds, \tag{4.16}$$

$$b^\alpha_{k,0}(x) = 0, \quad b^\alpha_{k,t_{i+1}}(x) = \frac{1}{\tau}\int_{t_i}^{t_{i+1}} b_k(s,x)\, ds, \tag{4.17}$$

$$F^\tau_{t_i}(x,p,r) := \frac{1}{\tau}\int_{t_i}^{t_{i+1}} F(s,x,p,r)\, ds,$$

$$G^\tau_{k,0}(x,r) := 0, \quad G^\tau_{k,t_{i+1}}(x,r) := \int_{t_i}^{t_{i+1}} G_k(s,x,r)\, ds.$$

Definition 4.5. A random vector $\{u^\tau(t_i) : i = 0,1,2,...,m\}$ is a called a generalized solution of the scheme (4.15) if $u^\tau(t_0) = u_0$, $u^\tau(t_i)$ is a $W^1_2(\mathbb{R}^d)$-valued $\mathcal{F}_{t_i}$-measurable random variable such that

$$E|u^\tau(t_i)|_1^2 < \infty$$

and almost surely

$$(u^\tau(t_i),\varphi)=\sum_{|\alpha|\le 1,|\beta|\le 1}(-1)^{|\alpha|}(a^{\alpha\beta}_{t_i}D^\beta u^\tau(t_i),D^\alpha\varphi)\tau+(F^\tau_{t_{i-1}}(\nabla u^\tau_{t_{i-1}},u^\tau_{t_{i-1}}),\varphi)\tau$$
$$+\sum_k\Big[\sum_{|\alpha|\le 1}(b^\alpha_{t_{i-1}}D^\alpha u^\tau_{t_{i-1}}+G_{k,t_{i-1}}(u^\tau_{t_{i-1}}),\varphi)(W^k(t_i)-W^k(t_{i-1})\Big]$$

for $i=1,2,...,m$ and all $\varphi\in C_0^\infty(\mathbb{R}^d)$, where $(\cdot,\cdot)$ is the inner product in $L_2(\mathbb{R}^d)$.

From this definition it is clear that, using the operators A, B_k defined by (4.9), we can cast the scheme (4.15) into the abstract scheme (2.14). Thus by applying Theorem 2.6 we get the following theorem.

Theorem 4.6. *Let (A1)-(A4) hold with $l=0$. Then there exists an integer m_0 such that* (4.15) *has a unique generalized solution $\{u^\tau(t_i):i=0,1,...,m\}$ for every $m\ge m_0$. Moreover, there exists a constant C such that*

$$E\max_{0\le i\le m}|u^\tau(t_i)|_0^2+E\sum_{i=1}^m|u^\tau(t_i)|_1^2\le C$$

for all integers $m\ge m_0$.

To ensure condition (T1) to hold we impose the following assumption.

Assumption (H) *There exists a constant C and a random variable ξ with finite first moment such that for $k=1,2,...,d_1$*

$$|D^\gamma(b^\alpha_k(t,x)-b^\alpha_k(s,x))|\le C|t-s|^{1/2}\quad\text{for all }\omega\in\Omega,x\in\mathbb{R}^d\text{ and }|\gamma|\le l,$$
$$|g_k(s)-g_k(s)|_l^2\le\xi|t-s|$$

for all $s,t\in[0,T]$.

Now applying Theorem 3.4 we obtain the following result.

Theorem 4.7. *Let (A1)-(A4) and (A6)-(A7) hold with $l=2$. Assume (H) with $l=0$. Then* (4.1)–(4.2) *and* (4.15) *have a unique generalized solution u and $u^\tau=\{u^\tau(t_i):i=0,1,2,...,m\}$, respectively, for all integers m larger than some integer m_0. Moreover, for all integers $m>m_0$*

$$E\max_{0\le i\le m}|u(i\tau)-u^\tau(i\tau)|_0^2+E\sum_{i=1}^m|u(i\tau)-u^\tau(i\tau)|_1^2\tau\le C\tau,\qquad(4.18)$$

where C is a constant, independent of τ.

Proof. By Theorems 4.2 and 4.6 the problems (4.1)–(4.2) and (4.15) have a unique solution u and u^τ, respectively. It is an easy exercise to verify that Assumption (H) ensures that condition (T1) holds. By virtue of Corollary 4.4 condition (T2) is valid with $\nu=1/2$. Condition (T3) clearly holds by statement (i) of Theorem 4.3. Now we can apply Theorem 3.4, which gives (4.18). □

4.2. *Linear stochastic PDEs*

Let Assumptions (A1)–(A3) and (A7) hold and impose also the following condition on F.

Assumption (A8) *There exist a $\mathcal{P}\otimes\mathcal{B}(\mathbb{R})$-measurable function f such that*

$$F(t,x,p,r)=f(t,x),\quad \text{for all } t,x,p,r,$$

and the derivatives in x of f up to order l are $\mathcal{P}\otimes\mathcal{B}(\mathbb{R})$-measurable functions such that

$$|f(t)|_l^2\le\xi,$$

for all (ω,t), where ξ is a random variable with finite first moment.

Now equation (4.13) has become the linear stochastic PDE

$$du(t,x)=\big[Lu(t,x)+f(t,x)\big]\,dt+\sum_{k=1}^{d_1}\big[M_ku(t,x)+g_k(t,x)\big]\,dW^k(t),\tag{4.19}$$

and by Theorem 3.4 we have the following result.

Theorem 4.8. *Let $r\ge0$ be an integer. Let Assumptions (A1)–(A3) and (A7)–(A8) hold with $l:=r+2$, and let Assumption (H) hold with $l=r$. Then there is an integer m_0 such that* (4.19)–(4.2) *and* (4.15) *have a unique generalized solution u and $u^\tau=\{u^\tau(t_i):i=0,1,2,...,m\}$, respectively, for all integers $m>m_0$. Moreover,*

$$E\max_{0\le i\le m}|u(i\tau)-u^\tau(i\tau)|_r^2+E\sum_{i=1}^{m}|u(i\tau)-u^\tau(i\tau)|_{r+1}^2\tau\le C\tau$$

holds for all $m>m_0$, where C is a constant independent of τ.

Proof. For $r=0$ the statement of this theorem follows immediately from Theorem 4.7. For $r>0$ set $H=W_2^r(\mathbb{R}^d)$ and $V=W_2^{r+1}(\mathbb{R}^d)$ and consider the normal triplet $V\hookrightarrow H\equiv H^*\hookrightarrow V^*$ based on the inner product $(\cdot,\cdot):=(\cdot,\cdot)_r$ in $W_2^r(\mathbb{R}^d)$, which determines the duality $\langle\cdot\,,\,\cdot\rangle$ between V and V^*. Using Assumptions (A3), (A7) and (A8) with $l=r$, one can easily show that there exist a constant C and a random variable ξ such that $E\xi^2<\infty$ and

$$\Big|\sum_{|\alpha|\le1,|\beta|\le1}(-1)^{|\alpha|}\big(a^{\alpha\beta}(t)\,D^\beta v\,,\,D^\alpha\varphi\big)_r\Big|\le C|v|_{r+1}|\varphi|_{r+1},$$

$$\sum_{k=1}^{d_1}|(b_k^\alpha(t)D^\alpha v\,,\,\varphi)_r|^2\le C|v|_{r+1}^2|\varphi|_r^2,$$

$$|(f(t)\,,\,\varphi)_r|^2\le\xi|\varphi|_r^2,\quad \sum_{k=1}^{d_1}|(g_k(t)\,,\,\varphi)_r|^2\le\xi|\varphi|_r^2$$

for all ω, $t \in [0,T]$ and $v, \varphi \in W_2^{r+1}(\mathbb{R}^d)$. Therefore the operators $A(t,\cdot)$, $B_k(t,\cdot)$ defined for $v, \varphi \in V$ by

$$\langle A(t,v), \varphi\rangle = \sum_{|\alpha|\le 1, |\beta|\le 1} (-1)^{|\alpha|} \big(a^{\alpha\beta}(t)\, D^\beta v\,,\, D^\alpha \varphi\big)_r + \big(f(t)\,,\,\varphi\big)_r,$$
$$\big(B_k(t,v)\,,\,\varphi\big) = \big(b_k^\alpha(t) D^\alpha v\,,\,\varphi\big)_r + \big(g_k(t)\,,\,\varphi\big)_r, \tag{4.20}$$

are mappings from V into V^* and H, respectively, for each k and ω, t, such that the growth conditions (2.6) and (2.7) hold. Thus we can cast the Cauchy problem (4.19)– (4.2) into the evolution equation (1.1), and it is an easy to verify that conditions (C1)–(C4) hold. Thus this evolution equation admits a unique solution u, which clearly a generalized solution to (4.19)– (4.2). Due to assumptions (A1)–(A3) and (A7)–(A8) by Theorem 1.1 in Ref. [10] u is a $W^{r+2}(\mathbb{R}^d)$-valued stochastic process such that

$$E \sup_{t\le T} |u(t)|_{r+2}^2 + E\int_0^T |u(t)|_{r+3}^2\, dt < \infty.$$

Hence it is obvious that (T3) holds, and it is easy to verify (T2) with $\nu = \frac{1}{2}$ like it is done in the proof of Corollary 4.4. Finally, it is an easy exercise to show that (T1) holds. Now we can finish the proof of the theorem by applying Theorem 3.4. □

From the previous theorem we obtain the following corollary by Sobolev's embedding from W_2^r to C^q.

Corollary 4.9. *Let q be any non-negative number and assume that the assumptions of Theorem 4.8 hold with $r > q + \frac{d}{2}$. Then there exist modifications $\bar u$ and $\bar u^\tau$ of u and u^τ, respectively, such that the derivatives $D^\gamma \bar u$ and $D^\gamma \bar u^\tau$ in x up to order q are functions continuous in x. Moreover, there exists a constant C independent of τ such that*

$$E \max_{0\le i\le m} \sup_{x\in\mathbb{R}^d} \sum_{|\gamma|\le q} |D^\gamma\big(\bar u(i\tau,x) - \bar u^\tau(i\tau,x)\big)|^2$$
$$+ E\sum_{i=1}^m \sup_{x\in\mathbb{R}^d} \sum_{|\gamma|\le q+1} \big|D^\gamma\big(\bar u(i\tau,x) - \bar u^\tau(i\tau,x)\big|^2 \tau \le C\tau. \tag{4.21}$$

Acknowledgments

This paper was written while I. Gyöngy was visiting the University of Paris 1. His research is partially supported by EU Network HARP. The research of A. Millet is partially supported by the research project BMF2003-01345 The authors wish to thank the referee for helpful comments.

References

[1] N. Krylov and B. Rosovskii, Stochastic evolution equations, *J.Soviet Mathematics.* **16**, 1233–1277, (1981).

[2] E. Pardoux. Équations aux dérivées partielles stochastiques nonlinéares monotones. Étude de solutions fortes de type Itô, (1975). Thése Doct. Sci. Math. Univ. Paris Sud.

[3] B. L. Rozovskiĭ, *Stochastic evolution systems.* vol. 35, *Mathematics and its Applications (Soviet Series)*, (Kluwer Academic Publishers Group, Dordrecht, 1990). ISBN 0-7923-0037-8. Linear theory and applications to nonlinear filtering, Translated from the Russian by A. Yarkho.

[4] I. Gyöngy and A. Millet, On discretization schemes for stochastic evolution equations, *Potential Anal.* **23**(2), 99–134, (2005). ISSN 0926-2601.

[5] N. V. Krylov, On L_p-theory of stochastic partial differential equations in the whole space, *SIAM J. Math. Anal.* **27**(2), 313–340, (1996). ISSN 0036-1410.

[6] C. Cardon-Weber, Cahn-Hilliard stochastic equation: existence of the solution and of its density, *Bernoulli.* **7**(5), 777–816, (2001). ISSN 1350-7265.

[7] N. V. Krylov and B. L. Rozovskiĭ, Conditional distributions of diffusion processes, *Izv. Akad. Nauk SSSR Ser. Mat.* **42**(2), 356–378, 470, (1978). ISSN 0373-2436.

[8] É. Pardoux. Filtrage non linéaire et équations aux dérivées partielles stochastiques associées. In *École d'Été de Probabilités de Saint-Flour XIX—1989*, vol. 1464, *Lecture Notes in Math.*, pp. 67–163. Springer, Berlin, (1991).

[9] E. Pardoux, Stochastic partial differential equations and filtering of diffusion processes, *Stochastics.* **3**(2), 127–167, (1979). ISSN 0090-9491.

[10] N. V. Krylov and B. L. Rozovskiĭ, The Cauchy problem for linear stochastic partial differential equations, *Izv. Akad. Nauk SSSR Ser. Mat.* **41**(6), 1329–1347, 1448, (1977). ISSN 0373-2436.

[11] I. Gyöngy and N. V. Krylov, On stochastics equations with respect to semimartingales. II. Itô formula in Banach spaces, *Stochastics.* **6**(3-4), 153–173, (1981/82). ISSN 0090-9491.

Chapter 12

Maximum Principle for SPDEs and Its Applications

Nicolai V. Krylov

School of Mathematics, University of Minnesota,
127 Vincent Hall, Minneapolis, MN, 55455, krylov@math.umn.edu.

The maximum principle for SPDEs is established in multidimensional C^1 domains. An application is given to proving the Hölder continuity up to the boundary of solutions of one-dimensional SPDEs.

Contents

1. Introduction

The maximum principle is one of the most powerful tools in the theory of second-order elliptic and parabolic partial differential equations. However, until now it did not play any significant role in the theory of SPDEs. In this paper we show how to apply it to one-dimensional SPDEs on the half line $\mathbb{R}_+ = (0, \infty)$ and prove the Hölder continuity of solutions on $[0, \infty)$. This result was previously known when the coefficients of the first order derivatives of solution appearing in the stochastic term in the equation obeys a quite unpleasant condition. On the other hand, if they just vanish, then the Hölder continuity was well known before (see, for instance, Ref. [1] and the references therein).

To the best of our knowledge the maximum principle was first proved in Ref. [2] (see also Ref. [3] for the case of random coefficients) for SPDEs in the whole space by the method of random characteristics introduced there and also in Ref. [4]. Later the method of random characteristics was used in many papers for various purposes, for instance, to prove smoothness of solutions (see, for instance, Refs. [5–8] and the references therein). It was very tempting to try to use this method for proving the maximum principle for SPDEs in domains. However, the implementation of the method turns out to become extremely cumbersome and inconvenient if the coefficients of the equation are random processes. Also, it requires more regularity of solutions than actually needed. Here in Section 2 we state the maximum principle in domains under minimal assumptions. We prove it in Section 4 by using methods taken from PDEs after we prepare some auxiliary results in Section 3.

Section 6 contains an application of the maximum principle to investigating the Hölder continuity up to the boundary of solutions of one-dimensional SPDEs. Note that, for instance, in Refs. [5, 6] and in many other papers that can be found from our list of references the regularity properties are proved only inside domains. Quite sharp regularity for solutions of SPDEs in multidimensional domains is established in Ref. [9], it is stated in terms of appropriate weighted Sobolev spaces and, unfortunately, do not imply even the pointwise continuity up to the boundary. It is worth saying that we only deal with one-dimensional case and coefficients independent of the space variable. In a subsequent paper we intend to treat the general case. In Section 5 we introduce some auxiliary functions used in Section 6.

We denote by $\mathbb{R}^d$ the Euclidean space of points $x = (x^1, ..., x^d)$,

$$D_i = \frac{\partial}{\partial x^i}.$$

For a domain $D \subset \mathbb{R}^d$ and we set $W_2^1(D)$ to be the closure of the set of infinitely differentiable functions φ having finite norm

$$\|\varphi\|^2_{W_2^1(D)} = \|\varphi\|^2_{L_2(D)} + \|\varphi_x\|^2_{L_2(D)}$$

with respect to this norm. Here φ_x is the gradient of φ. By $\overset{0}{W}{}_2^1(D)$ we denote the closure of $C_0^\infty(D)$ with respect to the norm $\|\cdot\|_{W_2^1(D)}$. Our way to say that $u \leq v$ on ∂D is that $(u-v)^+ \in \overset{0}{W}{}_2^1(D)$. As usual, the summation convention is enforced and writing $N(....)$ is to say that the constant N depends and depends only on the contents of the parentheses. Such constants may change from line to line.

2. The maximum principle

Let D be a domain in $\mathbb{R}^d$ of class C^1_{loc} and let $(\Omega, \mathcal{F}, P)$ be a complete probability space with a given filtration $(\mathcal{F}_t, t \geq 0)$ of σ-fields $\mathcal{F}_t \subset \mathcal{F}$ complete with respect to $\mathcal{F}, P$.

We are investigating some properties of a function $u_t(x) = u_t(\omega, x)$ satisfying

$$(\varphi, u_t) = (\varphi, u_0) + \int_0^t (\varphi, \sigma_s^{ik} D_i u_s + \nu_s^k u_s + g_s^k)\, dm_s^k$$

$$+ \int_0^t (\varphi, D_i(a_s^{ij} D_j u_s) + b_s^i D_i u_s + D_i(a_s^i u_s) - c_s u_s + f_s + D_i f_s^i)\, dV_s. \qquad (2.1)$$

for all $t \in [0, \infty)$ and any $\varphi \in C_0^\infty(D)$. Here m_t^k, $k = 1, 2, ...$, are one-dimensional continuous local $\mathcal{F}_t$-martingales, starting at zero, V_t is a nondecreasing continuous

$\mathcal{F}_t$-adapted process starting at zero, $(\varphi, \cdot)$ is the pairing between a generalized function on D and a test function φ, the summation convention over repeated indices is enforced, and the meaning of the remaining objects and further assumptions are described below. We need some real-valued functions $\xi^i_t(x)$, $K_1(t) > 0$, and $K_2(t) \geq 0$ defined for $i = 1, ..., d$, $t \in [0, \infty)$, $x \in \mathbb{R}^d$ and also depending on ω.

We assume that $a^{ij}_t(x)$, $b^i_t(x)$, $a^i_t(x)$, $c_t(x)$, $\sigma^{ik}_t(x)$, $\nu^k_t(x)$, and g^k_s are real-valued functions defined for $i, j = 1, ..., d$, $k = 1, 2, ...$, $t \in [0, \infty)$, $x \in \mathbb{R}^d$ and also depending on $\omega \in \Omega$.

Assumption 2.1. We suppose that, for any ω, $\langle m^i, m^j \rangle_t = 0$ if $i \neq j$, and for any k we have $d\langle m^k \rangle_t \leq dV_t$.

Assumption 2.2. For all values of the arguments

(i) $\sigma^i := (\sigma^{i1}, \sigma^{i2}, ...)$, $\nu := (\nu^1, \nu^2, ...)$, $g := (g^1, g^2, ...) \in \ell_2$;

(ii) for all $\lambda \in \mathbb{R}^d$

$$|\sum_i \lambda^i \xi^i|^2 \leq K_1(2a^{ij} - \alpha^{ij})\lambda^i \lambda^j,$$

where $\alpha^{ij} = (\sigma^i, \sigma^j)_{\ell_2}$.

The case $\xi \equiv 0$ is not excluded and in this case Assumption 2.2 (ii) is just the usual parabolicity assumption.

Assumption 2.3.

(i) The functions $a^{ij}_t(x)$, $b^i_t(x)$, $a^i_t(x)$, $c_t(x)$, $\sigma^{ik}_t(x)$, $\nu^k_t(x)$, $\xi^i_t(x)$, $K_1(t)$, and $K_2(t)$ are measurable with respect to (ω, t, x) and $\mathcal{F}_t$-adapted for each x;

(ii) the functions $a^{ij}_t(x)$, $b^i_t(x)$, $a^i_t(x)$, $c_t(x)$, $\sigma^{ik}_t(x)$, $\nu^k_t(x)$, and $\xi^i_t(x)$ are bounded;

(iii) for each ω, t the functions

$$\eta^i_t := a^i_t - b^i_t - (\sigma^i_t, \nu_t)_{\ell_2} - \xi^i_t$$

are once continuously differentiable on D, have bounded derivatives, and satisfy

$$D_i \eta^i - 2c + |\nu|^2_{\ell_2} \leq K_2 \tag{2.2}$$

for all values of arguments;

(iv) for each $\varphi \in C^\infty_0(D)$ the processes φf_t, φf^1_t,..., φf^d_t are $L_2(D)$-valued and φg_t is an $L_2(D, \ell_2)$-valued $\mathcal{F}_t$-adapted and jointly measurable; for all $t \in [0, \infty)$ and $\omega \in \Omega$

$$\int_0^t \left(\|\varphi f_s\|^2_{L_2(D)} + \sum_i \|\varphi f^i_s\|^2_{L_2(D)} + \|\varphi g_s\|^2_{L_2(D,\ell_2)} + K_1(s) + K_2(s)\right) dV_s < \infty.$$

Assumption 2.4. For each $\varphi \in C^\infty_0(D)$

(i) the process $\varphi u_t = \varphi u_t(\omega)$ is $L_2(D)$-valued, $\mathcal{F}_t$-adapted, and jointly measurable;

(ii) for any ω

$$\varphi u_t \in W^1_2(D) \quad (dV_t\text{-a.e.});$$

(iii) for each $t \in [0,\infty)$ and ω

$$\int_0^t \|\varphi u_s\|^2_{W^1_2(D)}\, dV_s < \infty.$$

The above assumptions are supposed to hold throughout this section. Here is the maximum principle saying, in particular, that if $g^k = f^i = 0$, $f \le 0$ and $u \le 0$ on the parabolic boundary of $[0,T] \times D$, then $u \le 0$ in $[0,T]$. By the way, our solutions are $L_{2,loc}(D)$-valued functions of ω and t, so that for each ω and t an equivalence class is specified. Naturally, if we write $u_t(\omega) \le 0$, or $u_t \le 0$ we mean that in the corresponding class there is a nonpositive function.

Theorem 2.5.
Let $\tau_2 \ge \tau_1$ be stopping times, $\tau_1 < \infty$ for any ω. Suppose that, for any ω, $i = 1,...,d$, $k = 1,2,...$,

$$I_{u_t>0} g^k_t = I_{u_t>0} f^i_t = 0, \quad u^+_t \in \overset{0}{W}{}^1_2(D), \quad I_{u_t>0} f_t \le 0$$

dV_t-almost everywhere on (τ_1,τ_2) and suppose that $u_{\tau_1} \le 0$ for any ω. Then almost surely $u_t \le 0$ for all $t \in [\tau_1,\tau_2] \cap [\tau_1,\infty)$.

The following comparison principle is a generalization of Theorem 2.5.

Theorem 2.6.
Let $\tau_2 \ge \tau_1$ be stopping times, $\tau_1 < \infty$ for any ω. Let $\rho_t \ge 0$, $t \in [0,\infty)$, be a nondecreasing continuous $\mathcal{F}_t$-adapted process and let $\bar f_t$, $\bar f^1_t$,..., $\bar f^d_t$, and $\bar g_t$ satisfy Assumption 2.3 (iv). Let $\bar u_t$ be a process satisfying Assumption 2.4 and such that equation (2.1) holds for all $t \in [0,\infty)$ and any $\varphi \in C^\infty_0(D)$ with $\bar f_t$, $\bar f^1_t$,..., $\bar f^d_t$, and $\bar g_t$ in place of f_t, f^1_t,..., f^d_t, and g_t, respectively.

Assume that, for any ω, (dV_t-a.e.) on $[\tau_1,\tau_2]$ we have

$$I_{u_t>\rho\bar u_t}(g_t - \rho_t \bar g_t) = I_{u_t>\rho_t\bar u_t}(f^i_t - \rho_t \bar f^i_t) = 0, \quad i = 1,...,d,$$

$$I_{u_t>\rho_t\bar u_t}(f_t - \rho_t \bar f_t) \le 0, \quad I_{u_t>\rho_t\bar u_t}\bar u_t \ge 0, \quad (u_t - \rho_t \bar u_t)^+ \in \overset{0}{W}{}^1_2(D).$$

Finally, assume that $u_{\tau_1} \le \rho_{\tau_1}\bar u_{\tau_1}$ for any ω.

Then almost surely $u_t \le \rho_t \bar u_t$ for all $t \in [\tau_1,\tau_2] \cap [\tau_1,\infty)$.

Corollary 2.7. *Assume that, for any ω, (dV_t-a.e.) on $(\tau_1,\tau_2) \times D$ we have*

$$I_{u_t>1}(\nu^k_t + g^k_t) = I_{u_t>1}(f^i_t + a^i_t) = 0, \quad i = 1,...,d, k = 1,2,...$$

$$I_{u_t>1} f \le I_{u_t>1} c, \quad (u_t - 1)^+ \in \overset{0}{W}{}^1_2(D).$$

Also assume that $u_{\tau_1} \le 1$ for any ω. Then almost surely $u_t \le 1$ for all $t \in [\tau_1,\tau_2] \cap [\tau_1,\infty)$.

Indeed, it suffices to take $\bar{u}_t \equiv 1$, $\rho_t \equiv 1$ and observe that $\bar{u}_t$ satisfies (2.1) with $\bar{f}^i_t = -a^i_t$, $\bar{f}_t = c_t$, and $\bar{g}_t = -\nu_t$ in place of f^i_t, f_t, and g_t, respectively.

This corollary generalizes the corresponding results of Refs. [2] and [3], where $\nu^k = g^k = f^i = a^i = 0$.

Remark 2.8. Our equation has a special structure, which may look quite restrictive. In particular, we assume that the martingales m^k_t are mutually orthogonal. The general case, actually, reduces to this particular one after using the fact that one can always orthogonalize the martingales by using, for instance, the Gramm-Schmidt procedure. This, of course, would change σ, ν, and g, and writing the corresponding general conditions would only obscure the matter. Then passing from m^k_t to (no summation in k)

$$\int_0^t \rho^k_s \, dm^k_s, \quad \rho^k_s = \Big(\frac{dt}{dt + d\langle m^k \rangle_t}\Big)^{1/2}$$

allows one to have $d\langle m^k \rangle_t \le dt$ and adding after that t to V_t allows one to have $d\langle m^k \rangle_t \le dV_t$. Again we should modify our coefficients but we will see in the proof of Theorem 2.6 that this modification does not affect Assumption 2.2, which is an assumption about parabolicity of our equation and not strict nondegeneracy.

3. Auxiliary results

In this section the notation u_t is sometimes used for different objects than in Section 2.

Denote by $\mathcal{R}$ the set of real-valued functions convex $r(x)$ on $\mathbb{R}$ such that
(i) r is continuously differentiable, $r(0) = r'(0) = 0$,
(ii) r' is absolutely continuous, its derivative r'' is bounded and left continuous, that is usual r'' which exists almost everywhere is bounded and there is a left-continuous function with which r'' coincides almost everywhere.

For $r \in \mathcal{R}$ by r'' we will always mean the left-continuous modification of the usual second-order derivative of r.

Remark 3.1. For each $r \in \mathcal{R}$ there exists a sequence $r_n \in \mathcal{R}$ of *infinitely* differentiable functions such that $|r_n(x)| \le N|x|^2$, $|r'_n(x)| \le N|x|$, and $|r''_n| \le N$ with $N < \infty$ independent of x and n, $r_n, r'_n, r''_n \to r, r'r''$ on $\mathbb{R}$.

Indeed, let $\zeta \in C^\infty_0(\mathbb{R})$ be a nonnegative function with support in $(0,1)$ and unit integral. For $\varepsilon > 0$ define $\zeta_\varepsilon(x) = \varepsilon^{-1}\zeta(x/\varepsilon)$ and $r_\varepsilon(x) = r * \zeta_\varepsilon(x) - r * \zeta_\varepsilon(0) - xr' * \zeta_\varepsilon(0)$. Then r_ε is infinitely differentiable, $r_\varepsilon(0) = r'_\varepsilon(0) = 0$,

$$|r''_\varepsilon| = |r'' * \zeta_\varepsilon| \le \sup |r''| < \infty.$$

In particular,

$$|r'_\varepsilon(x)| = |\int_0^x r''_\varepsilon(y)\, dy| \le N|x|, \quad |r_\varepsilon(x)| = |\int_0^x r'_\varepsilon(y)\, dy| \le N|x|^2.$$

Finally, the convergences $r_\varepsilon \to r$ and $r'_\varepsilon \to r'$ follow by the continuity of r and r' and the convergence $r''_\varepsilon \to r''$ follows from the dominated convergence theorem, the left continuity of r'' and the formula

$$r''_\varepsilon(x) = \int_0^1 r''(x - \varepsilon y)\zeta(y)\,dy.$$

In the following lemma the assumption that D is a locally smooth domain is not used.

Lemma 3.2.

Let $u_t = u_t(\omega)$ be an $L_2(D)$-valued process such that u_0 is $\mathcal{F}_0$-measurable. Let f_t and $g_t = (g_t^1, g_t^2, ...)$ be $\mathcal{F}_t$-adapted and jointly measurable processes with values $L_2(D)$ and $L_2(D, \ell_2)$, respectively. Assume that for each $t \in [0, \infty)$ we have

$$\int_0^t (\|f_s\|^2_{L_2(D)} + \|g_s\|^2_{L_2(D,\ell_2)})\,dV_s < \infty \tag{3.1}$$

and for any $\varphi \in C_0^\infty(D)$

$$(\varphi, u_t)_{L_2(D)} = (\varphi, u_0)_{L_2(D)} + \int_0^t (\varphi, f_s)_{L_2(D)}\,dV_s + \int_0^t (\varphi, g_s^k)_{L_2(D)}\,dm_s^k. \tag{3.2}$$

Then (i) u_t is a continuous $L_2(D)$-valued function (a.s.); (ii) for any $r \in \mathcal{R}$ (a.s.) for all $t \in [0, \infty)$

$$\|r^{1/2}(u_t)\|^2_{L_2(D)} = \|r^{1/2}(u_0)\|^2_{L_2(D)} + \int_0^t h_s\,dV_s + m_t, \tag{3.3}$$

where

$$h_s := (r'(u_s), f_s)_{L_2(D)} + (1/2)\|(r'')^{1/2}(u_s)\breve{g}_s\|^2_{L_2(D,\ell_2)}, \tag{3.4}$$

$$\breve{g}_s^k := \left(\frac{d\langle m^k\rangle_s}{dV_s}\right)^{1/2} g_s^k, \quad m_t := \int_0^t (r'(u_s), g_s^k)_{L_2(D)}\,dm_s^k$$

and m_t is a local martingale;

(iii) (a.s.) for $t \in [0, \infty)$

$$\|u_t^+\|^2_{L_2(D)} = \|u_0^+\|^2_{L_2(D)} + \int_0^t h_s\,dV_s + m_t, \tag{3.5}$$

where

$$h_s := 2(u_s^+, f_s)_{L_2(D)} + \|\breve{g}_s I_{u_s > 0}\|^2_{L_2(D,\ell_2)},$$

$$m_t := 2\int_0^t (u_s^+, g_s^k)_{L_2(D)}\,dm_s^k$$

and m_t is a local martingale.

Proof. (i) Recall that the operation of stochastic integration of Hilbert space valued processes is well defined. Therefore, the process

$$\widehat{u}_t = u_0 + \int_0^t f_s\,dV_s + \int_0^t g_s^k\,dm_s^k$$

is well defined as a continuous $L_2(D)$-valued process. We also recall how the scalar product interacts with integrals. Then it is seen that for any t and $\varphi \in C_0^\infty(D)$ we have $(\varphi, u_t) = (\varphi, \widehat{u}_t)$ (a.s.). Since both parts are continuous in t, the equality holds for all t at once (a.s.), and since $C_0^\infty(D)$ is dense in $L_2(D)$, we have that $u_t = \widehat{u}_t$ for all t (a.s.). This proves (i). As a corollary we obtain that

$$\sup_{t\le T} \|u_t\|_{L_2(D)} < \infty, \quad \forall T < \infty \quad \text{(a.s.)}. \tag{3.6}$$

(ii) It suffices to prove (3.3) for infinitely differentiable $r \in \mathcal{R}$. Indeed, for r_n from Remark 3.1, passing to the limit in all term in (3.3) apart from m_t presents no problem at all in light of (3.6) and the dominated convergence theorem. Also

$$m_t(n) := \int_0^t (r_n'(u_s), g_s^k)_{L_2(D)}\,dm_s^k \to m_t$$

uniformly in t on finite intervals in probability because

$$\langle m(n) - m\rangle_t = \int_0^t \sum_k ((r_n' - r')(u_s), g_s^k)^2_{L_2(D)}\,d\langle m^k\rangle_s$$

$$\le \int_0^t \sum_k ((r_n' - r')(u_s), g_s^k)^2_{L_2(D)}\,dV_s$$

$$\le \int_0^t \|(r_n' - r')(u_s)\|^2_{L_2(D)} \|g_s\|^2_{L_2(D,\ell_2)}\,dV_s \to 0$$

again owing to (3.6) and the dominated convergence theorem.

Thus, we may concentrate on the case that r is infinitely differentiable. Take a symmetric $\zeta \in C_0^\infty(\mathbb{R}^d)$ with support in the unit ball centered at the origin and unit integral. For $\varepsilon > 0$ set $\zeta_\varepsilon(x) = \varepsilon^{-d}\zeta(x/\varepsilon)$ and for functions $v = v(x)$ define $v^{(\varepsilon)} = v * \zeta_\varepsilon$. Also set

$$D_\varepsilon = \{x \in D : \text{dist}\,(x, \partial D) < \varepsilon\}.$$

According to (3.2) for any $x \in D_\varepsilon$ and $t \ge 0$ we have

$$u_t^{(\varepsilon)}(x) = u_0^{(\varepsilon)}(x) + \int_0^t f_s^{(\varepsilon)}(x)\,dV_s + \int_0^t g_s^{k(\varepsilon)}(x)\,dm_s^k.$$

By Itô's formula we have that on D_ε

$$r(u_t^{(\varepsilon)}) = r(u_0^{(\varepsilon)}) + \int_0^t [r'(u_s^{(\varepsilon)}) f_s^{(\varepsilon)} + (1/2) r''(u_s^{(\varepsilon)}) |\breve{g}_s^{(\varepsilon)}|^2_{\ell_2}]\,dV_s$$

$$+\int_0^t r'(u_s^{(\varepsilon)})g_s^{k(\varepsilon)}\,dm_s^k. \tag{3.7}$$

Here, for each $\varepsilon > 0$, the integrands are smooth functions of x and their magnitudes along with the magnitudes of each of their derivatives in D_ε are majorized by a constant (possibly depending on ε) times

$$\|f_s\|_{L_2(D)}\sup_{s\le t}\|u_s\|_{L_2(D)} \quad \text{or} \quad \|g_s\|^2_{L_2(D,\ell_2)} \quad \text{or} \quad \|g_s^k\|_{L_2(D)}\sup_{s\le t}\|u_s\|_{L_2(D)}.$$

This and (3.1) and (3.6) allow us to use Fubini's theorem while integrating through (3.7) and conclude

$$\|r^{1/2}(u_t^{(\varepsilon)})\|^2_{L_2(D_\varepsilon)} = \|r^{1/2}(u_0^{(\varepsilon)})\|^2_{L_2(D_\varepsilon)} + \int_0^t [(r'(u_s^{(\varepsilon)}), f_s^{(\varepsilon)})_{L_2(D_\varepsilon)}$$

$$+(1/2)\|(r'')^{1/2}(u_s^{(\varepsilon)})\check{g}_s^{(\varepsilon)}\|^2_{L_2(D_\varepsilon,\ell_2)}]\,dV_s$$

$$+\int_0^t (r'(u_s^{(\varepsilon)}), g_s^{k(\varepsilon)})_{L_2(D_\varepsilon)}\,dm_s^k. \tag{3.8}$$

Now we let $\varepsilon \downarrow 0$. We use that for any function $v \in L_2(D)$

$$\|v^{(\varepsilon)}\|_{L_2(D_\varepsilon)} \le \|(vI_D)^{(\varepsilon)}\|_{L_2(\mathbb{R}^d)} \le \|vI_D\|_{L_2(\mathbb{R}^d)} = \|v\|_{L_2(D)}$$

and $v^{(\varepsilon)}I_{D_\varepsilon} \to v$ in $L_2(D)$. In particular, $g_s^{k(\varepsilon)}I_{D_\varepsilon} \to g_s^k$ and $u_s^{(\varepsilon)}I_{D_\varepsilon} \to u_s$ implying that $r'(u_s^{(\varepsilon)})I_{D_\varepsilon} \to r'(u_s)$ in $L_2(D)$ and

$$(r'(u_s^{(\varepsilon)}), g_s^{k(\varepsilon)})_{L_2(D_\varepsilon)} \to (r'(u_s), g_s^k)_{L_2(D)}$$

for each k and $dP \times dV_s$-almost all (ω, s).

We also use (3.1) and (3.6) to assert that

$$\sum_{k=1}^{\infty}\int_0^t \sup_{\varepsilon\in(0,1)} |(r'(u_s^{(\varepsilon)}), g_s^{k(\varepsilon)})_{L_2(D_\varepsilon)}|^2\,dV_s$$

$$\le N\sup_{s\le t}\|u_s\|^2_{L_2(D)}\int_0^t \|g_s\|^2_{L_2(D,\ell_2)}\,dV_s < \infty.$$

As is easy to see this implies that the local martingale part in (3.8) converges to m_t as $\varepsilon \downarrow 0$ in probability locally uniformly with respect to t.

Similar manipulations with other terms in (3.8) allow us to get (3.3). Since (3.5) is just a particular case of (3.3), the lemma is proved.

Remark 3.3. Lemma 3.2 remains true if in the definition of $\mathcal{R}$ instead of requiring r'' to have a left-continuous modification we required it to have a right-continuous one, and of course, in (3.4) used this right-continuous modification. This is seen after replacing u with $-u$.

In case $r(x) = (x^+)^2$ the function r'' has both right- and left-continuous modifications, so that in the definition of m_t one can use $2I_{u_s>0}$ or $2I_{u_s\geq 0}$. It follows that (a.s.) for any t

$$\int_0^t \|\breve{g}_s I_{u_s=0}\|^2_{L_2(D,\ell_2)}\, dV_s = 0.$$

Furthermore, since, for any $v \in L_2(D)$, $u_t + v$ has the same form as u_t,

$$\int_0^t \|\breve{g}_s I_{u_s=v}\|^2_{L_2(D,\ell_2)}\, dV_s = 0.$$

Lemma 3.4. *Let D be an arbitrary domain. Let u_t be an $L_2(D)$-valued $\mathcal{F}_0$-measurable process such that for any ω*

$$u_t \in W^1_2(D)$$

(dV_t-a.e.) and for each $T \in [0,\infty)$ and ω

$$\int_0^T \|u_t\|^2_{W^1_2(D)}\, dV_t < \infty. \tag{3.9}$$

Let $f_t, f^1_t, ..., f^d_t$, and $g_t = (g^1_t, g^2_t, ...)$ be $\mathcal{F}_t$-adapted and jointly measurable processes with values in $L_2(D)$ and $L_2(D,\ell_2)$, respectively. Assume that for each $t \in [0,\infty)$ we have

$$\int_0^t (\|f_s\|^2_{L_2(D)} + \sum_i \|f^i_s\|^2_{L_2(D)} + \|g_s\|^2_{L_2(D,\ell_2)})\, dV_s < \infty, \tag{3.10}$$

and for each $t \in [0,\infty)$, $\varphi \in C^\infty_0(D)$, and ω

$$(\varphi, u_t) = (\varphi, u_0) + \int_0^t (\varphi, f_s + D_i f^i_s)\, dV_s + \int_0^t (\varphi, g^k_s)\, dm^k_s. \tag{3.11}$$

Finally, assume that there is a compact set $G \subset D$ such that

$$u_t(x) = f_t(x) = f^i_t(x) = g^k_t(x) = 0$$

outside G. Then

(a) u_t is a continuous $L_2(D)$-valued function (a.s.); (b) (a.s.) for all $t \in [0,\infty)$

$$\|u^+_t\|^2_{L_2(D)} = \|u^+_0\|^2_{L_2(D)} + \int_0^t h_s\, dV_s + m_t, \tag{3.12}$$

where

$$h_s := 2(u^+_s, f_s)_{L_2(D)} - 2(I_{u_s>0} D_i u_s, f^i_s)_{L_2(D)} + \|\breve{g}_s I_{\varphi u_s>0}\|^2_{L_2(D,\ell_2)},$$

$$m_t := 2\int_0^t (u^+_s, g^k_s)_{L_2(D)}\, dm^k_s.$$

Proof. Observe that (3.11) holds for all infinitely differentiable functions φ. Furthermore, since $u_t \in \overset{0}{W}{}^1_2(D)$ (dV_t-a.e.) assertion (a) is well known (see, for instance, Ref. [10], the references therein, and Remark 2.8).

To prove (b), take ε smaller than the distance between G and ∂D. Notice that, owing to the symmetry of ζ, for $\varphi \in C_0^\infty(D)$

$$(\varphi^{(\varepsilon)}, u_t) = (\varphi^{(\varepsilon)}, u_t)_{L_2(\mathbb{R}^d)} = (\varphi, u_t^{(\varepsilon)})_{L_2(\mathbb{R}^d)} = (\varphi, u_t^{(\varepsilon)})_{L_2(D)}.$$

Therefore, it follows from (3.11) that for any $\varphi \in C_0^\infty(D)$

$$(\varphi, u_t^{(\varepsilon)}) = (\varphi, u_0^{(\varepsilon)}) + \int_0^t (\varphi, \tilde{f}_s^\varepsilon)\, ds + \int_0^t (\varphi, g_s^{k(\varepsilon)})\, dm_s^k,$$

where

$$\tilde{f}_s^\varepsilon := f_s^{(\varepsilon)} + D_i f_s^{i(\varepsilon)}$$

is an $L_2(D)$-valued function with norm that is locally square integrable against dV_s. By Lemma 3.2 for any $r \in \mathcal{R}$

$$\|r^{1/2}(u_t^{(\varepsilon)})\|_{L_2(D)}^2 = \|r^{1/2}(u_0^{(\varepsilon)})\|_{L_2(D)}^2 + \int_0^t h_s^\varepsilon\, dV_s + m_t^\varepsilon, \tag{3.13}$$

where

$$m_t^\varepsilon := \int_0^t (r'(u_s^{(\varepsilon)}), \check{g}_s^{k(\varepsilon)})_{L_2(D)}\, dm_s^k,$$

$$h_s^\varepsilon := (r'(u_s^{(\varepsilon)}), \tilde{f}_s^\varepsilon)_{L_2(D)} + (1/2)\|(r'')^{1/2}(u_s^{(\varepsilon)})\check{g}_s^{(\varepsilon)}\|_{L_2(D,\ell_2)}^2$$

$$= (r'(u_s^{(\varepsilon)}), f_s^{(\varepsilon)})_{L_2(D)} - (r''(u_s^{(\varepsilon)}) D_i u_s^{(\varepsilon)}, f_s^{i(\varepsilon)})_{L_2(D)} + (1/2)\|(r'')^{1/2}(u_s^{(\varepsilon)})\check{g}_s^{(\varepsilon)}\|_{L_2(D,\ell_2)}^2.$$

If r is infinitely differentiable, then by using (3.9) and (3.10) one easily passes to the limit in (3.13) as $\varepsilon \to 0$. The argument is quite similar to the corresponding argument in the proof of Lemma 3.2 and, for smooth $r \in \mathcal{R}$, yields

$$\|r^{1/2}(u_t)\|_{L_2(D)}^2 = \|r^{1/2}(u_0)\|_{L_2(D)}^2 + \int_0^t h_s\, dV_s + m_t,$$

where

$$m_t = \int_0^t (r'(u_s), g_s^k)_{L_2(D)}\, dm_s^k,$$

$$h_s = (r'(u_s), f_s)_{L_2(D)} - (r''(u_s) D_i u_s, f_s^i)_{L_2(D)} + (1/2)\|(r'')^{1/2}(u_s)\check{g}_s\|_{L_2(D,\ell_2)}^2.$$

Finally, as in the proof of Lemma 3.2 one easily passes from smooth $r \in \mathcal{R}$ to arbitrary ones and gets (3.12) by taking $r(x) = (x^+)^2$. The lemma is proved.

Lemma 3.4 serves as an auxiliary tool to prove a deeper result.

Lemma 3.5. *Let D be an arbitrary domain. Assume that for each $\varphi \in C_0^\infty(D)$*

(i) φu_t is an $L_2(D)$-valued process such that φu_0 is $\mathcal{F}_0$-measurable;

(ii) for any ω

$$\varphi u_t \in W_2^1(D)$$

(dV_t-a.e.) and for each $T \in [0, \infty)$ and ω

$$\int_0^T \|\varphi u_t\|^2_{W_2^1(D)}\, dV_t < \infty.$$

(iii) Let $f_t, f_t^1, ..., f_t^d$, and $g_t = (g_t^1, g_t^2, ...)$ be $\mathcal{F}_t$-adapted and jointly measurable processes with values in $L_2(D)$ and $L_2(D, \ell_2)$, respectively. Assume that for each $t \in [0, \infty)$ and $\varphi \in C_0^\infty(D)$ we have

$$\int_0^t (\|\varphi f_s\|^2_{L_2(D)} + \sum_i \|\varphi f_s^i\|^2_{L_2(D)} + \|\varphi g_s\|^2_{L_2(D,\ell_2)})\, dV_s < \infty,$$

$$(\varphi, u_t) = (\varphi, u_0) + \int_0^t (\varphi, f_s + D_i f_s^i)\, dV_s + \int_0^t (\varphi, g_s^k)\, dm_s^k.$$

Then, for any $\varphi \in C_0^\infty(D)$,
(a) φu_t is a continuous $L_2(D)$-valued function (a.s.); (b) (a.s.) for all $t \in [0, \infty)$

$$\|(\varphi u_t)^+\|^2_{L_2(D)} = \|(\varphi u_0)^+\|^2_{L_2(D)} + \int_0^t h_s\, dV_s + m_t,$$

where

$$h_s := 2((\varphi u_s)^+, \varphi f_s - f_s^i D_i \varphi)_{L_2(D)} - 2(I_{\varphi u_s > 0} D_i(\varphi u_s), \varphi f_s^i)_{L_2(D)}$$

$$+\|\varphi \check{g}_s I_{\varphi u_s > 0}\|^2_{L_2(D,\ell_2)}, \quad m_t := 2\int_0^t (\varphi u_s^+, \varphi g_s^k)_{L_2(D)}\, dm_s^k.$$

Proof. Clearly, for any $\varphi, \eta \in C_0^\infty(D)$ we have

$$(\varphi, \eta u_t) = (\varphi, \eta u_0) + \int_0^t (\varphi, \eta f_s - f_s^i D_i \eta + D_i(\eta f_s^i))\, dV_s + \int_0^t (\varphi, \eta g_s^k)\, dm_s^k. \quad (3.14)$$

Therefore, ηu_t satisfies the assumptions of Lemma 3.4 with $\eta f_s - f_s^i D_i \eta$, ηf_s^i, and ηg_s^k in place of f_s, $D_i f_s^i$, and g_s^k, respectively.

By applying Lemma 3.4 to ηu_t in place of u_t we get the result with η in place of φ. This certainly proves the lemma.

4. Proof of Theorems 2.5 and 2.6

In this section the assumptions stated in Section 2 are supposed to be satisfied. We use the fact that due to our hypothesis that $D \in C^1$, there exist sequences ζ_n and $\bar{\zeta}_n$ of nonnegative $C_0^\infty(D)$-functions such that $0 \le \zeta_n, \bar{\zeta}_n \le 1$, $\zeta_n, \bar{\zeta}_n \to 1$ in D as $n \to \infty$ and for any $v \in \overset{0}{W}{}_2^1(D)$, $i = 1, ..., d$,

$$\|v D_i \zeta_n\|_{L_2(D)} \le N(\|(1 - \bar{\zeta}_n) v\|_{L_2(D)} + \|(1 - \bar{\zeta}_n) D v\|_{L_2(D)}), \quad (4.1)$$

where N is independent of n and v (see, for instance, the proof of Theorem 5.5.2 in Ref. [11]). We also know (see, for instance, the proof of Lemma 2.3.2 in Ref. [12] or Problem 17, Chapter 5 in Ref. [11]) that if $v \in W_2^1(D)$, then $v^+ \in W_2^1(D)$ and

$$D_i v^+ = I_{v>0} D_i v.$$

Proof of Theorem 2.5. Set

$$K = K_1 + K_2, \quad \varphi_t = \int_0^t K(s)\,ds.$$

Take the sequences of nonnegative $\zeta_n, \bar\zeta_n \in C_0^\infty(D)$ from above. By Itô's formula and Lemma 3.5

$$\|\zeta_n u_t^+\|_{L_2(D)}^2 e^{-\varphi_t} = \|\zeta_n u_0^+\|_{L_2(D)}^2 + \int_0^t h_s^n\,dV_s + m_t(n),$$

where

$$e^{\varphi_s} h_s^n = I_{1s} + I_{2s} + I_{3s} - K(s)\|\zeta_n u_s^+\|_{L_2(D)}^2,$$

$$I_{1s} = 2\big(\zeta_n u_s^+, \zeta_n[f_s + b_s^i D_i u_s - c_s u_s] - [u_s a_s^i + a_s^{ij} D_j u_s + f_s^i] D_i\zeta_n\big)_{L_2(D)},$$

$$I_{2s} = -2\big(I_{\zeta_n u_s>0} D_i(\zeta_n u_s), \zeta_n[u_s a_s^i + a_s^{ij} D_j u_s + f_s^i]\big)_{L_2(D)},$$

$$I_{3s} = \|\zeta_n I_{\zeta_n u_s>0}[\check\sigma_s^i D_i u_s + \check\nu_s u_s + \check g_s]\|_{L_2(D,\ell_2)}^2,$$

$$m_t(n) = 2\int_0^t e^{-\varphi_s}\big(\zeta_n u_s^+, \zeta_n[\sigma_s^{ik} D_i u_s + \nu_s^k u_s + g_s^k]\big)_{L_2(D)}\,dm_s^k.$$

Since $u_{\tau_1}^+ = 0$ we have

$$e^{-\varphi_{\tau_2\wedge t\vee\tau_1}}\|\zeta_n u_{\tau_2\wedge t\vee\tau_1}^+\|_{L_2(D)}^2 = \int_0^t I_{\tau_2>s>\tau_1} h_s^n\,dV_s + \bar m_t(n), \tag{4.2}$$

where

$$\bar m_t(n) := m_{\tau_2\wedge t\vee\tau_1}(n) - m_{\tau_1}(n)$$

is a local martingale.

Next we use the assumptions of the theorem and see that for dV_s-almost all $s \in (\tau_1, \tau_2)$ we have

$$I_{1s} \le 2(\zeta_n u_s^+, \zeta_n[b_s^i D_i u_s - c_s u_s] - [u_s a_s^i + a_s^{ij} D_j u_s] D_i\zeta_n)_{L_2(D)}$$

$$= 2(\zeta_n^2 u_s^+, b_s^i D_i u_s^+ - c_s u_s^+)_{L_2(D)} + I_{4s}$$

with

$$I_{4s} = -2(\zeta_n u_s^+ D_i\zeta_n, u_s^+ a_s^i + a_s^{ij} D_j u_s^+)_{L_2(D)}.$$

At this moment we recall (4.1) and observe that

$$(\zeta_n|D_i\zeta_n|,(u_s^+)^2)_{L_2(D)} \le N\|\zeta_n u_s^+\|_{L_2(D)}\|u_s^+ D_i\zeta_n\|_{L_2(D)}.$$

Then we see that

$$I_{4s} \le N\big(\|(1-\bar{\zeta}_n)u_s^+\|_{L_2(D)} + \|(1-\bar{\zeta}_n)Du_s^+\|_{L_2(D)}\big)\|u_s^+\|_{W_2^1(D)},$$

where and below by N we denote various finite constants.

In I_{2s}

$$I_{\zeta_n u_s>0}D_i(\zeta_n u_s) = D_i(\zeta_n u_s^+) = u_s^+ D_i\zeta_n + \zeta_n D_i u_s^+,$$

so that

$$I_{2s} = -2(\zeta_n^2 D_i u_s^+, u_s^+ a_s^i + a_s^{ij} D_j u_s^+)_{L_2(D)} + I_{4s}.$$

Next,

$$\zeta_n I_{\zeta_n u_s>0} = \zeta_n I_{u_s>0},$$

$$I_{3s} \le \|\zeta_n I_{u_s>0}[\sigma_s^i D_i u_s + \nu_s u_s]\|^2_{L_2(D,\ell_2)} = (\zeta_n^2 D_i u_s^+, \alpha_s^{ij} D_j u_s^+)_{L_2(D)}$$

$$+2(\zeta_n^2 D_i u_s^+, u_s^+(\sigma_s^i,\nu_s)_{\ell_2})_{L_2(D)} + \|\zeta_n|\nu_s|_{\ell_2} u_s^+\|^2_{L_2(D)}.$$

Also observe that certain parts of I_{2s} and I_{3s} can be combined if we use that

$$-2(\zeta_n^2 D_i u_s^+, a_s^{ij} D_j u_s^+)_{L_2(D)} + (\zeta_n^2 D_i u_s^+, \alpha_s^{ij} D_j u_s^+)_{L_2(D)}$$

$$\le -K_1^{-1}(s)\|\zeta_n \xi_s^i D_i u_s^+\|^2_{L_2(D)}.$$

It follows that for dV_s-almost all $s\in(\tau_1,\tau_2)$

$$e^{\varphi_s}h_s^n \le \int_D [\zeta_n^2(b_s^i - a_s^i + (\sigma_s^i,\nu_s)_{\ell_2})2u_s^+ D_i u_s^+ + \zeta_n^2(u_s^+)^2(|\nu_s|^2_{\ell_2} - 2c_s)]\,dx$$

$$+N\big(\|(1-\bar{\zeta}_n)u_s^+\|_{L_2(D)} + \|(1-\bar{\zeta}_n)Du_s^+\|_{L_2(D)}\big)\|u_s^+\|_{W_2^1(D)}$$

$$-K_1^{-1}(s)\|\zeta_n\xi_s^i D_i u_s^+\|^2_{L_2(D)} - K(s)\|\zeta_n u_s^+\|^2_{L_2(D)}.$$

Here

$$b_s^i - a_s^i + (\sigma_s^i,\nu_s)_{\ell_2} = -\xi_s^i - \eta_s^i$$

and we transform the integral of

$$\zeta_n^2(-\eta_s^i)2u_s^+ D_i u_s^+ = -\eta_s^i\zeta_n^2 D_i(u_s^+)^2$$

by integrating by parts. Then we get that for dV_s-almost all $s\in(\tau_1,\tau_2)$

$$e^{\varphi_s}h_s^n \le \int_D [-\zeta_n^2\xi_s^i 2u_s^+ D_i u_s^+ + \zeta_n^2(u_s^+)^2(|\nu_s|^2_{\ell_2} - 2c_s + D_i\eta_s^i)]\,dx$$

$$+2\int_D (u_s^+)^2\zeta_n\eta_s^i D_i\zeta_n\,dx + N\big(\|(1-\bar\zeta_n)u_s^+\|_{L_2(D)} + \|(1-\bar\zeta_n)Du_s^+\|_{L_2(D)}\big)\|u_s^+\|_{W_2^1(D)}$$

$$-K_1^{-1}(s)\|\zeta_n\xi_s^i D_i u_s^+\|^2_{L_2(D)} - K(s)\|\zeta_n u_s^+\|^2_{L_2(D)}.$$

We also use the fact that

$$|-\zeta_n^2\xi_s^i 2u_s^+ D_i u_s^+| \le K_1^{-1}(s)|\zeta_n\xi_s^i D_i u_s^+|^2 + K_1(s)\zeta_n^2(u_s^+)^2,$$

$$|\nu_s|^2_{\ell_2} - 2c_s + D_i\eta_s^i \le K_2(s).$$

Then we easily see that for dV_s-almost all $s\in(\tau_1,\tau_2)$

$$e^{\varphi_s}h_s^n \le N\big(\|(1-\bar\zeta_n)u_s^+\|_{L_2(D)} + \|(1-\bar\zeta_n)Du_s^+\|_{L_2(D)}\big)\|u_s^+\|_{W_2^1(D)}$$

$$+2\int_D (u_s^+)^2\zeta_n\eta_s^i D_i\zeta_n\,dx \le N\big(\|(1-\bar\zeta_n)u_s^+\|_{L_2(D)} + \|(1-\bar\zeta_n)Du_s^+\|_{L_2(D)}\big)\|u_s^+\|_{W_2^1(D)}.$$

Now (4.2) yields

$$e^{-\varphi_{\tau_2\wedge t\vee\tau_1}}\|u^+_{\tau_2\wedge t\vee\tau_1}\|^2_{L_2(D)} \le \bar m_t(n)$$

$$+N\int_0^t (\|(1-\bar\zeta_n)u_s^+\|_{L_2(D)} + \|(1-\bar\zeta_n)Du_s^+\|_{L_2(D)})\|u_s^+\|_{W_2^1(D)}\,dV_s. \tag{4.3}$$

The integrals against dV_s in (4.3) tend to zero as $n\to\infty$ by the dominated convergence theorem. Since the sum of them with continuous local martingales is nonnegative, the local martingales and the right-hand side of (4.3) tend to zero uniformly on finite time intervals in probability (see, for instance, Ref. [13]). So does the left-hand side and the theorem is proved.

Proof of Theorem 2.6. Obviously, $\widehat u_t = \rho_t\bar u_t$ satisfies

$$(\varphi,\widehat u_t) = (\varphi,\widehat u_0) + \int_0^t (\varphi, \sigma_s^{ik}D_i\widehat u_s + \nu_s^k\widehat u_s + \rho_s\bar g_s^k)\,dm_s^k + \int_0^t (\varphi,\bar u_s)\,d\rho_s$$

$$+\int_0^t (\varphi, D_i(a_s^{ij}D_j\widehat u_s) + b_s^i D_i\widehat u_s + D_i(a_s^i\widehat u_s) - c_s\widehat u_s + \rho_s\bar f_s + \rho_s D_i\bar f_s^i)\,dV_s. \tag{4.4}$$

We rewrite this equation introducing

$$\widehat V_t = V_t + \rho_t,\quad p_t = \frac{d\rho_t}{d\widehat V_t},\quad q_t = \frac{dV_t}{d\widehat V_t},$$

$$(\widehat a_t^{ij},\widehat a_t^i,\widehat b_t^i,\widehat c_t) = q_t(a_t^{ij},a_t^i,b_t^i,c_t),\quad (\widehat\sigma_t^{ik},\widehat\nu_t^k) = q_t^{1/2}(\sigma_t^{ik},\nu_t^k),$$

$$\widehat f_t = q_t\rho_t\bar f_t + p_t\bar u_t,\quad \widehat{f_t^i} = q_t\rho_t\bar f_t^i,\quad \widehat{g_t^k} = q_t^{1/2}\rho_t\bar g_t^k.$$

We also set

$$\widehat m_t^k = \int_0^t q_s^{-1/2}\,dm_s^k \qquad (0^{-1/2} := 0).$$

Notice that since $d\langle m^k\rangle_t \le dV_t = q_t d\widehat{V}_t$ the last integral makes sense.

In this notation (2.1) and (4.4) are rewritten as

$$(\varphi, u_t) = (\varphi, u_0) + \int_0^t (\varphi, \widehat{\sigma}_s^{ik} D_i u_s + \widehat{\nu}_s^k u_s + q_s^{1/2} g_s^k)\, d\widehat{m}_s^k$$

$$+ \int_0^t (\varphi, D_i(\widehat{a}_s^{ij} D_j u_s) + \widehat{b}_s^i D_i u_s + D_i(\widehat{a}_s^i u_s) - \widehat{c}_s u_s + q_s f_s + q_s D_i f_s^i)\, d\widehat{V}_s,$$

$$(\varphi, \widehat{u}_t) = (\varphi, \widehat{u}_0) + \int_0^t (\varphi, \widehat{\sigma}_s^{ik} D_i \widehat{u}_s + \widehat{\nu}_s^k \widehat{u}_s + \widehat{g}_s^k)\, d\widehat{m}_s^k$$

$$+ \int_0^t (\varphi, D_i(\widehat{a}_s^{ij} D_j \widehat{u}_s) + \widehat{b}_s^i D_i \widehat{u}_s + D_i(\widehat{a}_s^i \widehat{u}_s) - \widehat{c}_s \widehat{u}_s + \widehat{f}_s + D_i \widehat{f}_s^i)\, d\widehat{V}_s,$$

respectively. We subtract these equations, denote $v_t = u_t - \widehat{u}_t$, and observe that for any ω we have $d\widehat{V}_s$-almost everywhere on (τ_1, τ_2) that

$$I_{v_s>0}(q_s^{1/2} g_s^k - \widehat{g}_s^k) = I_{v_s>0}(q_s f_s^i - \widehat{f}_s^i) = 0,$$

$$I_{v_s>0}(q_s f_s - \widehat{f}_s) = q_s I_{v_s>0}(f_s - \rho_s \bar{f}_s) - p_s \bar{u}_s I_{v_s>0} \le 0.$$

We also use the fact that the above versions of equations (2.1) and (4.4) satisfy the same Assumptions 2.1, 2.2, 2.3, and 2.4 with $q_s \xi_s^i$ and $q_s K_i(s)$ in place of ξ_s^i and $K_i(s)$, respectively. Then we the desired result directly from Theorem 2.5. The theorem is proved.

5. Auxiliary functions

Let $C[0,\infty)$ be the set of real-valued continuous functions on $[0,\infty)$. For $x_\cdot \in C[0,\infty)$ set $x_s = x_0$ for $s < 0$ and for $n = 0, 1, 2, ...$ and $t \ge 0$ introduce

$$\Delta_n^-(x_\cdot, t) = 2^{n/2} \operatorname*{osc}_{[t-2^{-n},t]} x_\cdot\,.$$

If $c \in (0,\infty)$, then define

$$M_n^-(x_\cdot, c, t) = \#\{k = 0, ..., n : \Delta_k^-(x_\cdot, t) \le c\}.$$

For n negative we set $M_n^-(x_\cdot, c, t) := 0$. For $c \ge 0, d > 0, \delta > 0$ introduce

$$\gamma(c, d, \delta) = 1 - P(\min_{t \le \delta/2} w_t \le -c - d/\sqrt{2}, \max_{t \le \delta/2} w_t \le d - d/\sqrt{2}).$$

As is easy to see

$$\gamma(c, d, \delta) \ge P(w_t \quad \text{reaches} \quad d - d/\sqrt{2} \quad \text{before reaching} \quad -c - d/\sqrt{2})$$

$$= \frac{c + d/\sqrt{2}}{c + d} > 1/\sqrt{2},$$

so that $2\log_2\gamma(c,d,\delta) > -1$.

Set, for $m = 0, 1, 2, ...,$

$$Q_m := Q_m(x.) := \{(s,y) : s \ge 0, x_s < y < x_s + 2^{-m/2}\}.$$

Lemma 5.1. *For $m = 0, 1, 2, ..., t \ge 0$ and $x \in (0, 2^{-m/2})$ introduce*

$$r_m(t,x) = r_m(x., t, x) = P(x_t + x + w_\tau\sqrt{\delta} = x_{t-\tau} + 2^{-m/2}),$$

where $\tau = \inf\{s > 0 : (t-s, x_t + x + w_s\sqrt{\delta}) \notin Q_m\}$. Then

$$r_m(t,x) \le [\gamma(c,d,\delta)]^{M^-_{m+n}(x.,c,t) - M^-_{m-1}(x.,c,t) - k}, \tag{5.1}$$

where $n = n(2^{m/2}x/d)$, $k = k(c+d)$, and

$$n(y) = [(-2\log_2 y)_+], \quad k(d) = 2 + [(2\log_2 d)_+].$$

Proof. Define

$$\bar{t} = 2^m t, \quad \bar{x} = 2^{m/2}x, \quad \bar{w}_s = 2^{m/2}w_{s2^{-m}}, \quad \bar{x}_s = 2^{m/2}x_{s2^{-m}}.$$

Then as is easy to see $r_m(t,x)$ is rewritten as

$$P(\bar{x}_{\bar{t}} + \bar{x} + \bar{w}_{\bar{\tau}}\sqrt{\delta} = \bar{x}_{\bar{t}-\bar{\tau}} + 1), \tag{5.2}$$

where

$$\bar{\tau} = \inf\{s > 0 : (\bar{t} - s, \bar{x}_{\bar{t}} + \bar{x} + \bar{w}_s\sqrt{\delta}) \notin Q_0(\bar{x}.)\} = 2^m\tau.$$

Since $\bar{w}.$ is a Wiener process, by Corollary 3.4 in Ref. [14] expression (5.2) is less than

$$[\gamma(c,d,\delta)]^{M^-_{\bar{n}}(\bar{x}.,c,\bar{t}) - k},$$

where $\bar{n} = n(\bar{x}/d)$. Here

$$M^-_{\bar{n}}(\bar{x}.,c,\bar{t}) = \#\{j = 0, ..., \bar{n} : 2^{j/2} \operatorname*{osc}_{[\bar{t}-2^{-j},\bar{t}]} \bar{x}. \le c\}$$

$$= \#\{j = 0, ..., \bar{n} : 2^{(j+m)/2} \operatorname*{osc}_{[t-2^{-j-m},t]} x. \le c\}$$

$$= \#\{j = m, ..., m+\bar{n} : \Delta^-_j(x.) \le c\} = M^-_{m+\bar{n}}(x.,c,t) - M^-_{m-1}(x.,c,t)$$

and the result follows. The lemma is proved.

Lemma 5.2. *Let $T \in (0,\infty)$. Assume that*

$$\varliminf_{m\to\infty} \frac{1}{m+1} \inf_{t\in[0,T]} M^-_m(x.,c,t) > \alpha > 0.$$

Take constants $p > 0$ and ν so that

$$1 < \nu p < p\chi + 1 < 0, \tag{5.3}$$

where $\chi = -2\alpha\log_2\gamma(c,d,\delta)$. Then, for r_m from Lemma 5.1 it holds that

$$\sup_{m\ge 0}\sup_{t\in[0,T]} 2^{-m(\nu p-1)/(2\alpha)} \int_0^{2^{-m/2}} \frac{1}{x^{\nu p}} r^p_m(t,x)\,dx < \infty. \tag{5.4}$$

Proof. By Lemma 5.1 for a constant N and $\gamma = \gamma(c, d, \delta)$

$$r_m(t,x) \le N\gamma^{M^-_{m+n}-m},$$

where $x \le 2^{-m/2}$, $n = n(2^{m/2}x/d)$, $M^-_{m+n} = M^-_{m+n}(x.,c,t)$. Furthermore,

$$n(2^{m/2}x/d) \ge (-2\log_2(2^{m/2}x) + 2\log_2 d)_+ - 1$$

$$\le (-2\log_2(2^{m/2}x))_+ - N = -m - 2\log_2 x - N,$$

where N is a constant. Hence, $m + n \ge -2\log_2 x - N$. Since obviously $r_m \le 1$ we have that

$$r_m(t,x) \le 1 \wedge (N\gamma^{-m+M^-_{-2\log_2 x - N}}).$$

By the assumption if x is small enough

$$M^-_{-2\log_2 x - N} > \alpha(-2\log_2 x).$$

Therefore, for $x \in (0, 2^{-m/2}]$

$$r_m(t,x) \le 1 \wedge (N\gamma^{-m-2\alpha\log_2 x}) = 1 \wedge (N\gamma^{-m}x^{\chi}). \tag{5.5}$$

Next,

$$\int_0^{2^{-m/2}} \frac{1}{x^{\nu p}} r^p_m(t,x)\,dx \le \int_0^\infty \frac{1}{x^{\nu p}}(1 \wedge (N\gamma^{-m}x^{\chi}))^p\,dx$$

$$= \gamma^{m(1-\nu p)/\chi} \int_0^\infty \frac{1}{x^{\nu p}}(1 \wedge (Nx^{\chi}))^p\,dx,$$

where the last integral is finite owing to (5.3). This proves the lemma.

Let w_t be a Wiener process with respect to a filtration $\{\mathcal{F}_t, t \ge 0\}$ of complete σ-fields and let a_t and σ_t be bounded real-valued processes predictable with respect to $\{\mathcal{F}_t, t \ge 0\}$ and such that $a_t - \sigma_t^2 \ge \delta\sigma_t^2$, where $\delta \in (0,\infty)$ is a constant, $a_t - \sigma_t^2 > 0$ for all (ω, t) and for all ω

$$\int_0^\infty [a_t - \sigma_t^2]\,dt = \infty.$$

Set $D_x = \partial/\partial x$. For $m = 0, 1, 2, \ldots$ we will be dealing with the SPDE

$$dv(t,x) = (1/2)a_t D_x^2 v(t,x)\,dt + \sigma_t D_x v(t,x)\,dw_t$$

in $B_m = (0,\infty) \times (0, 2^{-m/2})$ with boundary conditions

$$v(t,0) = 0, \quad v(t, 2^{-m/2}) = 1, \quad t > 0, \tag{5.6}$$

$$v(0,x) = 0, \quad 0 < x < 1. \tag{5.7}$$

Recall that by Theorem 2.1 in Ref. [14] there is a deterministic function $\alpha_0(c)$, $c > 0$, such that $\alpha_0(c) \to 1$ as $c \to \infty$ and with probability one for any $T \in (0,\infty)$

$$\varliminf_{n\to\infty} \inf_{t\in[0,T]} \frac{1}{n+1} M^-_n(w.,c,t) = \alpha_0(c).$$

Theorem 5.3. *For each* $m = 0, 1, 2, ...$ *there is a function* $v_m(t,x) = v_m(\omega, t, x)$ *defined on* $\Omega \times \bar{B}_m$ *such that*

(i) $v_m(t,x)$ *is* $\mathcal{F}_t$*-measurable for each* $(t,x) \in \bar{B}_m$,

(ii) $v_m(t,x)$ *is bounded and continuous in* $\bar{B}_m \setminus \{(0, 2^{-m/2})\}$ *for each* ω,

(iii) derivatives of $v_m(t,x)$ *of any order with respect to* x *are continuous in* $B_m \cup (\{0\} \times (0, 2^{-m/2}))$ *for each* ω,

(iv) equations (5.6) *and* (5.7) *hold for each* ω,

(v) almost surely, for any $(t,x) \in B_m$

$$v_m(t,x) = \int_0^t (1/2) a_s D_x^2 v_m(s,x)\, ds + \int_0^t \sigma_s D_x v_m(s,x)\, dw_s,$$

(vi) for any $T \in (0,\infty)$, $c, d > 0$, $p > 0$, $\alpha > 0$ *such that* $\alpha_0(c\sqrt{\delta}) > \alpha$, *and* ν *satisfying*

$$1 < \nu p < \chi p + 1, \tag{5.8}$$

where $\chi = -2\alpha \log_2 \gamma(c, d, 1)$, *we have that with probability one*

$$\pi_T := \sup_{m \geq 0} \sup_{t \in [0,T]} 2^{-m(\nu p - 1)/(2\alpha)} \int_0^{2^{-m/2}} \frac{1}{x^{\nu p}} v_m^p(t,x)\, dx < \infty. \tag{5.9}$$

Proof. In Lemma 5.1 take $\delta = 1$ and set $\tilde{v}_m(x_\cdot, x_t + x, t) = r_m(x_\cdot, t, x)$, where r_m is introduced in that lemma. Set

$$\psi_t = \int_0^t (a_s - \sigma_s^2)\, ds, \quad \xi_t = \int_0^{\varphi_t} \sigma_s\, dw_s, \quad \tilde{\mathcal{F}}_t = \mathcal{F}_{\varphi_t},$$

$$\bar{v}_m(t,x) = \bar{v}_m(\omega, t, x) = \tilde{v}_m(\xi_\cdot, t, x), \quad v_m(t,x) = v_m(\omega, t, x) = \bar{v}_m(\psi_t, x + \xi_{\psi_t}),$$

where $\varphi_t = \inf\{s \geq 0 : \psi_s \geq t\}$ is the inverse function to ψ_t.

It is proved in Theorem 4.1 of Ref. [14] that v_0 possesses properties (i)-(v). The proof that this is also true for any m is no different.

Furthermore, it is well known that

$$\sqrt{\delta} \int_0^t \sigma_s\, dw_s = \tilde{w}_{\tilde{\psi}(t)},$$

where $\tilde{w}_t$ is a Wiener process and

$$\tilde{\psi}_t = \delta \int_0^t \sigma_s^2\, ds.$$

Hence $\xi_t = \delta^{-1/2} \tilde{w}_{\tilde{\psi}(\varphi_t)}$ with

$$(\tilde{\psi}(\varphi_t))' = \delta \sigma_s^2 / (a_s - \sigma_s^2)|_{s = \varphi_t} \leq 1.$$

It follows that for $n = 0, 1, 2, ...$ we have

$$M_n^-(\xi_\cdot, c, t) \geq M_n^-(\tilde{w}_\cdot, c\sqrt{\delta}, \tilde{\psi}(\varphi_t)),$$

$$\inf_{t\le T} M_n^-(\xi_\cdot, c, t) \ge \inf_{t\le T} M_n^-(\tilde{w}_\cdot, c\sqrt{\delta}, t),$$

and with probability one

$$\varliminf_{n\to\infty} \inf_{t\le T} \frac{1}{n+1} M_n^-(\xi_\cdot, c, t) \ge \alpha_0(c\sqrt{\delta}) > \alpha.$$

Finally, for $M = \sup_{\omega,t}(a_t - \sigma_t^2)$ we have

$$\sup_{t\le T} \int_0^{2^{-m/2}} \frac{1}{x^{\nu p}} v_m^p(t,x)\,dx \le \sup_{t\le MT} \int_0^{2^{-m/2}} \frac{1}{x^{\nu p}} \bar{v}_m^p(t, x+\xi_t)\,dx$$

$$= \sup_{t\le MT} \int_0^{2^{-m/2}} \frac{1}{x^{\nu p}} \tilde{v}_m^p(\xi_\cdot, t, x+\xi_t)\,dx = \sup_{t\le MT} \int_0^{2^{-m/2}} \frac{1}{x^{\nu p}} r_m^p(\xi_\cdot, t, x)\,dx.$$

After this it only remains to use Lemma 5.2. The theorem is proved.

Remark 5.4. Obviously, for any $\varepsilon \in (0, 2^{-(m+2)/2})$ we have

$$v_m(t,\cdot) \in W_2^1(\varepsilon, 2^{-m/2} - \varepsilon)$$

for any $t \in [0,\infty)$ and for any $T \in (0,\infty)$ we have

$$\int_0^T \|v_m(t,\cdot)\|_{W_2^1(\varepsilon, 2^{-m/2}-\varepsilon)}\,dt < \infty.$$

Furthermore, by using the deterministic and stochastic versions of Fubini's theorem one easily proves that for any $\varphi \in C_0^\infty(0, 2^{-m/2})$ with probability one for all $t \in [0,\infty)$

$$(\varphi, v_m(t,\cdot) = (1/2)\int_0^t (\varphi, a_s D_x^2 v_m(s,\cdot))\,ds + \int_0^t (\varphi, \sigma_s D_x v_m(s,\cdot))\,dw_s.$$

6. Continuity of solutions of SPDEs

We take the processes a_t, σ_t as before Theorem 5.3 but impose stronger assumptions on them.

Assume that there exist constants $\delta_0, \delta_1 \in (0,1]$ such that, for every (ω, t)

$$\delta_0 \le \delta_1 a_t \le a_t - \sigma_t^2 \le \delta_0^{-1},$$

We will be dealing with solutions $u_t(x)$ of

$$du_t = ((1/2)a_t D_x^2 u_t + f_t)\,dt + (\sigma_t D_x u_t + g_t)\,dw_t \tag{6.1}$$

on $\mathbb{R}_+$ with zero initial condition. To specify the assumptions on f, g and the class of solutions we borrow the Banach spaces $\mathbb{H}_{p,\theta}^\gamma(\tau)$ and $\mathbb{L}_{p,\theta}(\tau)$ from [15]. We also denote by M the operator of multiplying by x. Recall that, for $p \ge 2, 0 < \theta < p$, the norms in $\mathbb{H}_{p,\theta}^\gamma(\tau)$, $\gamma = 1, 2$, and $\mathbb{L}_{p,\theta}(\tau)$ are given by

$$\|v\|_{\mathbb{L}_{p,\theta}(\tau)}^p = E\int_0^\tau \int_0^\infty x^{\theta-1}|v(t,x)|^p\,dxdt,$$

$$\|v\|_{\mathbb{H}^1_{p,\theta}(\tau)} = \|v\|_{\mathbb{L}_{p,\theta}(\tau)} + \|MD_x v\|_{\mathbb{L}_{p,\theta}(\tau)},$$

$$\|v\|_{\mathbb{H}^2_{p,\theta}(\tau)} = \|v\|_{\mathbb{H}^1_{p,\theta}(\tau)} + \|M^2 D_x^2 v\|_{\mathbb{L}_{p,\theta}(\tau)}.$$

Given $p \geq 2$, $\theta \in [p-1, p)$, any stopping time τ, $f \in M^{-1}\mathbb{L}_{p,\theta}(\tau)$, and $g \in \mathbb{H}^1_{p,\theta}(\tau)$ by Theorem 3.2 in Ref. [15] equation (6.1) with zero initial condition has a unique solution $u \in M\mathbb{H}^2_{p,\theta}(\tau)$ and

$$\|M^{-1}u\|_{\mathbb{H}^2_{p,\theta}(\tau)} \leq N(\|Mf\|_{\mathbb{L}_{p,\theta}(\tau)} + \|g\|_{\mathbb{H}^1_{p,\theta}(\tau)}),$$

where $N = N(p, \theta, \delta_0, \delta_1)$.

We will also use Theorem 4.7 in Ref. [16], which implies that if u is a solution of (6.1) of class $M\mathbb{H}^2_{p,\theta}(\tau)$ with zero initial condition and $f \in M^{-1}\mathbb{L}_{p,\theta}(\tau)$, and $g \in \mathbb{H}^1_{p,\theta}(\tau)$ and if there are numbers $T \in (0, \infty)$ and β such that

$$2/p < \beta \leq 1, \quad \tau \leq T,$$

then for almost any ω the function $u_t(x)$ is continuous in (t, x) (that is, has a continuous modification) and

$$E \sup_{t\leq\tau} \sup_{x>0} |x^{\beta-1+\theta/p} u_t(x)|^p \leq NT^{\beta p/2}(\|M^{-1}u\|_{\mathbb{H}^2_{p,\theta}(\tau)} + \|Mf\|_{\mathbb{L}_{p,\theta}(\tau)} + \|g\|_{\mathbb{H}^1_{p,\theta}(\tau)}),$$

where $N = N(p, \theta, \beta, \delta_0)$.

Everywhere below we take

$$p > 2.$$

Theorem 6.1. *Let $T \in (0, \infty)$, $c > 0$, $\alpha \in (0, 1)$, $\theta > 0$, μ be some constants such that $\alpha_0(c\sqrt{\delta_1}) > \alpha$,*

$$\theta_0 < \theta < p, \quad \mu < p(1 + 2\log_2 \gamma(c)) - 2 = \theta_0 - 2 + 2p(1 - \alpha)\log_2 \gamma(c),$$

where

$$\gamma(c) = \gamma(c, 1, 1), \quad \theta_0 = p(1 + 2\alpha \log_2 \gamma(c)) \quad (> 0).$$

Let $f \in M^{-1}\mathbb{L}_{p,p-1}(T)$, $g \in \mathbb{H}^1_{p,p-1}(T)$, and let $u \in M\mathbb{H}^2_{p,p-1}(T)$ be a solution of (6.1) with zero initial condition.

Finally, assume that $f_t(x) = g_t(x) = 0$ for $x \geq 1$ and $f \in M^{-1}\mathbb{L}_{p,\mu}(T)$, $g \in \mathbb{H}^1_{p,\mu}(T)$. Then there exist stopping times $\tau_n \uparrow T$, defined independently of f and g such that, for each n, $u \in M\mathbb{H}^2_{p,\theta}(\tau_n)$ and

$$\|M^{-1}u\|^p_{\mathbb{H}^2_{p,\theta}(\tau_n)} \leq n(\|Mf\|^p_{\mathbb{L}_{p,\mu}(\tau_n)} + \|g\|^p_{\mathbb{H}^1_{p,\mu}(\tau_n)}) \tag{6.2}$$

Here is the result about the continuity of $u_t(x)$ we were talking about in the introduction.

Remark 6.2. By Theorem 4.7 in Ref. [16] and Theorem 6.1 if we have a number $\beta \in (2/p, 1]$, then there exists a sequence of stopping times $\tau_n \uparrow T$ such that

$$E \sup_{t\leq\tau_n} \sup_{x>0} |x^{-\varepsilon} u_t(x)|^p < \infty,$$

where $\varepsilon = 1 - \beta - \theta/p$. Due to the freedom of choosing α, β, and θ, the number ε can be made as close from the right as we wish to

$$1 - \lim_{\alpha \to \alpha_0(c\sqrt{\delta_1})} (2 + \theta_0)/p = -2/p - 2\alpha_0(c\sqrt{\delta_1}) \log_2 \gamma(c).$$

If we allow arbitrary p, then the rate of convergence of $u_t(x)$ to zero as $x \downarrow 0$ is almost

$$x^{\varepsilon_0}, \quad \varepsilon_0 = -2\alpha_0(c\sqrt{\delta_1}) \log_2 \gamma(c) > 0,$$

which is the same as we obtained for $v_m(t,x)$ (see (5.5)). Hence, the presence of f and g does not spoil the situation too much.

It is also worth noting that f and g still may blow up near zero even if p is large. When p is large we can take $(\mu - 1)/p$ as close to $1 + 2\log_2 \gamma(c)$ as we wish and then the integral

$$\int_0^1 x^{\mu-1} |x f_t(x)|^p \, dx$$

converges if $|f_t(x)|$ blows up near $x = 0$ slightly slower than $x^{-2(1+\log_2 \gamma(c))}$. Here $\log_2 \gamma(c) \to 0$ as $c \to \infty$ and one can allow $|f_t(x)|$ to blow up almost as x^{-2}.

However, when f and g become more irregular near 0, the rate with which the solution goes to zero at 0 deteriorates. In connection with this it is interesting to investigate what happens with ε_0 as $\delta_1 \downarrow 0$. Take an m so large that $\alpha_0(m) > 1/2$ and set $c = m\delta_1^{-1/2} - 1/2$. Then for δ_1 small we have $\alpha_0(c\sqrt{\delta_1}) > 1/2$ and

$$\varepsilon_0 \geq -\log_2[1 - P(\min_{s \leq 1/2} w_s \leq -c - 1/\sqrt{2}, \max_{s \leq 1/2} w_s \leq 1 - 1/\sqrt{2})),$$

$$\varepsilon_0 \ln 2 \geq -\ln[1 - P(\min_{s \leq 1/2} w_s \leq -c - 1/\sqrt{2}, \max_{s \leq 1/2} w_s \leq 1 - 1/\sqrt{2}))$$

$$\sim P(\min_{s \leq 1/2} w_t \leq -c - 1/\sqrt{2}, \max_{s \leq 1/2} w_s \leq 1 - 1/\sqrt{2})$$

$$= P(\min_{s \leq 1/2} w_t \leq -c - 1/\sqrt{2}) - P(\min_{s \leq 1/2} w_t \leq -c - 1/\sqrt{2}, \max_{s \leq 1/2} w_s \geq 1 - 1/\sqrt{2})$$

and

$$P(\min_{s \leq 1/2} w_t \leq -c - 1/\sqrt{2}, \max_{s \leq 1/2} w_s \geq 1 - 1/\sqrt{2}) \leq 2P(\min_{s \leq 1/2} w_t \leq -c - 1),$$

so that

$$P(\min_{s \leq 1/2} w_t \leq -c - 1/\sqrt{2}, \max_{s \leq 1/2} w_s \leq 1 - 1/\sqrt{2})$$

$$\geq P(\min_{s \leq 1/2} w_t \leq -c - 1/\sqrt{2}) - 2P(\min_{s \leq 1/2} w_t \leq -c - 1).$$

Next, as $a \to \infty$

$$P(\min_{s\leq 1/2} w_s \leq -a) = P(|w_{1/2}| \geq a) = \frac{2}{\sqrt{\pi}} \int_a^\infty e^{-x^2}\,dx \sim \frac{1}{\sqrt{\pi}} a^{-1} e^{-a^2}$$

and

$$\lim_{\delta_1 \downarrow 0} [P(\min_{s\leq 1/2} w_t \leq -c - 1/\sqrt{2})$$

$$-2P(\min_{s\leq 1/2} w_t \leq -c-1)](c+1/\sqrt{2})e^{(c+1/\sqrt{2})^2} = \frac{1}{\sqrt{\pi}}.$$

Hence

$$\varliminf_{\delta_1 \downarrow 0} [m\delta_1^{-1/2} e^{m^2/\delta_1} \varepsilon_0] \geq \frac{1}{\sqrt{\pi}\ln 2}.$$

This result may seem unsatisfactory since the guaranteed value of ε_0 is extremely small when δ_1 is small. However, recall that by Remark 4.2 in Ref. [14] the best possible rate with which the solutions go to zero for small δ_1 is less than

$$(1+\kappa)(2\pi\delta_1)^{-1/2} e^{-1/(2\delta_1)},$$

where $\kappa > 0$ is any number.

To prove Theorem 6.1, first we prove the following.

Lemma 6.3. *Assume that, for an $m = 0, 1, 2, \ldots$ we have $f_t(x) = g_t(x) = 0$ if $x \leq 2^{-m/2}$. Then almost surely for all $t \leq T$ and $x \in (0, 2^{-m/2})$*

$$|u_t(x)| \leq v_m(t,x) \sup_{s\leq t} |u_s(2^{-m/2})|. \tag{6.3}$$

Proof. By Theorem 4.7 in Ref. [16] the function $u_t(x)$ is continuous in $[0,T] \times (0, 2^{-m/2})$ (a.s.) and therefore to prove (6.3) it suffices to prove that for each $\varepsilon \in (0, 2^{-(m+2)/2})$ almost surely for all $t \leq T$ and $x \in D := (\varepsilon, 2^{-m/2} - \varepsilon)$

$$|u_t^\varepsilon(x)| \leq v_m(t,x) \sup_{s\leq t} |u_s^\varepsilon(2^{-m/2} - \varepsilon)| =: v_m(t,x)\rho_t^\varepsilon, \tag{6.4}$$

where $u_t^\varepsilon(x) = u_t(x-\varepsilon)$. The function u_t^ε satisfies (6.1) with $f = g = 0$ in $(0,T) \times (\varepsilon, 2^{-m/2} + \varepsilon)$ and in $(0,T) \times D$. Furthermore, (a.s.) for almost any $t \in (0,T)$ we have $D_x u_t \in L_p(D)$ implying that the limit of $u_t^\varepsilon(x)$ as $x \downarrow \varepsilon$ exists. Since (a.s.) for almost all $t \in (0,T)$ also $(x-\varepsilon)^{-1} u_t^\varepsilon \in L_p(D)$, the limit is zero. As $x \uparrow 2^{-m/2} - \varepsilon$ the situation is simpler and we see that (a.s.) for almost all $t \in (0,T)$ we have

$$\lim_{D \ni x \to \partial D} (u_t^\varepsilon(x) - v_m(t,x)\rho_t^\varepsilon)^+ = 0.$$

Furthermore, (a.s.) for almost all $t \in (0,T)$ it holds that $u_t^\varepsilon \in W_2^1(D)$ and

$$\int_0^T \|u_t^\varepsilon\|^2_{W_2^1(D)}\,dt < \infty.$$

Combining this with Remark 5.4 we see that (a.s.) for almost all $t \in (0,T)$ we have $(u^\varepsilon_t - v_m(t,\cdot)\rho^\varepsilon_t)^+ \in W^1_2(D)$, $(u^\varepsilon_t - v_m(t,\cdot)\rho^\varepsilon_t)^+ \in \overset{0}{W}{}^1_2(D)$ and

$$\int_0^T \|(u^\varepsilon_t - v_m(t,\cdot)\rho^\varepsilon_t)^+\|^2_{W^1_2(D)}\,dt < \infty.$$

By Theorem 2.6 we conclude that almost surely for all $t \le T$ and $x \in D$

$$u^\varepsilon_t(x) \le v_m(t,x)\rho^\varepsilon_t.$$

By combining this with similar inequality for $-u^\varepsilon_t$ we obtain (6.4). The lemma is proved.

Proof of Theorem 6.1. Clearly, we only need prove Theorem 6.1 for f and g such that $f_t(x) = g_t(x) = 0$ for all ω, t if x is small. Then

$$f \in M^{-1}\mathbb{L}_{p,\vartheta}(T), \quad g \in \mathbb{H}^1_{p,\vartheta}(T) \tag{6.5}$$

for any ϑ.

According to Lemma 3.6 in Ref. [15], for each stopping time $\tau_n \le T$, we have $u \in M\mathbb{H}^2_{p,\theta}(\tau_n)$ if $u \in M\mathbb{L}_{p,\theta}(\tau_n)$ and under this condition the left-hand side of (6.2) is dominated by a constant $N = N(\theta, p, \delta_0, \delta_1)$ times

$$\|M^{-1}u\|^p_{\mathbb{L}_{p,\theta}(\tau_n)} + E\int_0^{\tau_n}\int_0^\infty x^{\theta-1}|F_t(x)|^p\,dxdt, \tag{6.6}$$

where

$$F_t(x) := |xf_t(x)| + |g_t(x)| + |xD_xg_t(x)|.$$

Observe that obviously $(\alpha, \gamma(c) \le 1)$

$$\theta > \theta_0 > \mu \tag{6.7}$$

and since $f_t(x) = g_t(x) = 0$ for $x \ge 1$, the integral involving F_t will increase if we replace θ with μ. It follows that to prove the theorem, it suffices to estimate only the lowest norm of u, that is to prove the existence of $\tau_n \uparrow T$ such that

$$\|M^{-1}u\|^p_{\mathbb{L}_{p,\theta}(\tau_n)} \le n[\|Mf\|^p_{\mathbb{L}_{p,\mu}(\tau_n)} + \|g\|^p_{\mathbb{H}^1_{p,\mu}(\tau_n)}]. \tag{6.8}$$

Next, take a $\vartheta \in [p-1, p)$ such that $\vartheta > \theta$. For any stopping time $\tau \le T$, by Lemma 4.3 in Ref. [15] we have $u \in M\mathbb{H}^2_{p,\vartheta}(\tau)$ and by Theorem 3.2 of [15]

$$E\int_0^\tau\int_0^\infty x^{\vartheta-1}|u_t(x)/x|^p\,dxdt \le N[\|Mf\|^p_{\mathbb{L}_{p,\vartheta}(\tau)} + \|g\|^p_{\mathbb{H}^1_{p,\vartheta}(\tau)}], \tag{6.9}$$

where $N = N(p, \vartheta, \delta_0, \delta_1)$. As before on the right we can replace ϑ with μ. On the left one can replace ϑ with θ if one restricts the domain of integration with respect to x to $x \ge 1$. Therefore (6.8) will be proved if we prove the existence of appropriate stopping times τ_n such that

$$E\int_0^{\tau_n}\int_0^1 x^{\theta-1}|u_t(x)/x|^p\,dxdt \le n[\|Mf\|^p_{\mathbb{L}_{p,\mu}(\tau_n)} + \|g\|^p_{\mathbb{H}^1_{p,\mu}(\tau_n)}]. \tag{6.10}$$

Take a nonnegative $\eta \in C_0^\infty(\mathbb{R}_+)$ with support in $(1,4)$ such that the $(1/2)$-periodic function on $\mathbb{R}$

$$\sum_{k=-\infty}^{\infty} \eta(2^{x+k/2})$$

is identically equal to one. Introduce,

$$\eta_m(x) = \eta(2^{m/2}x), \quad (f_{mt}, g_{mt}) = (f_t, g_t)\eta_m.$$

Also introduce u_{mt} as solutions of class $M\mathbb{H}^2_{p,p-1}(T)$ of (6.1) with zero initial condition and f_{mt} and g_{mt} in place of f_t and g_t, respectively. Since only finitely many f_{mt} and g_{mt} are not zero, we have

$$u_t(x) = \sum_{m=1}^{\infty} u_{mt}(x) = I_1(t,x) + I_2(t,x),$$

where

$$I_1(t,x) := \sum_{m=1}^{\infty} u_{mt}(x) I_{x \le 2^{-m/2}}, \quad I_2(t,x) := \sum_{m=1}^{\infty} u_{mt}(x) I_{x > 2^{-m/2}}.$$

Estimating I_2. Take a ϑ as above, set $\varepsilon = (\vartheta - \theta)/(2p)$ and use Hölder's inequality to obtain

$$|I_2(t,x)|^p \le \sum_{m=1}^{\infty} 2^{\varepsilon p m} u^p_{mt}(x) J^{p/q}(x),$$

where

$$J(x) := \sum_{m=1}^{\infty} 2^{-\varepsilon q m} I_{x > 2^{-m/2}} \le N x^{2\varepsilon q}, \quad J^{p/q}(x) \le N x^{\vartheta - \theta}.$$

Then use (6.9) again to get

$$E\int_0^T \int_0^1 x^{\theta-1} |I_2(t,x)/x|^p \, dxdt$$

$$\le N \sum_{m=1}^{\infty} E\int_0^T \int_0^\infty 2^{m(\vartheta-\theta)/2} x^{\vartheta-1} |u_{mt}(x)/x|^p \, dxdt$$

$$\le N \sum_{m=1}^{\infty} E\int_0^T \int_0^\infty 2^{m(\vartheta-\theta)/2} x^{\vartheta-1} |F_{mt}(x)|^p \, dxdt,$$

where

$$F_{mt}(x) = |x f_{mt}(x)| + |g_{mt}(x)| + |x D_x g_{mt}(x)|.$$

Here we notice few facts, which will be also used in the future, that on the supports of $f_{mt}(x)$ and $g_{mt}(x)$ we have $x \sim 2^{-m/2}$, $2^{m(\vartheta-\theta)/2} F_{mt}(x) \sim x^{\theta-\vartheta} F_{mt}(x)$ and

$$F_{mt}(x) \le F_t(x) \bar{\eta}_m(x)$$

where $\bar{\eta}_m(x) = \eta_m(x) + x2^{m/2}|\eta'(2^{m/2}x)|$. Notice that the (1/2)-periodic function

$$\sum_{m=-\infty}^{\infty} \bar{\eta}_m^p(2^y)$$

is bounded on $\mathbb{R}$. Then we see that

$$E\int_0^\tau \int_0^1 x^{\theta-1}|I_2(t,x)/x|^p\,dxdt$$

$$\le NE\int_0^\tau \int_0^\infty x^{\theta-1}|F_t(x)|^p \sum_{m=1}^{\infty} \bar{\eta}_m^p(x)\,dxdt$$

$$\le N[\|Mf\|^p_{\mathbb{L}_{p,\theta}(\tau)} + \|g\|^p_{\mathbb{H}^1_{p,\theta}(\tau)}]$$

for any $\tau \le T$ with a constant N under control. As above we can reduce θ in the last expression to μ.

Estimating I_1. Here we will see how τ_n appear and how we get a substantial drop from θ to μ. We have seen above that the smaller μ is the weaker the statement of the theorem becomes. Therefore, we may concentrate on μ so close to $p(1 + 2\log_2 \gamma(c)) - 2$ from below that

$$2 < \beta p := p(1 + 2\log_2 \gamma(c)) - \mu \le p.$$

Then

$$2/p < \beta \le 1.$$

Observe that

$$|I_1(t,x)|^p \le \Big(\sum_{m=1}^{\infty} m^{-q}\Big)^{p/q} \sum_{m=1}^{\infty} m^p|u_{mt}(x)|^p I_{x\le 2^{-m/2}}$$

$$\le N|\log_2 x|^p \sum_{m=1}^{\infty} |u_{mt}(x)|^p I_{x\le 2^{-m/2}}.$$

It follows that for any $\theta' < \theta$

$$E\int_0^\tau \int_0^1 x^{\theta-1}|I_1(t,x)/x|^p\,dxdt$$

$$\le N\sum_{m=1}^{\infty} E\int_0^\tau \int_0^{2^{-m/2}} |\log_2 x|^p x^{\theta-1}|u_{mt}(x)/x|^p\,dxdt \le NJ(\tau),$$

where

$$J(\tau) := \sum_{m=1}^{\infty} E\int_0^\tau \int_0^{2^{-m/2}} x^{\theta'-1}|u_{mt}(x)/x|^p\,dxdt.$$

By Theorem 4.7 in Ref. [16] and Theorem 3.3 in Ref. [17], for any $\tau \leq T$

$$E \sup_{t\leq\tau} \sup_{x>0} |x^{\varepsilon} u_{mt}(x)|^p \leq NT^{\beta p/2} E \int_0^{\tau} \int_0^{\infty} x^{\theta_1 - 1} |F_{mt}(x)|^p \, dx dt, \tag{6.11}$$

where $N = N(p, \delta_0, \delta_1, \beta)$ and

$$\theta_1 := p - 1, \quad \varepsilon := \beta - 1 + \theta_1/p = \beta - 1/p > 0.$$

Therefore,

$$E \sup_{t\leq\tau} |u_{mt}(2^{-m/2})|^p \leq 2^{m(\beta p - p + \theta_1)/2} E \sup_{t\leq\tau} \sup_{x>0} |x^{\varepsilon} u_{mt}(x)|^p$$

$$\leq NE \int_0^{\tau} \int_0^{\infty} 2^{m(\beta p - p + \theta_1)/2} x^{\theta_1 - 1} |F_{mt}(x)|^p \, dx dt$$

$$\leq N 2^{m(\beta p - p + 1)/2} E \int_0^{\tau} \int_0^{\infty} |F_{mt}(x)|^p \, dx dt.$$

Next, observe that, by Lemma 6.3 for $x \in [0, 2^{-m/2}]$ and $t \leq T$,

$$|u_{mt}(x)| \leq v_m(t, x) \sup_{s\leq t} |u_{ms}(2^{-m/2})|.$$

Hence

$$J(\tau) \leq \sum_{m=1}^{\infty} E \sup_{t\leq\tau} |u_{mt}(2^{-m/2})| \int_0^{\tau} \int_0^{2^{-m/2}} x^{\theta' - 1} |v_m(t, x)/x|^p \, dx dt$$

$$\leq \sum_{m=1}^{\infty} 2^{m(\nu p - 1)/(2\alpha)} E \sup_{t\leq\tau} |u_{mt}(2^{-m/2})| \int_0^{\tau} \pi_t \, dt,$$

where ν is defined according to

$$\nu p = p - \theta' + 1$$

and π_t is introduced in Theorem 5.3. So far θ' was only restricted to $\theta' < \theta$, so that $\nu p > 1$. Due to the assumption that $\theta > \theta_0$ one can satisfy $\theta_0 < \theta' < \theta$ in which case (5.8) holds. Then in light of Theorem 5.3 one can find stopping times $\tau_n \uparrow T$ such that

$$\int_0^{\tau_n} \pi_t \, dt \leq n.$$

Then

$$J(\tau_n) \leq nN \sum_{m=1}^{\infty} 2^{m(\nu p - 1)/(2\alpha)} 2^{m(\beta p - p + 1)/2} E \int_0^{\tau_n} \int_0^{\infty} |F_{mt}(x)|^p \, dx dt.$$

As is easy to see the inequalities $\theta' > \theta_0$ and

$$\beta p - p + (\nu p - 1)/\alpha < -\mu$$

are equivalent. Hence,

$$E\int_0^{\tau_n}\int_0^1 x^{\theta-1}|I_1(t,x)/x|^p\,dxdt \le NJ(\tau_n)$$

$$\le nN\sum_{m=1}^{\infty} 2^{m(1-\mu)/2}E\int_0^{\tau_n}\int_0^{\infty}|F_{mt}(x)|^p\,dxdt$$

$$\le nN\sum_{m=1}^{\infty} E\int_0^{\tau_n}\int_0^{\infty}x^{\mu-1}|F_{mt}(x)|^p\,dxdt$$

$$\le nNE\int_0^{\tau_n}\int_0^{\infty}x^{\mu-1}|F_t(x)|^p\,dxdt.$$

By combining this estimate with the estimate of I_2, noticing that the above constants N are independent of f and g and, if necessary, renumbering the sequence τ_n we come to (6.10). This proves the theorem.

Acknowledgements

Several typos in the original version of the article were kindly pointed out by Kyeong-Hun Kim and Hongjie Dong. The author is sincerely grateful for that. Author's work was partially supported by NSF Grant DMS-0140405.

References

[1] K.-H. Kim, $L_q(L_p)$ theory and Hölder estimates for parabolic SPDEs, *Stochastic Process. Appl.* **114**(2), 313–330, (2004). ISSN 0304-4149.

[2] N. V. Krylov and B. L. Rozovskiĭ. On the first integrals and Liouville equations for diffusion processes. In *Stochastic differential systems (Visegrád, 1980)*, vol. 36, *Lecture Notes in Control and Information Sci.*, pp. 117–125. Springer, Berlin, (1981).

[3] N. V. Krylov and B. L. Rozovsky, On the characteristics of degenerate second order parabolic itô equations, *J. Soviet Math.* **32**(4), 336–348, (1986).

[4] H. Kunita, On backward stochastic differential equations, *Stochastics.* **6**(3-4), 293–313, (1981/82). ISSN 0090-9491.

[5] S. Bonaccorsi and G. Guatteri, Stochastic partial differential equations in bounded domains with Dirichlet boundary conditions, *Stoch. Stoch. Rep.* **74**(1-2), 349–370, (2002). ISSN 1045-1129.

[6] S. Bonaccorsi and G. Guatteri. Classical solutions for SPDEs with Dirichlet boundary conditions. In *Seminar on Stochastic Analysis, Random Fields and Applications, III (Ascona, 1999)*, vol. 52, *Progr. Probab.*, pp. 33–44. Birkhäuser, Basel, (2002).

[7] G. Da Prato and L. Tubaro, Fully nonlinear stochastic partial differential equations, *SIAM J. Math. Anal.* **27**(1), 40–55, (1996). ISSN 0036-1410.

[8] L. Tubaro, Some results on stochastic partial differential equations by the stochastic characteristics method, *Stochastic Anal. Appl.* **6**(2), 217–230, (1988). ISSN 0736-2994.

[9] K.-H. Kim, On stochastic partial differential equations with variable coefficients in C^1 domains, *Stochastic Process. Appl.* **112**(2), 261–283, (2004). ISSN 0304-4149.

[10] N. Krylov and B. Rosovskii, Stochastic evolution equations, *J.Soviet Mathematics.* **16**, 1233–1277, (1981).

[11] L. C. Evans, *Partial differential equations.* vol. 19, *Graduate Studies in Mathematics*, (American Mathematical Society, Providence, RI, 1998). ISBN 0-8218-0772-2.

[12] O. A. Ladyzhenskaya and N. N. Ural′tseva, *Linear and quasilinear elliptic equations.* Translated from the Russian by Scripta Technica, Inc. Translation editor: Leon Ehrenpreis, (Academic Press, New York, 1968).

[13] N. V. Krylov, *Introduction to the theory of diffusion processes.* vol. 142, *Translations of Mathematical Monographs*, (American Mathematical Society, Providence, RI, 1995). ISBN 0-8218-4600-0. Translated from the Russian manuscript by Valim Khidekel and Gennady Pasechnik.

[14] N. V. Krylov, One more square root law for Brownian motion and its application to SPDEs, *Probab. Theory Related Fields.* **127**(4), 496–512, (2003). ISSN 0178-8051.

[15] N. V. Krylov and S. V. Lototsky, A Sobolev space theory of SPDEs with constant coefficients on a half line, *SIAM J. Math. Anal.* **30**(2), 298–325 (electronic), (1999). ISSN 0036-1410.

[16] N. V. Krylov, Some properties of traces for stochastic and deterministic parabolic weighted Sobolev spaces, *J. Funct. Anal.* **183**(1), 1–41, (2001). ISSN 0022-1236.

[17] N. V. Krylov and S. V. Lototsky, A Sobolev space theory of SPDEs with constant coefficients in a half space, *SIAM J. Math. Anal.* **31**(1), 19–33 (electronic), (1999). ISSN 0036-1410.

Chapter 13

On Delay Estimation and Testing for Diffusion Type Processes

Yury A. Kutoyants

Laboratoire de Statistique et Processus, Facult des Sciences, Universit du Maine
Avenue Olivier Messiaen 72085 Le Mans CEDEX 9
kutoyants@univ-lemans.fr

We present a review of results concerning one example of parameter estimation problem, which is the same time regular and singular depending on the chosen asymptotics. The observed process satisfies the linear stochastic differential equation with the delayed trend coefficient and we are interested in the estimation of this delay. The trend coefficient depends on the unknown parameter (delay) as smooth as the Wiener process depends on time, i.e., we have a continuous but not differentiable w.r.t. parameter statistical model. We study the asymptotic behavior of the estimators and tests in two asymptotics: the first one corresponds to the *small noise perturbation*, when the diffusion coefficient tends to zero and the second is *large samples limit*, when the time of observation tends to infinity. We show that the first asymptotics corresponds well to regular statistical experiments situation and the second one is typical for non regular (discontinuous type) statistical models.

Contents

1. Introduction

We consider several statistical problems concerning a one dimensional parameter ϑ (called *delay parameter*) of the stochastic differential equation by the continuous time observations $X = \{X_t, 0 \leq t \leq T\}$ of the solution of stochastic differential equation

$$\mathrm{d}X_t = S(X_{t-\vartheta})\,\mathrm{d}t + \sigma(X_t)\,\mathrm{d}W_t, \quad X_s, s \leq 0, \qquad 0 \leq t \leq T.$$

Here $S(x)$ and $\sigma(x)^2$ are trend and diffusion coefficients, $W_t, 0 \leq t \leq T$ is standard Wiener process and X_s are independent of the Wiener process initial values. We are interested in asymptotic estimation and testing problems in two different asymptotics. The first one corresponds to the dynamical system with small noise ($\sigma \to 0$), also called *perturbed dynamical system* and the second asymptotics is typical for classical statistics *large samples asymptotics* ($T \to \infty$). The first type of asymptotics we realize on the linear model

$$\mathrm{d}X_t = -\gamma\, X_{t-\vartheta}\, \mathrm{d}t + \sigma\, \mathrm{d}W_t, \quad X_s = x_s, -\beta \leq s \leq 0, \quad 0 \leq t \leq T, \tag{1.1}$$

and the second asymptotics is studied for the linear ergodic model ($T \to \infty$)

$$\mathrm{d}X_t = -\gamma\, X_{t-\vartheta}\, \mathrm{d}t + \sigma\, \mathrm{d}W_t, \quad X_s = x_s, -\beta \leq s \leq 0, \quad 0 \leq t \leq T.$$

Therefore we have exactly the same statistical model in both cases. The unknown parameter (delay) $\vartheta \in (\alpha, \beta) = \Theta$, where $0 < \alpha < \beta$, $\gamma > 0$, and $x_s, -\beta \leq s \leq 0$ is a deterministic function (initial values). Note that this equation has a unique strong solution for all γ, ϑ and σ ([1], Theorem 4.6).

We study the properties of the maximum likelihood $\widehat{\vartheta}_{\sigma,T}$ and Bayesian $\tilde{\vartheta}_{\sigma,T}$ (for quadratic loss function and positive continuous prior density $p(\cdot)$) estimators defined by the usual equations

$$\sup_{\vartheta\in\Theta} L\left(\vartheta, X^T\right) = L\left(\widehat{\vartheta}_{\sigma,T}, X^T\right)$$

and

$$\tilde{\vartheta}_{\sigma,T} = \int_{\Theta} \vartheta p\left(\vartheta|X\right)\, \mathrm{d}\vartheta = \frac{\int_{\Theta} \vartheta p\left(\vartheta\right) L\left(\vartheta, X^T\right) \mathrm{d}\vartheta}{\int_{\Theta} p\left(\vartheta\right) L\left(\vartheta, X^T\right) \mathrm{d}\vartheta}$$

where the log-likelihood ratio is ([1], Theorem 7.7)

$$\ln L\left(\vartheta, X^T\right) = -\frac{\gamma}{\sigma^2} \int_0^T X_{t-\vartheta}\, \mathrm{d}X_t - \frac{\gamma^2}{2\sigma^2} \int_0^T X_{t-\vartheta}^2\, \mathrm{d}t.$$

The properties of the estimators are studied as $\sigma \to 0$ and $T \to \infty$ separately.

We present here a review of some results concerning delay parameter estimation obtained in the works [2, 3] and [4–9]. The further results on parameter estimation for linear stochastic differential equations with delay (in large samples asymptotics) can be found in the works by Gushchin and Küchler [10, 11]. The nonparametric estimation problems in the case of integral-type drift are treated by Reiss [12, 13].

What indeed is surprising with these problems is the possibility to have regular and singular estimation and testing problems for the same model. The first limit ($\sigma \to 0$) provides us statistically regular estimation and testing problems and the large samples asymptotics ($T \to \infty$) will be similar to the problems of change-point type.

Let us remind the difference between regular and singular statistical problems. We recall the well-known facts from classical estimation theory concerning regularity conditions and properties of estimators [6]. Let $X^n = \{X_1, \ldots, X_n\}$ be i.i.d. r.v.'s and X_j has density function $f(\vartheta, x)$, where $\vartheta \in \Theta = (\alpha, \beta)$. We denote by $\mathbf{P}_\vartheta^{(n)}$ the distribution of X^n. If we have to estimate $\vartheta \in \Theta$ by observations X^n and describe the properties of estimators as $n \to \infty$, then there are at least two situations: *regular* and *non regular*.

In *regular* situation the function $\sqrt{f(\vartheta, x)}$ is supposed to be differentiable w.r.t. ϑ in $\mathcal{L}_2$-sense and the Fisher information

$$\mathrm{I}(\vartheta) = \int_{-\infty}^{\infty} \left(\frac{\partial f(\vartheta, x)}{\partial \vartheta}\right)^2 f(\vartheta, x)^{-1} \, \mathrm{d}x$$

is positive continuous function of ϑ. Then the corresponding family of measures $\left\{\mathbf{P}_\vartheta^{(n)}, \vartheta \in \Theta\right\}$ is locally asymptotically normal (LAN), i.e., the normalized likelihood ratio admits the representation

$$Z_n(u) = \frac{\mathrm{d}\mathbf{P}^{(n)}_{\vartheta + \frac{u}{\sqrt{n}}}}{\mathrm{d}\mathbf{P}_\vartheta^{(n)}} (X^n) = \exp\left\{u\Delta_n(\vartheta) - \frac{u^2}{2}\,\mathrm{I}(\vartheta) + r_n\right\},$$

where $\Delta_n(\vartheta) \Rightarrow \Delta(\vartheta) \sim \mathcal{N}(0, \mathrm{I}(\vartheta))$ and $r_n \to 0$ (see Ref. [14], [15]). Hence

$$Z_n(u) \Longrightarrow Z(u) = \exp\left\{u\Delta(\vartheta) - \frac{u^2}{2}\,\mathrm{I}(\vartheta)\right\}.$$

The notion of LAN is equivalent to the notion of *regular statistical experiment*. If the family of measures of any statistical model is LAN, then we have immediately the minimax lower bound on the risk of all estimators (Hajek-Le Cam bound). For the quadratic loss function it is an asymptotically correct version of Rao-Cramér bound: for all estimators $\bar{\vartheta}_n$ and $\vartheta_0 \in \Theta$

$$\lim_{\delta \to 0} \lim_{n \to \infty} \sup_{|\vartheta - \vartheta_0| < \delta} n\, \mathbf{E}_\vartheta \left(\bar{\vartheta}_n - \vartheta\right)^2 \geq \mathrm{I}(\vartheta_0)^{-1}. \tag{1.2}$$

In this regular case (under some additional conditions) the maximum likelihood estimator (MLE) $\widehat{\vartheta}_n$ and the wide class of Bayesian estimators (BE) $\tilde{\vartheta}_n$ are consistent, asymptotically normal

$$\sqrt{n}\left(\widehat{\vartheta}_n - \vartheta\right) \Longrightarrow \widehat{u}, \qquad \sqrt{n}\left(\tilde{\vartheta}_n - \vartheta\right) \Longrightarrow \tilde{u} \tag{1.3}$$

where

$$\sup_u Z(u) = Z(\widehat{u}), \qquad \widehat{u} = \frac{\Delta(\vartheta)}{\mathrm{I}(\vartheta)} \sim \mathcal{N}\left(0, \mathrm{I}(\vartheta)^{-1}\right),$$

and

$$\tilde{u} = \frac{\int u Z(u)\, \mathrm{d}u}{\int Z(u)\, \mathrm{d}u} = \frac{\Delta(\vartheta)}{\mathrm{I}(\vartheta)} \sim \mathcal{N}\left(0, \mathrm{I}(\vartheta)^{-1}\right).$$

Moreover *both are asymptotically efficient* in the sense of the bound (1.2) (see Ref. [15]), i.e., for the MLE we have

$$\lim_{\delta \to 0} \lim_{n \to \infty} \sup_{|\vartheta - \vartheta_0| < \delta} n\, \mathbf{E}_\vartheta \left(\widehat{\vartheta}_n - \vartheta\right)^2 = \mathrm{I}(\vartheta_0)^{-1}$$

for all $\vartheta \in \Theta$.

In *non regular* cases the situations are different. Note that there are many types of *non regularity* (see, for example, Ref. [15], Chapters 5 and 6). Here we remind one case only. Suppose that we have the shift parameter $f(\vartheta, x) = f(x - \vartheta)$, where the function $f(\cdot)$ has a jump at point x_*, i.e., $f_+ = f(x_*+)$, $f_- = f(x_*-)$ and $f_+ - f_- \neq 0$. Then the normalized likelihood ratio (for $u \geq 0$)

$$Z_n(u) = \frac{\mathrm{d}\mathbf{P}^{(n)}_{\vartheta+\frac{u}{n}}}{\mathrm{d}\mathbf{P}^{(n)}_{\vartheta}}(X^n) \Longrightarrow Z(u) = \exp\left\{\ln\frac{f_-}{f_+}\,\pi_+(u) - u\,(f_- - f_+)\right\}$$

where $\pi_+(u), u \geq 0$ is Poisson process of intensity f_+. For $u \leq 0$ we have $\ln Z(u) = \ln\frac{f_+}{f_-}\,\pi_-(-u) - u\,(f_- - f_+)$, where $\pi_-(u), u \geq 0$ is Poisson process of intensity f_-. The MLE and a wide class of Bayesian estimators are consistent, their limit distributions are different

$$n\left(\hat{\vartheta}_n - \vartheta\right) \Longrightarrow \hat{u}, \qquad n\left(\tilde{\vartheta}_n - \vartheta\right) \Longrightarrow \tilde{u} \tag{1.4}$$

where the random variables $\hat{u}$ and $\tilde{u}$ are defined by the relations

$$Z(\hat{u}) = \sup_u Z(u), \qquad \tilde{u} = \frac{\int uZ(u)\,\mathrm{d}u}{\int Z(u)\,\mathrm{d}u}$$

and we have the convergence of their moments. The minimax lower bound for all estimators $\bar{\vartheta}_n$ and $\vartheta_0 \in \Theta$ is

$$\varliminf_{\delta\to 0}\varliminf_{n\to\infty}\sup_{|\vartheta-\vartheta_0|<\delta} n^2\mathbf{E}_\vartheta\left(\bar{\vartheta}_n - \vartheta\right)^2 \geq \mathbf{E}_{\vartheta_0}\tilde{u}^2. \tag{1.5}$$

Moreover, *asymptotically efficient are Bayesian estimators only* because we have the strict inequality $\mathbf{E}\hat{u}^2 > \mathbf{E}\tilde{u}^2$ (see Ref. [15], Chapter 5 for details).

Having the problem of delay parameter estimation by the observations (1.1) we can ask the first question: to what statistical situation corresponds it, *regular* or some *non regular*? Surprising, the answer depends on the type of the limit we have and is the following: *if we study the properties of estimators as* $\sigma \to 0$, *then the problem of estimation is regular (like* (1.2), (1.3)*) and if we are interested by large samples properties (*$T \to \infty$*) then the problem of parameter estimation is non regular in statistical sense (like* (1.4),(1.5)*).*

The particularity of this problem can be seen from the integral representation of the *trend coefficient*

$$X_{t-\vartheta} = X_0 - \gamma\int_0^{t-\vartheta} X_{s-\vartheta}\,\mathrm{d}s + \sigma\,W_{t-\vartheta}.$$

Hence $X_{t-\vartheta}$ it is as smooth w.r.t. ϑ as Wiener process w.r.t. time, i.e., *we have trend coefficient continuous but not differentiable w.r.t.* ϑ *for all* $t \in (0, T]$.

2. Estimation (small noise asymptotics)

Let us fix the time of observation T and consider the problem of delay estimation by observations

$$\mathrm{d}X_t = -\gamma\, X_{t-\vartheta}\,\mathrm{d}t + \sigma\,\mathrm{d}W_t, \quad X_s = x_s, -\beta \le s \le 0, \quad 0 \le t \le T \tag{2.1}$$

where the diffusion coefficient σ^2 is supposed to be small, i.e. we study the properties of estimators $\widehat{\vartheta}_{\sigma,T} = \widehat{\vartheta}_\sigma$ and $\tilde{\vartheta}_{\sigma,T} = \tilde{\vartheta}_\sigma$ as T is fixed and $\sigma \to 0$. The model (2.1) can be considered as a small perturbation of the deterministic linear system

$$\frac{\mathrm{d}x_t}{\mathrm{d}t} = -\gamma\, x_{t-\vartheta}, \qquad x_s, -\beta < s \le 0, \quad 0 \le t \le T. \tag{2.2}$$

by white Gaussian noise of intensity σ^2. The behavior of such stochastic systems (without delays) statistical problems for them are well studied (see, e.g., [16], [5]).

Remind that if the observed process is

$$\mathrm{d}X_t = S(\vartheta, X_t)\,\mathrm{d}t + \sigma\,\mathrm{d}W_t, \quad X_0 = x_0, \quad 0 \le t \le T \tag{2.3}$$

where $S(\vartheta, x)$ is a smooth function, then the corresponding family of measures is LAN. The lower (Hajek-Le Cam's) bound is

$$\varliminf_{\delta\to 0}\,\varliminf_{\sigma\to 0}\,\sup_{|\vartheta-\vartheta_0|<\delta} \mathbf{E}_\vartheta \left(\frac{\bar{\vartheta}_\sigma - \vartheta}{\sigma}\right)^2 \ge \mathrm{I}(\vartheta_0)^{-1}, \quad \mathrm{I}(\vartheta) = \int_0^T \dot{S}_\vartheta(\vartheta, x_t)^2\,\mathrm{d}t. \tag{2.4}$$

The behavior of estimators is entirely similar to that of the regular case (1.2),(1.3), i.e., the MLE $\widehat{\vartheta}_\sigma$ and Bayesian estimators $\tilde{\vartheta}_\sigma$ are consistent, asymptotically normal

$$\frac{\widehat{\vartheta}_\sigma - \vartheta}{\sigma} \Longrightarrow \mathcal{N}\left(0, \mathrm{I}(\vartheta)^{-1}\right), \qquad \frac{\tilde{\vartheta}_\sigma - \vartheta}{\sigma} \Longrightarrow \mathcal{N}\left(0, \mathrm{I}(\vartheta)^{-1}\right) \tag{2.5}$$

and asymptotically efficient in the sense of the bound (2.4).

If the trend coefficient is a discontinuous function of ϑ, say,

$$\mathrm{d}X_t = \left[S_1(X_t)\, 1_{\{t<\vartheta\}} + S_2(X_t)\, 1_{\{t\ge\vartheta\}}\right]\mathrm{d}t + \sigma \mathrm{d}W_t, \quad X_0 = x_0,\ 0 \le t \le T$$

where $S_2(x_t) - S_1(x_t) = V_t \ne 0$ and $S_1(x), S_2(x)$ are smooth functions of x, then the situation changes. The normalized likelihood ratio

$$Z_\sigma(u) = \frac{L\left(\vartheta + \frac{u\sigma^2}{V_\vartheta^2}, X^\sigma\right)}{L(\vartheta, X^\sigma)} \Longrightarrow Z(u) = \exp\left\{W(u) - \frac{|u|}{2}\right\}, \tag{2.6}$$

where $W(\cdot)$ is two-sided Wiener process. Let us introduce two random variables $\widehat{u}$ and $\tilde{u}$ by the equations

$$Z(\widehat{u}) = \sup_{u\in\mathbb{R}} Z(u), \qquad \tilde{u} = \frac{\int_{\mathbb{R}} u\, Z(u)\,\mathrm{d}u}{\int_{\mathbb{R}} Z(u)\,\mathrm{d}u}. \tag{2.7}$$

The lower bound on the risk of all estimators is

$$\varliminf_{\delta\to 0}\varliminf_{\sigma\to 0}\sup_{|\vartheta-\vartheta_0|<\delta}\sigma^{-4}\,V_{\vartheta_0}^4\,\mathbf{E}\left(\bar\vartheta_\sigma-\vartheta\right)^2\geq \mathbf{E}\tilde u^2. \tag{2.8}$$

The MLE $\widehat\vartheta_\sigma$ and BE $\tilde\vartheta_\sigma$ are consistent, have the following limit distributions

$$\frac{V_\vartheta^2}{\sigma^2}\left(\widehat\vartheta_\sigma-\vartheta\right)\Longrightarrow\widehat u,\qquad \frac{V_\vartheta^2}{\sigma^2}\left(\tilde\vartheta_\sigma-\vartheta\right)\Longrightarrow\tilde u, \tag{2.9}$$

and BE are asymptotically efficient in the sense of this bound (see Ref. [5], Section 5.1).

Note, that the random process $Z(u)$ appeared as well in the change point problem for the model of *signal in white Gaussian noise*

$$\mathrm{d}X_t=S(t-\vartheta)\,\mathrm{d}t+\sigma\,\mathrm{d}W_t,\quad X_0=0,\quad 0\leq t\leq T$$

where $S(t)$ has a jump $S(t_*+)-S(t_*-)\neq 0$. In the asymptoics of small noise $(\sigma\to 0)$ it was shown that the bound like (2.8) is valid and the estimators have behavior similar (2.9) (see Ref. [15], Section 7.2). The values $\mathbf{E}\widehat u^2$ and $\mathbf{E}\tilde u^2$ were calculated by several authors. Terent'ev [17]showed that $\mathbf{E}\widehat u^2=26$ (see also Ref. [15], Section 7.3). Ibragimov and Khasminski [15] obtained the first approximation of the value $\mathbf{E}\tilde u^2$ by Monte-Carlo simulation. Then Golubev [18] found an integral representation of $\mathbf{E}\tilde u^2$, which provided by numerical calculation the value $\mathbf{E}\tilde u^2$ with higher precision. Finally, Rubin and Song [19] showed that

$$\mathbf{E}\tilde u^2=16\,\zeta(3),$$

where $\zeta(s)=\sum_{n=1}^\infty n^{-s}$ is Riemann's zeta function. As $16\,\zeta(3)=19,276\pm 0,006$ we see that $\mathbf{E}\widehat u^2>\mathbf{E}\tilde u^2$ and the MLE is not asymptotically efficient.

In our case of observations (2.1) the derivative w.r.t. ϑ of the trend coefficient $-\gamma X_{t-\vartheta}$ does not exist. Nevertheless, the family of measures is LAN, i.e.,

$$Z_\sigma(u)=\frac{L(\vartheta+u\sigma,X^\sigma)}{L(\vartheta,X^\sigma)}=\exp\left\{u\,\Delta_\sigma(\vartheta)-\frac{u^2}{2}\,\mathrm{I}(\vartheta)+r_\sigma\right\},$$

where $\Delta_\sigma(\vartheta)\sim\mathcal{N}(0,\mathrm{I}(\vartheta))$ and $r_\sigma\to 0$ and the following quantity plays the role of Fisher information:

$$\mathrm{I}(\vartheta)=\gamma^2\int_0^T x_{t-2\vartheta}^2\,\mathrm{d}t$$

where we suppose for simplicity that the initial values $x_s, s\leq 0$ are given for $s\in[-2\beta,0]$.

As the family is LAN we have the following minimax lower bound on the mean square error of all estimators $\bar\vartheta_\sigma$ (like (1.2),(2.4)).

$$\varliminf_{\delta\to 0}\varliminf_{\sigma\to 0}\sup_{|\vartheta-\vartheta_0|<\delta}\mathbf{E}\left(\frac{\bar\vartheta_\sigma-\vartheta}{\sigma}\right)^2\geq\mathrm{I}(\vartheta_0)^{-1}. \tag{2.10}$$

Comparison with (1.2), (2.6) shows that $\mathrm{I}(\vartheta)$ here indeed plays the role of Fisher information. Then we define the asymptotically efficient estimators as estimators for which we have equality in this inequality for all $\vartheta_0 \in \Theta$.

Theorem 2.1. *The MLE $\widehat{\vartheta}_\sigma$ and BE $\tilde{\vartheta}_\sigma$ are consistent, asymptotically normal*

$$\frac{\widehat{\vartheta}_\sigma - \vartheta}{\sigma} \Longrightarrow \mathcal{N}\left(0, \mathrm{I}(\vartheta)^{-1}\right), \qquad \frac{\tilde{\vartheta}_\sigma - \vartheta}{\sigma} \Longrightarrow \mathcal{N}\left(0, \mathrm{I}(\vartheta)^{-1}\right), \tag{2.11}$$

we have the convergence of all moments of these estimators and both estimators are asymptotically efficient.

Proof can be found in Ref. [4], (see also Ref. [5], Section 2.4). The proofs of this theorem and the Theorems 4 and 5 below consist in the verification of the conditions of two general theorems by Ibragimov and Khasminski [15]. These conditions are given in terms of the normalized likelihood ratios $Z_\sigma(\cdot)$ and $Z_T(\cdot)$ respectively.

Comparison of (2.6),(2.7) with (2.4),(2.5) shows that we have the regular problem of parameter estimation.

2.1. *Asymptotic expansion*

The asymptotic normality (2.6) can be proved *in probability* too, i.e., the estimators can be written as

$$\widehat{\vartheta}_\sigma = \vartheta + \frac{\xi_1}{\sqrt{\mathrm{I}(\vartheta)}}\,\sigma + o(\sigma), \qquad \tilde{\vartheta}_\sigma = \vartheta + \frac{\xi_1}{\sqrt{\mathrm{I}(\vartheta)}}\,\sigma + o(\sigma),$$

where $\sigma^{-1} o(\sigma) \to 0$ in probability and the r.v.

$$\xi_1 = -\frac{\gamma}{\sqrt{\mathrm{I}(\vartheta)}} \int_0^T x_{t-2\vartheta}\, \mathrm{d}W_t \sim \mathcal{N}(0,1)$$

is defined on the same probability space. The proof of these representations can be obtained using the method developed in Ref. [5], Section 3.1.

It is interesting to see the next after Gaussian terms of these expansions by the powers of σ. In regular problems of parameter estimation by observations of the process (2.3) with sufficiently smooth w.r.t. ϑ and x trend coefficient $S(\vartheta, x)$ such expansion for the MLE has the form

$$\widehat{\vartheta}_\sigma = \vartheta + \frac{\xi_1}{\sqrt{\mathrm{I}(\vartheta)}}\,\sigma + \left(a_1\xi_1^2 + a_2\xi_1\xi_2 + a_3\xi_1\xi_3 + a_4\xi_4\right)\,\sigma^2 + o\left(\sigma^2\right),$$

where ξ_1, ξ_2, ξ_3 are Gaussian r.v.'s and ξ_4 is an Itô integral of a Gaussian process. We see that the next after Gaussian term have to be of order σ^2 (see Ref. [5], Section 3.1).

The following theorem shows that this term in the case of the model (2.3) is of the order $\sigma^{3/2}$ only.

Theorem 2.2. *The MLE $\widehat{\vartheta}_\sigma$ admits the representation*

$$\widehat{\vartheta}_\sigma = \vartheta + \frac{\xi_1}{\sqrt{\mathrm{I}(\vartheta)}}\,\sigma - \frac{\gamma\sqrt{T}}{2\,\mathrm{I}(\vartheta)^{\frac{3}{4}}}\,\frac{\zeta_\sigma\,\mathrm{sgn}(\xi_1)}{\sqrt{|\xi_1|}}\,\sigma^{3/2} + o\left(\sigma^{3/2}\right), \tag{2.12}$$

where $\zeta_\sigma \sim \mathcal{N}(0,1)$ and is asymptotically independent of $\xi_1 \sim \mathcal{N}(0,1)$.

Proof. The proof can be found in Ref. [6]. It is based on the expansion of the likelihood ratio by the powers of σ

$$\begin{aligned}\ln Z_\sigma(u) &= \ln\frac{L\left(\vartheta+u\sigma, X^T\right)}{L\left(\vartheta, X^T\right)}\\ &= -\frac{\gamma}{\sigma}\int_0^T \left[X_{t-\vartheta-u\sigma} - X_{t-\vartheta}\right]\mathrm{d}W_t - \frac{\gamma^2}{2\sigma^2}\int_0^T \left[X_{t-\vartheta-u\sigma} - X_{t-\vartheta}\right]^2\mathrm{d}t\\ &= u\,\sqrt{\mathrm{I}(\vartheta)}\,\xi_1 - \frac{u^2}{2}\,\mathrm{I}(\vartheta) - \gamma\sqrt{|u|\,T}\,\zeta_\sigma\,\sigma^{1/2} + O(\sigma),\end{aligned}$$

where

$$\zeta_\sigma = \int_0^\tau \eta(t,u,\sigma)\,\mathrm{d}W_t, \qquad \eta(t,u,\sigma) = \frac{W_{t-\vartheta-u\sigma} - W_{t-\vartheta}}{\sqrt{T\,|u|\,\sigma}}$$

and τ is a stopping time

$$\tau = \inf\left\{t:\ \int_0^t \eta(s,u,\sigma)^2\,\mathrm{d}s \geq 1\right\}.$$

For construction of the r.v. ζ_σ we define $\eta(s,u,\sigma) = 1$ for $s \in [T, T+1]$ and we introduce an independent Wiener process W_s for $s \in [T, T+1]$ (see Ref. [6], proof of the theorem 1.19). Note that in Ref. [7] we used in the expansion the random variable

$$\zeta = \int_0^T \eta(t,u,\sigma)\,\mathrm{d}W_t,$$

which is asymptotically normal, but using the convergence

$$\int_0^T \eta(t,u,\sigma)^2\,\mathrm{d}t \longrightarrow 1$$

proved there it is easy to see that $\zeta - \zeta_\sigma \to 0$ in probability.

The similar expansion can be found for Bayesian estimator as well, but the final expression is more cumbersome.

2.2. *Generalizations*

There are several direct generalizations of Theorem 2.1. We can call the model (2.3) as *shift-delay* and consider as well the problem of **scale-delay** estimation for the model of observations

$$\mathrm{d}X_t = -\gamma\,X_{\vartheta t}\,\mathrm{d}t + \sigma\,\mathrm{d}W_t, \quad X_0 = x_0, \quad 0 \leq t \leq T,$$

where $\vartheta \in (0,1)$. The study is quite close to (2.3) and yields the same properties of MLE and BE with the Fisher information

$$\mathrm{I}(\vartheta) = \gamma^2 \int_0^T t^2\, x_{\vartheta^2 t}^2\, \mathrm{d}t,$$

i.e., these estimators are consistent, asymptotically normal (2.7) and asymptotically efficient. The proof can be found in Ref. [4], [5], Section 2.4. The asymptotic expansion like (2.8) can be obtained as well.

Another generalization can be done for **nonlinear system**

$$\mathrm{d}X_t = S\left(X_{t-\vartheta}, t\right) \mathrm{d}t + \sigma\, b\left(X_t, t\right) \mathrm{d}W_t, \quad X_s = x_s, -\beta < s \le 0, \quad 0 \le t \le T$$

where $S(\cdot)$ is a smooth function, then we have for the same estimators the asymptotic normality (2.7) with the Fisher information

$$\mathrm{I}(\vartheta) = \int_0^T \frac{S\left(x_{t-2\vartheta}, t-\vartheta\right)^2 S'\left(x_{t-\vartheta}, t\right)^2}{b\left(x_t, t\right)^2}\, \mathrm{d}t.$$

Here x_t is solution of the equation

$$\frac{\mathrm{d}x_t}{\mathrm{d}t} = S\left(x_{t-\vartheta}\right), \qquad x_s, -2\beta < s \le 0, \quad 0 \le t \le T$$

and $S'(x,t) = \frac{\partial}{\partial x} S(x,t)$. The proof can be found in Ref. [2] (see also Ref. [5]).

It is interesting to see what happens if we have **multiple delays** in the linear system

$$\mathrm{d}X_t = \sum_{j=1}^k \lambda_j\, X_{t-\tau_j}\, \mathrm{d}t + \sigma\, \mathrm{d}W_t, \quad X_s = x_s, \ -\beta < s \le 0, \quad 0 \le t \le T,$$

where $\boldsymbol{\vartheta} = (\lambda_1, \ldots, \lambda_k, \tau_1, \ldots, \tau_k)$ is unknown parameter.

It is shown that the MLE $\widehat{\boldsymbol{\vartheta}}_T$ is consistent, asymptotically normal and asymptotically efficient [20].

Let us consider the linear stochastic differential equation with **integral-type delay**

$$\mathrm{d}X_t = \left(\int_0^\delta X_{t-s}\, \mu(\mathrm{d}s)\right) \mathrm{d}t + \sigma\, \mathrm{d}W_t, \quad 0 \le t \le T. \tag{2.13}$$

We suppose that the deterministic initial values $X_s = x_s$, $-\beta < s \le 0$ are given. Here the measure $\mu(\cdot)$ is unknown. Then we can study the nonparametric estimation of the function

$$f(t) = \int_0^\delta x_{t-s}\, \mu(\mathrm{d}s), \quad 0 \le t \le T$$

where $x_t = x_t(\mu)$ is solution of the equation (2.13) with $\sigma = 0$. We show that the kernel-type estimator

$$\widehat{f}_\sigma(t) = \frac{1}{\psi_\sigma}\int_0^T K\left(\frac{\tau - t}{\psi_\sigma}\right) \mathrm{d}X_t$$

with $\psi_\sigma = \sigma^{2/3}$ is uniformly consistent

$$\lim_{\sigma\to 0}\ \sup_{\mu\in\Theta_\delta(L)}\ \sup_{a\leq t\leq b} \mathbf{E}_\mu\left(\widehat{f}_\sigma(t) - f(t)\right)^2 = 0$$

and asymptotically normal (see Ref. [9] for details). Here $0 < a < b < T$ and

$$\Theta_\delta(L) = \{\mu:\ \operatorname{supp}(\mu) \subset [0,\delta] \quad \|\mu\| \leq L\}.$$

Using this nonparametric estimator we can estimate the measure $\mu(\cdot)$ in the case (2.4), i.e., when $\mu = \sum_{i=1}^k \lambda_i \delta_{\tau-i}$. Note that the minimum distance estimator of the parameter $\boldsymbol{\vartheta} = (\lambda_1, \ldots, \lambda_k, \tau_1, \ldots, \tau_k)$ is consistent and asymptotically normal (see Refs. [8, 9]).

3. Hypotheses Testing (small noise asymptotics)

The observed process is always

$$\mathrm{d}X_t = -\gamma\, X_{t-\vartheta}\, \mathrm{d}t + \sigma\, \mathrm{d}W_t, \quad X_s = x_s, s\leq 0 \quad 0 \leq t \leq T$$

and we have to test the following two hypotheses

$$\begin{aligned} &\mathcal{H}_0: &&\vartheta = 0, &&\text{(no delay)}\\ &\mathcal{H}_1: &&\vartheta > 0. \end{aligned}$$

Therefor the observed process under hypothesis $\mathcal{H}_0$ is Ornstein-Uhlenbeck. The contiguous alternatives correspond to $\vartheta = u\sigma$ and we can rewrite the hypotheses testing problem as :

$$\begin{aligned} &\mathcal{H}_0: &&u = 0, &&\text{(no delay)}\\ &\mathcal{H}_1: &&u > 0, \end{aligned}$$

Let us fix some $\varepsilon \in (0,1)$ and denote by $\mathcal{K}_\varepsilon$ the class of tests of asymptotic level $1-\varepsilon$. The direct calculation shows that the limit ($\sigma \to 0$) of the power function of Neyman-Pearson test $\varphi_\sigma \in \mathcal{K}_\varepsilon$ is

$$\beta_{NP}(u) = \mathbf{P}\left\{\zeta > z_\varepsilon - u\, \mathrm{I}_0^{1/2}\right\},$$

where $\zeta \sim \mathcal{N}(0,1)$, z_ε is $1-\varepsilon$ quantil of $\mathcal{N}(0,1)$ and the Fisher information $\mathrm{I}_0 = \mathrm{I}(0)$ is

$$\mathrm{I}(0) = \frac{x_0^2\,\gamma^3}{2}\left(1 - e^{-2\gamma T}\right).$$

We call a test $\varphi_\sigma^\circ \in \mathcal{K}_\varepsilon$ *locally asymptotically uniformly most powerful* if its power function $\beta_\sigma^\circ(u) = \mathbf{E}_u\varphi_\sigma^\circ(X^\sigma)$ satisfies the relation: for any test $\bar\varphi_\sigma \in \mathcal{K}_\varepsilon$ and any $K > 0$

$$\lim_{\sigma\to 0}\inf_{0\le u\le K}\left[\beta_\sigma^\circ(u) - \bar\beta_\sigma(u)\right] \ge 0.$$

Here $\bar\beta_\sigma(u) = \mathbf{E}_u\bar\varphi_\sigma(X^\sigma)$.

Let us introduce three tests. The first one is *score-function test*

$$\varphi_\sigma^*(X^\sigma) = \chi_{\{\Delta_\sigma > z_\varepsilon\}},$$

where

$$\Delta_\sigma = -\frac{\gamma}{\sigma\sqrt{\mathrm{I}_0}}\int_0^T x_t\,\left[\mathrm{d}X_t + \gamma X_t\,\mathrm{d}t\right].$$

Note that under $\mathcal{H}_0$ the statistic $\Delta_\sigma \sim \mathcal{N}(0,1)$.

The second test is the *likelihood ratio test*

$$\tilde\varphi_\sigma(X^\sigma) = \chi_{\{\delta_\sigma > b_\varepsilon\}},\qquad \delta_\sigma(X^\sigma) = \sup_{\vartheta>0} L(\vartheta, X^\sigma),$$

where $b_\varepsilon = \exp\left\{z_\varepsilon^2/2\mathrm{I}_0\right\}$.

The third is *Wald's test* based on the MLE $\widehat\vartheta_\sigma$:

$$\widehat\varphi_\sigma(X^\sigma) = \chi_{\left\{\widehat\delta_\sigma > z_\varepsilon \mathrm{I}_0^{-1/2}\right\}},\qquad \widehat\delta_\sigma(X^\sigma) = \frac{\widehat\vartheta_\sigma}{\sigma}.$$

Theorem 3.1. *The tests $\varphi_\sigma^*, \tilde\varphi_\sigma, \widehat\varphi_\sigma$ belong to $\mathcal{K}_\varepsilon$ and are locally asymptotically uniformly most powerful.*

Proof. The proof follows from the local asymptotical normality of the family of measures at the point $\vartheta = 0$ and the standard arguments applied to regular statistical experiments (see, e.g. [21]). It consists in the verification that the power functions of these tests converge uniformly on compacts $[0, K]$ to the limit power function $\beta_{NP}(u)$ of the Neyman-Pearson test.

4. Estimation (large samples asymptotics)

Let us consider the same linear stochastic differential equation

$$\mathrm{d}X_t = -\gamma\, X_{t-\vartheta}\,\mathrm{d}t + \sigma\,\mathrm{d}W_t,\quad X_s = x_s,\ -\beta < s \le 0,\quad 0 \le t \le T$$

but now we study the properties of MLE $\widehat\vartheta_{\sigma,T} = \widehat\vartheta_T$ and BE $\tilde\vartheta_{\sigma,T} = \tilde\vartheta_T$ of the delay ϑ in the asymptotics of large samples, i.e., as $T \to \infty$. Suppose that $\gamma > 0$, $\vartheta \in \left(0, \frac{\pi}{2\gamma}\right) = \Theta$, ($\beta = \pi/2\gamma$), then the process X_t has ergodic properties [22].

Remind the properties of parameter estimators for ergodic diffusion processes

$$\mathrm{d}X_t = S(\vartheta, X_t)\,\mathrm{d}t + \mathrm{d}W_t,\quad X_0,\quad 0 \le t \le T. \tag{4.1}$$

If the function $S(\vartheta, x)$ is smooth w.r.t. ϑ, then the family of measures is LAN and we have the lower bound on the risks of all estimators :

$$\varliminf_{\delta\to 0}\varliminf_{T\to\infty}\sup_{|\vartheta-\vartheta_0|<\delta} T\,\mathbf{E}_\vartheta\left(\bar\vartheta_T-\vartheta\right)^2\geq \mathrm{I}(\vartheta_0)^{-1},\quad \mathrm{I}(\vartheta)=\int \dot S(\vartheta,x)^2 f(\vartheta,x)\,\mathrm{d}x,$$

where $f(\vartheta, x)$ is the density of invariant law. The properties of estimators are typical for regular experiments, i.e., the MLE $\widehat\vartheta_T$ and BE $\tilde\vartheta_T$ are consistent, asymptotically normal

$$\sqrt{T}\left(\widehat\vartheta_T-\vartheta\right)\Longrightarrow\mathcal{N}\left(0,\mathrm{I}(\vartheta)^{-1}\right),\quad \sqrt{T}\left(\tilde\vartheta_T-\vartheta\right)\Longrightarrow\mathcal{N}\left(0,\mathrm{I}(\vartheta)^{-1}\right),$$

and asymptotically efficient (see Ref. [6]).

The properties of estimators change if the trend coefficient is not a smooth function. For example, if $S(\vartheta, x) = S(x-\vartheta)$, where $S(x)$ has a finite jump at some point x_*, i.e.; $S(x_*+) - S(x_* - +) \neq 0$, then the rate of convergence of estimators is T and not $\sqrt{T}$. To illustrate this let us consider a simple switching ergodic diffusion process

$$\mathrm{d}X_t=-\mathrm{sgn}\left(X_t-\vartheta\right)\,\mathrm{d}t+\mathrm{d}W_t,\quad X_0,\quad 0\leq t\leq T. \tag{4.2}$$

To describe the properties of estimators we reintroduce the random process (2.6)

$$Z(u)=\exp\left\{W(u)-\frac{1}{2}\,|u|\right\},\qquad u\in\mathbb{R},$$

where $W(\cdot)$ is two-sided Wiener process and corresponding two random variables $\widehat u$ and $\tilde u$ defined by the equations (2.7).

Note that the normalized likelihood ratio

$$Z_T(u)=\frac{L\left(\vartheta+\frac{u}{4T},X^T\right)}{L\left(\vartheta,X^T\right)}\Longrightarrow Z(u) \tag{4.3}$$

and this (together with some other estimates on the likelihood ratio) provides us the lower bound like (2.8)

$$\varliminf_{\delta\to 0}\varliminf_{T\to\infty}\sup_{|\vartheta-\vartheta_0|<\delta} 16\,T^2\,\mathbf{E}_\vartheta\left(\bar\vartheta_T-\vartheta\right)^2\geq \mathbf{E}\tilde u^2. \tag{4.4}$$

Moreover, the MLE and BE are consistent, have the following limit distributions

$$4T\left(\widehat\vartheta_T-\vartheta\right)\Longrightarrow\widehat u,\qquad 4T\left(\tilde\vartheta_T-\vartheta\right)\Longrightarrow\tilde u, \tag{4.5}$$

and the BE are asymptotically efficient in the sense of this bound (see [6], Section 3.4).

Let us return to delay estimation problem. First we verify that the normalized likelihood ratio

$$Z_T(u)=\frac{L\left(\vartheta+\frac{u}{\gamma^2 T},X^T\right)}{L\left(\vartheta,X^T\right)}\Longrightarrow Z(u)=\exp\left\{W(u)-\frac{|u|}{2}\right\}. \tag{4.6}$$

Then we obtain the lower bound: for all estimators $\bar{\vartheta}_T$ we have

$$\underline{\lim}_{\delta\to 0}\ \underline{\lim}_{T\to\infty}\ \sup_{|\vartheta-\vartheta_0|<\delta} \gamma^4\, T^2\, \mathbf{E}_\vartheta \left(\bar{\vartheta}_T - \vartheta\right)^2 \geq \mathbf{E}\tilde{u}^2. \tag{4.7}$$

Therefore we can define the asymptotically efficient estimators as estimators for which we have equality in this inequality for all $\vartheta_0 \in \Theta$.

Theorem 4.1. *The MLE $\widehat{\vartheta}_T$ and BE $\tilde{\vartheta}_T$ are consistent, have the following limits in distribution*

$$\gamma^2\, T\left(\widehat{\vartheta}_T - \vartheta\right) \Longrightarrow \widehat{u}, \qquad\qquad \gamma^2\, T\left(\tilde{\vartheta}_T - \vartheta\right) \Longrightarrow \tilde{u} \tag{4.8}$$

and BE are asymptotically efficient.

For the proof see Küchler and Kutoyants [3] and Kutoyants [6], Section 3.3. Remind how the weak convergence of the random function $Z_T(\cdot)$ to $Z(\cdot)$ yields the weak convergence (4.8) of the MLE [15]. Let us denote ϑ_0 the true value of the parameter. Then we can write

$$\begin{aligned}
&\mathbf{P}_{\vartheta_0}^{(T)}\left\{\gamma^2\, T\left(\widehat{\vartheta}_T - \vartheta_0\right) < z\right\} = \mathbf{P}_{\vartheta_0}^{(T)}\left\{\widehat{\vartheta}_T < \vartheta_0 + \frac{z}{\gamma^2\, T}\right\}\\
&\quad = \mathbf{P}_{\vartheta_0}^{(T)}\left\{\sup_{\vartheta<\vartheta_0+\frac{z}{\gamma^2 T}} L\left(\vartheta, X^T\right) > \sup_{\vartheta\geq\vartheta_0+\frac{z}{\gamma^2 T}} L\left(\vartheta, X^T\right)\right\}\\
&\quad = \mathbf{P}_{\vartheta_0}^{(T)}\left\{\sup_{\vartheta<\vartheta_0+\frac{z}{\gamma^2 T}} \frac{L\left(\vartheta, X^T\right)}{L\left(\vartheta_0, X^T\right)} > \sup_{\vartheta\geq\vartheta_0+\frac{z}{\gamma^2 T}} \frac{L\left(\vartheta, X^T\right)}{L\left(\vartheta_0, X^T\right)}\right\}\\
&\quad = \mathbf{P}_{\vartheta_0}^{(T)}\left\{\sup_{u<z} Z_T(u) > \sup_{u\geq z} Z_T(u)\right\} = \mathbf{P}_{\vartheta_0}^{(T)}\left\{\sup_{u<z} Z_T(u) - \sup_{u\geq z} Z_T(u) > 0\right\}\\
&\quad \longrightarrow \mathbf{P}\left\{\sup_{u<z} Z(u) - \sup_{u\geq z} Z(u) > 0\right\} = \mathbf{P}\left\{\widehat{u} < z\right\},
\end{aligned}$$

where we changed the variable $\vartheta = \vartheta_0 + \frac{u}{\gamma^2 T}$ and used the convergence (4.6).

Comparison of (4.7), (4.8) with (4.4), (4.5) shows that the delay estimation problem is singular and similar to space change point estimation problem.

Let us suppose that the observed process

$$\mathrm{d}X_t = -\gamma\, X_{t-\tau}\,\mathrm{d}t + \sigma\,\mathrm{d}W_t, \quad X_s = x_s, \ -\frac{2\pi}{\beta} < s \leq 0, \quad 0 \leq t \leq T$$

has two unknown parameters $\gamma \in (\alpha, \beta)$, $\alpha > 0$ and $\tau \in \left(0, \frac{2\pi}{\beta}\right)$ and we have to estimate the both parameters simultaneously, i.e. we have two-dimensional parameter $\boldsymbol{\vartheta} = (\gamma, \tau)$. We need the further notation.

- normalizing matrix

$$\varphi_T = \begin{pmatrix} T^{-1/2}, & 0 \\ 0 & , T^{-1} \end{pmatrix}.$$

- Fisher information of the component γ

$$\mathrm{I}(\boldsymbol{\vartheta}) = \int_0^\infty x_t^2 \, \mathrm{d}t$$

where x_t is *fundamental solution* of the deterministic equation

$$\frac{\mathrm{d}x_t}{\mathrm{d}t} = -\gamma\, x_{t-\tau}, \qquad x_s = 0,\ -\frac{2\pi}{\beta} < s < 0, \quad x_0 = 1,$$

- the random vectors $\widehat{\boldsymbol{w}} = \left(\widehat{v}, \frac{\widehat{u}}{\gamma^2}\right)$ and $\tilde{\boldsymbol{w}} = \left(\widehat{v}, \frac{\tilde{u}}{\gamma^2}\right)$ where the random variables $\widehat{u}$ and $\tilde{u}$ are the same as before and $\widehat{v}$ is independent of $\widehat{u}$ and $\tilde{u}$ Gaussian r.v. $\mathcal{N}\left(0, \mathrm{I}(\boldsymbol{\vartheta})^{-1}\right)$

The lower bound is given by the following inequality: for any $\boldsymbol{\vartheta}_0 \in \boldsymbol{\Theta} = (\alpha, \beta) \times \left(0, \frac{2\pi}{\beta}\right)$ and any estimator $\bar{\boldsymbol{\vartheta}}_T$

$$\varliminf_{\delta\to 0}\ \varliminf_{T\to\infty}\ \sup_{|\boldsymbol{\vartheta}-\boldsymbol{\vartheta}_0|<\delta} \mathbf{E}_{\boldsymbol{\vartheta}} \left|\varphi_T^{-1}\left(\bar{\boldsymbol{\vartheta}}_T - \boldsymbol{\vartheta}\right)\right|^2 \geq \mathrm{I}(\boldsymbol{\vartheta})^{-1} + \frac{\mathbf{E}\tilde{u}^2}{\gamma^4}.$$

We call an estimator $\boldsymbol{\vartheta}_T^*$ asymptotically efficient if for all $\boldsymbol{\vartheta}_0 \in \boldsymbol{\Theta}$

$$\lim_{\delta\to 0}\ \lim_{T\to\infty}\ \sup_{|\boldsymbol{\vartheta}-\boldsymbol{\vartheta}_0|<\delta} \mathbf{E}_{\boldsymbol{\vartheta}} \left|\varphi_T^{-1}\left(\boldsymbol{\vartheta}_T^* - \boldsymbol{\vartheta}\right)\right|^2 = \mathrm{I}(\boldsymbol{\vartheta})^{-1} + \frac{\mathbf{E}\tilde{u}^2}{\gamma^4}.$$

The asymptotic properties of the MLE $\widehat{\boldsymbol{\vartheta}}_T = \left(\widehat{\gamma}_T, \widehat{\tau}_T\right)$ and BE $\tilde{\boldsymbol{\vartheta}}_T = (\tilde{\gamma}_T, \tilde{\tau}_T)$ are described in the following theorems.

Theorem 4.2. *The MLE $\widehat{\boldsymbol{\vartheta}}_T$ and BE $\tilde{\boldsymbol{\vartheta}}_T$ of the parameter $\boldsymbol{\vartheta}$ are consistent, have different limit distributions*

$$\varphi_T^{-1}\left(\widehat{\boldsymbol{\vartheta}}_T - \boldsymbol{\vartheta}\right) \Longrightarrow \widehat{\boldsymbol{w}}, \qquad \varphi_T^{-1}\left(\tilde{\boldsymbol{\vartheta}}_T - \boldsymbol{\vartheta}\right) \Longrightarrow \tilde{\boldsymbol{w}},$$

for any $p > 0$ the moments $\mathbf{E}_{\boldsymbol{\vartheta}}|\varphi_T^{-1}\left(\widehat{\boldsymbol{\vartheta}}_T - \boldsymbol{\vartheta}\right)|^p$ and $\mathbf{E}_{\boldsymbol{\vartheta}}|\varphi_T^{-1}\left(\tilde{\boldsymbol{\vartheta}}_T - \boldsymbol{\vartheta}\right)|^p$ converge. Moreover, the BE are asymptotically efficient.

For the proof see Ref. [6], Section 3.3.

From this theorem it follows that the parameters γ and τ are estimated with the different rates, say,

$$\sqrt{T}\left(\widehat{\gamma}_T - \gamma\right) \Longrightarrow \widehat{v}, \qquad T\left(\widehat{\tau}_T - \tau\right) \Longrightarrow \frac{\widehat{u}}{\gamma^2}.$$

5. Hypotheses Testing (large samples asymptotics)

The observed process is

$$\mathrm{d}X_t = -\gamma\, X_{t-\vartheta}\, \mathrm{d}t + \sigma\, \mathrm{d}W_t, \quad X_s = x_s, s \leq 0 \quad 0 \leq t \leq T$$

and we have to test the following two hypotheses

$$\begin{aligned} \mathcal{H}_0 : \qquad & \vartheta = 0, \qquad \text{(no delay)} \\ \mathcal{H}_1 : \qquad & \vartheta > 0, \end{aligned}$$

The contiguous alternatives correspond to $\vartheta = u/\gamma T$:

$$\begin{aligned} \mathcal{H}_0 : \qquad & u = 0, \qquad \text{(no delay)} \\ \mathcal{H}_1 : \qquad & u > 0, \end{aligned}$$

We have no asymptotically uniformly most powerful test and describe the likelihood ratio test only. This test is defined as follows

$$\tilde{\varphi}_T\left(X^T\right) = \chi_{\left\{\delta_T > \frac{1}{\varepsilon}\right\}}, \qquad \delta_T\left(X^T\right) = \sup_{\vartheta>0} \frac{L\left(\vartheta, X^T\right)}{L\left(0, X^T\right)}.$$

Its properties are given in the following theorem.

Theorem 5.1. *The likelihood ratio test $\tilde{\varphi}_T$ belongs to $\mathcal{K}_\varepsilon$ and its power function*

$$\beta_T\left(u, \widehat{\varphi}_T\right) = \mathbf{P}\left\{\zeta + \max\left[\eta, \xi\right] > \ln\frac{1}{\varepsilon} - \frac{u}{2}\right\} + o\left(1\right), \tag{5.1}$$

where the random variables ζ, η and ξ are independent and

$$\zeta \sim \mathcal{N}\left(0, u\right), \qquad F_\xi\left(x\right) = 1 - \mathrm{e}^{-x}, \quad x \geq 0 \tag{5.2}$$

$$F_\eta\left(x\right) = \Phi\left(\frac{x}{\sqrt{u}} + \frac{\sqrt{u}}{2}\right) + \mathrm{e}^{-x}\left[1 - \Phi\left(\frac{x}{\sqrt{u}} - \frac{\sqrt{u}}{2}\right)\right], \tag{5.3}$$

where $\Phi\left(x\right)$ is Gaussian $\mathcal{N}\left(0, 1\right)$ distribution function.

Proof. Remind that the normalized likelihood ratio $Z_T\left(u\right)$ converges to the process $Z\left(u\right)$ (see (4.6)), hence

$$\begin{aligned} \mathbf{P}_0\left\{\delta_T\left(X^T\right) > \frac{1}{\varepsilon}\right\} &= \mathbf{P}_0\left\{\sup_{u\geq 0} Z_T\left(u\right) > \frac{1}{\varepsilon}\right\} \\ \longrightarrow \mathbf{P}\left\{\sup_{u\geq 0} Z\left(u\right) > \frac{1}{\varepsilon}\right\} &= \mathbf{P}\left\{\sup_{u\geq 0}\left[W\left(u\right) - \frac{u}{2}\right] > \ln\left(\frac{1}{\varepsilon}\right)\right\} = \varepsilon \end{aligned}$$

because the random variable $\xi = \sup_{u\geq 0}\left[W\left(u\right) - \frac{u}{2}\right]$ has exponential distribution. Therefore $\tilde{\varphi}_T \in \mathcal{K}_\varepsilon$. The proof of (5.1)-(5.3) can be found in Ref. [6], Section 5.2.

6. Discussion

The difference of the properties of estimators and tests in these two different limits can be explain as follows. The properties of estimators depend strongly of the local structure of the likelihood ratio. Let us write it as

$$Z_{\sigma,T}(u) = \frac{L(\vartheta + \varphi_{\sigma,T} u, X)}{L(\vartheta, X)} = \exp\left\{ -\gamma \int_0^T \frac{\left[X_{t-\vartheta-\varphi_{\sigma,T} u} - X_{t-\vartheta}\right]}{\sigma} \mathrm{d}W_t \right.$$
$$\left. - \frac{\gamma^2}{2} \int_0^T \left[\frac{X_{t-\vartheta-\varphi_{\sigma,T} u} - X_{t-\vartheta}}{\sigma} \right]^2 \mathrm{d}t \right\},$$

where $\varphi_{\sigma,T} \to 0$.

Using equation (1.1) we obtain the representation

$$\frac{X_{t-\vartheta-\varphi_{\sigma,T} u} - X_{t-\vartheta}}{\sigma} = -\frac{\gamma}{\sigma} \int_{t-\vartheta}^{t-\vartheta-\varphi_{\sigma,T} u} X_{s-\vartheta}\, \mathrm{d}s + W_{t-\vartheta-\varphi_{\sigma,T} u} - W_{t-\vartheta}.$$

If we put $\varphi_{\sigma,T} = \sigma \to 0$, then

$$\frac{X_{t-\vartheta-\varphi_{\sigma,T} u} - X_{t-\vartheta}}{\sigma} \longrightarrow \gamma\, u\, x_{t-2\vartheta}$$

because the stochastic process X_t converges uniformly in $t \in [0, T]$ to the deterministic function $x_t = x_t(\vartheta)$, which is solution of the differential equation

$$\frac{\mathrm{d}x_t}{\mathrm{d}t} = -\gamma\, x_{t-\vartheta}, \qquad x_s,\ s \leq 0.$$

Therefore the normalized likelihood ratio converges

$$Z_\sigma(u) \Longrightarrow Z(u) = \exp\left\{ u\gamma \int_0^T x_{t-2\vartheta}\, \mathrm{d}W_t - \frac{u^2\gamma^2}{2} \int_0^T x_{t-2\vartheta}^2\, \mathrm{d}t \right\}$$

and we have regular estimation and testing problems. The contribution of the term $W_{t-\vartheta-\varphi_{\sigma,T} u} - W_{t-\vartheta}$ in this case is asymptotically negligible.

If we fix σ and consider $T \to \infty$ with $\varphi_{\sigma,T} = T^{-1}$, then

$$\frac{X_{t-\vartheta-\frac{u}{T}} - X_{t-\vartheta}}{\sigma} = -\frac{\gamma}{\sigma} \int_{t-\vartheta}^{t-\vartheta-\frac{u}{T}} X_{s-\vartheta}\, \mathrm{d}s + W_{t-\vartheta-\frac{u}{T}} - W_{t-\vartheta}$$
$$\sim -\frac{\gamma u}{T} X_{t-2\vartheta} + \sqrt{\frac{u}{T}}\, \eta(t, \vartheta, u),$$

where $\mathbf{E}_\vartheta\, \eta(t, \vartheta, u)^2 = 1$. We see that the main part of the contribution is due to the Wiener process and as the Wiener process is not differentiable w.r.t. time, the statistical problems became non regular.

References

[1] R. S. Liptser and A. N. Shiryaev, *Statistics of random processes. I.* vol. 5, *Applications of Mathematics (New York)*, (Springer-Verlag, Berlin, 2001), expanded edition. ISBN 3-540-63929-2. General theory, Translated from the 1974 Russian original by A. B. Aries, Stochastic Modelling and Applied Probability.

[2] G. T. Apoyan, An estimate for the parameter of a nondifferentiable drift coefficient, *Erevan. Gos. Univ. Uchen. Zap. Estestv. Nauki.* **161**(1), 33–42, (1986). ISSN 0132-0173.

[3] U. Küchler and Y. A. Kutoyants, Delay estimation for some stationary diffusion-type processes, *Scand. J. Statist.* **27**(3), 405–414, (2000). ISSN 0303-6898.

[4] Y. A. Kutoyants, An example of an estimate for the parameter of a nondifferentiable drift coefficient, *Teor. Veroyatnost. i Primenen.* **33**(1), 188–192, (1988). ISSN 0040-361X.

[5] Y. Kutoyants, *Identification of dynamical systems with small noise.* vol. 300, *Mathematics and its Applications*, (Kluwer Academic Publishers Group, Dordrecht, 1994). ISBN 0-7923-3053-6.

[6] Y. A. Kutoyants, *Statistical inference for ergodic diffusion processes.* Springer Series in Statistics, (Springer-Verlag London Ltd., London, 2004). ISBN 1-85233-759-1.

[7] Y. A. Kutoyants, On delay estimation for stochastic differential equations, *Stoch. Dyn.* **5**(2), 333–342, (2005). ISSN 0219-4937.

[8] Y. Kutoyants and T. Mourid, Estimation dans un modèle autorégressif avec retards, *C. R. Acad. Sci. Paris Sér. I Math.* **315**(4), 455–458, (1992). ISSN 0764-4442.

[9] Y. Kutoyants and T. Mourid, Estimation par la distance minimale pour un processus de type diffusion avec retards, *Publ. Inst. Statist. Univ. Paris.* **38**(2), 3–18, (1994).

[10] A. A. Gushchin and U. Küchler, Asymptotic inference for a linear stochastic differential equation with time delay, *Bernoulli.* **5**(6), 1059–1098, (1999). ISSN 1350-7265.

[11] A. A. Gushchin and U. Küchler, On parametric statistical models for stationary solutions of affine stochastic delay differential equations, *Math. Methods Statist.* **12** (1), 31–61, (2003). ISSN 1066-5307.

[12] M. Reiß, Minimax rates for nonparametric drift estimation in affine stochastic delay differential equations, *Stat. Inference Stoch. Process.* **5**(2), 131–152, (2002). ISSN 1387-0874.

[13] M. Reiss, Adaptive estimation for affine stochastic delay differential equations, *Bernoulli.* **11**(1), 67–102, (2005). ISSN 1350-7265.

[14] J. Hájek. Local asymptotic minimax and admissibility in estimation. In *Proceedings of the Sixth Berkeley Symposium on Mathematical Statistics and Probability (Univ. California, Berkeley, Calif., 1970/1971), Vol. I: Theory of statistics*, pp. 175–194, Berkeley, Calif., (1972). Univ. California Press.

[15] I. A. Ibragimov and R. Z. Has′minskiĭ, *Statistical estimation.* vol. 16, *Applications of Mathematics*, (Springer-Verlag, New York, 1981). ISBN 0-387-90523-5. Asymptotic theory, Translated from the Russian by Samuel Kotz.

[16] M. I. Freidlin and A. D. Wentzell, *Random perturbations of dynamical systems.* vol. 260, *Grundlehren der Mathematischen Wissenschaften [Fundamental Principles of Mathematical Sciences]*, (Springer-Verlag, New York, 1984). ISBN 0-387-90858-7. Translated from the Russian by Joseph Szücs.

[17] A. S. Terent'yev, Probability distribution of a time location of an absolute maximum at the output of a sinchronized filter, *Radioengineering and Electronics.* **13**(4), 652–657, (1968).

[18] G. K. Golubev, Computation of efficiency of maximum-likelihood estimate when observing a discontinuous signal in white noise, *Problems Inform. Transmission.* **15**(3), 61–69, (1979).

[19] H. Rubin and K. S. Song, Exact computation of the asymptotic efficiency of maximum likelihood estimators of a discontinuous signal in a Gaussian white noise, *Ann. Statist.* **23**(3), 732–739, (1995). ISSN 0090-5364.

[20] Y. A. Kutoyants, T. Mourid, and D. Bosq, Estimation paramétrique d'un processus de diffusion avec retards, *Ann. Inst. H. Poincaré Probab. Statist.* **28**(1), 95–106, (1992). ISSN 0246-0203.

[21] G. G. Roussas, *Contiguity of probability measures: some applications in statistics.*

(Cambridge University Press, London, 1972). Cambridge Tracts in Mathematics and Mathematical Physics, No. 63.

[22] U. Küchler and B. Mensch, Langevin's stochastic differential equation extended by a time-delayed term, *Stochastics Stochastics Rep.* **40**(1-2), 23–42, (1992). ISSN 1045-1129.

Chapter 14

On Cauchy-Dirichlet Problem for Linear Integro-Differential Equation in Weighted Sobolev Spaces

Remigijus Mikulevicius and Henrikas Pragarauskas*

Department of Mathematics, University of Southern California
Los Angeles, CA 90089-2532
mikulvcs@usc.edu

We study the Cauchy-Dirichlet problem in a smooth bounded domain for linear parabolic integro-differential equations. Sufficient conditions are derived under which the problem has a unique solution in weighted Sobolev classes. The result can be used in the regularity analysis of certain functionals arising in the theory of Markov processes.

Contents

1. Introduction

The paper is devoted to the Cauchy-Dirichlet problem for integro-differential equations associated with d-dimensional Markov process $X_t^{s,x}$, $t \geq s$, defined by Ito stochastic differential equation

$$dX_t = \sigma(t, X_t)\, dW_t + b(t, X_t)\, dt + \int_{|y|\leq 1} y\, q(dt, dy) + \int_{|y|>1} y\, p(dt, dy), \quad (1.1)$$

$$X_s = x,$$

where W_t is a standard d-dimensional Wiener process, $p(dt, dy)$ is the jump measure of X_t with the compensator $\pi(t, X_t, dy)dt$, and $q(dt, dy) = p(dt, dy) - \pi(t, X_t, dy)dt$ is the corresponding martingale measure. A very simple example is the process X_t

*Institute of Mathematics and Informatics, Akademijos 4, Vilnius, Lithuania, hpraga@ktl.mii.lt satisfying the equation

$$dX_t = \sigma(t, X_t)\, dW_t + b(t, X_t)\, dt + dZ_t, \qquad (1.2)$$
$$X_s = x,$$

where Z_t is a d-dimensional α-stable process, $\alpha \in (0, 2)$. In this case $\pi(t, x, dy) = dy/|y|^{d+\alpha}$.

In many problems arising in the theory of Markov processes it is important to study the smoothness properties of the functionals

$$v(s,x) = \mathbf{E} \int_s^{\tau^{s,x} \wedge T} f(t, X_t^{s,x})\, dt,$$

where $\tau^{s,x}$ is the first exit time of the process $X_t^{s,x}$ from a domain $G \subseteq \mathbf{R}^d$. If v is a sufficiently smooth function, it is a solution to Cauchy-Dirichlet problem

$$\partial_t u + Au + f = 0, \text{ in } (0,T) \times G,$$
$$u(T, x) = 0, \ x \in \mathbf{R}^d, \qquad (1.3)$$
$$u(t, x) = 0, \ t \in [0, T], \ x \notin G,$$

where

$$Au(t,x) = a^{ij}(t,x)\partial_{ij}^2 u(t,x) + b^i(t,x)\partial_i u(t,x)$$
$$+ \int [u(t, x+y) - u(t,x) - \partial_i u(t,x) y_i \mathbf{1}_{\{|y|\leq 1\}}]\ \pi(t,x,dy),$$

$a^{ij}(t,x) = (1/2)\sigma(t,x)\sigma^*(t,x)$, $\mathbf{1}_{\{|y|\leq 1\}}$ is the indicator function of $\{y \in \mathbf{R}^d : |y| \leq 1\}$, and the implicit summation convention over repeated indices is assumed.

The problem of such type (including the case of nonlinear equations) was considered by a number of authors (see e.g. Refs. [1–4], and references therein) in Sobolev and Hölder spaces under certain restrictive assumptions on $\pi(t, x, dy)$. In [3], only finite number of jumps can occur outside of D in finite time, and in Ref. [2] very restrictive assumptions on the order of jumps are imposed. In Ref. [5], general results were obtained in weighted Sobolev classes for a differential operator L. We extend some results in Ref. [5] for the integro-differential operator Lu and show for a bounded domain $G \subseteq \mathbf{R}^d$ that the assumptions on $\pi(t, x, dy)$ can be considerably relaxed. The results in Refs. [3] and [2] do not apply for the equation (1.2) which is obviously covered by our Theorem 2.1.

The results obtained can be also used in the analysis of non-linear equations arising in the optimal control theory of Markov processes.

The main result of the paper is presented in Section 2 and proved in Section 3.

2. Notation and main result

We will consider a bounded domain G in $\mathbf{R}^d$ of class C^1_u. In other words (see Ref. [5]), there are two numbers $r_0, K_0 \in (0,\infty)$ and an increasing function $\omega_0(\varepsilon)$ on $[0,\infty)$ such that $\omega_0(\varepsilon) \downarrow 0$ as $\varepsilon \downarrow 0$ so that for each $x_0 \in \partial G$, there exists a one-to-one continuously differentiable mapping Ψ of $B_{r_0}(x_0) = \{y \in \mathbf{R}^d : |y - x_0| < r_0\}$ onto a domain $D \subseteq \mathbf{R}^d$ having the following properties:

(i) $D_+ = \Psi\left(B_{r_0}(x_0) \cap G\right) \subseteq \mathbf{R}^d_+ = \{x \in \mathbf{R}^d : x_1 > 0\}$ and $\Psi(x_0) = 0$;

(ii) $\Psi\left(B_{r_0}(x_0) \cap \partial G\right) = D \cap \{x \in \mathbf{R}^d : x_1 = 0\}$;

(iii) $|\Psi(x)| + |\partial \Psi(x)| \leq K_0$ on $B_{r_0}(x_0)$ and $|\Psi^{-1}(y_1) - \Psi^{-1}(y_2)| \leq K_0 |y_1 - y_2|$ for all $y_1, y_2 \in D$;

(iv) for $x_1, x_2 \in B_{r_0}(x_0)$, we have $|\partial\Psi(x_1) - \partial\Psi(x_2)| \leq \omega_0(|x_1 - x_2|)$.

Denote $\rho(x) = \text{dist}(x, \partial G), x \in G$; $\rho(x) = 0$, if $x \notin G$. For $p > 1$, $\theta \in \mathbf{R}$, define $L_{p,\theta}(G) = L_p(G; \rho^{\theta-d}(x)\, dx)$ the space of all measurable functions $u(x)$ on G such that

$$\int_G |u(x)|^p \rho^{\theta-d}(x)\, dx < \infty.$$

Let $H^\gamma_{p,\theta}(G)$ be the space of all measurable function $u(x)$ on G such that $u, \rho\partial u, \ldots \rho^\gamma \partial^\gamma u \in L_{p,\theta}(G)$, where $\gamma = 0, 1, 2$.

Denote $\delta = \frac{\theta - d}{p}$, $|u|^p_{p,G} = \int_G |u(x)|^p\, dx$. We mostly are interested in the spaces $H^0_{p,\theta+p}(G), H^1_{p,\theta}(G), H^2_{p,\theta-p}(G)$ with the norms

$$|u|_{H^0_{p,\theta+p}(G)} = |u\rho^{1+\delta}|_{p,G},$$

$$|u|_{H^1_{p,\theta}(G)} = |u\rho^{\delta}|_{p,G} + |\partial u \rho^{1+\delta}|_{p,G},$$

$$|u|_{H^2_{p,\theta-p}(G)} = |u\rho^{-1+\delta}|_{p,G} + |\partial u \rho^{\delta}|_{p,G} + |\partial^2 u \rho^{1+\delta}|_{p,G}.$$

Let $T > 0$. We introduce the following spaces $H^\gamma_{p,\theta}(T, G), \gamma = 0, 1, 2$, of functions $u = u(t,x)$ with the norm

$$|u|_{H^\gamma_{p,\theta}(T,G)} = \left\{ \int_0^T |u(t,\cdot)|^p_{H^\gamma_{p,\theta}(G)}\, dt \right\}^{1/p}.$$

Let $\mathcal{H}^1_{p,\theta}(T, G)$ be the space of all functions $u \in H^2_{p,\theta-p}(T, G)$ such that

$$u(t,x) = \int_0^t f(s,x)\, ds, 0 \leq t \leq T,$$

in a generalized sense, where $f \in H^0_{p,\theta+p}(T, G)$, which means that for all $\varphi \in C_0^\infty(G)$,

$$\int u(t,x)\varphi(x)\, dx = \int_0^t \int f(s,x)\varphi(x)\, dx ds,\ t \leq T.$$

For $u \in \mathcal{H}^1_{p,\theta}(T,G)$, we define

$$|u|_{\mathcal{H}^1_{p,\theta}(T,G)} = |u|_{H^2_{p,\theta-p}(T,G)} + |f|_{H^0_{p,\theta+p}(T,G)}.$$

Let us introduce the operators

$$Lu(t,x) = a^{ij}(t,x)\partial^2_{ij}u(t,x) + b^i(t,x)\partial_i u(t,x) - r(t,x)u(t,x), (t,x) \in [0,T] \times G,$$

where $a = (a^{ij})_{1\le i,j\le d}$ is symmetric and nonnegative definite, and

$$Iu(t,x) = \int \nabla^2_y \tilde{u}(t,x)\pi(t,x,dy),\ (t,x) \in [0,T] \times G,$$

where

$$\tilde{u}(t,x) = \begin{cases} u(t,x),\ (t,x) \in [0,T] \times G, \\ 0, \quad 0 \le t \le T, x \notin G, \end{cases}$$

and

$$\nabla^2_y \tilde{u}(t,x) = \tilde{u}(t,x+y) - \tilde{u}(t,x) - \partial_i \tilde{u}(t,x) y_i \mathbf{1}_{\{|y|\le 1\}}.$$

The summation convention that repeated indices indicate summation from 1 to d is followed here as it will throughout.

We will need the following assumptions.

A1. The domain G is bounded of class C^1_u (see Ref. [5])

A2. There exist two positive numbers $0 < c < C$ such that

$$c|\xi|^2 \le a^{ij}\xi_i\xi_j \le C|\xi|^2$$

for all $\xi \in \mathbf{R}^d$, and

$$\lim_{|x-y|\to 0} \sup_{0\le t\le T} |a^{ij}(t,x) - a^{ij}(t,y)| = 0.$$

A3. The functions $b^i(t,x), r(t,x)$ are measurable and bounded;

A4. Let $p > 1,\ d-1 < \theta < d-1+p$.

The following two types of assumptions will be used for the measurable family of non-negative measures $\pi(t,x,dy)$ on $\mathbf{R}^d\backslash\{0\}$.

B. (The case of a dominating measure). There are measures $\bar{\pi}(t,dy), t \in [0,T]$ such that $\pi(t,x,dy) \le \bar{\pi}(t,dy), t \in [0,T], x \in G$, and

$$\sup_{0\le t\le T} \int |y|^2 \wedge 1\bar{\pi}(t,dy) < \infty,$$

$$\lim_{\varepsilon\to 0} \sup_{0\le t\le T} \int_{|y|\le\varepsilon} |y|^2\bar{\pi}(t,dy) = 0.$$

C. ("general case") There is a constant N such that

$$\int |y|^2 \wedge 1\ \pi(t,x,dy) \le N$$

and

$$\lim_{\varepsilon\to 0}\sup_{0\le t\le T, x\in G}\int_{|y|\le\varepsilon}|y|^2\pi(t,x,dy)=0.$$

We consider a linear equation

$$\begin{aligned}\partial_t u(t,x) &= Lu(t,x)+Iu(t,x)+h(t,x),(t,x)\in[0,T]\times G,\\ u(0,x)&=0.\end{aligned}\tag{2.1}$$

A function $u\in\mathcal{H}^1_{p,\theta}(T,G)$ is called a solution of (2.1), if for every $\varphi\in C_0^\infty(G)$ and all $t\in[0,T]$,

$$\int u(t,x)\varphi(x)\,dx=\int_0^t\int[Lu(s,x)\ +Iu(s,x)+h(s,x)]\varphi(x)\,dxds.$$

Our main results are the following two statements. The first one holds for the case of a dominating measure.

Theorem 2.1. *Assume the assumptions A1-A4 and B are satisfied. Then for each $h\in H^0_{p,\theta+p}(T,G)$ there is a unique solution $u\in\mathcal{H}^1_{p,\theta}(T,G)$ of (2.1). Moreover, there is a constant C independent of h such that*

$$|u|_{H^2_{p,\theta-p}(T,G)}\le C|h|_{H^0_{p,\theta+p}(T,G)}.$$

The next statement holds under the assumption C for $\pi(t,x,dy)$.

Theorem 2.2. *Assume $p>d,\theta\le p$ and the assumptions A1-A4 and C hold. Then for each $h\in H^0_{p,\theta+p}(T,G)$ there is a unique solution $u\in\mathcal{H}^1_{p,\theta}(T,G)$ of (2.1). Moreover, there is a constant C independent of h such that*

$$|u|_{H^2_{p,\theta-p}(T,G)}\le C|h|_{H^0_{p,\theta+p}(T,G)}.$$

3. Proof of the main results

Consider a partial differential equation

$$\begin{aligned}\partial_t u(t,x) &= Lu(t,x)+h(t,x),\ (t,x)\in[0,T]\times G,\\ u(0,x)&=0,x\in G.\end{aligned}\tag{3.1}$$

According to Theorem 2.10 in Ref. [5], the following statement holds.

Theorem 3.1. *Assume A1- A4 are satisfied. Then for each $h\in H^0_{p,p+\theta}(T,G)$ there is a unique solution $u\in\mathcal{H}^1_{p,\theta}(T,G)$ of (3.1). Moreover,*

$$|u|_{H^2_{p,\theta-p}(T,G)}\le Ne^{NT}|h|_{H^0_{p,\theta+p}(T,G)},$$

where N is independent of T and h.

Assume $V^t : H^2_{p,\theta-p}(G) \to H^0_{p,p+\theta}(G), t \geq 0$, is a family of linear operators such that the following assumption holds.

L. (i) For any $u \in H^2_{p,\theta-p}(T,G)$, the function $Vu(t,x) = V^t u(t,x)$ is a measurable function on $G_T = [0,T] \times G$

(ii) For each $\varepsilon > 0$ there is a constant μ_ε such that

$$|V^t u|_{H^0_{p,\theta+p}(T,G)} \leq \varepsilon |u|_{H^2_{p,\theta-p}(G)} + \mu_\varepsilon |u|_{H^1_{p,\theta}(G)},$$

for all $u \in H^2_{p,\theta-p}(G), t \geq 0$.

We will show that under the assumptions A1-A4 and L the equation

$$\begin{aligned} \partial_t u(t,x) &= Lu(t,x) + V^t u(t,x) + h(t,x),\ (t,x) \in [0,T] \times G, \\ u(0,x) &= 0, x \in G. \end{aligned} \tag{3.2}$$

has a unique solution $u \in u \in \mathcal{H}^1_{p,\theta}(T,G)$.

According to Lemma 2.6 in Ref. [5], there is a function $\psi \in C^\infty(G)$ such that

(i) ψ and $\partial\psi$ are continuous on $\bar{G}$, $|\partial\psi(x)| \geq 1$ on ∂G;

(ii) $\psi(x) > 0$ for any $x \in G$, and $\psi = 0$ on ∂G; for each $\varepsilon > 0$ the function ψ is bounded away from zero on the set $\{x \in G : \rho(x) \geq \varepsilon\}$;

(iii) for any integer $k \geq 0$ we have

$$\sup_G \rho^k(x)|\partial^{k+1}\psi(x)| < \infty. \tag{3.3}$$

(iv) there are two constants $0 < N_1 < N_2$ such that

$$N_1\rho(x) \leq \psi(x) \leq N_2\rho(x), x \in G. \tag{3.4}$$

Remark 3.2. a) Since ρ and ψ are comparable ((3.4) holds), we obtain equivalent norms of the spaces if the distance ρ is replaced by its regularized version ψ.

b) The inequality (3.4) implies that for each $\varepsilon > 0$ there is a constant C_ε such that

$$1 \leq \frac{\varepsilon}{\psi} + C_\varepsilon \psi. \tag{3.5}$$

We will need the following estimate of $H^1_{p,\theta}(G)$ norm.

Lemma 3.3. *For each $\varepsilon > 0$ there is a constant C_ε such that for all $v \in H^2_{p,\theta-p}(G)$*

$$|v|_{H^1_{p,\theta}(G)} \leq \varepsilon |v|_{H^2_{p,\theta-p}(G)} + C_\varepsilon |v\rho^{1+\delta}|_{p,G}.$$

Proof. Let $v \in H^2_{p,\theta-p}(G)$ and $u = v\psi^{1+\delta}$. We have

$$\partial_i u = \partial_i v \psi^{1+\delta} + (1+\delta) v\psi^\delta \partial_i \psi,$$

$$\partial^2_{ij} u = \partial^2_{ij} v\psi^{1+\delta} + (1+\delta)\partial_i v\psi^\delta \partial_j \psi + (1+\delta)\partial_j v\psi^\delta \partial_i \psi$$

$$+(1+\delta)\delta v\psi^{\delta-1}\partial_j\psi\partial_i\psi + (1+\delta)v\psi^\delta\partial^2_{ij}\psi.$$

For each $\varepsilon > 0$ there is a constant C_ε such that $|\partial u|_{p,G} \le \varepsilon|\partial^2 u|_{p,G} + C_\varepsilon|u|_{p,G}$. Therefore, using the properties of ψ, for each ε we can find a constant $C_\varepsilon > 0$ such that

$$|\partial_i u|_{p,G} \le \varepsilon \left(|\partial^2 v\psi^{1+\delta}|_{p,G} + |\partial_i v\psi^{\delta}|_{p,G} + |v\psi^{\delta-1}|_{p,G}\right)$$

$$+C_\varepsilon|v\psi^{1+\delta}|_{p,G}.$$

By (3.5), for each $\varepsilon > 0$ there is a constant C_ε such that

$$|v|_{H^1_{p,\theta}(G)} \le C(|v\psi^{\delta}|_{p,G} + |\partial v\psi^{1+\delta}|_{p,G}) \le C(|v\psi^{\delta}|_{p,G} + |\partial u|_{p,G})$$

$$\le \varepsilon|u|_{H^2_{p,\theta-p}(G)} + C_\varepsilon|u\psi^{1+\delta}|_{p,G},$$

and the statement follows. □

Now we prove the existence and uniqueness of the solution to (3.2).

Proposition 3.4. *Assume A1- A4 and L are satisfied. Then for each $h \in H^0_{p,p+\theta}(T,G)$ there is a unique solution $u \in \mathcal{H}^1_{p,\theta}(T,G)$ of (3.2). Moreover,*

$$|u|_{H^2_{p,\theta-p}(T,G)} \le C|h|_{H^0_{p,\theta+p}(T,G)},$$

where C is independent of h.

Proof. For $\tau \in [0.1]$ define the operators $A_\tau u = Lu + \tau Vu$. Assume $u \in \mathcal{H}^1_{p,\theta}(T,G)$ solves the equation

$$\begin{aligned} \partial_t u(t,x) &= A_\tau u(t,x) + h(t,x),\ (t,x) \in [0,T]\times G, \\ u(0,x) &= 0, x \in G, \end{aligned} \tag{3.6}$$

where $h \in H^0_{p,p+\theta}(T,G)$. We will prove that there is a constant independent of h, u and τ such that

$$|u|_{H^2_{p,\theta-p}(T,G)} \le C|h|_{H^0_{p,\theta+p}(T,G)}, \tag{3.7}$$

Then by Theorem 3.1 , there is a constant C independent of τ, h and u such that for all $t \le T$

$$|u|^p_{H^2_{p,\theta-p}(t,G)} \le C(|h|^p_{H^0_{p,\theta+p}(t,G)} + |Vu|^p_{H^0_{p,\theta+p}(t,G)}).$$

Therefore, by Lemma 3.3, there is a constant C such that for all $t \le T$,

$$|u|^p_{H^2_{p,\theta-p}(t,G)} \le C(|h|^p_{H^0_{p,\theta+p}(t,G)} + |\rho^{1+\delta}u|^p_{p,G_t}), \tag{3.8}$$

where $|f|^p_{p,G_t} = \int_0^t \int_G |f(s,x)|^p dxds$. On the other hand, multiplying both sides of the equation (3.6) by $\rho^{1+\delta}$ we have, by Lemma 3.3, (3.8) and assumption L,

$$\begin{aligned}
&|\rho^{1+\delta}u(t,\cdot)|^p_{p,G} \\
&\quad\leq C\int_0^t [\sum_{k=0}^2 |\rho^{1+\delta}\partial^k u(s,\cdot)|^p_{p,G} + |\rho^{1+\delta}V^s u(s,\cdot)|^p_{p,G} + |\rho^{1+\delta}h(s,\cdot)|^p_{p,G}]\ ds \\
&\quad\leq C\int_0^t [u(s,\cdot)|^p_{H^2_{p,\theta-p}(G)} + |\rho^{1+\delta}u(s,\cdot)|^p_{p,G} + |\rho^{1+\delta}h(s,\cdot)|^p_{p,G}]\ ds \\
&\quad\leq C\int_0^t [|\rho^{1+\delta}u(s,\cdot)|^p_{p,G} + |\rho^{1+\delta}h(s,\cdot)|^p_{p,G}]\ ds,
\end{aligned}$$

$t \leq T$. Therefore (3.7) follows by Gronwall's lemma and (3.8).

Consider the mappings $T_\tau : \mathcal{H}^1_{p,\theta}(T,G) \to H^0_{p,\theta+p}(T,G)$ defined by

$$u(t,x) = \int_0^t f(s,x)\ ds \mapsto f - A_\tau u,$$

where $f \in H^0_{p,\theta+p}(T,G)$. Obviously, there is a constant independent of τ such that

$$|T_\tau u|_{H^0_{p,\theta+p}(T,G)} \leq C|u|_{\mathcal{H}^1_{p,\theta}(T,G)}.$$

On the other hand, there is a constant C independent of τ such that for all $u \in \mathcal{H}^1_{p,\theta}(T,G)$,

$$|u|_{\mathcal{H}^1_{p,\theta}(T,G)} \leq C|T_\tau u|_{H^0_{p,\theta+p}(T,G)}.$$

Indeed, for $u(t,x) = \int_0^t f(s,x)\ ds$ ($f \in H^0_{p,\theta+p}(T,G), u \in \mathcal{H}^1_{p,\theta}(T,G)$) we have $\partial_t u = A_\tau u + (f - A_\tau u)$ and by (3.7),

$$\begin{aligned}
|u|_{\mathcal{H}^1_{p,\theta}(T,G)} &\leq C(|f - A_\tau u|_{H^0_{p,\theta+p}(T,G)} + |A_\tau u|_{H^0_{p,\theta+p}(T,G)}) \\
&\leq C(|f - A_\tau u|_{H^0_{p,\theta+p}(T,G)} + |u|_{H^2_{p,\theta-p}(T,G)}) \\
&\leq C|f - A_\tau u|_{H^0_{p,\theta+p}(T,G)} = C|T_\tau u|_{H^0_{p,\theta+p}(T,G)}.
\end{aligned}$$

Since by Theorem 3.1, T_0 is an onto map, all $T_\tau, \tau \in [0,1]$ are onto maps as well by Theorem 5.2 in Ref. [6]. □

We will show that under the assumption B or C,

$$I^t u(x) = Iu(t,x) = \int \nabla^2_y \tilde{u}(x)\pi(t,x,dy) \in H^0_{p,\theta+p}(G),$$

if $u \in H^2_{p,\theta-p}(G)$, and $I^t u, t \geq 0$, satisfies the assumption L.

3.1. *Proof of Theorem 2.1*

The following lemma shows that under B, the assumptions L for holds for Iu. Therefore Theorem 2.1 follows immediately by Proposition 3.4.

Lemma 3.5. *Assume B holds and $u \in H^2_{p,\theta-p}(G)$. Then for each $\varepsilon > 0$ there is C_ε such that for all $t \geq 0$*

$$|\rho^{1+\delta} I^t u|_{p,G} \leq \varepsilon |u|_{H^2_{p,\theta-p}(G)} + C_\varepsilon |u|_{H^1_{p,\theta}(G)}.$$

Proof. Assume $u \in H^2_{p,\theta-p}(G)$. Then

$$Iu(x) = \int_{\Gamma^1_\varepsilon} \ldots + \int_{\Gamma^2_\varepsilon} \ldots + \int_{|y|>\varepsilon} \ldots = (I_1 + I_2 + I_3)u(x),$$

where

$$\Gamma^1_\varepsilon = \{y \in \mathbf{R}^d : x + y \in G, |y| \leq \varepsilon\},$$

$$\Gamma^2_\varepsilon = \{y \in \mathbf{R}^d : x + y \notin G, |y| \leq \varepsilon\}.$$

1^0. We start with $I_1 u$ and split it into two integrals

$$I_1 u(x) = \int_{\Gamma^1_\varepsilon} \ldots = \int_{\Gamma^{11}_\varepsilon} \ldots + \int_{\Gamma^{12}_\varepsilon} \ldots = (I_{11} + I_{12})u(x),$$

where $\Gamma^{11}_\varepsilon = \Gamma^1_\varepsilon \cap \{y : |y| \leq \rho(x)\}$, $\Gamma^{12}_\varepsilon = \Gamma^1_\varepsilon \cap \{y : |y| > \rho(x)\}$.

For any $\tau \in (0,1), y \in \Gamma^{11}_\varepsilon$ we have $\rho(x + \tau y) \geq \rho(x) - \tau|y| \geq (1-\tau)\rho(x)$. Therefore

$$\frac{\rho(x)}{\rho(x+\tau y)} \leq (1-\tau)^{-1}.$$

If ε is sufficiently small,

$$\begin{aligned}\rho^{1+\delta}(x)|\nabla^2_y u(t,x)| &= \rho^{1+\delta}(x)\Big|\int_0^1 (1-\tau)\partial^2_{ij} u(x+\tau y) y_i y_j \, d\tau\Big| \\ &\leq N|y|^2 \int_0^1 (1-\tau)|(\rho^{1+\delta}\partial^2 u)(x+\tau y)| \left[\frac{\rho(x)}{\rho(x+\tau y)}\right]^{1+\delta} d\tau.\end{aligned}$$

By Hölder inequality,

$$\begin{aligned}
&|\rho^{1+\delta} I_{11} u(x)|_{p,G}^p \\
&\leq N \int_G \left\{ \int_{\Gamma_\varepsilon^{11}} \int_0^1 (1-\tau)^{-\delta} |(\rho^{1+\delta}\partial^2 u)(x+\tau y)| \, d\tau \, |y|^2 \bar{\pi}(t,dy) \right\}^p dx \\
&\leq N \int_G \int_{\Gamma_\varepsilon^{11}} \int_0^1 |(\rho^{1+\delta}\partial^2 u)(x+\tau y)|^p \, d\tau \, |y|^2 \bar{\pi}(t,dy) \\
&\times \left\{ \int_{\Gamma_\varepsilon^{11}} \int_0^1 (1-\tau)^{-\delta q} d\tau |y|^2 \bar{\pi}(t,dy) \right\}^{\frac{p}{q}} dx \\
&\leq N |\rho^{1+\delta}\partial^2 u|_{p,G}^p \left\{ \int_{|y|\leq\varepsilon} |y|^2 \bar{\pi}(t,dy) \right\}^{1+\frac{p}{q}} \left\{ \int_0^1 (1-\tau)^{-\delta q} d\tau \right\}^{\frac{p}{q}},
\end{aligned}$$

where $\delta q = \frac{\theta-d}{p}$ and $\frac{p}{p-1} = \frac{\theta-d}{p-1} < 1$, because $\theta < d-1+p$. Thus

$$|\rho^{1+\delta} I_{11} u|_{p,G} \leq N\gamma(\varepsilon) |\rho^{1+\delta}\partial^2 u|_{p,G} \leq N\gamma(\varepsilon) |u|_{H^2_{p,\theta-p}(G)},$$

where

$$\gamma(\varepsilon) = \sup_{t>0} \int_{|y|\leq\varepsilon} |y|^2 \bar{\pi}(t,dy) \left\{ \int_0^1 (1-\tau)^{-\delta q} d\tau \right\}^{\frac{p}{q}} \to 0,$$

as $\varepsilon \to 0$, by assumption B.

If $y \in \Gamma_\varepsilon^{12}$, we have $\rho(x+y) \leq \rho(x) + |y| \leq 2|y|$, and the estimate is straightforward:

$$\begin{aligned}
\rho^{1+\delta}(x)|u(x+y)| &\leq \rho^{1+\delta}(x)\rho^{1-\delta}(x+y)|(\rho^{\delta-1}u)(x+y)| \\
&\leq 2|y|^2 |(\rho^{\delta-1}u)(x+y)|,
\end{aligned}$$

$$\rho^{1+\delta}(x)|u(x)| \leq \rho^2(x)|(\rho^{\delta-1}u)(x)| \leq |y|^2 |(\rho^{\delta-1}u)(x)|,$$

$$\rho^{1+\delta}(x)|\partial u(x)||y| \leq |y|^2 |(\rho^{\delta}\partial u)(x)|,$$

and

$$|\rho^{1+\delta} I_{12} u(x)|_{p,G}^p \leq C |u|_{H^2_{p,\theta-p}(G)}^p \left(\int_{|y|\leq\varepsilon} |y|^2 \bar{\pi}(t,dy) \right)^p.$$

2^0. Consider now $I_2 u$. Obviously,

$$|I_2 u(x)| \leq |u(x)| \bar{\pi}(t,\Gamma_\varepsilon^2) + |\partial u(x)| \int_{\Gamma_\varepsilon^2} |y| \, \bar{\pi}(t,dy).$$

If $y \in \Gamma^2_\varepsilon$, then $\varepsilon \geq |y| > \rho(x)$ and

$$|\left(\rho^{1+\delta} I_2 u\right)(x)| \leq C\left(|(\rho^{-1+\delta} u)(x)| + |(\rho^\delta \partial u)(x)|\right) \sup_t \int_{|y|\leq \varepsilon} |y|^2 \bar{\pi}(t, dy).$$

Thus

$$|\rho^{1+\delta} I_2 u|_{p,G} \leq C\kappa(\varepsilon)|u|_{H^2_{p,\theta-p}(G)},$$

where $\kappa(\varepsilon) = \sup_t \int_{|y|\leq\varepsilon} |y|^2 \bar{\pi}(t, dy)$.

3^0. Finally, we handle $I_3 u$. It follows that

$$|I_3 u(x)| \leq \int_{|y|>\varepsilon} |u(x+y)|\bar{\pi}(t, dy) + |u(x)|\bar{\pi}(t, \{|y| > \varepsilon\})$$

$$+|\partial u(x)| \int_{1\geq|y|>\varepsilon} |y|\bar{\pi}(t, dy).$$

First we estimate

$$Ju(x) = \int_{|y|>\varepsilon} |u(x+y)|\bar{\pi}(t, dy).$$

For $\varepsilon_0 > 0$ and $z \in G$ we have

$$|u(z)| = |u(z)\rho^{-1+\delta}(z)|\rho^{1-\delta}(z)\mathbf{1}_{\{\rho(z)\leq\varepsilon_0\}} + |u(z)\rho^\delta(z)|\rho^{-\delta}(z)\mathbf{1}_{\{\rho(z)>\varepsilon_0\}}$$

$$\leq \varepsilon_0^{1-\delta}|u(z)\rho^{-1+\delta}(z)| + \varepsilon_0^{-\delta}|u(z)\rho^\delta(z)|.$$

Therefore,

$$Ju(x) \leq \int_{|y|>\varepsilon} |u(x+y)|\left(\varepsilon_0^{1-\delta}\rho^{-1+\delta}(x+y) + \varepsilon_0^{-\delta}\rho^\delta(x+y)\right)\bar{\pi}(t, dy).$$

By Minkowski's inequality,

$$|Ju|_{p,G} \leq \sup_{t>0} \int_{|y|>\varepsilon} \bar{\pi}(t, dy)\left(\varepsilon_0^{1-\delta}|u\rho^{-1+\delta}|_{p,G} + \varepsilon_0^{-\delta}|u\rho^\delta|_{p;G}\right)$$

$$\leq N\varepsilon^{-2}\left(\varepsilon_0^{1-\delta}|u\rho^{-1+\delta}|_{p,G} + \varepsilon_0^{-\delta}|u\rho^\delta|_{p;G}\right).$$

Since G is bounded,

$$|\rho^{1+\delta} Ju|_{p,G} \leq N|Ju|_{p,G}.$$

The remaining terms

$$J_2 u(x) = |u(x)|\bar{\pi}(t, \{|y| > \varepsilon\}, \; J_3 u(x) = |\partial u(x)| \int_{1\geq|y|>\varepsilon} |y|\bar{\pi}(t, dy)$$

are estimated in the following way:

$$\rho^{1+\delta}(x) J_2 u(x) \leq N\rho^\delta(x)|u(x)|\bar{\pi}(t, \{|y| > \varepsilon\}) \leq N\varepsilon^{-2}|(\rho^\delta u)(x)|,$$

$$\rho^{1+\delta}(x) J_3 u(x) \leq N\varepsilon^{-1}|(\rho^{1+\delta}\partial u)(x)|.$$

Therefore,

$$|\rho^{1+\delta} J_2 u|_{p,G} \leq N\varepsilon^{-2}|u|_{H^1_{p,\theta}(G)},$$

$$|\rho^{1+\delta} J_3 u|_{p;G} \leq N\varepsilon^{-1}|u|_{H^1_{p,\theta}(G)}.$$

By choosing in these estimates a small $\varepsilon > 0$ and then a small ε_0 we obtain the statement □

So, the statement of Theorem 2.1 follows by Proposition 3.4.

3.2. *Proof of Theorem 2.2*

We will use the maximal functions. So, the function u needs to be bounded. The following embedding theorem holds in weighted Sobolev spaces.

Proposition 3.6. *(see Proposition 2.2 in Ref. [7]) If $\gamma p > d$, then*

$$|\rho^{\frac{\theta}{p}} u|_{\infty,G} \leq N|u|_{H^\gamma_{p,\theta}(G)}.$$

In particular,

$$|\rho^{\frac{\theta}{p}-1} u|_{\infty,G} \leq N|u|_{H^2_{p,\theta-p}(G)},$$

and u is bounded on $\bar{G}$, if $p \geq \theta$.

Denote

$$T_y u(x) = \frac{|\nabla^2_y u(x)|}{|y|^2}, x \in G, y \in \mathbf{R}^d,$$

and define the maximal function

$$\mathcal{M}f(x) = \sup_{R>0} R^{-d} \int_{|z|\leq R} f(x+z)\, dz.$$

Lemma 3.7. *Let $u \in C_0^\infty(G), p > \frac{d}{2} \vee 1$. Then there is a constant N such that for all $x \in G$*

$$\sup_{|y|\leq \frac{1}{4}\rho(x)} |T_y u(x)|^p \leq N\rho^{-(1+\delta)p}(x)[\mathcal{M}\left(|\partial^2 u\rho^{1+\delta}|^p\right)(x) + \mathcal{M}\left(|\partial u\rho^{\delta}|^p\right)(x)$$

$$+\mathcal{M}\left(|u\rho^{-1+\delta}|^p\right)(x)].$$

Proof. According to [4] and [1], for all $x, y \in \mathbf{R}^d$,

$$(T_y u(x))^p \leq N\mathcal{M}\left(|\partial^2 u|^p\right)(x). \tag{3.9}$$

Fix $x \in G$. Let $\varphi \in C_0^\infty(\mathbf{R}^d)$, $\varphi(z) = 1$, if $|z| \leq \frac{1}{2}$, $\varphi(z) = 0$, if $|z| > 1$ and

$$\varphi_r(z) = \varphi(\frac{z-x}{r}).$$

Obviously, $T_y u(x) = T_y(u\varphi_r)(x)$, if $|y| \le \frac{1}{2}r$. By (3.9), we have for $|y| \le \frac{1}{2}r \le \frac{1}{4}\rho(x)$

$$(T_y u(x))^p = (T_y(u\varphi_r)(x))^p \le N\mathcal{M}\left(|\partial^2(u\varphi_r)|^p\right)(x)$$

$$\le N\mathcal{M}\left(|\partial^2 u\varphi_r|^p + |\partial u \cdot \partial\varphi_r|^p + |u\partial^2\varphi_r|^p\right)(x).$$

For $r = \frac{1}{2}\rho_G(x)$,

$$\mathcal{M}\left(|\partial^2 u\varphi_r|^p\right)(x) = \sup_{R>0} R^{-d}\int_{|z|\le R} |\partial^2 u(x+z)|^p\varphi_r^p(x+z)\,dz$$

$$\le \sup_{R>0} R^{-d}\int_{|z|\le R\wedge r} |(\partial^2 u\rho^{1+\delta})(x+z)|^p\rho^{-(1+\delta)p}(x+z)\,dz$$

$$\le N\rho^{-(1+\delta)p}(x)\mathcal{M}\left(|\partial^2 u\rho^{1+\delta}|^p\right)(x),$$

because $\rho(x+z) \ge \frac{1}{2}\rho(x)$, $|z| \le r = \frac{1}{2}\rho(x)$.

Similarly we get

$$\mathcal{M}\left(|\partial u \cdot \partial\varphi_r|^p\right)(x) \le Nr^{-p}\sup_{R>0}\int_{|z|\le R\wedge r} |(\partial u\rho^{\delta})(x+z)|^p\rho^{-\delta p}(x+z)\,dz$$

$$\le N\rho^{-(1+\delta)p}(x)\mathcal{M}\left(|\partial u\rho^{\delta}|^p\right)(x),$$

and

$$\mathcal{M}\left(|u\partial^2\varphi_r|^p\right)(x) \le Nr^{-2p}\sup_{R>0} R^{-d}\int_{|z|\le R\wedge r} |(u\rho^{-1+\delta})(x+z)|^p\rho^{(1-\delta)p}(x+z)\,dz$$

$$\le N\rho^{-(1+\delta)p}(x)\mathcal{M}\left(|u\rho^{-1+\delta}|^p\right)(x).$$

□

Using the properties of maximal functions (see Ref. [8]) we obtain the following obvious statement.

Corollary 3.8. *If $u \in H^2_{p,\theta-p}(G), p > \frac{d}{2} \vee 1$, then*

$$|\rho^{1+\delta}Tu|_{p,G} \le N|u|_{H^2_{p,\theta-p}(G)},$$

where

$$Tu(x) = \sup_{|y|\le\frac{1}{4}\rho(x)} T_y u(x).$$

Proof. For each $p > \frac{d}{2} \vee 1$, there are $\bar{p} > \frac{d}{2} \vee 1$ and $\varepsilon > 0$ such that $p = \bar{p}(1+\varepsilon)$. By Lemma 3.7,

$$\rho^{(1+\delta)\bar{p}}(x)\sup_{|y|\le\frac{1}{4}\rho(x)}|T_y u(x)|^{\bar{p}} \le N[\mathcal{M}\left(|\partial^2 u\rho^{1+\delta}|^{\bar{p}}\right)(x) + \mathcal{M}\left(|\partial u\rho^{\delta}|^{\bar{p}}\right)(x)$$

$$+\mathcal{M}\left(|u\rho^{-1+\delta}|^{\bar{p}}\right)(x)].$$

Therefore

$$\int_G (\rho^{(1+\delta)\bar{p}}(x) \sup_{|y|\le \frac{1}{4}\rho(x)} |T_y u(x)|^{\bar{p}})^{1+\varepsilon} dx$$

$$\le C(\int_G |\partial^2 u \rho^{1+\delta}|^{\bar{p}(1+\varepsilon)} dx + \int_G |\partial u \rho^{\delta}|^{\bar{p}(1+e)}\, dx + \int_G |u\rho^{-1+\delta}|^{\bar{p}(1+\varepsilon)} dx).$$

□

In order to prove that the assumption L holds for $I^t u$, the following estimate is needed as well.

Lemma 3.9. *Assume $\theta \le p, r > 0, u \in C_0^\infty(G)$. Then there is a constant N such that for all $x \in G$*

$$\sup_{\substack{|y|>r,\\ \rho(x+y)>0}} \frac{|u(x+y)|^p}{|y|^{2p}} \le N r^{-(1+\delta)p} [\mathcal{M}\left(|\partial^2 u \rho^{1+\delta}|^p\right)(x)$$

$$+\mathcal{M}\left(|\partial u \rho^{\delta}|^p\right)(x) + \mathcal{M}\left(|u\rho^{-1+\delta}|^p\right)(x)].$$

Proof. Fix $x \in G$ and denote $U(y) = u(x+y)$. Then

$$\sup_{\substack{|y|>r,\\ \rho(x+y)>0}} \frac{|u(x+y)|}{|y|^2} = \sup_{t>r} \frac{1}{t^2} \sup_{\substack{y\in\Gamma_t\\ |y|=1}} |U(ty)|, \tag{3.10}$$

where

$$\Gamma_t = \{y : \rho(x+ty) > 0\}.$$

For $y \in \Gamma_t$, let $\rho_t(y) = dist(y, \partial\Gamma_t)$. Notice that

$$\begin{aligned} dist(y, \partial\Gamma_t) &= \inf_{z,\rho(x+tz)=0} |y-z| = \frac{1}{t} \inf_{z,\rho(x+tz)=0} |ty-tz| \\ &= \frac{1}{t} \inf_{z,\rho(x+tz)=0} |(x+ty)-(x+tz)| \\ &= \frac{1}{t}\rho(x+ty). \end{aligned}$$

Let $\varphi \in C_0^\infty(\mathbf{R}^d), \varphi \geq 0, \varphi(y) = 1$, if $|y| \leq 1$, and $\varphi(y) = 0$, if $|y| > 2$. By Proposition 3.6 for $U(ty)\varphi(y), y \in \Gamma_t$, we have

$$
\begin{aligned}
|U(t\cdot)|_{\infty;B_1\cap\Gamma_t} &\leq |U(t\cdot)\varphi|_{\infty,\Gamma_t} \leq N|U(t\cdot)\varphi|_{H^2_{p,\theta-p}(\Gamma_t)} \\
&\leq N[|\partial^2(U(t\cdot)\varphi)\rho_t^{1+\delta}|_{p,\Gamma_t} + |\partial(U(t\cdot)\varphi)\rho_t^{\delta}|_{p,\Gamma_t} \\
&\quad + |U(t\cdot)\varphi\rho_t^{-1+\delta}|_{p,\Gamma_t}] \\
&\leq N[|\partial^2 U(t\cdot)\rho_t^{1+\delta}|_{p,B_2\cap\Gamma_t} + |\partial U(t\cdot)\rho_t^{\delta}|_{p,B_2\cap\Gamma_t} \\
&\quad + |U(t\cdot)\rho_t^{-1+\delta}|_{p,B_2\cap\Gamma_t}],
\end{aligned}
$$

where $B_R = \{y : |y| < R\}$. Using the change of variable $ty = z$, we obtain

$$
\begin{aligned}
|\partial^2 U(t\cdot)\rho_t^{1+\delta}|^p_{p,B_2\cap\Gamma_t} &= \int_{B_2\cap\Gamma_t} t^{2p}|\partial^2 u(x+ty)|^p t^{-(1+\delta)p}\rho^{(1+\delta)p}(x+ty)\, dy \\
&= t^{(1-\delta)p}t^{-d}\int_{B_{2t}\cap G} |\partial^2 u(x+z)|^p\rho^{(1+\delta)p}(x+z)\, dz \\
&\leq Nt^{(1-\delta)p}\mathcal{M}\left(|\partial^2 u\rho^{1+\delta}|^p\right)(x),
\end{aligned}
$$

and

$$
\begin{aligned}
|\partial U(t\cdot)\rho_t^{\delta}|^p_{p,B_2\cap\Gamma_t} &= \int_{B_2\cap\Gamma_t} t^p|\partial u(x+ty)|^p t^{-\delta p}\rho^{\delta p}(x+ty)\, dy \\
&= t^{(1-\delta)p}t^{-d}\int_{B_{2t}\cap G} |\partial u(x+z)|^p\rho^{\delta p}(x+z)\, dz \\
&\leq Nt^{(1-\delta)p}\mathcal{M}\left(|\partial u\rho^{\delta}|^p\right)(x).
\end{aligned}
$$

Also,

$$
\begin{aligned}
|U(t\cdot)\rho_t^{-1+\delta}|^p_{p,B_2\cap\Gamma_t} &= \int_{B_2\cap\Gamma_t} |u(x+ty)|^p t^{(1-\delta)p}\rho^{(-1+\delta)p}(x+ty)\, dy \\
&= t^{(1-\delta)p}t^{-d}\int_{B_{2t}\cap G} |u(x+z)|^p\rho^{(-1+\delta)p}(x+z)\, dz \\
&\leq Nt^{(1-\delta)p}\mathcal{M}\left(|u\rho^{-1+\delta}|^p\right)(x).
\end{aligned}
$$

These estimates and (3.10) imply that

$$\sup_{\substack{|y|>r,\\ \rho(x+y)>0}} \frac{|u(x+y)|^p}{|y|^{2p}} = \sup_{t>r} \frac{1}{t^{2p}} |U(t\cdot)|^p_{\infty;S_1\cap\Gamma_t}$$

$$\leq \sup_{t>r} \frac{1}{t^{2p}} N t^{(1-\delta)p} (\mathcal{M}... + \mathcal{M}... + \mathcal{M}...)$$

$$\leq N r^{-(1+\delta)p} (\mathcal{M}... + \mathcal{M}... + \mathcal{M}...)$$

and the statement follows. □

The Lemma below completes the proof of Theorem 2.2. It shows that $I^t u$ satisfies the assumption L.

Lemma 3.10. *Assume $u \in H^2_{p,\theta-p}(G), p > d, \theta \leq p$ and the assumption C holds. Then for each $\varepsilon > 0$ there is a constant $C_\varepsilon > 0$ such that*

$$|\rho^{1+\delta} I u|_{p,G} \leq \varepsilon |u|_{H^2_{p,\theta-p}(G)} + C_\varepsilon |u|_{H^1_{p,\theta}(G)}.$$

Proof. Fix $\varepsilon > 0$ and split the integral

$$Iu(x) = \int_{|y|\leq\varepsilon\wedge\frac{1}{4}\rho(x)} \nabla^2_y u(x)\pi(t,x,dy) + \int_{\varepsilon\geq|y|>\varepsilon\wedge\frac{1}{4}\rho(x)} ... + \int_{|y|>\varepsilon} ...$$

$$= I_1 u(x) + I_2 u(x) + I_3 u(x).$$

1^0. Estimate of $I_1 u$. Obviously,

$$|I_1 u(x)| \leq \gamma(\varepsilon) \sup_{|y|\leq\varepsilon\wedge\frac{1}{4}\rho(x)} \frac{|\nabla^2_y u(x)|}{|y|^2},$$

where $\gamma(\varepsilon) = \sup_{t>0, x\in G} \int_{|y|\leq\varepsilon} |y|^2\ \pi(t,x,dy)$. By Lemma 3.7,

$$|\rho^{1+\delta}(x) I_1 u(x)|^p \leq (N\gamma(\varepsilon))^p \left[\mathcal{M}\left(|\partial^2 u \rho^{1+\delta}|^p\right)(x) + \mathcal{M}\left(|\partial u \rho^{\delta}|^p\right)(x)\right.$$

$$\left. + \mathcal{M}\left(|u\rho^{-1+\delta}|^p\right)(x)\right]$$

for $p > \frac{d}{2} \vee 1$. Therefore using properties of the maximal functions (see Ref. [8] and the proof of Corollary 3.8 above)

$$|\rho^{1+\delta} I_1 u|_{p;G} \leq N\gamma(\varepsilon) |u|_{H^2_{p,\theta-p}(G)}.$$

2^0. Estimate of $I_2 u$. We have

$$|I_2 u(x)| \leq \int_{\Gamma_\varepsilon(x)} |\tilde{u}(x+y)| \pi(t,x,dy)| + |u(x)|\pi(t,x,\Gamma_\varepsilon(x))$$

$$+ |\partial u(x)| \int_{\Gamma_\varepsilon(x)} |y| \pi(t,x,dy)$$

$$= I_{21} u(x) + I_{22} u(x) + I_{23} u(x),$$

where $\Gamma_\varepsilon(x) = \{y : \varepsilon \geq |y| > \frac{1}{4}\rho(x)\}$.

We have

$$I_{21}u(x) \leq \sup_{\substack{y\in\Gamma_\varepsilon(x),\\ \rho(x+y)>0}} \frac{|u(x+y)|}{|y|^2} \int_{|y|\leq\varepsilon} |y|^2\pi(t,x,dy)$$

$$\leq N\gamma(\varepsilon) \sup_{\substack{|y|>\frac{1}{4}\rho(x),\\ \rho(x+y)>0}} \frac{|u(x+y)|}{|y|^2}.$$

Therefore by Lemma 3.9,

$$|\rho^{1+\delta}(x)I_{21}u(x)|^p \leq (N\gamma(\varepsilon))^p \left[\mathcal{M}\left(|\partial^2 u\rho^{1+\delta}|^p\right)(x) + \mathcal{M}\left(|\partial u\rho^{\delta}|^p\right)(x)\right.$$

$$\left.+\mathcal{M}\left(|u\rho^{-1+\delta}|^p\right)(x)\right],$$

and by properties of maximal functions (see Ref. [8] and the proof of Corollary 3.8 above))

$$|\rho^{1+\delta}I_{21}u|_{p;G} \leq N\gamma(\varepsilon)|u|_{H^2_{p,\theta-p}(G)}.$$

Obviously,

$$|I_{22}u(x)| + |I_{23}u(x)| \leq N\gamma(\varepsilon)\left(\rho^{-2}(x)|u(x)| + \rho^{-1}(x)|\partial u(x)|\right)$$

and therefore

$$|\rho^{1+\delta}I_{22}u|_{p,G} + |\rho^{1+\delta}I_{23}u|_{p,G} \leq N\gamma(\varepsilon)\left(|\rho^{-1+\delta}u|_{p,G} + |\rho^{\delta}\partial u|_{p,G}\right)$$

$$\leq N\gamma(\varepsilon)|u|_{H^2_{p,\theta-p}(G)}.$$

Thus

$$|\rho^{1+\delta}I_2u|_{p;G} \leq N\gamma(\varepsilon)|u|_{H^2_{o,\theta-p}(G)}.$$

3^0. Estimate of I_3u. We have

$$|I_3u(x)| \leq \int_{|y|>\varepsilon} |u(x+y)|\pi(t,x,dy) + |u(x)|\pi(t,x,\{|y|>\varepsilon\})$$

$$+|\partial u(x)|\int_{1\geq|y|>\varepsilon} |y|\pi(t,x,dy) \tag{3.11}$$

$$\leq N(\varepsilon^{-2}|u|_{\infty;G} + \varepsilon^{-1}|\partial u(x)|).$$

For $\varepsilon_0 > 0$ and $z \in G$,

$$|u(z)| = |u(z)\rho^{\frac{\theta}{p}-1}(z)|\rho^{1-\frac{\theta}{p}}(z)\mathbf{1}_{\{\rho(z)\leq\varepsilon_0\}}$$

$$+|u(z)\rho^{\frac{\theta}{p}}(z)|\rho^{-\frac{\theta}{p}}(z)\mathbf{1}_{\{\rho(z)>\varepsilon_0\}} \tag{3.12}$$

$$\leq \varepsilon_0^{1-\frac{\theta}{p}}|u\rho^{\frac{\theta}{p}-1}|_{\infty;G} + \varepsilon_0^{-\frac{\theta}{p}}|u\rho^{\frac{\theta}{p}}|_{\infty;G}.$$

It follows from (3.11) and (3.12) that

$$|\rho^{1+\delta}(x)I_3u(x)| \leq N\varepsilon^{-2}\varepsilon_0^{1-\frac{\theta}{p}}|u\rho^{\frac{\theta}{p}-1}|_{\infty;G}\rho^{1+\delta}(x)$$

$$+N\varepsilon^{-2}\varepsilon_0^{-\frac{\theta}{p}}|u\rho^{\frac{\theta}{p}}|_{\infty;G}\rho^{1+\delta}(x)$$

$$+N\varepsilon^{-1}|\partial u(x)|\rho^{1+\delta}(x).$$

By embedding theorem (we use $p > d$ when $\gamma = 1$),

$$|\rho^{1+\delta}I_3u|_{p;G} \leq N\varepsilon^{-2}\varepsilon_0^{1-\frac{\theta}{p}}|u|_{H^2_{p,\theta-p}(G)} + N\varepsilon^{-2}\varepsilon_0^{-\frac{\theta}{p}}|u|_{H^1_{p,\theta}(G)}$$

$$+N\varepsilon^{-1}|u|_{H^1_{p,\theta}(G)}.$$

Choosing a small $\varepsilon > 0$ and then a small ε_0 we obtain our statement. □

Acknowledgement

We are very grateful to our referee for valuable comments.

References

[1] J.-M. Bony, Problème de Dirichlet et semi-groupe fortement fellérien associés à un opérateur intégro-differentiel, *C. R. Acad. Sci. Paris Sér. A-B.* **265**, A361–A364, (1967).

[2] F. Gimbert and P.-L. Lions, Existence and regularity results for solutions of second-order, elliptic integro-differential operators, *Ricerche Mat.* **33**(2), 315–358, (1984).

[3] M. G. Garroni and J.-L. Menaldi, *Green functions for second order parabolic integro-differential problems.* vol. 275, *Pitman Research Notes in Mathematics Series*, (Longman Scientific & Technical, Harlow, 1992). ISBN 0-582-02156-1.

[4] R. Mikulevičius and H. Pragarauskas, On the Cauchy problem for certain integro-differential operators in Sobolev and Hölder spaces, *Liet. Mat. Rink.* **32**(2), 299–331, (1992). ISSN 0132-2818.

[5] K.-H. Kim and N. V. Krylov, On the Sobolev space theory of parabolic and elliptic equations in C^1 domains, *SIAM J. Math. Anal.* **36**(2), 618–642 (electronic), (2004). ISSN 0036-1410.

[6] D. Gilbarg and N. S. Trudinger, *Elliptic partial differential equations of second order.* vol. 224, *Grundlehren der Mathematischen Wissenschaften [Fundamental Principles of Mathematical Sciences]*, (Springer-Verlag, Berlin, 1983), second edition. ISBN 3-540-13025-X.

[7] S. V. Lototsky, Dirichlet problem for stochastic parabolic equations in smooth domains, *Stochastics Stochastics Rep.* **68**(1-2), 145–175, (1999). ISSN 1045-1129.

[8] E. M. Stein, *Singular integrals and differentiability properties of functions.* Princeton Mathematical Series, No. 30, (Princeton University Press, Princeton, N.J., 1970).

Chapter 15

Strict Solutions of Kolmogorov Equations in Hilbert Spaces and Applications

Giuseppe Da Prato

Scuola Normale Superiore
56126, Pisa, Italy
g.daprato@sns.it

Consider a stochastic differential equation in a Hilbert space H, with the associated differential operator K_0 determined by the coefficients in the equation, and the corresponding Markov semigroup P_t. The paper investigates the problem of constructing a core for the (weak) generator K of P_t in the space of real, uniformly continuous, and bounded functions on H, and studies the relation between K and K_0. The paper proposes an approach based on the concept of a strict solution of the corresponding Kolmogorov equation: a suitable assumption of existence of strict solutions imply good results on the core, and the assumption is verified in an SPDE example.

Contents

1. Introduction

We are concerned with a stochastic differential equation in a separable Hilbert space H (norm $|\cdot|$, inner product $\langle\cdot,\cdot\rangle$) of the following form

$$\begin{cases} dX(t) = (AX(t) + b(X(t))dt + \sigma(X(t))dW(t), \quad t \geq 0, \\ \\ X(0) = x \in H, \end{cases} \tag{1.1}$$

where $A\colon D(A) \subset H \to H$ is the infinitesimal generator of a strongly continuous semigroup e^{tA} in H and the mappings $b\colon D(b) \subset H \to H$ and $\sigma : H \to L(H)$ are (generally) nonlinear. Moreover, $W(t)$ is a cylindrical Wiener process in H.

We shall assume that problem (1.1) has a unique mild solution $X(t,x)$ (a concept that has to be made precise in any specific situation) which is adapted and continuous in mean square. We shall denote by P_t the transition semigroup

$$P_t\varphi(x) = \mathbb{E}[\varphi(X(t,x))], \quad \varphi \in C_b(H),\ x \in H, \tag{1.2}$$

where $C_b(H)$ denotes the Banach space of all real uniformly continuous and bounded functions in H endowed with the norm

$$\|\varphi\|_0 = \sup_{x\in H} |\varphi(x)|, \quad \varphi \in C_b(H).$$

The Kolmogorov equation corresponding to (1.1) looks like

$$\begin{cases} u_t(t,x) = \dfrac{1}{2}\text{Tr}\ [(\sigma\sigma^*)(x)u_{xx}(t,x)] + \langle Ax + b(x), u_x(t,x)\rangle, \\ \qquad\qquad \text{for } x \in D(A) \cap D(b), \\ u(0,x) \ = \varphi(x), \quad x \in H. \end{cases} \tag{1.3}$$

As well known, a candidate for the solution of (1.3) is given by

$$u(t,x) = \mathbb{E}[\varphi(X(t,x))], \quad t \geq 0,\ x \in H.$$

Let us give some notations. We shall denote by $C_b^1(H)$ the space of all real functions in H which are continuous together with their Fréchet derivatives of the first order. It is a Banach space with the norm $\|\varphi\|_1 = \|\varphi\|_0 + \|\varphi_x\|_0$. Spaces $C_b^k(H), k > 1$, are defined similarly. Moreover we denote by $\{e_k\}$ a complete orthonormal system in H.

Definition 1.1. A *strict* solution of (1.3) is a function $u\colon [0,\infty)\times H \to \mathbb{R}$ such that

(i) u is continuous in $[0,\infty) \times H$ and $u(0,x) = \varphi(x),\ x \in H$.
(ii) For all $t \geq 0$, $u(t,\cdot) \in C_b^1(H)$ and there exist the second derivatives of u, $u_{x,x}(t,\cdot)(h,k)$ in all directions $h,k \in H$.
(iii) For all $t > 0$ the function

$$H \to \mathbb{R},\ x \mapsto \text{Tr}\ [(\sigma\sigma^*)(x)u_{xx}(t,x)] = \sum_{k=1}^{\infty} u_{xx}(t,x)(\sigma(x)e_k, \sigma(x)e_k)$$

belongs to $C_b(H)$.
(iv) For all $x \in H$, $u(\cdot,x) \in C^1((0,+\infty))$.
(v) For all $(t,x) \in (0,+\infty) \times D(A) \cap D(b)$, equation (1.3) is fulfilled.

We are also interested in the elliptic equation

$$\lambda\varphi - \frac{1}{2}\text{Tr}\ [(\sigma\sigma^*)(x)\varphi_{xx}(t,x)] - \langle Ax + b(x), \varphi_x(t,x)\rangle = f(x) \tag{1.4}$$

where $\lambda > 0$ and $f \in C_b(H)$ are given. The following definition is the elliptic counterpart of Definition 1.1.

Definition 1.2. A *strict* solution of (1.4) is a function $\varphi : H \to \mathbb{R}$ such that

(i) $\varphi \in C_b^1(H)$ and there exist the second derivatives of φ, $\varphi_{x,x}(x)(h,k)$ in all directions $h,k \in H$.
(ii) For all $x \in D(A) \cap D(b)$, equation (1.4) is fulfilled.

For this equation a candidate solution is given, as well known, by

$$\varphi(x) = \int_0^{+\infty} e^{-\lambda t} P_t f(x) dt, \quad x \in H. \tag{1.5}$$

Finally, we define the Kolmogorov operator K_0 setting

$$\begin{cases} K_0\varphi(x) = \frac{1}{2}\mathrm{Tr}\,[(\sigma\sigma^*)(x)\varphi_{xx}(x)] + \langle Ax + b(x), \varphi_x(x)\rangle, \quad \varphi \in D(K_0), \\ D(K_0) \;= \left\{\varphi \in C_b^1(H) : \textit{ conditions (i) of Definition } 1.2 \textit{ holds}\right\}. \end{cases} \tag{1.6}$$

It is well known that the semigroup P_t is not strongly continuous in $C_b(H)$ in general (in fact in all interesting situations). However, we can define its infinitesimal generator by proceeding as in Ref. [1]. For any $\lambda > 0$ and any $f \in C_b(H)$ we set

$$F_\lambda(f)(x) = \int_0^{\infty} e^{-\lambda t} P_t f(x) dt, \quad x \in H. \tag{1.7}$$

Then the following result holds.

Proposition 1.3. *For any $f \in C_b(H)$ and any $\lambda > 0$ we have $F_\lambda(f) \in C_b(H)$ and*

$$\|F_\lambda(f)\|_0 \le \frac{1}{\lambda}\, \|f\|_0. \tag{1.8}$$

Moreover there exists a unique closed operator $K\colon D(K) \subset C_b(H) \to C_b(H)$ such that for any $\lambda > 0$ and any $f \in C_b(H)$ we have $F_\lambda(f) = (\lambda - K)^{-1} f$.

Proof. Let first take f in $C_b^1(H)$. Then for all $x, y \in H$ we have

$$|F_\lambda(f)(x) - F_\lambda(f)(y)| \le \int_0^{\infty} e^{-\lambda t}\, \mathbb{E}(|f(X(t,x)) - f(X(t,y))|) dt$$

$$\le \|f\|_1 \int_0^{\infty} e^{-\lambda t}\, \mathbb{E}|X(t,x) - X(t,y)| dt.$$

Since $X(t,x)$ is mean square continuous, we see that $F_\lambda(f) \in C_b(H)$. Moreover, it is obvious that (1.8) holds. Since $C_b^1(H)$ is dense in $C_b(H)$ we can conclude, by a straightforward argument, that $F_\lambda(f) \in C_b(H)$ for all $f \in C_b(H)$ and that (1.8) holds.

Now, by a direct computation, we see that F_λ fulfills the resolvent identity

$$F_\lambda - F_\mu = (\mu - \lambda) F_\lambda F_\mu, \quad \lambda, \mu > 0.$$

Since for every $f \in C_b(H)$

$$\lim_{\lambda\to\infty} \lambda F_\lambda(f)(x) = \lim_{\lambda\to\infty} \int_0^{+\infty} e^{-\tau} P_{\frac{\tau}{\lambda}} f(x) d\tau = f(x), \quad x \in H,$$

F is one-to-one. So, by a classical result, see e.g. Ref. [2], there exists a unique closed operator $K: D(K) \subset C_b(H) \to C_b(H)$ such that for any $\lambda > 0$ and any $f \in C_b(H)$ we have $F_\lambda(f) = (\lambda - K)^{-1} f$. □

The goal of this paper is to study strict solutions both of equations (1.3) and (1.4) and to investigate the relationship between the "abstract" generator K and the "concrete" differential operator K_0 defined by (1.6).

We shall consider situations where K is an extension of K_0 (see Remark 2.4 below) and K_0 determines K, more precisely that $D(K_0)$ is a *core* for K (in a sense to be made precise). This fact is important in order to prove several properties of the operator K (starting from the analogous ones for the more accessible operator K_0), as we shall explain in section 3 below.

A natural idea to find strict solutions of equation (1.4) when $f \in C_b^3(H)$ is to check, by successive differentiations, that the candidate function φ given by (1.5) fulfills (1.4). This method works easily provided b and σ are of class C^3, since in this case $X(t,\cdot)$ is of class C^2 for any $t \geq 0$, see Ref. [3]. But in all interesting applications b and σ are not even of class C^1 so that, proving that u is a strict solution, can be much more involved. In section 4 we shall present an application to a very simple but non trivial stochastic partial differential equation. We plan to take in consideration more interesting stochastic PDEs in successive papers. We quote in this direction the paper [4] where this method was used for the $3D$-stochastic Navier-Stokes. However, in that case a much weaker notion (than Definition 1.1) of strict solution was used.

We end this section by giving a useful characterization of $D(K)$ similar to that proved in Ref. [5].

Proposition 1.4. *Let $\varphi \in D(K)$. Then we have*

$$\lim_{h\to 0^+} \frac{1}{h}\,(P_h\varphi(x) - \varphi(x)) = K\varphi(x), \quad \text{for all } x \in H \tag{1.9}$$

and

$$\sup_{h\in(0,1]} \left\| \frac{1}{h}\,(P_h\varphi - \varphi) \right\|_0 < +\infty. \tag{1.10}$$

Conversely, if there exists $\varphi, g \in C_b(H)$ such that

$$\lim_{h\to 0^+} \frac{1}{h}\,(P_h\varphi(x) - \varphi(x)) = g(x), \quad \text{for all } x \in H \tag{1.11}$$

and (1.10) holds we have $\varphi \in D(K)$ and $K\varphi = g$.

Proof. Assume that $\varphi \in D(K)$. Fix $\lambda > 0$ and set $f = \lambda\varphi - K\varphi$. Then $F_\lambda(f) = \varphi$ and for any $h > 0$ and any $x \in H$ we have

$$P_h\varphi(x) = P_hF_\lambda(f)(x) = e^{\lambda h}\int_h^{+\infty} e^{-\lambda s}P_sf(x)ds.$$

It follows that

$$D_hP_h\varphi(x)|_{h=0} = \lambda\int_0^{+\infty} e^{-\lambda s}P_sf(x)ds - f(x)$$

$$= \lambda F_\lambda(f)(x) - f(x) = KF_\lambda(f)(x) = K\varphi(x),$$

and so (1.9) follows. Moreover, since

$$|P_h\varphi(x) - \varphi(x)| \leq (e^{\lambda h} - 1)\left|\int_h^{+\infty} e^{-\lambda s}P_sf(x)ds\right|$$

$$+\left|\int_0^h e^{-\lambda s}P_sf(x)ds\right| \leq \|f\|_0\left[\frac{e^{\lambda h}-1}{\lambda}\,e^{-\lambda h} + \frac{1-e^{-\lambda h}}{\lambda}\right] \leq ch,$$

where c is a suitable positive constant, we see that (1.10) follows as well.

Assume now that there exists $\varphi, g \in C_b(H)$ such that (1.11) is fulfilled. Since clearly for any $x \in H$

$$\frac{d}{dt}\,P_t\varphi(x) = \lim_{h\to 0}\frac{1}{h}\,(P_{t+h}\varphi(x) - P_t\varphi(x)) = P_tg(x),$$

we have

$$F_\lambda(\varphi)(x) = -\frac{1}{\lambda}\int_0^t P_t\varphi(x)de^{-\lambda t}$$

$$= \frac{1}{\lambda}\,\varphi(x) + \frac{1}{\lambda}\int_0^t e^{-\lambda t}P_tg(x)dt$$

$$= \frac{1}{\lambda}\,\varphi(x) + \frac{1}{\lambda}\,F_\lambda(g)(x).$$

Therefore

$$F_\lambda(\varphi)(x) = \frac{1}{\lambda}\,\varphi(x) + \frac{1}{\lambda}\,F_\lambda(g)(x),$$

which implies $\varphi \in D(K)$ and $K\varphi = g$. □

2. Existence of cores

When the dimension of H is infinite it is well known that $C_b^2(H)$ is not dense in $C_b(H)$, see Ref. [6]. For this reason we introduce a notion of convergence weaker than that of $C_b(H)$, see Ref. [5]. We say that a sequence $\{\varphi_n\} \subset C_b(H)$ is π-*convergent* to $\varphi \in C_b(H)$ if $\lim_{n\to\infty} \varphi_n(x) = \varphi(x)$ for all $x \in H$ and there exists $c > 0$ such that $|\varphi_n(x)| \le c$ for all $n \in \mathbb{N}$ and $x \in H$.

Proposition 2.1. *Let $\{f_n\} \subset C_b(H)$ be π-convergent to a function f in $C_b(H)$. Then $\{F_\lambda(f_n)\}$ is π-convergent to f.*

Proof. The conclusion follows from the mean square continuity of $X(t,x)$ and the dominated convergence theorem. □

In the next definition we follow Ref. [5].

Definition 2.2.

(i) A subspace Z of $C_b(H)$ is said to be π-*dense* in $C_b(H)$ if for any $\varphi \in C_b(H)$ there exists a sequence $\{\varphi_n\} \subset Z$ which is π–convergent to φ.
(ii) A subspace Y of $D(K)$ is a π-*core* for K if for any $\varphi \in D(K)$ there exists a sequence $\{\varphi_n\} \subset Y$ which is π–convergent to φ and such that the sequence $\{K\varphi_n\}$ is π–convergent to $K\varphi$.

It is easy to see that for any $k \in \mathbb{N}$, $C_b^k(H)$ is π-dense in $C_b(H)$, see e.g. Refs. [5] and [3].

We notice that there are many π-cores for K. It is enough to choose a π-dense linear subspace $\mathcal{R}(H)$ in $C_b(H)$ and set

$$\Gamma := \bigcup_{\lambda>0} (\lambda - K)^{-1}\mathcal{R}(H). \tag{2.1}$$

In fact Γ is π-dense in $C_b(H)$ and consequently in $D(K)$, endowed with the graph norm, by Proposition 2.1.

Concerning existence of π-cores, we shall make the following basic assumption.

Hypothesis 2.3.

(i) K is an extension of K_0.
(ii) There exists a linear subspace $\mathcal{R}(H)$ which is π-dense in $C_b(H)$ and such that for any $f \in \mathcal{R}(H)$, $\varphi = F_\lambda(f)$ is a strict solution of (1.4).

Then we define Γ by (2.1).

Remark 2.4. *Assumption 2.3-(i) holds provided the following Itô formula is fulfilled for any $\varphi \in C_b^2(H)$,*

$$\mathbb{E}[\varphi(X(t,x))] = \varphi(x) + \mathbb{E}\int_0^t K_0\varphi(X(s,x))ds, \quad x \in H,\ t \ge 0.$$

In fact, from Proposition 1.4 it follows that $D(K_0) \subset D(K)$ and

$$K_0\varphi = K\varphi.$$

Let $\varphi \in \Gamma$; we do not know if $\varphi^2 \in \Gamma$; however we can show that $\varphi^2 \in D(K_0)$. We have in fact

Lemma 2.5. *Assume that Hypothesis 2.3 is fulfilled. Then for all $\varphi \in \Gamma$ we have $\varphi^2 \in D(K_0)$ and*

$$K_0(\varphi^2)(x) = 2\varphi(x)\ K_0\varphi(x) + |\sigma^*(x)D\varphi(x)|^2, \quad x \in D(A) \cap D(b). \tag{2.2}$$

Proof. Let $\varphi \in \Gamma$, $\lambda > 0$. Then $f = \lambda\varphi - K\varphi$ belongs to $\mathcal{R}(H)$ by assumption. We know by Hypothesis 2.3 that $\varphi \in D(K_0)$. Let now $x \in D(A) \cap D(b)$. Then we have

$$\langle Ax, D(\varphi^2(x))\rangle = 2\varphi(x)\langle Ax, D\varphi(x)\rangle,$$

$$\langle b(x), D(\varphi^2(x))\rangle = 2\varphi(x)\langle b(x), D\varphi(x)\rangle,$$

and

$$\text{Tr}\ [(\sigma\sigma^*)(x)D^2(\varphi^2(x))] = 2\varphi\ \text{Tr}\ [(\sigma\sigma^*)(x)D^2(\varphi(x))] + 2|\sigma^*(x)D\varphi(x)|^2.$$

Consequently $\varphi^2 \in D(K_0)$ and (2.2) follows from a straightforward computation. □

3. Invariant measures

In this section we assume, besides Hypothesis 2.3 that there exists an invariant measure ν for P_t, that is a probability measure in $(H, \mathcal{B}(H))$ [a] such that

$$\int_H P_t f d\nu = \int_H f d\nu, \quad f \in C_b(H). \tag{3.1}$$

We shall prove the basic integration by parts formula (3.4) below. We recall that this formula is the main tool in order to define the Sobolev space $W^{1,2}(H,\nu)$, for studying regularity properties of the domain $D(K)$ of K and also to prove the Poincarè and log-Sobolev inequalities. We will not deal with these problems in the present paper; several results in this direction can be found in Refs. [7], [8] and references therein.

Let $\lambda > 0$. Multiplying both sides of (3.1) by $e^{-\lambda t}$ and integrating with respect to t over $(0, +\infty)$, yields

$$\int_H R(\lambda, K) f d\nu = \frac{1}{\lambda} \int_H f d\nu. \tag{3.2}$$

Let now $\varphi \in D(K)$ and $\lambda > 0$. Set $f = \lambda\varphi - K\varphi$ so that $\varphi = R(\lambda, K)f$. Then, taking into account (3.2), it follows that

$$\int_H \varphi d\nu = \int_H R(\lambda, K) f d\nu = \frac{1}{\lambda} \int_H f d\nu = \frac{1}{\lambda} \int_H (\lambda\varphi - K\varphi) d\nu,$$

[a] $\mathcal{B}(H)$ is the σ-algebra of all Borel subsets of H.

which implies

$$\int_H K\varphi d\nu = 0, \quad \text{for all } \varphi \in D(K). \tag{3.3}$$

It is well known that P_t can be uniquely extendible to a strongly continuous semigroup of contractions in $L^2(H,\nu)$ still denoted P_t; let K_2 be its infinitesimal generator.

Remark 3.1. *Assume that Hypothesis 2.3 is fulfilled. Then, by the dominated convergence theorem and a standard monotone classes argument, it follows that $\mathcal{R}(H)$ is dense in $L^2(H,\nu)$. Consequently, Γ is a core for K_2.*

We now consider the mapping

$$\gamma \colon \Gamma \subset D(K_2) \to L^2(H,\nu),\ \varphi \mapsto |\sigma^* D\varphi|^2,$$

where $D(K_2)$ is endowed with the graph norm.

Proposition 3.2. *The mapping γ is continuous and uniquely extendible to $D(K_2)$. Moreover the following identity holds*

$$\int_H \varphi\ K\varphi d\nu = -\frac{1}{2}\int_H |\sigma^* D\varphi|^2 d\nu, \quad \varphi \in D(K_2). \tag{3.4}$$

Proof. Let first $\varphi \in \Gamma$. Since the measure ν is invariant for P_t we have by Lemmas 2.5 and identity (3.3)

$$\int_H K_0(\varphi^2) d\nu = 0.$$

Then, integrating both sides of identity (2.2) yields

$$\int_H \varphi\ K_0\varphi d\nu = -\frac{1}{2}\int_H |\sigma^* D\varphi|^2 d\nu, \quad \varphi \in \Gamma. \tag{3.5}$$

Now let $\varphi \in D(K_2)$. Since Γ is a core for K_2, there exists a sequence $\{\varphi_n\} \subset \Gamma$ convergent to φ in $L^2(H,\nu)$ and such that $\{K\varphi_n\}$ is convergent to $K\varphi$ in $L^2(H,\nu)$. Then by (3.5) we have

$$\int_H |\sigma^*(x) D(\varphi_n - \varphi_m)|^2 d\nu = -2\int_H (\varphi_n - \varphi_m)\ K_0(\varphi_n - \varphi_m) d\nu,$$

so the sequence $\{\sigma^* D\varphi_n\}$ is Cauchy in $L^2(H,\nu)$ and the conclusion follows. □

4. Application

We consider here the following reaction-diffusion equation in the Hilbert space $H = L^2(0, 2\pi)$.

$$\begin{cases} dX(t,\xi) = [(D_\xi^2 X(t,\xi) - X(t,\xi)) + g(X(t,\xi))]dt + dW(t,\xi), \\ \qquad\qquad t \geq 0,\ \xi \in [0, 2\pi], \\ X(t,0) \ = X(t,2\pi), \quad D_\xi X(t,0) = D_\xi X(t,2\pi), \quad t \geq 0, \\ X(0,\xi) \ = x(\xi), \quad \xi \in [0,2\pi], \end{cases} \tag{4.1}$$

where $g \in C_b^3(\mathbb{R})$ and W is a cylindrical Wiener process in H which will be specified below.

Let us write problem (4.1) in the abstract form (1.1). Define

$$\begin{cases} Ax(\xi) = D_\xi^2 x(\xi) - x(\xi), \quad \xi \in [0,2\pi],\ x \in D(A) \\ D(A) = \{x \in H^2(0,2\pi) : \ x(0) = x(2\pi),\ D_\xi x(0) = D_\xi x(2\pi)\} \end{cases}$$

and

$$b(x)(\xi) = g(x(\xi)), \quad x \in H,\ \xi \in [0,2\pi].$$

Denote by $(e_k)_{k\in\mathbb{Z}}$ the complete orthonormal system of $L^2(0,2\pi)$ [b],

$$e_k(\xi) = \frac{1}{\sqrt{2\pi}}\, e^{ik\xi}, \quad \xi \in [0,2\pi],\ k \in \mathbb{Z}$$

and define

$$W(t) = \sum_{k\in\mathbb{Z}} \beta_k(t) e_k,$$

where $\{\beta_k(t)\}_{k\in\mathbb{Z}}$ is a family of standard Brownian motions mutually independent in a filtered probability space $(\Omega, \mathcal{F}, (\mathcal{F}_t)_{t\geq 0}, \mathbb{P})$. We have

$$Ae_k = -(1+k^2)e_k, \quad k \in \mathbb{Z}$$

and

$$\|e^{tA}\| \leq e^{-t}, \quad t \geq 0.$$

It is well known that problem (4.1) has a unique mild solution $X(t,x)$, see e.g. Ref. [9]. We recall that a *mild solution* of equation (4.1) is a stochastic process

[b] Here we are dealing with the usual complexification of $L^2(0,2\pi)$.

$X \in C_W([0,T],H)$ such that

$$X(t,x) = e^{tA}x + \int_0^t e^{(t-s)A} b(X(s,x))dW(s) + \int_0^t e^{(t-s)A} dW(s), \quad t \geq 0,\ x \in H. \tag{4.2}$$

Here $C_W([0,T],H)$ is the space of all mean square continuous adapted stochastic process $X(\cdot)$ defined in $[0,T]$ and taking values in H, endowed with the norm

$$\|X\|_{C_W([0,T],H)} = \left(\sup_{t\in[0,T]} \mathbb{E}(|X(t)|^2) \right)^{1/2}.$$

Notice that, in spite of the fact that the function g is of class C^3, the Nemitskii operator b is not of class C^1 (except when g is constant). However, b is differentiable in all directions of $L^2(0,2\pi)$ and it is twice differentiable in all directions of $L^4(0,2\pi)$. So, the Kolmogorov equation cannot be solved by the standard method from Chapter 7 in Ref. [3].

We notice that the existence of a strict solutions of the Kolmogorov equation corresponding to (4.1) have been studied (in a more general setting) in Ref. [10] using the Bismut-Elworthy formula.

4.1. *Estimates for $X_x(t,x)$*

For any $h \in H, t \geq 0, x \in H$, we denote by $X_x(t,x)h := \eta^h(t,x)$ the directional derivative of $X_x(t,x)$ in the direction of h. It is well known that $\eta^h(t,x)$ does exist, (see Ref. [11]) and that it is the mild solution of the problem

$$\frac{d}{dt}\,\eta^h(t,x) = A\eta^h(t,x) + f'(X(t,x))\eta^h(t,x), \quad \eta^h(0,x) = h, \tag{4.3}$$

that is

$$\eta^h(t,x) = e^{tA}h + \int_0^t e^{(t-s)A} f'(X(s,x))\eta^h(s,x)ds. \tag{4.4}$$

Proposition 4.1. *We have*

$$|\eta^h(t,x)|^2_{L^2(0,2\pi)} \leq e^{-2(1-\|g\|_1)t}|h|^2_{L^2(0,2\pi)}, \quad t \geq 0,\ h \in H. \tag{4.5}$$

Proof. In fact, multiplying scalarly both sides of (4.3) by $\eta^h(t,x)$ and integrating over $[0,2\pi]$ yields,

$$\begin{aligned}\frac{1}{2}\,\frac{d}{dt}\,|\eta^h(t,x)|^2_{L^2(0,2\pi)} &= \langle A\eta^h(t,x), \eta^h(t,x)\rangle_{L^2(0,2\pi)} \\ &\quad + \langle g'(X(t,x))\eta^h(t,x), \eta^h(t,x)\rangle_{L^2(0,2\pi)} \\ &\leq (-1 + \|g\|_1)|\eta^h(t,x)|^2_{L^2(0,2\pi)}\end{aligned}$$

and so, the conclusion follows from a classical comparison result. □

Assume that one has to estimate the series

$$\sum_{k=1}^{\infty} |\eta^h(t,x)|^2_{L^2(0,2\pi)}.$$

To this purpose the estimate (4.5) will not be enough. For this reason we prove another estimate.

Proposition 4.2. *We have*

$$|\eta^h(t,x)|^2_{L^2(0,2\pi)} \le |e^{tA}h|_{L^2(0,2\pi)} + \|g\|_1 \int_0^t e^{(\|g\|_1-1)(t-s)}|e^{sA}h|_{L^2(0,2\pi)}ds. \quad (4.6)$$

Proof. By (4.4) we have

$$\begin{aligned}|\eta^h(t,x)|_{L^2(0,2\pi)} &\le |e^{tA}h|_{L^2(0,2\pi)} + \int_0^t |e^{(t-s)A}g'(X(s,x))\eta^h(s,x)|_{L^2(0,2\pi)}ds \\ &\le |e^{tA}h|_{L^2(0,2\pi)} + \|g\|_1 \int_0^t e^{-(t-s)}|\eta^h(s,x)|_{L^2(0,2\pi)}ds.\end{aligned}$$

Now, setting

$$\varphi(t) = e^t|\eta^h(t,x)|_{L^2(0,2\pi)},$$

we have

$$\varphi(t) \le e^t|e^{tA}h|_{L^2(0,2\pi)} + \|g\|_1 \int_0^t \varphi(s)ds.$$

The conclusion follows from the Gronwall lemma. □

We are going now to estimate, for further use, the L^2 norm of $[\eta^h(t,x)]^2$. Write

$$\frac{1}{2}\frac{d}{dt}[\eta^h(t,x)]^2 = A\eta^h(t,x)\,\eta^h(t,x) + g'(X(s,x))[\eta^h(s,x)]^2.$$

Since

$$A([\eta^h(t,x)]^2) = 2A\eta^h(t,x)\,\eta^h(t,x) + |D\eta^h(t,x)|^2,$$

we deduce that

$$\begin{aligned}\frac{d}{dt}[\eta^h(t,x)]^2 &\le A([\eta^h(t,x)]^2) - |D\eta^h(t,x)|^2 + 2\|g\|_1[\eta^h(t,x)]^2 \\ &\le A([\eta^h(t,x)]^2) + 2\|g\|_1[\eta^h(t,x)]^2.\end{aligned}$$

Therefore

$$[\eta^h(t,x)]^2 \le e^{tA}(h^2) + 2\|g\|_1 \int_0^t e^{(t-s)A}[\eta^h(s,x)]^2ds.$$

Arguing as in the proof of Proposition 4.2, we find the following result.

Proposition 4.3. *We have*

$$|[\eta^h(t,x)]^2|_{L^2(0,2\pi)} \le |e^{tA}(h^2)|_{L^2(0,2\pi)} + \int_0^t e^{(2\|g\|_1-1)(t-s)}|e^{sA}(h^2)|_{L^2(0,2\pi)}ds. \quad (4.7)$$

Using the well known ultracontractivity type estimate

$$|e^{tA}f|_{L^2(0,2\pi)} \le ct^{-1/4}e^{-t}|f|_{L^1(0,2\pi)}, \quad (4.8)$$

we find the result

Corollary 4.4. *There exist $c_1>0, \kappa_1>0$ such that*

$$|[\eta^h(t,x)]^2|_{L^2(0,2\pi)} \le c_1(1+t^{-1/4})e^{\kappa_1 t}\,|h|^2_{L^2(0,2\pi)}. \quad (4.9)$$

Proof. We have in fact by (4.8)

$$|e^{tA}(h^2)|_{L^2(0,2\pi)} \le ct^{-1/4}e^{-t}|h^2|_{L^1(0,2\pi)}$$

$$= ct^{-1/4}e^{-t}|h|^2_{L^2(0,2\pi)},$$

and the conclusion follows from (4.7). □

4.2. *Estimates for $X_{xx}(t,x)$*

We set now

$$\zeta^h(t,x) = \langle X_{xx}(t,x)h,h\rangle, \quad t\ge 0,\ x,h\in H.$$

ζ^h does exist and fulfills the equation (see e.g. Ref. [11])

$$\begin{cases} \dfrac{d}{dt}\,\zeta^h(t,x) = A\zeta^h(t,x) + g'(X(t,x))\zeta^h(t,x) + g''(X(t,x))[\eta^h(t,x)]^2, \\ \zeta^h(0,x) = 0. \end{cases} \quad (4.10)$$

Proposition 4.5. *There exist $c_2>0$ and $\kappa_2>0$ such that*

$$|\zeta^h(t,x)|^2_{L^2(0,2\pi)} \le c_2 e^{\kappa_2 t}|h|^4_{L^2(0,2\pi)}. \quad (4.11)$$

Proof. Multiplying both sides of (4.10) by $\zeta^h(t,x)$ and integrating with respect to ξ over $[0,2\pi]$ yields

$$\frac{1}{2}\,\frac{d}{dt}\,|\zeta^h(t,x)|^2_{L^2(0,2\pi)} \le \|g\|_1|\zeta^h(t,x)|^2_{L^2(0,2\pi)}$$

$$+\|g\|_2\int_0^1 [\eta^h(t,x)]^2|\zeta^h(t,x)|d\xi$$

$$\le (\|g\|_1| + \frac{1}{2}\,\|g\|_2^2)|\zeta^h(t,x)|^2_{L^2(0,2\pi)}$$

$$+\frac{1}{2}\,|[\eta^h(t,x)]^2|_{L^2(0,2\pi)}.$$

It follows that

$$|\zeta^h(t,x)|^2_{L^2(0,2\pi)} \le \int_0^t e^{(2\|g\|_1|+\|g\|_2^2)(t-s)}|[\eta^h(s,x)]^2|^2_{L^2(0,2\pi)}ds.$$

Now, using (4.9), we find

$$|\zeta^h(t,x)|^2_{L^2(0,2\pi)} \le c_1^2 \int_0^t e^{(2\|g\|_1|+\|g\|_2^2)(t-s)}(1+s^{-1/4})^2 e^{2\kappa_1 t}\, |h|^4_{L^2(0,2\pi)}ds$$

and the conclusion follows.

4.3. *Estimates of $TR[X_{x,x}(t,x)]$*

We want here to estimate the norm of the vector trace of $X_{x,x}(t,x)$, namely

$$\mathrm{TR}[X_{x,x}(t,x)]:\ = \sum_{k=1}^\infty \langle X_{x,x}(t,x)e_k, e_k\rangle = \sum_{k=1}^\infty \zeta^{e_k}(t,x).$$

Let us set

$$T(t,x) = \mathrm{TR}[X_{x,x}(t,x)], \quad t \ge 0,\ x \in H.$$

Setting in (4.10) $h = e_k$ and summimg up on k we see that $T(t,x)$ fulfills the equation

$$\begin{cases} \dfrac{d}{dt}\,T(t,x) = AT(t,x) + g'(X(t,x))T(t,x) + g''(X(t,x))Z(t,x), \\ \\ T(0,x) = 0, \end{cases} \tag{4.12}$$

where

$$Z(t,x) = \sum_{k=1}^\infty [\eta^{e_k}(t,x)]^2.$$

Proposition 4.6. *There exist $c_3 > 0$ and $\kappa_3 > 0$ such that*

$$|\mathrm{TR}\ (X_{x,x}(t,x))|^2_{L^2(0,2\pi)} \le c_3 e^{\kappa_3 t}. \tag{4.13}$$

Proof. By (4.7) we have, taking into account that $e_k^2 = \sqrt{2\pi}\ e_{2k}$,

$$\begin{aligned} |[\eta^{e_k}(t,x)]^2|_{L^2(0,2\pi)} &\le |e^{tA}(e_k^2)|_{L^2(0,2\pi)} + \int_0^t e^{(2\|g\|_1-1)(t-s)}|e^{sA}(e_k^2)|_{L^2(0,2\pi)}ds \\ &\le (2\pi)^2 e^{-4k^2 t} + (2\pi)^2 \int_0^t e^{(2\|g\|_1-1)(t-s)} e^{-4k^2 s}ds. \end{aligned}$$

Summing up on k yields

$$|Z(t,x)|_2 \le (2\pi)^2 \sum_{k=1}^\infty e^{-4k^2 t} + (2\pi)^2 \int_0^t e^{(2\|g\|_1-1)(t-s)} \sum_{k=1}^\infty e^{-4k^2 s}ds.$$

Since, for a suitable constant $c > 0$,

$$\sum_{k=1}^{\infty} e^{-4k^2 t} \le ct^{-1/2},$$

the conclusion follows. □

4.4. *Estimates of $P_t\varphi$*

In this subsection we show that Hypothesis 2.3 is fulfilled by choosing $\mathcal{R}(H) = \mathcal{E}_A(H)$, where $\mathcal{E}_A(H)$ is the linear span of all real and imaginary parts of the *exponential functions* φ_h,

$$\varphi_h(x) := e^{i\langle h,x\rangle}, \quad x \in H, \quad h \in D(A).$$

It is easy to see that the space $\mathcal{E}_A(H)$ is dense in $L^2(H,\nu)$, see e.g. Ref. [8].

Let $\varphi \in \mathcal{E}_A(H)$ and set $u(t,x) = P_t\varphi(x)$. Then for any $h \in H$ we have

$$\langle u_x(t,x), h\rangle = \mathbb{E}\left[\langle D\varphi(X(t,x)), \eta^h(t,x)\rangle\right], \tag{4.14}$$

$$\begin{aligned}\langle u_{xx}(t,x)h, h\rangle &= \mathbb{E}\left[\langle D^2\varphi(X(t,x))\eta^h(t,x), \eta^h(t,x)\rangle\right]\\ &+\mathbb{E}\left[\langle D\varphi(X(t,x)), \zeta^h(t,x)\rangle\right].\end{aligned} \tag{4.15}$$

By (4.5) it follows that

$$|\langle u_x(t,x), h\rangle| \le \|\varphi\|_1 e^{-(1-\|g\|_1)t}|h|, \tag{4.16}$$

for all $h \in H$. Moreover by (4.11) we deduce that

$$|\langle u_{xx}(t,x)h, h\rangle| \le c_2^{1/2} e^{\frac{1}{2}\,\kappa_2 t}|h|^2, \tag{4.17}$$

for all $h \in H$. Now by (4.17) it follows easily that $\varphi \in C_b^1(H)$ and

$$|u_x(t,x)| \le \|\varphi\|_1 e^{-(1-\|g\|_1)t}. \tag{4.18}$$

It remains to estimate $\text{TR}[u_{xx}(t,x)]$. We have

$$\begin{aligned}\text{TR}\,[u_{xx}(t,x)] &= \text{Tr}\,[\mathbb{E}(X_x^*(t,x)D^2\varphi(X(t,x))X_x^*(t,x))]\\ &+\mathbb{E}\left[\langle D\varphi(X(t,x)), \text{TR}[X_{xx}(t,x)\rangle\right].\end{aligned} \tag{4.19}$$

It follows, by (4.5) and (4.13) that

$$\text{TR}\,[u_{xx}(t,x)] = \|\text{Tr}\,D^2\varphi\|_0 + \|\varphi\|_1 c_3 e^{\kappa_3 t}. \tag{4.20}$$

Moreover, recalling that for all $\varphi \in \mathcal{E}(H)$ we have that $D\varphi(x) \in D(A)$ for any $x \in H$ and that $AD\varphi \in C_b(H)$, we have

$$T_A(u(t,x)) = \langle Ax, u_x(t,x\rangle = \mathbb{E}\left[\langle AD\varphi(X(t,x)), \eta^h(t,x)\rangle\right], \quad x \in D(A).$$

So $u(t,x)$ fulfills conditions (i)–(iii) of Definition 1.1. We can now prove

Proposition 4.7. *$u(t,x)$ is a strict solution of* (1.3).

Proof. Let us introduce Galerkin approximations. For any $n \in \mathbb{N}$ we denote by P_n the projector

$$P_n x = \sum_{k=1}^{n} \langle x, e_k \rangle e_k, \quad x \in H$$

and set $A_n = AP_n$. Then we consider the approximating equation

$$\begin{cases} dX^n(t,x) = (P_n A X^n(t,x) + P_n g(P_n X^n(t,x)))dt + P_n dW(t), \\ \\ X^n(0,x) = P_n x, \end{cases} \tag{4.21}$$

which has a unique solution $X^n(t,x)$. We define also

$$\eta_n^h(t,x) = X_x^n(t,x)h, \quad \zeta_n^h(t,x) = X_{xx}^n(t,x)(h,h).$$

The following results are standard.

$$\lim_{n\to\infty} X^n(\cdot,x) = X(\cdot,x), \quad \text{in } C_W([0,T],H),$$

$$\lim_{n\to\infty} \eta_n^h(t,x) = \eta^h(t,x), \quad \text{in } C_W([0,T],H),$$

and

$$\lim_{n\to\infty} \zeta_n^h(t,x) = \zeta_n(t,x), \quad \text{in } C_W([0,T],H),$$

for any $h \in H$.

Moreover $u^n(t,x) = \mathbb{E}[\varphi(X^n(t,x))]$ is a strict solution of the approximating Kolmogorov equation

$$\begin{cases} u_t^n(t,x) = \dfrac{1}{2} \operatorname{Tr}\, [u_{xx}^n(t,x)] + \langle AP_n x, u_x^N(t,x)\rangle + \langle P_n g(P_n x), u_x^N(t,x)\rangle, \\ \\ u^n(0,x) = \varphi(P_n x), \quad x \in H. \end{cases} \tag{4.22}$$

Letting $n\to\infty$ we see that

$$\lim_{n\to\infty} u^n(t,x) = u(t,x), \quad \text{uniformly in } t,\ x \in H,$$

$$\lim_{n\to\infty} \langle P_n A x, u^n(t,x)\rangle = \langle x, Au(t,x)\rangle, \quad \text{uniformly in } t,\ x \in H,$$

and

$$\lim_{n\to\infty} \operatorname{Tr}\, [u_{xx}^n(t,x)] = \operatorname{Tr}\, u_{xx}(t,x), \quad \text{uniformly in } t,\ x \in H.$$

Thus

$$\lim_{n\to\infty} \frac{d}{dt}\, u^n(t,x) = \frac{d}{dt} u(t,x) \quad \text{uniformly in } t,\ x \in H$$

and (1.3) is fulfilled. □

In an analogous way, using inequalities (4.16), (4.17), (4.20), we can prove the result.

Proposition 4.8. *Set*

$$\varphi(x) = \int_0^\infty e^{-\lambda t} u(t,x)dt.$$

Then φ is a strict solution of (1.4).

References

[1] S. Cerrai, A Hille-Yosida theorem for weakly continuous semigroups, *Semigroup Forum.* **49**(3), 349–367, (1994). ISSN 0037-1912.

[2] K. Yosida, *Functional analysis.* Die Grundlehren der Mathematischen Wissenschaften, Band 123, (Academic Press Inc., New York, 1965).

[3] G. Da Prato and J. Zabczyk, *Second order partial differential equations in Hilbert spaces.* vol. 293, *London Mathematical Society Lecture Note Series*, (Cambridge University Press, Cambridge, 2002). ISBN 0-521-77729-1.

[4] G. Da Prato and A. Debussche, Ergodicity for the 3D stochastic Navier-Stokes equations, *J. Math. Pures Appl. (9).* **82**(8), 877–947, (2003). ISSN 0021-7824.

[5] E. Priola, On a class of Markov type semigroups in spaces of uniformly continuous and bounded functions, *Studia Math.* **136**(3), 271–295, (1999). ISSN 0039-3223.

[6] A. S. Nemirovski and S. M. Semenov, The polynomial approximation of functions in hilbert spaces, *Mat. Sb. (N.S.).* **92**, (1973).

[7] G. Da Prato, A. Debussche, and B. Goldys, Some properties of invariant measures of non symmetric dissipative stochastic systems, *Probab. Theory Related Fields.* **123** (3), 355–380, (2002). ISSN 0178-8051.

[8] G. Da Prato, *Kolmogorov equations for stochastic PDEs.* Advanced Courses in Mathematics. CRM Barcelona, (Birkhäuser Verlag, Basel, 2004). ISBN 3-7643-7216-8.

[9] G. Da Prato and J. Zabczyk, *Ergodicity for infinite-dimensional systems.* vol. 229, *London Mathematical Society Lecture Note Series*, (Cambridge University Press, Cambridge, 1996). ISBN 0-521-57900-7.

[10] S. Cerrai. Classical solutions for Kolmogorov equations in Hilbert spaces. In *Seminar on Stochastic Analysis, Random Fields and Applications, III (Ascona, 1999)*, vol. 52, *Progr. Probab.*, pp. 55–71. Birkhäuser, Basel, (2002).

[11] S. Cerrai, *Second order PDE's in finite and infinite dimension.* vol. 1762, *Lecture Notes in Mathematics*, (Springer-Verlag, Berlin, 2001). ISBN 3-540-42136-X. A probabilistic approach.

Author Index

Subject Index